T0238504

Communications in Computer and Information Science   1265

Commenced Publication in 2007
Founding and Former Series Editors:
Simone Diniz Junqueira Barbosa, Phoebe Chen, Alfredo Cuzzocrea,
Xiaoyong Du, Orhun Kara, Ting Liu, Krishna M. Sivalingam,
Dominik Ślęzak, Takashi Washio, Xiaokang Yang, and Junsong Yuan

Editorial Board Members

More information about this series at http://www.springer.com/series/7899

Haijun Zhang · Zhao Zhang ·
Zhou Wu · Tianyong Hao (Eds.)

Neural Computing for Advanced Applications

First International Conference, NCAA 2020
Shenzhen, China, July 3–5, 2020
Proceedings

Springer

Editors
Haijun Zhang (iD)
Harbin Institute of Technology
Shenzhen, China

Zhou Wu (iD)
Chongqing University
Chongqing, China

Zhao Zhang (iD)
Hefei University of Technology
Hefei, China

Tianyong Hao (iD)
South China Normal University
Guangzhou, China

ISSN 1865-0929 ISSN 1865-0937 (electronic)
Communications in Computer and Information Science
ISBN 978-981-15-7669-0 ISBN 978-981-15-7670-6 (eBook)
https://doi.org/10.1007/978-981-15-7670-6

This Springer imprint is published by the registered company Springer Nature Singapore Pte Ltd.
The registered company address is: 152 Beach Road, #21-01/04 Gateway East, Singapore 189721, Singapore

Preface

This volume contains the accepted papers presented at the International Conference on Neural Computing for Advanced Applications (NCAA 2020). Neural computing and artificial intelligence (AI) have become hot topics in recent years. Most people, with little knowledge in such areas, are expecting to follow the AI mainstream for pursuing possible changes of business, education, and so on. For promoting multidisciplinary development and application of neural computing, the NCAA conference series was initiated around the theme "make academia more practical," to provide an open platform of academic discussions, industrial showcases, and basic training tutorials.

NCAA 2020 was organized by Harbin Institute of Technology, Chongqing University, Hefei University of Technology, South China Normal University, and was supported by Springer. Due to effects of the coronavirus, the mainstream of NCAA 2020 transfered to an online event, in which people could freely connect to live broadcasts of keynote speeches and presentations. NCAA 2020 attracted more than 1,000 participants from all over the world. Most AI experts invited to the event were young scholars from universities or active researchers from industries, who were greatly enthusiastic about sharing fruitful thoughts on the Internet. This first attempt won overwhelming applauses from both presenters and audiences, whose feedback showed that an open platform is an urgent and effective way to break barriers between academia and industry.

In this volume, 43 high-quality papers were selected for publication among 113 submissions after a double-blind peer-review process, which indicated that the acceptance rate of NCAA 2020 was less than 40%. These papers were categorized into eight technical tracks, i.e., Neural network theory, cognitive sciences, neuro-system hardware implementations, and NN-based engineering applications; Machine learning, data mining, data security and privacy protection, and data-driven applications; Computational intelligence, nature-inspired optimizers, and their engineering applications; Cloud/edge/fog computing, the Internet of Things/Vehicles (IoT/IoV), and their system optimization; Control systems, network synchronization, system integration, and industrial artificial intelligence; Fuzzy logic, neuro-fuzzy systems, decision making, and their applications in management sciences; Computer vision, image processing, and their industrial applications; and Natural language processing, machine translation, knowledge graphs, and their applications.

In this volume, the authors of each paper reported their novel results of computing theory or application. The volume cannot cover all aspects of neural computing and advanced applications, but must inspire insightful thoughts of readers. With these proceedings, we wish for the secrets of AI to be unveiled and that academia is made practical.

June 2020

Haijun Zhang
Zhao Zhang
Zhou Wu
Tianyong Hao

Organization

Honorary Chairs

John MacIntyre   University of Sunderland, UK
Tommy W. S. Chow   City University of Hong Kong, Hong Kong

General Chair

Haijun Zhang   Harbin Institute of Technology, China

Technical Program Committee Co-chairs

Zhou Wu   Chongqing University, China
Zhao Zhang   Hefei University of Technology, China

Demo Co-chairs

Weiwei Wu   Southeast University, China
Kai Liu   Chongqing University, China

Panel Co-chairs

Mingbo Zhao   Donghua University, China
Wei Huang   Zhejiang University of Technology, China

Organizing Committee Co-chairs

Qiang Jia   Jiangsu University, China
Yu Wang   Xi'an Jiaotong University, China

Local Arrangement Co-chairs

Xiaofeng Zhang   Harbin Institute of Technology, China
Ke Wang   Harbin Institute of Technology, China

Registration Co-chairs

Jingjing Cao   Wuhan University of Technology, China
Shi Cheng   Shaanxi Normal University, China

Publication Co-chairs

Tianyong Hao   South China Normal University, China
Cuili Yang   Beijing University of Technology, China

Publicity Co-chairs

Liang Feng   Chongqing University, China
Xiangping Zhai   Nanjing University of Aeronautics and Astronautics,
                      China
Xianghua Chu   Shenzhen University, China

Finance Co-chairs

Yuzhu Ji   Nanyang Technological University, Singapore
Qun Li   Harbin Institute of Technology, China

Sponsor Co-chairs

Choujun Zhan   South China Normal University, China
Biao Yang   Harbin Institute of Technology, China

Web Chair

Ziliang Yin   Harbin Institute of Technology, China

NCAA Steering Committee Liaison

Xiaoying Zhong   Harbin Institute of Technology, China

Track Co-chairs

Track 1: Neural Network Theory, Cognitive Sciences, Neuro-System Hardware Implementations, and NN-Based Engineering Applications

Zenghui Wang   University of South Africa, South Africa
Shi Cheng   Shaanxi Normal University, China
Cuili Yang   Beijing University of Technology, China
Jianghong Ma   City University of Hong Kong, Hong Kong

Track 2: Machine Learning, Data Mining, Data Security and Privacy Protection, and Data-Driven Applications

Li Zhang	Soochow University, China
Zhao Kang	University of Electronic Science and Technology, China
Yimin Yang	Lakehead University, Canada
Jicong Fan	Cornell University, USA

Track 3: Computational Intelligence, Nature-Inspired Optimizers, and their Engineering Applications

Sheng Li	University of Georgia, USA
Yaqing Hou	Dalian University of Technology, China
Zhile Yang	Chinese Academy of Sciences, China
Xiaozhi Gao	University of Eastern Finland, Finland

Track 4: Cloud/Edge/Fog Computing, the Internet of Things/Vehicles (IoT/IoV), and their System Optimization

Cheng-Kuan Lin	Fuzhou University, China
Kai Liu	Chongqing University, China
Zhiying Tu	Harbin Institute of Technology, China
Xiangping Zhai	Nanjing University of Aeronautics and Astronautics, China
Weizhi Meng	Technical University of Denmark, Denmark

Track 5: Control Systems, Network Synchronization, System Integration, and Industrial Artificial Intelligence

Chengqing Li	Hunan University, China
Qiang Jia	Jiangsu University, China
Yongji Wang	Huazhong University of Science and Technology, China
Ning Sun	Nankai University, China
Wenkai Hu	China University of Geosciences, China

Track 6: Fuzzy Logic, Neuro-Fuzzy Systems, Decision Making, and their Applications in Management Sciences

Zhigang Zeng	Huazhong University of Science and Technology, China
Dongrui Wu	Huazhong University of Science and Technology, China
Zhaohong Deng	Jiangnan University, China

Liang Bai	Shanxi University, China
Jingjing Cao	Wuhan University of Technology, China
Xin Zhang	Tianjin Normal University, China

Track 7: Computer Vision, Image Processing, and their Industrial Applications

Zhong Ji	Tianjin University, China
Bineng Zhong	Huaqiao University, China
Jiayi Ma	Wuhan University, China
Xiangyuan Lan	Hong Kong Baptist University, Hong Kong
Mingbo Zhao	Donghua University, China

Track 8: Natural Language Processing, Machine Translation, Knowledge Graphs, and their Applications

Lap-Kei Lee	The Open University of Hong Kong, Hong Kong
Haitao Wang	China National Institute of Standardization, China
Tianyong Hao	South China Normal University, China
Xiaolei Lu	City University of Hong Kong, Hong Kong

Contents

Adaptive Multiple-View Label Propagation for Semi-supervised Classification

Lei Jia[1,2], Huan Zhang[1,2], and Zhao Zhang[1,3(✉)]

[1] School of Computer Science and Technology,
Soochow University, Suzhou 215006, China
cszzhang@gmail.com
[2] Provincial Key Laboratory for Computer Information Processing Technology,
Soochow University, Suzhou 215006, China
[3] School of Computer and Information Science,
Hefei University of Technology, Hefei, China

Abstract. We propose a novel Adaptive Multiple-view Label Propagation (MLP) framework for semi-supervised classification. MLP performs classification on multiple views rather than on the single view, and can exploit the complementarity of multiple views in the label prediction process. Moreover, MLP integrates the multi-view label propagation and the adaptive multiple graph weight learning into a unified model, where a linear transformation is used to enforce different weights form different view spaces. Thus, an optimal graph weight matrix can be constructed from each view. Due to the adaptive manner for defining the weight matrices from multiple views, MLP can avoid the complex and tricky process to select the neighborhood size or kernel parameter. Extensive results on real data show that MLP can deliver the enhanced performances, by comparing with other related techniques.

Keywords: Semi-supervised classification · Multiple-view label propagation · Adaptive multiview graph weight learning

1 Introduction

Graph based semi-supervised learning [1, 10] for classification by label propagation (LP) [14, 25–29] has been attracting many interests in recent years. Most existing LP algorithms focus on predicting labels from a single view, e.g., Gaussian Fields and Harmonic Function (GFHF) [2], Learning with Local and Global Consistency (LLGC) [3], Special LP (SLP) [6], Neighborhood Propagation (LNP) [4], Linear Adaptive Neighborhood Propagation (AdaptiveNP) [8], Prior Class Dissimilarity based LNP (CD-LNP) [5], Laplacian Linear Discriminant Analysis [13], Projective Label Propagation [23] and Nonnegative Sparse Neighborhood Propagation (SparseNP) [7]. But in various data mining applications, a sample may have multiple-view representations of in various subspaces, which has inspired a new research direction, i.e., *multiple-view learning* [12, 17, 31, 32]. To extend the real application areas from the single-view to multiple-view, Auto-weighted Multiple Graph Learning (AMGL) [9] and Multiple Locality Preserving Projections with Cluster-based LP (MLPP-CLP) [23] incorporated

© Springer Nature Singapore Pte Ltd. 2020
H. Zhang et al. (Eds.): NCAA 2020, CCIS 1265, pp. 1–11, 2020.
https://doi.org/10.1007/978-981-15-7670-6_1

the idea of multiple-view learning into LP. But their learning processes are not straightforward, and still suffer from certain problems. First, they separate the multiple graph weight construction process from the multiple-view LP process, so the pre-calculated weights cannot be guaranteed as optimal for subsequent classification. Second, MLPP-CLP and AMGL suffer from the same tricky issue in selecting an optimal neighbor number or kernel width when defining the weights. Moreover, the number of neighbors is usually set as the same number for each view artificially, but such operation fails to consider the actual distributions of various real data. The above suffered issues may result in the decreased label estimation results in reality.

We, therefore, in this paper propose a novel Adaptive Multiview Label Propagation (MLP) model for enhancing the classification. The main contributions of this paper are shown as follows. (1) MLP enhances the performance by discovering the effects of multiple views rather than single view. (2) MLP integrates the multiview LP and adaptive multiple graph weight learning into a unified model to exploit the complementation using a linear transformation to make different adaptive weights form different view spaces. Note that the general process of MLP is shown in Fig. 1, where a three-view set (i.e., X^1, X^2, X^3) with different poses is used as an example, each view X^v, $v = 1, 2, 3$ has six samples, W^v is the adaptive weight matrix over each view v and $L^v = (I - W^v)(I - W^v)^{\mathrm{T}}$. Clearly, the label formation of each unlabeled data is partly from its initial state and is partly from its neighbors. Based on mining valuable knowledge from different views in an adaptive manner, the label prediction results can be potentially enhanced.

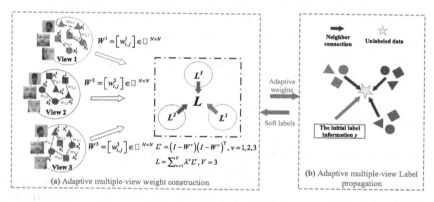

Fig. 1. The learning framework of the proposed adaptive multiple-view label propagation.

2 Related Work

We first review the closely related AMGL algorithm to our formulation. Let V be the number of views. AMGL defines the following framework for learning:

$$\min_{F} \sum_{v=1}^{V} \sqrt{tr(FL^{v}F^{\mathrm{T}})}, \tag{1}$$

where $L^{v} = D^{v} - W^{v}$, F is a class indicator matrix and D^{v} is a diagonal matrix [9]. The Lagrange function to the above optimization problem can be written as

$$\sum_{v=1}^{V} \sqrt{tr(FL^{(v)}F^{\mathrm{T}})} + \varphi(\tau, F), \tag{2}$$

where τ is a Lagrange multiplier, $\varphi(\tau, F)$ is the formalized term derived from the constraints. By taking the derivative of Eq. (2) subject to F, we have

$$\sum_{v=1}^{V} \alpha^{v} \left(\partial tr(F^{\mathrm{T}}L^{v}F)/\partial F \right) + \partial \varphi(\tau, F)/\partial F = 0, \tag{3}$$

where $\alpha^{v} = 1/\left(2\sqrt{tr(F^{\mathrm{T}}L^{v}F)}\right)$. Note that α^{v} is dependent on the target variable F, so the problem in Eq. (3) cannot be solved directly. According to [9], if α^{v} is set to be stationary, Eq. (3) can be considered as the solution to following problem:

$$\min_{F} \sum_{v=1}^{V} \alpha^{v} tr(F^{\mathrm{T}}L^{v}F). \tag{4}$$

Note that AMGL can be used for semi-supervised classification. Specifically, suppose that there are l $(1 \leq l \leq n)$ labeled data in each view and let y_i be the initial label vector of x_i, AMGL performs semi-supervised classification by solving

$$\underset{\alpha^{v}, F}{Min} \sum_{v=1}^{V} \alpha^{v} tr(F^{\mathrm{T}}L^{v}F), \quad s.t. \, f_i = y_i, \, \forall i = 1, 2, \ldots, l. \tag{5}$$

3 Adaptive Multiple-View Label Propagation

3.1 Proposed Formulation

Give a dataset $X = \{X^{v}\}_{v=1}^{V}$ and each sample has a single label in $\{1, 2, \ldots, c\}$. If each X^{v} is formed by an unlabeled set $X_{U}^{v} \in \mathbb{R}^{d^{v} \times U}$ and a labeled set $X_{L}^{v} \in \mathbb{R}^{d^{v} \times L}$, then our MLP aims at exploiting the complementarity of multiple views. Clearly, V graphs can be easily constructed based on the V view data by finding the nearest neighbor information within each view. Specifically, MLP integrates the adaptive multiple weight learning and adaptive multi-view LP into a unified model to exploit the complementation using a linear transformation sufficiently to make different weights form different view spaces, where w_i^{v} is the i-th column vector of weights W^{v} within the v-th view data. Thus, we have the following problem MLP:

$$\min_{F,W^v} \sum_{v=1}^{V} \sqrt{\sum_{i=1}^{N} \|f_i - Fw_i^v\|_2^2} + \sum_{v=1}^{V} \left[\alpha \sum_{i=1}^{d^v} \left\| (X^v)^i - (X^v W^v)^i \right\|_2 + \beta \sum_{i=1}^{N} \left\| (W^v)^i \right\|_2 \right] + \sum_{i=1}^{N} \mu_i \|f_i - y_i\|_2^2, \tag{6}$$

where $Y \in \mathbb{R}^{c \times N}$ is the initial label matrix, $(X^v)^i$ denotes the i-th row vector of X^v and $F \in \mathbb{R}^{c \times N}$ denotes the estimated soft labels. Both α and β are two parameters to balance two terms in constructing the reconstruction weights. μ_i is a trade-off parameter for weighting both labeled and unlabeled samples. Note that MLP with different μ_i can detect and erase the noisy labels in labeled data effectively [6]. In above problem, the first term is a multiple-view manifold smoothing term. The second term is the $L_{2,1}$-norm based sparse regularization for seeking the weights over different views as adaptive weights for adaptive multiple-view label prediction. The last term is the label fitness term to characterize the difference between predicted soft labels F and initial labels Y across all views. Due to the definition of $L_{2,1}$-norm, the above problem can be easily rewritten using matrix form as

$$\min_{F,W^v} \sum_{v=1}^{V} \sqrt{\|F - FW^v\|_F^2} + \sum_{v=1}^{V} \left(\alpha \|X^v - X^v W^v\|_{2,1} + \beta \|W^v\|_{2,1} \right) + tr(F - Y)^T U(F - Y), \tag{7}$$

where U denotes a diagonal matrix with the adjustable parameters μ_i as its diagonal entries. In this paper, we fix $\mu_i = 0.999$ for the labeled data and $\mu_i = 0$ for unlabeled data. Clearly, the L2,1-norm based $\|W^v\|_{2,1}$ on the weights can ensure vectors of W^v to be sparse and the L2,1-norm based reconstruction $\|X^v - X^v W^v\|_{2,1}$ can also enable the distance metric to have robust properties to noise in data [11].

3.2   Optimization Procedure

It is noted that the variables depend on each other, so we solve them alternately.

(1) Given W^v, update the label matrix F:

Note that graph weight matrix W^v over each view is simply initialized as the identity matrix. Given W_t^v at the i-th iteration, the label matrix F_{t+1} can be inferred as

$$F_{t+1} = \arg\min_{F} \sum_{v=1}^{V} \sqrt{\|F - FW_t^v\|_F^2} + tr(F - Y)^T U(F - Y). \tag{8}$$

The Lagrange function of the above formulation Eq. (8) can be constructed as

$$\sum_{v=1}^{V} \sqrt{\|F - FW_t^v\|_F^2} + tr(F - Y)^T U(F - Y) + \psi(\delta, F), \tag{9}$$

where δ denotes a Lagrange multiplier, $\psi(\delta, F)$ is also the formalized term derived from constraints. From the derivative of the above problem w.r.t. F, we have

$$\sum_{v=1}^{V} \lambda^v \frac{\partial \|F - FW_t^v\|_F^2}{\partial F} + \frac{\partial tr(F - Y)^T U(F - Y)}{\partial F} + \frac{\partial \psi(\delta, F)}{\partial F} = 0, \quad (10)$$

where λ^v is a weighting factor that trading-off the contributions of various view data on minimizing the reconstruction error, which can be calculated by

$$\lambda_{t+1}^v = 1/\left[2\sqrt{\|F - FW_t^v\|_F^2}\right] = 1/\left[2\sqrt{tr(FL_t^v F^T)}\right], \quad (11)$$

where $L_t^v = (I - W_t^v)(I - W_t^v)^T$ is an adaptive affinity matrix of view v, I is an identity matrix in $\mathbb{R}^N$, $I - W_t^v$ can be treated as the graph Laplacian, and the values of λ^v is similarly initialized as $1/V$, i.e., treating each view equally. Since λ^v is dependent F, we use a similar alternating strategy as [9] to solve F and λ^v iteratively. Specifically, when λ^v is known, the problem in Eq. (10) can be considered as

$$\min_F \ tr\left(F\left(\sum_{v=1}^{V} \lambda^v L_t^v\right) F^T\right) + tr(F - Y)^T U(F - Y). \quad (12)$$

From the derivative of the above problem w.r.t. F, we can update F as follows:

$$F_{t+1} = YU\left(\sum_{v=1}^{V} \lambda_t^v L_t^v + U\right)^{-1} = YU(L + U)^{-1}, \quad (13)$$

where $L = \sum_{v=1}^{V} \lambda^v L_t^v$. After getting F, we update $\lambda_{t=1}^v$ by Eq. (11). With F_{t+1} and $\lambda_{t=1}^v$ obtained, we can use them to update the graph weight matrix W^v.

(2) **Given F and λ^v, update the adaptive reconstruction weights W^v:**

We aim to update the graph weight matrix W^v over various views and sufficiently exploit the complementarity of multiple views. Given F_{t+1} and λ_{t+1}^v, the adaptive multiple-view weight matrices W_{t+1}^v can be updated as follows:

$$\begin{aligned}
W_{t+1}^v = \min_{W^v} \ &\sum_{v=1}^{V} \lambda_{t+1}^v \ tr\left(F_{t+1}(I - W_{t+1}^v)(I - W_{t+1}^v)^T F_{t+1}^T\right) \\
&+ \sum_{v=1}^{V} \left(\alpha\|X^v - X^v W^v\|_{2,1} + \beta\|W^v\|_{2,1}\right).
\end{aligned} \quad (14)$$

Due to the involvement of $L_{2,1}$-norm, it is very hard to obtain its closed solution directly. Although the reconstruction error $\sum_{v=1}^{V} \left(\alpha\|X^v - X^v W^v\|_{2,1} + \beta\|W^v\|_{2,1}\right)$ is generally convex, its derivative does not exist suppose that w_i^v and h_i^v are zeros, where h_i^v denotes the i-th vector of $H^v = X^v - X^v W^v$. As each $w_i^v \neq 0$ and $h_i^v \neq 0$, the above formulation can be approximated as follows:

$$W_{t+1}^v = \min_{W^v} \ tr\left(F_{t+1}\left(\sum_{v=1}^{V} \lambda_{t+1}^v L_{t+1}^v\right)F_{t+1}^T\right)$$
$$+ \sum_{v=1}^{V}\left[\alpha\left(tr(X^v - X^v W^v)^T G^v(X^v - X^v W^v)\right) + \beta tr\left(W^{vT}M^v W^v\right)\right], \quad (15)$$

where $G^v \in \mathbb{R}^{d^v \times d^v}$ and $M^v \in \mathbb{R}^{N \times N}$ denote two diagonal matrices defined as

$$G_{jj}^v = 1/\left[2\left\|(X^v)^j - (X^v W^v)^j\right\|_2\right], \quad M_{ii}^v = 1/\left[2\left\|(W^v)^i\right\|_2\right], \quad (16)$$

where $j = 1, \ldots, d^v$, $i = 1, 2, \ldots, N$, $(X^v - X^v W_{t+1}^v)^j$ is the j-th row of $X^v - X^v W_{t+1}^v$, and $(W_{t+1}^v)^i$ is the i-th row of W_{t+1}^v. When G^v and M^v are all fixed, the derivative in Eq. (14) can be regarded as a derivative of the optimization problem in Eq. (15). By setting the derivative of Eq. (15) w.r.t. each W^v to zero, we can have

$$W_{t+1}^v = \left(F_{t+1}^T F_{t+1} + \sum_{v=1}^{V} \alpha X^{vT} G_t^v X^v + \beta M_t^v\right)^{-1}\left(F_{t+1}^T F_{t+1} + \sum_{v=1}^{V} \alpha X^{vT} G_t^v X^v\right). \quad (17)$$

It should be noted that minimizing $tr(W^{vT}M^v W^v) = \|W^v\|_{2,1}/2$ will add explicit sparse constraint on W^v. After the weight matrix W_{t+1}^v is obtained, we can obtain the diagonal matrices G_{t+1}^v and M_{t+1}^v by Eq. (23). Finally, we can get an optimal W^{v*} and F^*. Finally, the hard label of each x_i can be obtained as $\arg\max_{i \le c}(f_i)_i$.

3.3   Convergence Analysis

As we solve the variables W^v, λ^v and F alternately, we need to describe the convergence of our method. Firstly, a lemma [15] for helping the proof is shown.

Lemma 1. For any nonzero vectors Φ and Ω, we have the following inequality:

$$\|\Phi\|_2 - \frac{\|\Phi\|_2^2}{2\|\Omega\|_2} \le \|\Omega\|_2 - \frac{\|\Omega\|_2^2}{2\|\Omega\|_2^2}. \quad (18)$$

Proposition 1. The objective function value of our MLP is non-increasing based in the proposed optimization process.

Proof: Suppose we fix F as F_t to compute W_{t+1}, we have the following inequality:

$$J\left(W_{t+1}^v, F_t\right) \le J\left(W_t^v, F_t\right). \quad (19)$$

Therefore, recalling the inequality in Lemma 1, it is easy to obtain the following two inequalities:

$$\sum \frac{\left\| \left(X^v - X^v W^v_{t+1} \right)^i \right\|_2}{\left\| \left(X^v - X^v W^v_{t+1} \right)^i \right\|_2} - \sum \left\| \left(X^v - X^v W^v_{t+1} \right)^i \right\|_2 \geq \sum \frac{\left\| \left(X^v - X^v W^v_t \right)^i \right\|_2}{\left\| \left(X^v - X^v W^v_t \right)^i \right\|_2}$$
$$- \sum \left\| \left(X^v - X^v W^v_t \right)^i \right\|_2. \tag{20}$$

$$- \left[\sum \left\| \left(W^v_{t+1} \right)^i \right\|_2 - \sum \frac{\left\| \left(W^v_{t+1} \right)^i \right\|_2}{\left\| \left(W^v_t \right)^i \right\|_2} \right] \geq - \left[\sum \left\| \left(W^v_t \right)^i \right\|_2 - \sum \frac{\left\| \left(W^v_t \right)^i \right\|_2}{\left\| \left(W^v_t \right)^i \right\|_2} \right]. \tag{21}$$

By combining the inequality in Eq. (19) with those in Eqs. (20)–(21), we can obtain

$$\begin{aligned} &\sum_{v=1}^{V} \sqrt{\left\| F_t - F_t W^v_{t+1} \right\|_F^2} + \sum_{v=1}^{V} \left(\alpha \left\| X^v - X^v W^v_{t+1} \right\|_{2,1} + \beta \left\| W^v_{t+1} \right\|_{2,1} \right) + tr(F_t - Y)^T U (F_t - Y) \\ &\leq \sum_{v=1}^{V} \sqrt{\left\| F_t - F_t W^v_t \right\|_F^2} + \sum_{v=1}^{V} \left(\alpha \left\| X^v - X^v W^v_t \right\|_{2,1} + \beta \left\| W^v_t \right\|_{2,1} \right) + tr(F_t - Y)^T U (F_t - Y) \end{aligned}. \tag{22}$$

It is clear that the above Proposition 1, can indicate that the objective function is non-increasing, but we still need to know F also converges. Thus, we can measure the difference between two sequential Fs based on the following metric:

$$Error(t) = \sum_{i=1}^{N} \left| \left\| f^i_{t+1} \right\|_2 - \left\| f^i_t \right\|_2 \right|. \tag{23}$$

We also provide two experiments for illustration. Figure 2 (a) is about the objective function and Fig. 2 (b) is the divergence between two consecutive F based on evaluation metric. The convergence analysis results of our method are shown in Fig. 2. ORL face image database (http://www.cl.cam.ac.uk/research/dtg/attarchive/facedatabase.html) is used as an exasmple. As can be seen, the produced objective function values of MLP are not increasing in the iterative optimizations and can converge to a fixed value. The divergence also converges to zero, i.e., the final result will not be changed drastically. The convergence speed of our method is relatively fast, and the number of iterations is usually less than 20.

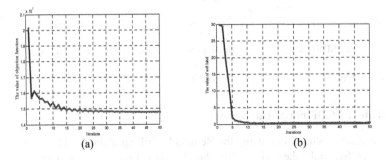

(a)　　　　　　　　　(b)

Fig. 2. Convergence behavior of our proposed MLP algorithm.

4 Experimental Results and Analysis

Experiments are conducted to examine MLP. For each database, we split it into a labeled set and an unlabeled set, where the unlabeled samples are to be classified. Compared methods use their own parameters for fair comparison. All simulations were performed on a PC with Intel (R) Core (TM) i5-4590 @ 3.30 Hz 8.00 GB.

4.1 Face Image Databases

1) *Dataset Descriptions and Setting:* Three face image databases, including ORL, UMIST and Georgia Tech face (http://www.face-rec.org/databases/) are used. We extract five visual features from each image, including the global principal component features (V1) [18] (500 dimensions), neighborhood preserving features (V2) [19], Fourier transform coefficients (V3) [20] (1024 dimensions), LBP features (V4) [21] (900 dimensions) and Relief features (V5) [24] (800 dimensions).

2) *Experimental results:* We mainly compared with those of multiple-view based AMGL and recent single-view based AdaptiveNP. Between AMGL and MLP, we provide the results of them based on all the views. While for AdaptiveNP, we present its results over each single view. For each method, we vary the number of labeled images from 1 to 9. The test results are illustrated in Fig. 3, where the abscissa denotes the randomly labeled number of images. From the results we can find: (1) The performance of each method can be enhanced by the increasing number of labeled face images in each class; (2) MLP delivers higher accuracies than other criteria in most cases, including the recently proposed AMGL, which can be attributed to the reasonable formulation of integrating the multiple-view label propagation and adaptive multiple graph weight learning into a unified model that can sufficiently explore the complementation over all views.

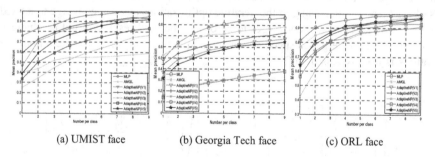

(a) UMIST face   (b) Georgia Tech face   (c) ORL face

Fig. 3. Classification results on different face recognition datasets.

4.2 Reuters Multilingual Database

We also evaluate each method using the Reuters Multilingual dataset [17] that has 1200 samples and five published views for classification. Firstly, we project the data by Latent Semantic Analysis (LSA) [22] to a 100-dimensional subspace and compute similarities in lower dimensional space, which is similar to compute the topic based similarity of documents. The results are illustrated in Fig. 4. We see that the classification results of each method can be improved by the increasing numbers of labeled data. MLP obtains better results under each setting. AMGL achieves the second best accuracy behind MLP; (3) For single-view models, their results over all views are worse than the multiple-view based AMGL and MLP, as the single-view models cannot exploit the common space of all views, and hence the connections of different views cannot be discovered accurately.

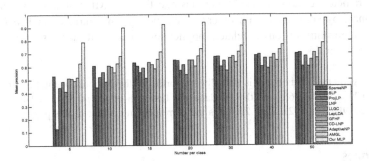

Fig. 4. Classification results of each method on Reuters Multilingual database.

4.3   Handwritten Numerals Database

The handwritten numerals dataset [16] contains 2,000 samples of 10 handwritten digit classes (i.e., 0 to 9). Following the common evaluations [18], six features are used for classification. In this study, 5, 10, 15, 20, 30, 40 and 50 samples from each digital data are randomly selected as labeled and set the remaining ones as unlabeled. The test results are illustrated in Fig. 5. We can also find that our proposed MLP algorithm can still produce more promising results by exploring the relations within each view and the correlations across all the views.

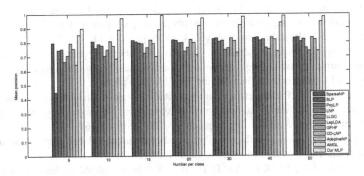

Fig. 5. Classification results of each method on handwritten numerals database.

5 Conclusion

We proposed an adaptive multiple-view label propagation framework for semi-supervised classification. MLP performs the label prediction adaptively based on the complementary space of multiple-views by mining the connections of multiple-views. For learning, MLP integrates multiview LP and adaptive weight learning into a unified framework, where a linear transformation is introduced to make different weights form different view spaces. Due to the adaptive way in defining the weight matrices from multiple views, MLP can avoid the tricky process to choose the neighbor number or kernel parameter. Extensive results demonstrate that our MLP obtains the enhanced performance, by comparing with other related models. In future, we will extend our proposed algorithm to other related application areas and deep semi-supervised learning scenario [30].

Acknowledgements. This paper is partially supported by the National Natural Science Foundation of China (61672365) and the Fundamental Research Funds for Central Universities of China (JZ2019HGPA0102). Zhao Zhang is the corresponding author of this paper.

References

1. Chapelle, O., Scholkopf, B., Zien, A.: Semi-Supervised Learning. MIT Press, Cambridge (2006)
2. Zhu, X., Ghahramani, Z., Lafferty, J.D.: Semi-supervised learning using gaussian fields and harmonic functions. In: Proceedings of International Conference on Machine Learning, pp. 912–919 (2003)
3. Zhou, D., Bousquet, O., Lal, T., Weston, J., Scholkopf, B.: Learning with local and global consistency. In: Neural Information Processing Systems, pp. 321–328 (2004)
4. Wang, F., Zhang, C.S.: Label propagation through linear neighborhoods. IEEE Trans. Knowl. Data Eng. **20**(1), 55–67 (2007)
5. Zhang, C., Wang, S., Li, D.: Prior class dissimilarity based linear neighborhood propagation. Knowl. Based Syst. **83**, 58–65 (2015)
6. Nie, F., Xiang, S., Liu, Y.: A general graph-based semi-supervised learning with novel class discovery. Neural Comput. Appl. **19**, 549–555 (2010)
7. Zhang, Z., Zhang, L., Zhao, M., Jiang, W., Liang, Y., Li, F.: Semi-supervised image classification by nonnegative sparse neighborhood propagation. In: ACM International Conference on Multimedia Retrieval, vol. 4, no. 6, pp. 139–146 (2015)
8. Jia, L., Zhang, Z., Wang, L., Zhao, M.: Adaptive neighborhood propagation by joint L2,1-norm regularized sparse coding for representation and classification. In: Proceedings of IEEE International Conference on Data Mining, pp. 201–210 (2016)
9. Nie, F., Li, J., Li, X.: Parameter-free auto-weighted multiple graph learning: a framework for multiview clustering and semi-supervised classification. In: Proceedings of International Joint Conference on Artificial Intelligence, pp. 1881–1887 (2016)
10. Zoidi, O., Tefas, A., Nikolaidis, N., Pitas, I.: Person identity label propagation in stereo videos. IEEE Trans. Multimedia **16**(5), 1358–1368 (2014)
11. Hou, C., Nie, F.P., Li, X.L., Yi, D.Y., Wu, Y.: Joint embedding learning and sparse regression: a framework for unsupervised feature selection. IEEE Trans. Cybern. **44**(6), 793–804 (2014)

12. Muslea, I., Minton, S., Knoblock, C.A.: Active learning with multiple views. J. Artif. Intell. Res. **27**, 203–233 (2006)
13. Tang, H., Fang, T., Shi, P.: Laplacian linear discriminant analysis. Pattern Recogn. **39**(1), 136–139 (2006)
14. Zhang, Z., Li, F. Zhao, M.: Transformed neighborhood propagation. In: Proceedings of International Conference on Pattern Recognition, Sweden, pp. 3792–3797 (2014)
15. Nie, F., Huang, H., Cai, X., Ding, C.: Efficient and robust feature selection via joint l2,1-norms minimization. In: Neural Information Processing Systems (NIPS) (2010)
16. Asuncion, A., Newman, D.J.: UCI Machine Learning Repository. University of California, Irvine (2007)
17. Kumar, A., Rai, P., Daume, H.: Co-regularized multi-view spectral clustering. In: Proceedings of Neural Information Processing Systems, pp. 1413–1421 (2011)
18. Jolliffe, I.: Principal Component Analysis, Wiley Online Library (2002)
19. He, X., Cai, D., Yan, S. Zhang, H.: Neighborhood preserving embedding. In: Proceeding of the IEEE International Conference on Computer Vision, vol.1, pp. 1208–1213 (2005)
20. Kuo, S.M., Lee, B.H.: Fast fourier transform and its applications, pp: 27–34 (1988)
21. Ojala, T., Pietikainen, M., Maenpaa, T.: Multiresolution gray-scale and rotation invariant texture classification with local binary patterns. IEEE Trans. Pattern Anal. Mach. Intell. **24** (7), 971–987 (2002)
22. Evangelopoulos, N.E.: Latent semantic analysis. Wiley Interdisc. Rev. Cogn. Sci. **4**(6), 683–692 (2013)
23. Zhang, Z., Jiang, W., Li, F., Zhang, L., Zhao, M., Jia L.: Projective label propagation by label embedding. In: Computer Analysis of Images and Patterns, Valetta, Malta (2015)
24. Aha, D., Kibler, D.: Instance-based learning algorithms. Mach. Learn. **6**(1), 37–66 (1991)
25. Zhang, Z., Li, F., Jia, L., Qin, J., Zhang, L., Yan, S.: Robust adaptive embedded label propagation with weight learning for inductive classification. IEEE Trans. Neural Netw. Learn. Syst. **29**(8), 3388–3403 (2018)
26. Zhang, Z., Jia, L., Zhao, M., Liu, G., Wang, M., Yan, S.: Kernel-induced label propagation by mapping for semi-supervised classification. IEEE Trans. Big Data **5**(2), 148–165 (2019)
27. Jia, L., Zhang, Z., Jiang, W.: Transductive classification by robust linear neighborhood propagation. In: Chen, E., Gong, Y., Tie, Y. (eds.) PCM 2016. LNCS, vol. 9916, pp. 296–305. Springer, Cham (2016). https://doi.org/10.1007/978-3-319-48890-5_29
28. Zhang, H., Zhang, Z., Zhao, M., Ye, Q., Zhang, M., Wang, M: Robust triple-matrix-recovery-based auto-weighted label propagation for classification. IEEE Trans. Neural Netw. Learn. Syst. (2019). https://doi.org/10.1109/TNNLS.2019.2956015
29. Zhang, Z., Jia, L., Zhao, M., Ye, L., Zhang, M., Wang, M.: Adaptive non-negative projective semi-supervised learning for inductive classification. Neural Netw. **108**, 128–145 (2018)
30. Iscen, Z., Tolias, G., Avrithis, Y., Chum, Q.: Label propagation for deep semi-supervised learning. In: Proceedings of IEEE CVPR, pp. 5070–5079 (2019)
31. Lin, G., Liao, K., Sun, B., Chen, Y., Zhao, F.: Dynamic graph fusion label propagation for semi-supervised multi-modality classification. Pattern Recogn. **68**, 14–23 (2017)
32. Karasuyama, M., Mamitsuka, H.: Multiple graph label propagation by sparse integration. IEEE Trans. Neural Netw. Learn. Syst. **24**(12), 1999–2012 (2013)

Container Damage Identification Based on Fmask-RCNN

Xueqi Li[✉], Qing Liu[✉], Jinbo Wang, and Jiwei Wu

School of Automation Wuhan University of Technology, Wuhan 430070, China
719488621@qq.com

Abstract. The inspection of container body damage is an inevitable detection work for containers entering the terminal of the port area. In the past, this work was manually recorded, with hidden safety risks and inaccurate handling. To solve this problem, the paper proposes a Fmask-RCNN model. It is based on Mask-RCNN, introducing Res2Net101 framework and adding path fusion augmentation, multiple fully connected layers, fusion upsampling or enhancement. Fmask-RCNN model is applied to the identification of port container damage, and 4407 samples are trained and tested, including 2537 training sets and 1870 testing sets which are untrained. The miss rate of the final damage identification was 4.599% and the error rate was 18.887%.

Keywords: Fmask-RCNN · Deep learning · Container damage identification · Port application

1 Problem Raising

In recent years, Chinese port container transport industry has been developing rapidly. In fully automated terminals, information perception system, operation control system, dispatching command system and other systems are becoming more and more perfect, but the automatic container inspection module has not been fully developed. The inspection of damaged container body is an inevitable inspection work of container entering the wharf of port area, mainly to prevent the disputes caused by damaged container body between transportation enterprises and wharf. The common types of port container damage include dented, convex, hole, broken, distorted, scratch, mark broken, damage of container door, deformation of container door railing, etc. At present, there is no corresponding research result based on visual technology to realize the detection and identification of the damaged state of containers, but the application scope of the damaged state detection is wide. The research and application of non-destructive testing and health testing have been carried out in construction, power system, automobile and other fields, including non-contact method based on vision to detect the damaged state. For example, in 2017, Deep Learning-Based Crack Damage Detection Using Convolutional Neural Networks [1] was proposed to detect concrete cracks by Convolutional Neural Networks. Ashley Varghese et al. published Power infrastructure monitoring and damage detection using drone captured images [2] to detect Power infrastructure damage using GoogLeNet. Vehicle Damage based on Mask

H. Zhang et al. (Eds.): NCAA 2020, CCIS 1265, pp. 12–22, 2020.
https://doi.org/10.1007/978-981-15-7670-6_2

r-cnn was proposed in 2018 [3], in which Mask r-cnn was used to detect the damaged parts of vehicles and it achieved ideal detection results.

As the proportion of damaged containers in the total number of containers in the port is small, the sample size is less under such working conditions, and the location, size and category of damage are uncertain. In addition, due to the change of background, it is easy to have missed and false detection. The traditional method and previous neural network framework are not very effective in detecting such conditions. Therefore, based on the idea of detecting the damaged parts of automobiles [3], the paper proposes the Fmask-RCNN network algorithm to identify the damaged parts of containers, which increases the multi-scale features, enhances the utilization rate of the location information, reduces the upsampling information loss and increases the efficiency of existing sample. Fmask-RCNN aims to decrease the omission and error detection of damage identification, to replace manual testing and can check the surface condition of container quickly and accurately in the unmanned operation mode.

2   Container Damage Detection Based on Mask-RCNN

Mask RCNN is a two-stage detector, which generates candidate regions for detection in the first stage, and performs classification of regional objects and regression of boundary boxes in the second stage. Such models can achieve a high accuracy rate, reaching 35.7 AP in the standard COCO test set, which is better than all similar models [4]. In addition, the classification, regression and segmentation of the model are parallel to each other. Compared with the traditional image processing algorithm of "segmentation first and classification later", this parallel design is simple and efficient. Mask r-cnn is used to detect the damaged containers. The framework of Mask r-cnn is shown in Fig. 1.

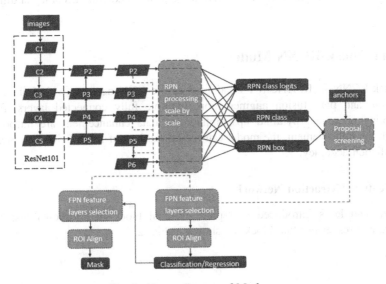

Fig. 1. Frame diagram of Maskrcnn

As shown in Fig. 1, features C1–C5 of different depths are extracted in the input image through the ResNet101 network, and FPN is used to combine feature maps C2, C3, C4 and C5 to form new feature maps P2, P3, P4, P5 and P6, so that feature information of different depths is included in P2–P6 at the same time. For each point in p2–p6, multiple anchors are preset with different widths and heights, centered on its coordinates. According to RPN network, the center, width and height of each Anchor are modified, and the generated box is screened. ROI Align is used to cut out the features corresponding to the candidate box from the feature map, transform the features into uniform size to classify, regress and segment.

Mask-RCNN is used to train and test the sample set. Its omission rate was 12.246% and the error rate was 20.267%, which was still unsatisfactory. Since the container body is large, the damage area is small, and the background in the port environment is variable, small targets are easy to be missed and the background is easy to be wrongly detected when Mask-RCNN is used for damage identification. Therefore, Fmask-RCNN model is proposed to improve the model. Fmask-RCNN is based on Mask-RCNN and is improved as follows:

(1) Optimize the feature extraction network and increase the information utilization to reduce the loss of information.

(2) P2–P6 layer in Mask-RCNN increase the utilization of high-level information, but lose low-level information. Therefore, this part is improved to increase the utilization rate of low-level information without reducing the utilization rate of high-level information.

(3) Parameters need to be set artificially when bilinear interpolation is adopted for upsampling, which will cause the loss of partial information. This part is improved to reduce the influence of artificial parameter setting.

(4) Since the full connection layer predicts different positions of the feature graph through a large number of parameters, the full connection layer is position sensitive. Multiple fully connected layers are used to increase the prediction accuracy at different scales.

3   The Fmask-RCNN Model

In Fmask-RCNN n, Res2Net101 is adopted as the CNN network with feature map extraction, and path fusion augmentation, multiple fully connected layers, fusion upsampling or spatial pyramid upsampling and data enhancement are introduced. Through the improvement, the model can reduce the loss of information and increase the prediction accuracy.

3.1   Feature Extraction Network

Res2Net module is introduced to replace residual block in traditional resnet. Its structure and resnet residual block are shown in Fig. 2.

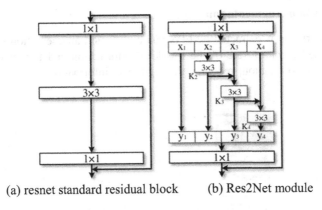

(a) resnet standard residual block   (b) Res2Net module

Fig. 2. Res2Net module and resnet residuals

In Fig. 2, resnet standard residuals are serialized by 3 convolutional layers (1 * 1, 3 * 3, 1 * 1) as a residual module, while Res2Net divides the input feature map into several groups. A set of convolution kernels first extracts features from a group of input feature maps. Then the output feature graph is combined with another group of input feature maps and features are extracted through another set of convolution kernels. This process is repeated several times until all the input feature maps are processed. Finally, the feature maps of all groups are spliced together, and then the feature fusion is carried out through the convolution kernel of 1 × 1. There is no convolution operation for the first split. One is to reduce the number of parameters, and the other is to reuse features. Res2Net represents features at multiple scales and increases the receptive field range for each network layer.

The 101-layer Res2Net module is used to extract the features of the input image. The architecture is shown in Table 1.

Table 1. Res2Net101 framework.

Res2Net -101 -layer					
Layer	Conv1	Conv2	Conv3	Conv4	Conv5
para	$[3 \times 3, 64] \times 3$ 3×3 max pool	$\begin{bmatrix} 1 \times 1, 64 \\ 3 \times 3, 64 \\ 1 \times 1, 256 \end{bmatrix} \times 3$	$\begin{bmatrix} 1 \times 1, 128 \\ 3 \times 3, 128 \\ 1 \times 1, 512 \end{bmatrix} \times 4$	$\begin{bmatrix} 1 \times 1, 256 \\ 3 \times 3, 256 \\ 1 \times 1, 1024 \end{bmatrix} \times 23$	$\begin{bmatrix} 1 \times 1, 512 \\ 3 \times 3, 512 \\ 1 \times 1, 2048 \end{bmatrix} \times 3$

The first layer of Res2Net101 model adopts three 3 * 3 convolution kernels, channel number is 64, stride is 2, and input to the second layer after standardization and pooling. The second layer sets 3 Res2Net modules, increasing the number of channels from 64 to 256. The third layer sets 4 Res2Net modules, increasing the number of channels to 512. The fourth layer sets 23 Res2Net modules, increasing the number of channels to 1024. The last layer sets up 3 Res2Net modules, increasing the number of channels to 2048.

3.2   Path Fusion Augmentation

The candidate region is extracted by FPN and path fusion augmentation is introduced to improve the utilization rate of the low-level information and the accuracy of the location information without reducing the high-level information.

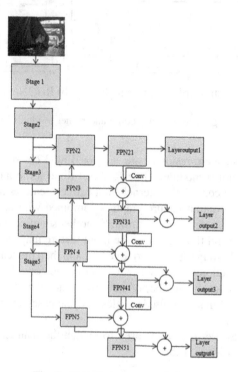

Fig. 3. Path fusion augmentation.

As shown in Fig. 3, the original Maskrcnn uses FPN to generate FPN2-FPN5 for the next conversion, and here the FPN2-FPN5 is enhanced. FPN2, one feature map of FPN output, is converted to FPN21, and it is convoluted to get FPN31 after adding it to FPN3. Similarly, FPN31 is convolved with the convolution kernel of 3×3, and FPN4 is added to obtain FPN41. FPN51 is obtained from FPN41. In addition to the direct output of the first layer, the next three layers fuse the input and output of the previous steps and then output, that is, FPN3 and FPN31, FPN4 and FPN41, FPN5 and FPN51 are fused respectively to obtain the output of the layer.

3.3   Multiple Fully Connected Layers

In order to improve the quality of mask prediction, multiple fully connected layers are introduced into the mask prediction branch. Because fully connected layer predicts different positions of the feature map by a large number of parameters, it is position sensitive. Moreover, it can use global information to predict each subregion, which is helpful to distinguish different instances and identify different parts belonging to the same object. Multiple fully connected layers are used to increase the prediction accuracy at different scales.

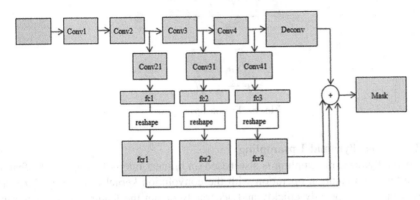

Fig. 4. Multiple fully connected layers.

The structure of multiple fully connected layers is shown in Fig. 4. Three branches are introduced into the original mask prediction branch, and convolutional layer, fully connected layer and dimension modification part are set in the branch. The feature maps obtained by conv2, conv 3 and conv 4 through this branch are fused with the feature map obtained by deconvolution with conv 4 to get the final output.

3.4 Upsampling Improvement

Bilinear interpolation is one of the most widely used methods in upsampling. Its advantages are no need to train, fast running speed and simple operation. The disadvantage is that it relies on artificial parameters. The advantage of deconvolution is that it can learn to optimize its parameters, but may have checkerboard artifacts.

Based on this, an improved upsampling method is proposed in the paper. Two schemes are proposed, namely, the fusion upsampling and spatial pyramid upsampling.

3.4.1 Fusion Upsampling
Based on bilinear interpolation and deconvolution, the fusion upsampling can not only optimize the parameters through deconvolution learning, but also quickly and accurately obtain the feature graph of upsampling.

Figure 5 shows the schematic of fusion upsampling. Deconvolution and bilinear interpolation are carried out for the feature map that needs to be sampled up at the same time. The obtained feature maps are fused to obtain the final result.

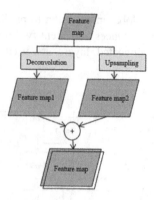

Fig. 5. Fusion upsampling

3.4.2 Spatial Pyramid Upsampling

The spatial pyramid upsampling is also based on bilinear interpolation, and the feature maps after deconvolution operation with the convolution kernel of different sizes are fused, which can not only quickly and accurately obtain the feature graph of upsampling, but also optimize the spatial information through deconvolution learning parameters of different sizes.

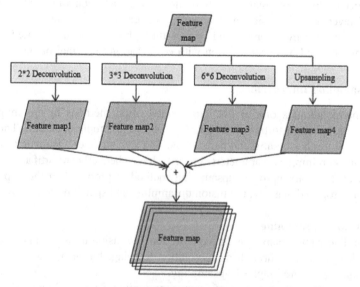

Fig. 6. spatial pyramid upsampling

Figure 6 is the schematic diagram of spatial pyramid upsampling. Bilinear interpolation and deconvolution of different convolution kernel sizes are performed on the feature graphs that need to be sampled. The obtained feature maps are fused to obtain the final result.

3.5   Data Enhancement

In order to identify such conditions as container damage, data enhancement are used for the existing samples to increase the number of training samples and improve the generalization ability of the network. At the same time, noise data is introduced to improve the robustness of the model.

The paper proposes two data enhancement schemes. The first is to flip only 50% of the dataset horizontally and vertically and introduce a random gaussian blur. The second data enhancement operation is applied to 50% of the data set. 10 enhancement operations are set. When data enhancement is used, choose from 1 to 5 of the 10 enhancement operations to process the image. Here are 10 enhancements: (1) flip horizontal; (2) flip vertical; (3) random gaussian blur introduction; (4) image scaling; (5) image contrast change, including contrast raising and lowering; (6) image sharpening and the mix of the sharpened image and input image using a random mixing factor between 35% and 60%; (7) input pixel value reverse, that is, the original value is v, and the changed value is 255-v; (8) random deformation of the image; (9) image shifting to left, right, up and down; (10) image rotating clockwise and counter clockwise.

4   Experiment and Analysis

The sample set is trained and tested by Mask-RCNN. The improved methods proposed above are respectively introduced into the model to test the effects of different methods on the identification of container damage under the conditions of the same number of samples, training steps and parameter settings, and to calculate the rate of omission and error detection respectively.

Table 2. Comparison between different improved methods and the original model for omission and error detection.

Model	Omission rate	Error rate
Mask-RCNN	12.246%	20.267%
Mask-RCNN, with the introduction of path fusion augmentation	9.733%	18.556%
Mask-RCNN, with the introduction of fusion upsampling	10.214%	17.166%
Mask-RCNN, with the introduction of spatial pyramid upsampling	10.214%	21.390%
Mask-RCNN, with the introduction of flip fuzzy data enhancement	10.909%	17.166%
Mask-RCNN, with the introduction of multi-processing data enhancement	11.711%	15.294%

As can be seen from Table 2, the omission rate of Mask-RCNN was 12.246%, and the error rate was 20.267%. When path fusion augmentation is introduced, the miss rate and error rate are reduced by 2.513% and 1.711% respectively. With the introduction of fusion upsampling, the missed detection rate decreases by 2.032% and the false detection rate decreases by 3.101%. When spatial pyramid upsampling is introduced, the omission rate decreases by 2.032% and the false rate increases by 1.123%. With the introduction of the flip fuzzy data enhancement, the missed detection rate decreases by 1.337% and the false detection rate decreases by 3.101%. As multi-processing data enhancement is introduced, the omission rate decreased by 0.535% and the error rate decreased by 4.973%. It can be seen that the proposed improvement method can be introduced into the model to effectively improve the model's omission and error detection of container damage identification to a certain extent.

Multiple methods proposed above are introduced into the original Mask-RCNN model to detect the omission rate and error rate respectively under the same samples. The results are shown in Table 3.

Table 3. Comparison of omission and error detection in different improved methods.

Model	Omission rate	Error rate
Mask-RCNN, introducing path fusion augmentation and multiple fully connected layers	8.449%	19.840%
Mask-RCNN, introducing path fusion augmentation, multiple fully connected layers and fusion upsampling	7.112%	19.572%
Mask-RCNN, introducing path fusion augmentation, multiple fully connected layers, fusion upsampling and multi-processing	9.198%	12.941%
Fmask-RCNN	4.599%	18.887%
Fmask-RCNN, training steps increased to 150,000 times	6.364%	12.032%

As can be seen from Table 3, with the introduction of path fusion augmentation and multiple fully connected layers, the missed detection rate decreases by 3.797% and the false detection rate decreases by 0.427%, when compared with the original model. The missed detection rate is 5.134% lower than the original model and the false detection rate is 0.695% lower when path fusion augmentation, multiple fully connected layers and fusion upsampling are introduced. With the introduction of path fusion augmentation, multiple fully connected layers, fusion upsampling and multi-processing, the omission rate decreases by 3.048% and the error rate decreases by 7.326%. While Fmask-RCNN, that is, the CNN network using Res2Net as feature map extraction, introduced path fusion augmentation, multiple fully connected layers, fusion upsampling and flip fuzzy data enhancement, is used, after screening the positioning results, the miss rate is 7.647% lower than the original model, and the error rate is 1.38% lower. The Fmask-RCNN model is adopted, and the training steps are increased. The omission rate decreases by 5.882% and the error rate decreases by 8.235%. Based on the condition of container damage identification, the scheme with a low omission rate is

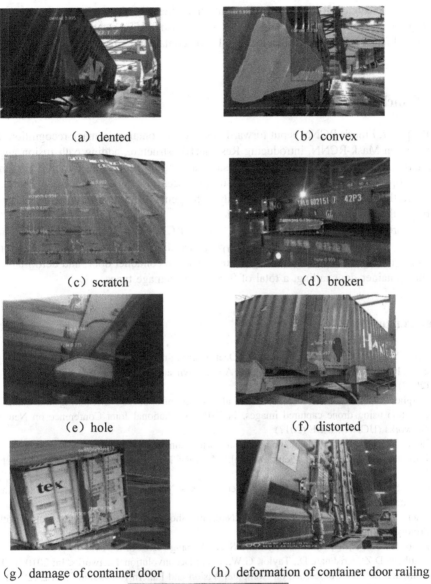

(a) dented   (b) convex

(c) scratch   (d) broken

(e) hole   (f) distorted

(g) damage of container door   (h) deformation of container door railing

(i) mark broken

Fig. 7. Identification results of container damage

preferred, so Fmask-RCNN is finally chosen to identify the samples of container damage, with an omission rate of 4.599% and a misdetection rate of 18.887%.

The Fmask-RCNN identification results of container damage are shown in the Fig. 7.

5   Conclusion

In the paper, Fmask-RCNN is put forward to solve the container damage recognition. It is based on Mask-RCNN, introducing Res2Net101 structure, adding path fusion augmentation, multiple connection layer, fusion upsampling or spatial pyramid upsampling, flip fuzzy data enhancement or multi-processing data enhancement. It can enhance the robustness of model, strengthen the generalization ability of model, and improve the positioning accuracy of model.

According to the experimental data, Fmask-RCNN can accurately identify and display a variety of container damage types, including dented, convex, hole, deformation, broken, scratch, mark damage, damage to the container door, and deformation of the container door railing, a total of 9 common damage types.

References

1. Cha., Y.J., Choi., W., Büyükztürk., O.: Deep learning-based crack damage detection using convolutional neural networks. Comput. Aided Civil and Infrastruct. Eng. **32**(5), 361–378 (2017)
2. Varghese, A., Gubbi, J., Sharma, H., et al.: Power infrastructure monitoring and damage detection using drone captured images. In: 2017 International Joint Conference on Neural Networks (IJCNN). IEEE (2017)
3. Priya, D., et al.: Vehicle Damage based on Mask r-cnn[J/OL] (2018)
4. Kaiming., H., Georgia., G., Piotr., D., et al.: Mask R-CNN. IEEE Trans. Pattern Anal. Mach. Intell. **1**, 1 (2018)
5. Gao, S.H., Cheng, M.M., Zhao, K., et al.: Res2Net: A New Multi-scale Backbone Architecture (2019)
6. Wang, K., Liew, J.H., Zou, Y., et al.: PANet: Few-Shot Image Semantic Segmentation with Prototype Alignment (2019)
7. Zhao, H., Shi, J., Qi, X., et al.: Pyramid Scene Parsing Network (2016)
8. Matthew, D.Z., Krishnan, D., Taylor, G.W., et al.: Deconvolutional networks. In: 2010 IEEE Computer Society Conference on Computer Vision and Pattern Recognition. IEEE (2010)
9. Ankit, V., Monika, S., Ramya, H., et al.: Automatic container code recognition via spatial transformer networks and connected component region proposals. Int. Conf. Mach. Learn. Appl. IEEE (2016)

Multiagent Reinforcement Learning for Combinatorial Optimization

Yifan Gu, Qi Sun, and Xinye Cai[✉]

Nanjing University of Aeronautics and Astronautics, Nanjing 210016, Jiangsu, China
alvinguyf@163.com, sunshinesq@163.com, xinye@nuaa.edu.cn

Abstract. In this paper, we combine multiagent reinforcement learning (MARL) with grid-based Pareto local search for combinatorial multiobjective optimization problems (CMOPs). In the multiagent system, each agent (grid) maintains at most one solution after the MARL-guided selection for local search. MARL adaptively adjusts the selection strategy for conducting better collaborative Pareto local search. In the experimental studies, the MARL-guided grid Pareto local search (MARL-GPLS) is compared with the Pareto local search (PLS), two decomposition-based multiobjective local search approaches, a grid-based approach (ϵ-MOEA), and one state-of-the-art hybrid approach on benchmark CMOPs. The results show that the MARL-GPLS outperforms the other six algorithms on most instances.

Keywords: Combinatorial optimization · Multiagent reinforcement learning · Grid-based dominance · Local search

1 Introduction

The combinatorial optimization problems (COPs), such as traveling salesman problem [19], knapsack problem [4], flow shop scheduling problem [5], software next release problem [8], vehicle routing problem [16], have been researched for decades due to their wide applications in real life. When such COPs have multiple objective to be optimized simultaneously, they are called combinatorial multi-objective optimization problems (CMOPs).

CMOPs are usually NP-hard by nature. Local search (LS) is a typical method to solve COPs which iteratively tries to improve the current solution by exploring a better soluin its neighborhood. Pareto local search (PLS) [12] can be considered as an extension of LS for CMOPs. The process of PLS can be described as searching the neighborhood of non-dominated solution sets and selecting more efficient ones based on Pareto dominance for the next iteration. As PLS needs to reserve all the nondominated solutions for LS, it is usually time-consuming.

There already exist several works of adopting classic or deep reinforcement learning techniques in multi-objective optimization algorithms. An $(1 + 1)$ evolution strategy by controlling the step-size such as standard deviation through reinforcement learning [15]. Later work [9] extended the use of reinforcement

© Springer Nature Singapore Pte Ltd. 2020
H. Zhang et al. (Eds.): NCAA 2020, CCIS 1265, pp. 23–34, 2020.
https://doi.org/10.1007/978-981-15-7670-6_3

learning to the simultaneous control of multiple numerical evolution parameters. According to [1], reinforcement learning is used to select from a set of handcrafted auxiliary fitness functions that can be added to the main objective function, in order to reshape the fitness landscape. Reinforcement learning is used to control two numerical parameters per individual in the local search strategy of a memetic algorithm [2]. It should also be noted that all the aforementioned works are concerned with learning during the execution of an evolutionary algorithm for a specific problem instance, whereas our method is designed to learn throughout the evolutionary algorithm in multiple runs. To the best of our knowledge, there has been no previous work on combining multi-agent reinforcement learning with PLS for multi-objective combinatorial optimization.

In this paper, multiagent reinforcement learning (MARL) is adopted to enhance the Pareto local search in a grid system. More specifically, each agent (grid) maintains one representative solution in a grid system for Pareto local search. MARL is used for adaptively selecting a diversely-populated nondominated set that enables better collaborative local search.

The rest of this paper is organized as follows. Section 2 introduces the relative backgrounds. Some basic knowledge of CMOPs, grid-based Pareto dominance, and multiagent reinforcement learning (MARL) are presented in this section. The MARL-based grid Pareto local search (MARL-GPLS) is introduced in Sect. 3. The experimental studies are given in Sect. 4 which includes the introductions of two benchmark CMOPs, parameter settings, and the experimental results and analysis. Finally Sect. 5 concludes this paper.

2 Background

2.1 Definitions

A *multiobjective optimization problem* (MOP) can be formally stated as follows:

$$
\begin{aligned}
\text{minimize } & F(x) = (f_1(x), \ldots, f_m(x)) \\
\text{subject to } & x \in \Omega
\end{aligned}
\tag{1}
$$

where Ω denotes the *decision space*, $F : \Omega \to R^m$ means m real-valued objective functions. $\{F(x) | x \in \Omega\}$ is known as the *attainable objective set*. Equation 1 is called a combinatorial MOP (CMOP) when Ω is a finite set.

Let $p, q \in R^m$, q is said to be *dominated* by p, denoted by $p \prec q$, if and only if $p_i \leq q_i$ for every $i \in \{1, \ldots, m\}$ and $p_j < q_j$ for at least one index $j \in \{1, \ldots, m\}$. Given a set P in R^m, if no other solution in P can dominate the solution, then the solution is called non-dominated in P. In other words, a solution $x^* \in \Omega$ is called as *Pareto-optimal* if and only if $F(x^*)$ is non-dominated in the attainable objective set.

The *ideal* and *nadir points* are used to bound the PFs and construct the grid system. The ideal objective vector $z^* = (z_1^*, \ldots, z_m^*)^T$ can be computed by

$$
z_j^* = \min_{x \in \Omega} f_j(x), j \in \{1, \ldots, m\}.
\tag{2}
$$

The nadir point $z^{nad} = (z_1^{nad}, \ldots, z_m^{nad})^T$ can be computed by

$$z_j^{nad} = \max_{x \in PS} f_j(x), j \in \{1, \ldots, m\} \tag{3}$$

2.2 Grid Pareto Dominance

To integrate MARL into PLS, this paper adopts multi-agent system in a grid system. The following part is the definition of grid system.

In the grid system [4], L is division parameter and D is the width of each grid interval for each objective

$$D = (d_1, ..., d_m)^T$$
$$\text{where} \quad d_i = (z_i^{nad} - z_i^*)/L \tag{4}$$

Theorem 1 (Index of Grid). *For each solution y, if $k_i = \lfloor (f_i(y) - z_i^*)/d_i \rfloor$, $\forall x \in 1, .., m$, then $K = (k_1, .., k_m) \in 0, ..., (L-1)$ denotes the grid index of y.*

Theorem 2 (Corner of Grid). *For each solution y with grid index $(k_1, ..., k_m)$, its grid corner point z^{gcp} denotes as follow:*

$$z^{gcp} = (z_1^{gcp}, ..., z_m^{gcp})^T$$
$$\text{where} \quad z_i^{gcp} = z_i^* + d_i \times k_i \tag{5}$$

Theorem 3 (Grid Strong dominance). *Let $y^a, y^b \in \Omega$, grid index $K^a = (k_1^a, ..., k_m^a)$ and $K^b = (k_1^b, ..., k_m^b)$.*

$$y^a \prec_{sg} y^b \Leftrightarrow \quad \forall i \in (1, ..., m), k_i^a < k_i^b \tag{6}$$

where $y^a \prec_{sg} y^b$ denotes that y^a strongly grid-dominates y^b.

Theorem 4 (Grid Pareto Dominance (GPD)). *Let $y^a, y^b \in \Omega$, $K^a = (k_1^a, ..., k_m^a)$, and $K^b = (k_1^b, ..., k_m^b)$ are their grid index, respectively.*

$$y^a \prec_{gpd} y^b \Leftrightarrow \begin{cases} y^a \prec y^b, & K^a = K^b \\ y^a \prec_{sg} y^b, & otherwise \end{cases} \tag{7}$$

2.3 Model and Problem Formulation

Multiagent Model in a Grid System. In this section, the selection of non-dominated solutions by MARL for PLS is described as follows.

As illustrated in Fig. 1, there are N agents $\mathcal{N} = \{1, ..., i, ..., N\}$ in the considered model. Each agent can represent a set of grids. In a particular iteration, each agent operates only in one grid as solutions in other grids have been grid Pareto dominated.

The space of each grid is divided into K subregions, according to [11], and each agent selects one of the K subregions according to its strategy. A solution closest to the grid corner point within the selected subregion by the agent is selected for PLS.

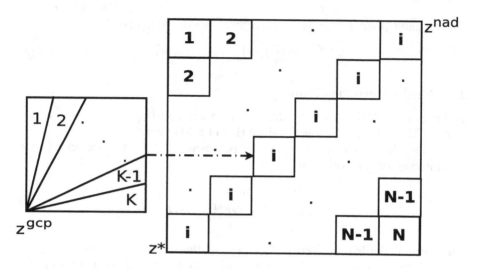

Fig. 1. The demonstration of agents in a grid-based objective space.

Solution Selection Problem Formulation. The problem can be formulated as a Markov game [3]. Mathematically, it can be described as $\mathcal{G} = (\mathcal{S}, \mathcal{N}, \Theta, F, \mathcal{R})$, where $\mathcal{S}$ is the finite set of state with $\mathcal{S} = \mathcal{S}_1 \times ... \times \mathcal{S}_n$, $\mathcal{S}_i$ is a binary vector representing whether the subregion has solutions. $\mathcal{N}$ is the set of agents, Θ is the finite joint action set and Θ_i is agent i's action set, $F : S \times \Theta \times S$ is the probability function of state transition. $\mathcal{R} = \{R_1, ..., R_n\}$, R_i is agent i's reward calculated by a real value function.

At any iteration t, the feasible subregions which contain solutions in the grid represented by the agent n can be observed. Hence, the state s_n^t for each agent n, $n \in \mathcal{M}$, is fully observed. $s_n^t(i), i \in \{1, ..., K\}$, denotes whether the subregion Ω_i, the definition of Ω is the same as [11], have solutions. The definition of $s_n^t(i)$ is:

$$s_n^t(i) = \begin{cases} 1, & \exists x \in \Omega_i \\ 0, & otherwise \end{cases} \tag{8}$$

For combinatorial multiobjective optimization, hypervolume (HV) [22] is a comprehensive quality indicator that can represent the quality of a non-dominated set. It is adopted as the reward function for MARL in this paper. Therefore, the reward of agent n at iteration t can be described as:

$$R_n^t = HV(gpls(pop_n^t)) - HV(pop^t) \tag{9}$$

where pop^t denotes the population at iteration t, pop_n^t denotes the solution in the grid which agent n represents, $gpls$ denotes the grid Pareto local-search operator, $gpls(pop_n^t)$ denotes the non-dominated solutions generated by the grid Pareto local search from pop_n^t.

2.4 MARL

This paper uses an IL-based MARL [13] to solve the area selection among agents. Each agent adopts a Q-learning algorithm to update its Q-value and determines the best strategy for MDP at the same time. The update rule [20] is

$$
\begin{aligned}
Q_n^{t+1}(s_n, \theta_n) = {} & Q_n^t(s_n, \theta_n) + \alpha^t \{ R_n^t \\
& + \delta \max_{\theta_n'} Q_n^t(s_n', \theta_n') - Q_n^t(s_n, \theta_n) \}
\end{aligned}
\tag{10}
$$

with $s_n^t = s_n$, $\theta_n^t = \theta_n$, where s_n' and θ_n' correspond to s_n^{t+1} and θ_n^{t+1}. α_t denotes the learning rate and Q_n^t denotes the action-value of agent n at iteration t.
ϵ-greedy method [20] at state s_n is

$$
\pi_n(s_n, \theta_n) = \begin{cases} \epsilon, & \text{if } Q_n \text{ of } \theta_n \text{ is the highest} \\ 1 - \epsilon, & \text{otherwise} \end{cases}
\tag{11}
$$

where $\epsilon \in (0, 1)$. According to [6], the learning rate α_t is given by

$$
\alpha_t = \frac{1}{(t + c_\alpha)^{\varphi_\alpha}}
\tag{12}
$$

where $c_\alpha > 0$, $\varphi_\alpha \in (\frac{1}{2}, 1]$. Each agent n runs the Q learning process independently in IL-based MARL.

3 MARL-GPLS

MARL-GPLS has four major steps: 1) initialization; 2) grid Pareto local search; 3) updating agents' Q-value; 4) solution selection based on MARL.

3.1 The Framework of MARL-GPLS

As illustrated in Fig. 2 and Algorithm 1, the population P is initialized, and then TP is generated by processing local search and grid Pareto dominance based on P. After that, the process of MARL-based selection generates state and transfers them to the model of MARL based on population TP, and then the model gives a set of actions which tell PLS from which subregions to select solutions for next generation.

The population size, the subregion number of each agent, and the division parameter L are set as N, AN, L. The output of this algorithm is the final population P obtained by the whole process, which means the optimal solutions for this problem. Due to fixed-parameter L and only one solution allowed in each grid in MARL-GPLS, the size of the population adaptively changes during evolution. It's worth noting that the update of agents of MARL starts from the second iteration.

Algorithm 1. MARL-GPLS

   Input :
         N, L
         AN: the subregion number
   Output:
         P: final population
   `// Step 1: Initialization:`
1   $P = Initialize\_population(N)$;
2   update ideal and nadir points based on Eq. 2 and Eq. 3;
3   set grid system based on z^*, z^{nad} and L;
4   **for** *each* $n \in \mathcal{N}$ **do**
5   Initialize the action-value $Q_n^t(s_n, \theta_n)$, strategy $\pi_n(s_n, \theta_n) = \frac{1}{|\Theta_n|}$, for all
       $\theta_n \in \Theta_n$;
6   Initialize the state $s_n = s_n^t = 0$ according to AN;
7   **end**
8   $t = 1$;
9   **while** *terminate criterian is not met* **do**
10     $TP = \emptyset$;
11     $PopHV = HV(P)$;
      `// Step 2: Pareto Local Search`
12   **for** *each* $x \in P$ **do**
13   **for** *each* $y \in N(x)$ **do**
14   **if** $x \not\succ_{gpd} y$ **then**
15           $TP =$ UPDATEPOP(y, TP);
16   **end**
17   **end**
18       $tHV =$ HV(TP);
19       $r = tHV - PopHV$;
20   obtain agent n corresponding to solution x;
21       $R_n = r$;
22   **end**
23   use Eq. 2 and Eq.3 to update ideal and nadir points ;
24   update grid system;
      `// Step 3: Update Q-value`
25   **if** $t > 1$ **then**
26   **for** *each* $n \in \mathcal{N}$ **do**
27   update the learning rate α_t according to Eq. 12;
28   observe the state vector s_n^{t+1} according to Eq. 8;
29   update the action-value $Q_n^{t+1}(s_n, \theta_n)$ according to Eq. 10;
30   update the strategy $\pi_n(s_n, \theta_n)$ according to Eq. 11;
31         $s_a = s_n^{t+1}$;
32   **end**
33   **end**
      `// Step 4: MARL-based selection`
34     $[P, \theta] = $ MARL_Selection(TP, A);
35     $t + +$;
36   **end**
37   **return** P;

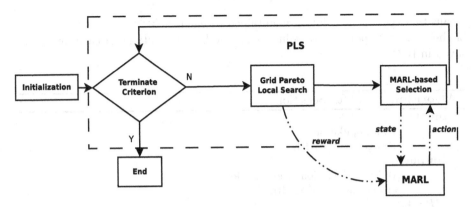

Fig. 2. The interactive process of GPD-based PLS and MARL

3.2 Initialization

Algorithm 1 line $1-7$ is the process of initialization. A population P is generated at random or by the heuristic way. Nadir and ideal points are set according to Eq. 2 and Eq. 3. The update of grid system is based on ideal and nadir points which are described in Sect. 2.1. Finally, the agents of MARL are initialized with division parameter L and the state of each agent is initialized by P and AN as described in Sect. 2.3.

3.3 Pareto Local Search

In the process of Pareto local search (Algorithm 1 line $12-22$), population P is used to store the solutions after local search, r is the reward of the agent n corresponding to the solution x and HV is the hypervolume value of the original population. In this paper, we use the grid Pareto dominance instead of Pareto dominance to update P.

3.4 Update Q-Value

The procedure of updating Q-value is given in Algorithm 1 line $25-33$. The learning rate α_t is updated during the iteration. After that state vectors are generated by observing the population P. Then action-value is updated through rewards r and actions θ. Finally, the strategy π and current state s are updated.

3.5 MARL-Based Selection

In the process of MARL Selection (Algorithm 2), each agent obtains its action θ_n through the strategy π_n. For each agent, solution with better convergence is maintained in the designated subregion. Note that θ_n means the area index selected by agent, MAXINT means the largest integer, TP_n equal to NULL means that there is no solution maintained by agent n, $z^{gcp}(x)$ means the corner of the grid corresponding to solution x. The better solutions of each agent are saved in WP.

Algorithm 2. MARL_Selection

Input :
$\quad\quad\quad$ P: a population
$\quad\quad\quad$ A: agents of MARL

Output:
$\quad\quad\quad$ WP: a population after selection
$\quad\quad\quad$ θ: joint action of MARL

1   $WP = \emptyset$;
2   **foreach** $n \in \mathcal{N}$ **do**
3      $minD_n = MAXINT$;
4      $TP_n = NULL$;
5   select action θ_n according to strategy $\pi_n(s_n)$;
6   **end**
7   **for** *each* $x \in P$ **do**
8   Find the corresponding region ω and agent n according to x;
9   **if** $\theta_n == \omega$ *and* $minDist_n > ||x - z^{gcp}(x)||_2$ **then**
10        $minDist_n = ||x - z^{gcp}(x)||_2$;
11        $TP_n = x$;
12   **end**
13   **end**
14   **for** *each* $n \in \mathcal{N}$ **do**
15   **if** $TP_n \neq NULL$ **then**
16        $WP = WP \cup x$;
17   **end**
18   **end**
19   **return** WP, θ;

---

4 Experiment and Discussion

4.1 Experiment Setup

There are seven bi-objective instances considered in the experiments. Four instances of MOTSP are from [12,17,18] and three bi-objective MOKPs are used in this paper [22]. Multiobjective knapsack problem (MOKP) instances are named in the following way. "Knapsack500-2" denotes that the instance has 500 variables with two objectives.

The parameter of MOEA/D-LS, GWS-PLS [4], MOGLS [7] are set according to the paper. The initial size of population is set to 300. The penalty parameter θ is set to 5.0. For MARL-GPLS, the parameter L is set as 200. The number of Agents for bi-objective instances is set to $2 * L - 1$. The number of weight vectors for each grid is set to 15, the parameters of reinforcement learning is set as follow: ϵ is set to 0.9; discount δ is set to 0.9; learning rate α_t is set according to Sect. 2.4; For ϵ-MOEA-LS, in order to maintain the same population size with MARL-GPLS, ϵ in ϵ-MOEA-LS is set to 100.

Stopping Criterion: Each algorithm runs 20 times independently on each instance. Each algorithm is stopped when reaching the specified number of iteration or no new solutions generated for the process of LS. The iteration number of MOKPs, MOTSPs, and MOTSPs is set to 100, 200, 500 respectively. For MARL-GPLS, agents are trained 20 times continuously, and then agents are extracted. Finally, The algorithm runs 20 times through these extracted agents without training them.

Methods of Initialization: For MOTSP, the initial population is randomly generated. For MOKP, the initial population is generated by a greedy method [10]. This paper uses two metric: Inverted Generational Distance (IGD) [14] and Hypervolume Indicator (HV) [22].

4.2 Effectiveness of MARL

In this subsection, in order to show the effectiveness of MARL, we compare MARL-GPLS with GCPD [21] which always selects the solution closest to its grid corner.

Table 1 shows the HV and IGD values of MARL-GPLS and GCPD-PLS. It's obvious that MARL-GPLS outperforms GCPD-PLS in most CMOPs instances. The experimental results can confirm the effectiveness of MARL on PLS, and provide experimental proof for our motivation.

Table 1. IGD and HV values of the non-dominated solutions obtained by GCPD-PLS and MARL-GPLS on various multiobjective instances

instance	HV			IGD	
	GCPD-PLS	MARL-GPLS		GCPD-PLS	MARL-GPLS
clusterAB100	9.222e-01(9.0e-04)-	9.254e-01(3.3e-04)	1.562e+03(1.2e+02)-	1.274e+03(6.1e+01)
clusterAB300	9.444e-01(2.0e-01)-	9.512e-01(3.0e-04)	3.804e+03(1.1e+05)-	1.815e+03(1.9e+02)
euclidAB100	8.761e-01(8.6e-04)-	8.791e-01(6.8e-04)	1.519e+03(1.0e+02)-	1.225e+03(7.0e+01)
euclidAB300	9.062e-01(7.5e-04)-	9.151e-01(4.5e-04)	4.201e+03(2.7e+02)-	1.781e+03(1.3e+02)
kroAB100	9.128e-01(7.8e-04)-	9.157e-01(5.4e-04)	1.593e+03(9.8e+01)-	1.355e+03(2.9e+04)
kroAB200	9.228e-01(8.3e-04)-	9.282e-01(3.6e-04)	2.281e+03(2.0e+02)-	1.262e+03(8.8e+01)
kroAB300	9.312e-01(1.1e-03)-	9.388e-01(3.2e-04)	4.040e+03(3.4e+02)-	1.812e+03(1.4e+02)
knapsack250-2	8.525e-01(8.4e-06)-	8.525e-01(1.9e-05)	4.867e+00(2.0e-02)-	4.491e+00(8.7e-02)
knapsack500-2	8.376e-01(1.7e-05)+	8.371e-01(1.3e-04)	6.745e+00(1.1e-01)+	8.068e+00(4.3e-01)
knapsack750-2	8.397e-01(1.5e-04)-	8.410e-01(2.0e-04)	1.568e+01(4.1e-01)-	1.268e+01(4.3e-01)

Wilcoxons rank sum test at a 0.05 significance level is performed to HV/IGD values.
" + " means the HV/IGD value of the algorithm on this problem is significantly better than that of MARL-GPLS
" − " means the HV/IGD value of the algorithm on this problem is significantly worse than that of MARL-GPLS
" ≈ " means there is no significant difference between the compared results

4.3   Comparisons with Other Algorithms

In this part, MARL-GPLS is compared with two MOEA/D based algorithms and two well-known algorithms (MOGLS, NSGA-II) on seven MOTSPs and MOKPs.

This paper calculates the IGD and HV value for all compared algorithms and tabulates the median value in Table 2 and 3. It can be seen that the performance of MARL-GPLS is better than other comparison algorithms in most CMOPS instances. The performance of MARL-GPLS is worse than GWS and ϵ-MOEA-LS for MOKP instances. The reason is that the greedy method can generate the initial population for MOKP with great performance in both convergence and diversity in the first place. MARL-GPLS can not perform well in finding a strategy to guide PLS under this circumstance due to rough representation of the state s_n. It is worth noting that MARL-GPLS proposes a novel idea of applying multiagent reinforcement learning to multiobjective combinatorial optimization.

Table 2. Median and standard deviation values of HV obtained by MARL-GPLS, GWS-PLS, MOEA/D-LS (TCH, PBI), ϵ-MOEA, MOGLS, NSGA-II-LS.

instance	MARL	GWS	MOEAD-LS(PBI)	MOEAD-LS(TCH)	eMOEA-LS	MOGLS	NSGAII
clusterAB100	9.254e-01(3.3e-04)	9.231e-01(9.8e-04)-	8.322e-01(6.1e-03)-	8.847e-01(9.4e-03)-	8.822e-01(6.6e-02)-	9.241e-01(6.7e-03)-	9.228e-01(4.2e-02)-
clusterAB300	9.512e-01(3.0e-04)	9.492e-01(4.3e-04)-	8.348e-01(9.6e-03)-	9.016e-01(2.2e-02)-	8.026e-01(2.5e-02)-	9.504e-01(4.0e-04)-	8.819e-01(2.6e-01)-
euclidAB100	8.791e-01(6.8e-04)	8.762e-01(1.2e-03)-	8.081e-01(7.4e-03)-	8.456e-01(7.0e-03)-	8.396e-01(5.1e-03)-	8.428e-01(9.6e-03)-	8.764e-01(4.3e-03)-
euclidAB300	9.151e-01(4.5e-04)	9.118e-01(3.6e-04)-	8.114e-01(8.1e-03)-	8.910e-01(6.5e-03)-	8.198e-01(3.5e-01)-	9.149e-01(4.4e-04)=	8.198e-01(3.5e-02)-
kroAB100	9.157e-01(5.4e-04)	9.136e-01(7.5e-04)-	8.419e-01(8.8e-03)-	8.736e-01(7.8e-03)-	8.763e-01(7.5e-03)-	9.150e-01(5.9e-04)=	9.179e-01(5.7e-02)=
kroAB200	9.282e-01(3.6e-04)	9.249e-01(5.3e-04)-	8.363e-01(6.3e-03)-	8.865e-01(1.2e-02)-	8.235e-01(1.5e-02)-	9.279e-01(5.4e-04)-	9.219e-01(3.4e-03)-
kroAB300	9.388e-01(3.2e-04)	9.362e-01(4.3e-04)-	8.353e-01(1.0e-02)-	9.134e-01(1.1e-02)-	7.931e-01(2.4e-02)-	9.382e-01(4.2e-04)-	8.577e-01(3.8e-01)-
knapsack250-2	8.525e-01(1.9e-05)	8.525e-01(2.2e-06)-	7.732e-01(4.0e-03)-	7.822e-01(7.1e-03)-	8.538e-01(5.9e-05)+	8.417e-01(3.5e-04)-	8.415e-01(1.2e-03)-
knapsack500-2	8.371e-01(1.3e-04)	6.674e+00(1.8e-01)+	1.892e+02(8.4e+00)-	8.331e-01(1.0e-03)-	8.369e-01(2.0e-04)-	8.289e-01(5.4e-04)-	8.285e-01(8.7e-04)-
knapsack750-2	8.410e-01(2.0e-04)	8.403e-01(1.4e-05)-	7.034e-01(4.5e-03)-	8.333e-01(1.6e-03)-	8.410e-01(2.0e-04)=	8.356e-01(4.5e-04)-	8.341e-01(6.8e-04)-

Wilcoxons rank sum test at a 0.05 significance level is performed to HV values.
" + " means the HV value of the algorithm on this problem is significantly better than that of MARL-GPLS
" − " means the HV value of the algorithm on this problem is significantly worse than that of MARL-GPLS
" = " means there is no significant difference between the compared results

Table 3. Median and standard deviation values of IGD obtained by MARL-GPLS, GWS-PLS, MOEA/D-LS (TCH, PBI), ϵ-MOEA, MOGLS, NSGA-II-LS.

instance	MARL	GWS	MOEAD-LS(PBI)	MOEAD-LS(TCH)	eMOEA-LS	MOGLS	NSGAII
clusterAB100	1.274e+01(6.1e+01)	1.443e+02(1.3e+02)-	3.607e+04(1.6e+03)-	1.555e+04(3.4e+03)-	1.414e+04(4.6e+03)-	1.593e+03(2.0e+02)-	1.603e+03(3.8e+02)-
clusterAB300	1.815e+03(1.9e+02)	2.982e+03(2.8e+02)-	7.839e+04(3.1e+03)-	2.329e+04(1.1e+04)-	7.178e+04(1.3e+04)-	2.427e+03(3.7e+02)-	3.948e+04(1.4e+05)-
euclidAB100	1.225e+03(7.0e+01)	1.456e+03(1.0e+02)-	2.102e+04(1.8e+03)-	6.936e+03(1.3e+03)-	7.343e+03(1.6e+03)-	7.192e+03(2.2e+03)-	1.540e+03(3.9e+02)-
euclidAB300	1.781e+03(1.3e+02)	3.092e+03(1.7e+02)-	7.342e+04(3.4e+03)-	9.084e+03(3.3e+03)-	7.378e+01(1.9e+02)-	2.098e+03(1.5e+02)-	4.105e+04(1.9e+06)-
kroAB100	1.355e+03(2.9e+04)	1.496e+03(1.2e+02)-	2.865e+04(1.9e+03)-	1.477e+04(2.9e+03)-	1.395e+04(3.0e+03)-	1.669e+03(2.4e+02)-	1.157e+03(5.7e+02)=
kroAB200	1.262e+03(8.8e+01)	1.824e+03(1.3e+02)-	4.890e+04(1.7e+03)-	1.963e+04(4.2e+03)-	3.854e+04(9.1e+03)-	1.857e+03(2.5e+02)-	3.765e+03(6.7e+02)-
kroAB300	1.812e+03(1.4e+02)	2.877e+03(2.2e+02)-	7.875e+04(4.4e+03)-	1.552e+04(5.7e+03)-	7.627e+04(7.2e+03)-	3.102e+03(3.1e+02)-	4.142e+04(2.1e+05)-
knapsack250-2	4.491e+00(8.7e-02)	4.724e+00(1.9e-03)-	1.663e+02(1.5e+01)-	1.209e+02(1.5e+01)-	3.994e+00(1.2e-01)+	3.360e+01(9.3e-01)-	2.135e+01(1.7e+00)-
knapsack500-2	8.068e+00(4.3e-01)	8.378e-01(7.1e-05)+	7.636e-01(2.9e-03)-	1.686e+01(1.4e+00)-	7.490e+00(2.3e-01)+	2.727e+01(2.5e+00)-	2.847e+01(1.9e+00)-
knapsack750-2	1.268e-01(4.3e-01)	1.529e-01(1.2e-01)-	7.489e+02(2.9e+01)-	3.873e+01(4.7e+00)-	2.459e+01(4.4e+00)-	3.988e+01(1.7e+00)-	4.417e+01(2.2e+00)-

Wilcoxons rank sum test at a 0.05 significance level is performed to IGD values.
" + " means the IGD value of the algorithm on this problem is significantly better than that of MARL-GPLS
" − " means the IGD value of the algorithm on this problem is significantly worse than that of MARL-GPLS
" = " means there is no significant difference between the compared results

5 Conclusion

This paper proposed a novel way to combine multiagent reinforcement learning with multiobjective combinatorial optimization. In this work, MARL-GPLS can achieve better results in most instances. The future research directions include making MARL-GPLS more efficient for tri- and many-objective problems and using a more elaborate method to describe the states of the solution set.

Acknowledgment. This work was supported in part by the Aeronautical Science Foundation of China under grant 20175552042, by the National Natural Science Foundation of China (NSFC) under grant 61300159, by the Natural Science Foundation of Jiangsu Province of China under grant BK20181288 and by China Postdoctoral Science Foundation under grant 2015M571751.

References

1. Afanasyeva, A., Buzdalov, M.: Choosing best fitness function with reinforcement learning. In: 2011 10th International Conference on Machine Learning and Applications and Workshops, vol. 2, pp. 354–357, December 2011. https://doi.org/10.1109/ICMLA.2011.163
2. Bhowmik, P., Rakshit, P., Konar, A., Kim, E., Nagar, A.K.: DE-TDQL: an adaptive memetic algorithm. In: 2012 IEEE Congress on Evolutionary Computation, pp. 1–8, June 2012. https://doi.org/10.1109/CEC.2012.6256573
3. Busoniu, L., Babuska, R., De Schutter, B.: A comprehensive survey of multiagent reinforcement learning. IEEE Trans. Syst. Man Cybern. Part C (App. Rev.) **38**(2), 156–172 (2008). https://doi.org/10.1109/TSMCC.2007.913919
4. Cai, X., Sun, H., Zhang, Q., Huang, Y.: A grid weighted sum pareto local search for combinatorial multi and many-objective optimization. IEEE Trans. Cybern. **49**(9), 3586–3598 (2019). https://doi.org/10.1109/TCYB.2018.2849403
5. Hisao, I., Tadahiko, M.: Multi-objective genetic local search algorithm. In: Fukuda, T., Furuhashi, T. (eds.) Proceedings of the 1996 International Conference on Evolutionary Computation, Nagoya, Japan, pp. 119–124. IEEE (1996)
6. Jaakkola, T., Jordan, M.I., Singh, S.P.: On the convergence of stochastic iterative dynamic programming algorithms. Neural Comput. **6**(6), 1185–1201 (2014)
7. Jaszkiewicz, A.: On the performance of multiple-objective genetic local search on the 0/1 knapsack problem - a comparative experiment. IEEE Trans. Evol. Comput. **6**(4), 402–412 (2002). https://doi.org/10.1109/TEVC.2002.802873
8. Juan, J.D., Zhang, Y., Enrique, A., Mark, H., Antonio, J.N.: A study of the bi-objective next release problem. Empirical Softw. Eng. **16**(1), 29–60 (2011)
9. Karafotias, G., Eiben, A.E., Hoogendoorn, M.: Generic parameter control with reinforcement learning. In: Proceedings of the 2014 Annual Conference on Genetic and Evolutionary Computation, pp. 1319–1326 (2014)
10. Ke, L., Zhang, Q., Battiti, R.: Hybridization of decomposition and local search for multiobjective optimization. IEEE Trans. Cybern. **44**(10), 1808–1820 (2014). https://doi.org/10.1109/TCYB.2013.2295886
11. Liu, H., Gu, F., Zhang, Q.: Decomposition of a multiobjective optimization problem into a number of simple multiobjective subproblems. IEEE Trans. Evol. Comput. **18**(3), 450–455 (2014). https://doi.org/10.1109/TEVC.2013.2281533

12. Lust, T., Teghem, J.: Two-phase pareto local search for the biobjective traveling salesman problem. J. Heuristics **16**(3), 475–510 (2010). https://doi.org/10.1007/s10732-009-9103-910.1007/s10732-009-9103-9

13. Matignon, L., Laurent, G.J., Fort-Piat, N.L.: Independent reinforcement learners in cooperative Markov games: a survey regarding coordination problems. Knowl. Eng. Rev. **27**(01), 1–31 (2012)

14. Sierra, M.R., Coello, C.A.: A new multi-objective particle swarm optimizer with improved selection and diversity mechanisms (2004)

15. Muller, S.D., Schraudolph, N.N., Koumoutsakos, P.D.: Step size adaptation in evolution strategies using reinforcement learning. In: Proceedings of the 2002 Congress on Evolutionary Computation, CEC 2002 (Cat. No. 02TH8600), vol. 1, pp. 151–156, May 2002. https://doi.org/10.1109/CEC.2002.1006225

16. Papadimitriou, C.H., Steiglitz, K.: Combinatorial Optimization: Algorithms and Complexity. Dover, New York (1998)

17. Paquete, L.F.: Stochastic Local Search Algorithms for Multiobjective Combinatorial Optimizations: Methods and Analysis. IOS Press, Inc., Amsterdam (2006)

18. Reinelt, G.: TSPLIB: traveling salesman problem library. Orsa J. Comput. **3**(1), 376–384 (1991)

19. Shim, V.A., Tan, K.C., Cheong, C.Y.: A hybrid estimation of distribution algorithm with decomposition for solving the multiobjective multiple traveling salesman problem. IEEE Trans. Syst. Man Cybern. Part C **42**(5), 682–691 (2012)

20. Sutton, R.S., Barto, A.G.: Introduction to Reinforcement Learning, vol. 135. MIT Press, Cambridge (1998)

21. Yang, S., Li, M., Liu, X., Zheng, J.: A grid-based evolutionary algorithm for many-objective optimization. IEEE Trans. Evol. Comput. **17**(5), 721–736 (2013)

22. Zitzler, E., Thiele, L.: Multiobjective evolutionary algorithms: a comparative case study and the strength pareto approach. IEEE Trans. Evol. Comput. **3**(4), 257–271 (1999). https://doi.org/10.1109/4235.797969

Extended Kalman Filter-Based Adaptively Sliding Mode Control with Dead-Zone Compensator for an Anchor-Hole Driller

Zhen Zhang[1], Yi-Nan Guo[1(✉)] (ID), Xi-Wang Lu[1], Dun-Wei Gong[1], and Yang Zhang[2]

[1] School of Information and Control Engineering,
China University of Mining and Technology, Xuzhou 221116, China
nanfly@126.com
[2] School of Electrical and Power Engineering,
China University of Mining and Technology, Xuzhou 221116, China

Abstract. To improve the control performance of the swing angle for an anchor-hole, an extended Kalman filter-based adaptively sliding mode control with dead-zone compensator is developed, with the purpose of tracking the pre-set swing angle of an anchor-hole driller as soon as possible without steady-state error. Taking the load disturbance and the dead-zone with the uncertain parameters of a proportional reversing valve into consideration, the rotation part of an anchor-hole driller is modeled. Based on this, a dead-zone compensator is designed by introducing its smooth inverse model. Following that, an adaptively sliding mode controller is designed. Finally, extended Kalman filter is employed to predict the swing angle in the next control period, as well as filter the noises derived from the measurement of the swing angle. The experimental results show that the proposed controller has the capability of rapidly tracking the pre-set swing angle without overshoot and chattering.

Keywords: Positioning accuracy · Anchor-hole driller · Extended Kalman filter · Adaptively sliding mode control · Dead-zone compensator

1 Introduction

To support a roadway by accurately installing bolts and anchor cables as soon as possible has a direct influence on its tunneling speed and stability [1,2]. Based on this, moving an anchor-hole driller to the pre-designed location exactly and carry out the drilling process is the necessary issue for supporting.

In an anchor-hole driller, the inherent nonlinearity caused by dead-zone with uncertain parameters, as well as unknown load disturbances not only make the exact mathematical model of the system difficultly built, but also complicate

© Springer Nature Singapore Pte Ltd. 2020
H. Zhang et al. (Eds.): NCAA 2020, CCIS 1265, pp. 35–46, 2020.
https://doi.org/10.1007/978-981-15-7670-6_4

the control design [3]. Based on this, Sliding mode control (SMC) [4], which is insensitive to parameters perturbation and external disturbances, as well as independence on the exact model of the system, had been successfully applied in the real-world [5,6]. Qian et al. proposed SMC with disturbance observer to improve the system control performance [7,8]. For a turbine governing system with internal disturbance and external noise, a fuzzy SMC show the strong robustness [9]. Also, a fuzzy adaptively sliding mode controller was designed for a nonlinear system with unknown dynamics and bounded disturbances [10]. However, the inherent nonlinearity, the overall un-modeled disturbances and frictions did not fully take into account.

In a proportional reversing valve, the dead-zone will deteriorate the performance of the whole system, even causes unstable. To solve the problem, Deng et al. [11] designed a linear controller based on dead-zone inverse model for adjusting a rotation system. However, the parameters of dead-zone can not be detected exactly. Hence, Hu et al. [12] proposed a nonlinear adaptive control method, it is noteworthy that, the employed discontinuous dead-zone inverse model may cause chattering of the control output, even instability of the motion control system. Based on this, a smooth dead-zone inverse model was constructed and an adaptively controller was designed for a nonlinear system [13]. Obviously, the reasonable compensation technology based on the estimated parameters of dead-zone is of great significance to weaken its negative effects.

Without loss of generality, noise produced during the measurement will provide the wrong information on the swing angle of an anchor-hole driller, and finally aggravate the chattering of the output, which deteriorates the control precision. Effectively predicting the swing angle is an important issue for designing a controller by extended Kalman filter (EKF) [14–16]. Based on this, we propose an EKF-based adaptively sliding mode control method with dead-zone compensator (ASMC-DC), with the purpose of tracking the pre-set swing angle accurately. This paper has the following threefold contributions:

(1) Establishing the novel mathematic model for the rotation motion. As a key component, the hydraulic valve has the obvious nonlinear characteristic caused by dead-zone with uncertain parameters. Moreover, the system is inevitably affected by un-modeled disturbances and frictions. Until now, less reports on the model of electro-hydraulic anchor-hole driller were given for its swing angle control. Based on this, we model the above-mentioned electro-hydraulic rotation system by considering the dead-zone and un-modeled factors.

(2) Designing an adaptively sliding mode controller with dead-zone compensator for adjusting the swing angle. A compensation strategy is introduced to weaken the negative influence of dead-zone. Moreover, two adaptive laws are designed for a sliding mode controller to respectively estimate the uncertain parameters of dead-zone and un-modeled factors, with the purpose of tracking angle as soon as possible without overshoot and chattering.

(3) Predicting the swing angle of an anchor-hole driller by extended Kalman filter. The noise produced during the measurement is seldom considered,

however, the above-mentioned noise provides the wrong information about the swing angle, and then causes the chattering of the output, which makes the following drilling process of bolts or anchor cables unstable, even break the drill pipe. Based on this, we introduced EKF to predict the swing angle and suppress the noise, with the purpose of further improving the control precision.

2 Modeling the Rotating System of an Anchor-Hole Driller

In the rotating system shown in Fig. 1, the emulsion, pumped from a quantitative pump which is driven by an asynchronous motor. The input flow of the hydraulic motor is adjusted by a proportional reversing valve to control the swing angle.

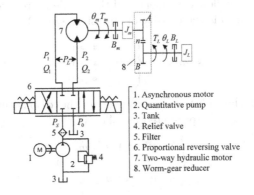

Fig. 1. The schematic diagram of the rotating system (θ_m is the angle, T_L is the load torque, T_m is the motor torque, P_1 is the oil-in pressure, P_2 is the oil-return pressure, J_L is the load inertia, Q_1 is the inflow, B_L is the load coefficient, J_m is the motor inertia, P_S is the oil-in pressure, n is the transmission ratio, Q_2 is the outflow, P_0 is the oil-return pressure, θ_L is the swing angle, B_m is the coefficient).

2.1 Dynamic Characteristics of the Hydraulic Motor

Taking the external load torque, un-modeled disturbances and frictions into consideration, the torque of the hydraulic motor can be formulated as follows [17].

$$D_m P_L = J_t \ddot{\theta}_L + B_t \dot{\theta}_L + G_t \theta_L + T_t + \Delta_f \tag{1}$$

where, $J_t = nJ_m + J_L/n$, $B_t = nB_m + B_L/n$, $T_t = T_L/n$, $n = \theta_m/\theta_L$, $P_L = P_1 - P_2$. D_m is the displacement; J_t is the total inertia; B_t is the equivalent viscous damping coefficient; G_t is the equivalent stiffness; T_t is the equivalent external load torque; Δ_f represents the total un-modeled disturbances and frictions.

The outflow Q_L of the proportional reversing valve described as follows [18].

$$Q_L = nD_m\dot{\theta}_L + \frac{V_m}{4\beta_e}\dot{P}_L + C_{tm}P_L \tag{2}$$

where, C_{tm} is the total leakage coefficient; V_m is the total volume; β_e is the effective bulk modulus.

2.2 Formulation of the Proportional Reversing Valve

The mathematical model of a proportional reversing valve is built under the condition of $P_0 = 0$ [19] as follows.

$$Q_L = C_d\omega x_v\sqrt{\frac{P_S - sign(x_v)P_L}{\rho}} \tag{3}$$

where C_d and ω are the flow coefficient and the gradient area, respectively. x_v is the displacement, and ρ is the density of emulsion. $sign(\cdot)$ is defined as follows.

$$sign(\bullet) = \begin{cases} 1 & \bullet > 0 \\ 0 & \bullet = 0 \\ -1 & \bullet < 0 \end{cases} \tag{4}$$

The voltage of the valve is denoted as u, the displacement of a spool, expressed with x_v, can be represented as follows [19].

$$x_v = \begin{cases} m(u - \delta_l) & u \le \delta_l \\ 0 & \delta_l < u < \delta_r \\ m(u - \delta_r) & u \ge \delta_r \end{cases} \tag{5}$$

where, m is the proportional ratio, δ_l and δ_r mean the turning points that decide the width of the dead-zone. Normally, δ_l and δ_r are bounded and their lower and upper limits are known in advance, suggesting that $\delta_{l\,min} \le \delta_l \le \delta_{l\,max} < 0$, $0 < \delta_{r\,min} \le \delta_r \le \delta_{r\,max}$. The Eq. (5) can be transformed as follows.

$$x_v = m(u - \delta(u)) \tag{6}$$

$$\delta(u) = \begin{cases} \delta_l & u \le \delta_l \\ u & \delta_l < u < \delta_r \\ \delta_r & u \ge \delta_r \end{cases} \tag{7}$$

2.3 Formulation of the Rotating System of an Anchor-Hole Driller

Denote $\mathbf{x} = [x_1, x_2, x_3]^T = [\theta_L, \dot{\theta}_L, \ddot{\theta}_L]^T$, $F = -\frac{1}{J_t}(\dot{T}_t + \dot{\Delta}_f) - \frac{4\beta_e C_{tm}}{J_t V_m}(T_t + \Delta_f)$, and y is chosen as the output, the rotating system can be formulated as follows.

$$\begin{cases} \dot{x}_1 = x_2 \\ \dot{x}_2 = x_3 \\ \dot{x}_3 = f(\mathbf{x}) + g(\mathbf{x})(u - \delta(u)) + F \\ y = x_1 \end{cases} \tag{8}$$

Denote $f(\mathbf{x}) = \theta_1 x_1 + \theta_2 x_2 + \theta_3 x_3$, $g(\mathbf{x}) = \theta_4 R(\mathbf{x})$, where, $\theta_1 = -\frac{4\beta_e G_t C_{tm}}{J_t V_m}$, $\theta_2 = -(\frac{G_t}{J_t} + \frac{4\beta_e n D_m^2}{J_t V_m} + \frac{4\beta_e B_t C_{tm}}{J_t V_m})$, $\theta_3 = -(\frac{B_t}{J_t} + \frac{4\beta_e C_{tm}}{V_m})$, $\theta_4 = \frac{4\beta_e D_m C_d \omega m}{J_t V_m \sqrt{\rho}}$, $R(\mathbf{x}) = \sqrt{P_S - sign(x_v)P_L}$. The oil-in pressure of a pump satisfies $P_S > P_L$ [17]. Based on this, $0 < R(\mathbf{x}) < \sqrt{2P_S}$. F is usually unknown and bounded in actual control, indicating that $|F| \leq \bar{F}$.

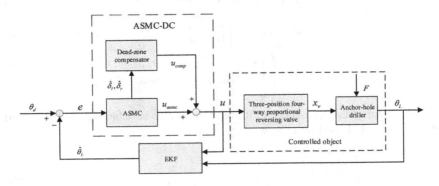

Fig. 2. The proposed control method.

3   EKF-Based ASMC-DC of an Anchor-Hole Driller

To overcome the negative affect caused by the inherent nonlinearity and un-modeled disturbances, an EKF-based ASMC-DC is designed mainly includes ASMC and dead-zone compensator shown in Fig. 2.

3.1   Dead-Zone Compensator

The smooth dead-zone inverse model [11] is introduced to effectively eliminate the negative influence of dead-zone on the stability of the rotating system. Denote $\phi(x_v) = e^{x_v/\varepsilon}(e^{x_v/\varepsilon} + e^{-x_v/\varepsilon})^{-1}$ is a smooth continuous exponential function, $\varepsilon > 0$. Let $N^{-1}(x_v)$ be the smooth anti-dead-zone function, one has

$$u = N^{-1}(x_v) = \left(\delta_r + \frac{x_v}{m}\right)\phi(x_v) + \left(\delta_l + \frac{x_v}{m}\right)(1 - \phi(x_v)) \tag{9}$$

The turning points of the valve's dead-zone, expressed with $\vartheta = [\delta_r, \delta_l]^T$. Their estimations, denoted as $\hat{\vartheta} = [\hat{\delta}_r, \hat{\delta}_l]^T$, and the estimation errors are $\tilde{\vartheta} = \vartheta - \hat{\vartheta}$. Based on the above-mentioned smooth exponential function $\phi(x_v)$, the output of the dead-zone compensator is designed as follows.

$$u_{comp} = \hat{\delta}_r \phi(x_v) + \hat{\delta}_l(1 - \phi(x_v)) = \hat{\vartheta}^T \phi \tag{10}$$

where $\phi = [\phi(x_v), 1 - \phi(x_v)]^T$.

As shown in Fig. 2, the input of proportional reversing valve is the sum of the output from the ASMC and the dead-zone compensator, denoted as $u = u_{asmc} + u_{comp} = u_{asmc} + \hat{\vartheta}^{\mathrm{T}}\phi$. After integrating it into Eq. (6), one has

$$x_v = m\left(u_{asmc} - \tilde{\vartheta}^{\mathrm{T}}(\phi - \varphi)\right) \tag{11}$$

$$\varphi = \left[\chi\left(\tilde{\delta}_r\right), \chi\left(\tilde{\delta}_l\right)\right]^{\mathrm{T}} \phi\left(u_{asmc}\right) sat\left(u_{asmc}\right) \tag{12}$$

$$\phi\left(u_{asmc}\right) = \begin{cases} 1 & u_{asmc} \geq 0 \\ 0 & else \end{cases} \tag{13}$$

$$sat\left(u_{asmc}\right) = \begin{cases} 1 - u_{asmc}/\delta_l & \delta_l \leq u_{asmc} < 0 \\ 1 - u_{asmc}/\delta_r & 0 \leq u_{asmc} < \delta_r \\ 0 & else \end{cases} \tag{14}$$

In above formulas, $\chi(\cdot)$ is the unit step function. φ satisfies $\|\varphi\| \leq 1$ [17], and $|sat\left(u_{asmc}\right)| \leq 1$. Based on this, we have

$$\begin{cases} \dot{x}_1 = x_2 \\ \dot{x}_2 = x_3 \\ \dot{x}_3 = f(\mathbf{x}) + g(\mathbf{x})u_{asmc} - g(\mathbf{x})\tilde{\vartheta}^{\mathrm{T}}(\phi - \varphi) + F \\ y = x_1 \end{cases} \tag{15}$$

3.2   Adaptively Sliding Mode Control with Dead-Zone Compensation

In this section, an adaptively sliding mode controller with dead-zone compensation is designed, with the purpose of effectively tracking the pre-set swing angle accurately.

Let $x_d = \theta_d$ be the pre-set swing angle, and e is the tracking error.

$$e = x_d - x_1 \tag{16}$$

Based on this, the switching function is designed as follows.

$$s = k_1 e + k_2 \dot{e} + \ddot{e} \tag{17}$$

where, $k_1 > 0$, $k_2 > 0$. Let $\hat{F}$ is the estimation of F and the corresponding estimation error is $\tilde{F} = F - \hat{F}$. The adaptively sliding mode controller is formulated.

$$u_{asmc} = \frac{1}{g(\mathbf{x})}\left(k_1\dot{e} + k_2\ddot{e} + \ddot{x}_d - f(\mathbf{x}) - \hat{F} + \tau \mathrm{sgn}(s)\right) \tag{18}$$

Two adaptive laws are designed to estimate the uncertain parameters of dead-zone and the un-modeled factors, expressed by δ_r, δ_l and F, respectively, with the purpose of improving the steady-state performance of the rotating system.

$$\dot{\hat{\vartheta}} = \gamma_1 g(\mathbf{x})(\phi - \varphi)s \tag{19}$$

$$\dot{\hat{F}} = -\gamma_2 s \tag{20}$$

In the above formulas, $\tau > 0$, $\gamma_1 > 0$, $\gamma_2 > 0$.

Theorem 1. *Under adaptively sliding mode controller with dead-zone compensation, the rotating system of an anchor-hole driller is stable and the tracking error of the swing angle can converge to the sliding surface, denoted by s.*

Proof. According to Eq. (15) and Eq. (16), the three-order derivative of e is

$$\dddot{e} = \dddot{x}_d - \dddot{e}_1 = \dddot{x}_d - \dot{x}_3 \tag{21}$$

Thus, the derivative of s is

$$\dot{s} = k_1\dot{e} + k_2\ddot{e} + \dddot{x}_d - f(\mathbf{x}) - g(\mathbf{x})\left(u_{asmc} + \tilde{\vartheta}^{\mathrm{T}}(\phi - \varphi)\right) - F \tag{22}$$

By incorporating Eq. (18) with Eq. (22), one has

$$\dot{s} = g(\mathbf{x})\tilde{\vartheta}^{\mathrm{T}}(\phi - \varphi) - \tilde{F} - \tau\mathrm{sgn}(s) \tag{23}$$

Considering the following Lyapunov function

$$V = \frac{1}{2}s^2 + \frac{1}{2\gamma_1}\tilde{\vartheta}^{\mathrm{T}}\tilde{\vartheta} + \frac{1}{2\gamma_2}\tilde{F}^2 \tag{24}$$

By incorporating Eq. (23), the derivative of V is

$$\dot{V} = \tilde{\vartheta}^{\mathrm{T}}[g(\mathbf{x})(\phi - \varphi)s - \frac{1}{\gamma_1}\dot{\tilde{\vartheta}}] - \tilde{F}(s + \frac{1}{\gamma_2}\dot{\tilde{F}}) - \tau|s| \tag{25}$$

By substituting Eq. (19) and Eq. (20) into Eq. (25), one has

$$\dot{V} = -\tau|s| \leq 0 \tag{26}$$

Equation (26) indicates that the system is stable and its integral formula is

$$\lim_{t\to\infty}\int_0^t -\tau|s|\,d\eta \leq \lim_{t\to\infty}[V(0) - V(t)] \leq V(0) < \infty \tag{27}$$

According to Barbalat's lemma [13] that when $t \to \infty$, we have $s \to 0$, which shows that the asymptotical convergence of the tracking error will be ensured.

3.3   Extended Kalman Filter

Extended Kalman filter is employed to estimate the current swing angle of an anchor-hole driller and predict the swing angle in the next control period, as well as filter the noises derived from the measurement of the swing angle.

For the sake of simplicity, the state and output equations of the rotating system in Eq. (15) can be transformed to the following form.

$$\begin{cases} \dot{\mathbf{x}}(t) = \bar{f}(\mathbf{x}(t)) + b(t)u(t) + \bar{g}(t) \\ y(t) = h(\mathbf{x}(t)) \end{cases} \tag{28}$$

In the above formula, $b(t) = [0, 0, \theta_4 R(\mathbf{x})]^T$ is the control matrix, $\bar{g}(t) = [0, 0, F]^T$ is the disturbance matrix, $h(\mathbf{x}(t)) = [x_1, 0, 0]^T$ is the output matrix, $\bar{R}(\mathbf{x}) = \theta_4 R(\mathbf{x}) \tilde{\vartheta}^T (\phi - \varphi)$, and the system state matrix is denoted as

$$\bar{f}(\mathbf{x}(t)) = \begin{bmatrix} x_2 \\ x_3 \\ \theta_1 x_1 + \theta_2 x_2 + \theta_3 x_3 - \theta_4 R(\mathbf{x}) \tilde{\vartheta}^T (\phi - \varphi) \end{bmatrix} \tag{29}$$

The measured swing angle satisfies the assumption of Gaussian white noise, expressed by $v(k)$, and $w(k)$ represents the uncertain parameters satisfying the Gaussian distribution. Thus, $p(w(k)) \sim N(0, Q)$ and $p(v(k)) \sim N(0, R)$. Based on this, Eq. (28) is discretized as follows.

$$\begin{cases} X(k+1) = \bar{f}(\mathbf{x}(k)) + b(k)u(k) + \bar{g}(k) + w(k) \\ Y(k) = h(\mathbf{x}(k)) + v(k) \end{cases} \tag{30}$$

After locally linearizing Eq. (30), one has

$$\begin{cases} X(k+1) = A(k)X(k) + b(k)u(k) + \bar{g}(k) + w(k) \\ Y(k) = H(k)X(k) + v(k) \end{cases} \tag{31}$$

Denote $A(k) = I + F(k)T$ is the state transition matrix, I is the unit matrix, T is the sampling period. $H(k) = \frac{\partial h}{\partial \mathbf{x}} |_{\mathbf{x}=\hat{\mathbf{x}}_{k|k-1}} = [1\,0\,0]^T$ is the measurement matrix, and

$$F(k) = \frac{\partial \bar{f}}{\partial \mathbf{x}} |_{\mathbf{x}=\hat{\mathbf{x}}_k} = \begin{bmatrix} 0 & 1 & 0 \\ 0 & 0 & 1 \\ \theta_1 + \frac{\partial \bar{R}(\mathbf{x})}{\partial x_1} & \theta_2 + \frac{\partial \bar{R}(\mathbf{x})}{\partial x_2} & \theta_3 + \frac{\partial \bar{R}(\mathbf{x})}{\partial x_3} \end{bmatrix} \tag{32}$$

Suppose that $\hat{X}(k-1)$ and $P(k-1)$ are the estimation value of the state variables and the covariance matrix of the error at the $(k-1)-th$ sampling period. Based on this, the swing angle of an anchor-hole driller is estimated in the following two steps.

Step 1: Time update.

$$\hat{X}(k|k-1) = \hat{X}(k-1) + [\bar{f}(\hat{X}(k-1)) + b(k-1)u(k-1)]T \tag{33}$$

$$P(k|k-1) = A(k|k-1)P(k-1)A^T(k|k-1) + \bar{g}(k-1)Q(k-1)\bar{g}^T(k-1) \tag{34}$$

Step 2: State update.

$$K(k) = P(k|k-1)H^T(k)(H(k)P(k|k-1)H^T(k) + R(k))^{-1} \tag{35}$$

$$\hat{X}(k) = \hat{X}(k|k-1) + K(k)[Y(k) - h(\hat{X}(k|k-1))] \tag{36}$$

$$P(k) = (I - K(k)H(k))P(k|k-1) \tag{37}$$

In the above-mentioned estimation strategy, the expectation of the initial state variables and covariance matrix are set as follows: $\hat{X}(0) = E(X(0))$, $P(0) = E[(X(0) - \hat{X}(0))(X(0) - \hat{X}(0))^T]$.

4   Experimental Results and Discussion

The effectiveness of the proposed control method is verified by comparative experiments. All experiments are done under the joint simulation environment composed of AMEsim and Matlab, shown as Fig. 3(a). The parameter settings as follows: $P_s = 160$ bar, $J_t = 4.71$ kg/m^2, $C_d = 0.61$, $T = 0.01$ s, $\omega = 8\pi e - 3$ m^2, $D_m = 100$ ml/r, $m = 0.25$ m/A, $C_{tm} = 2e - 8$ m^3/s/MPa, $\varepsilon = 0.1$, $B_t = 0.01$ mNs/rad, $V_m = 8e - 5$ m^3, $\beta_e = 700$ MPa, $\delta_{rmax} = 1.0$ V, $\delta_{lmin} = -1.1$, $G_t = 100$ N/m, $n = 30$, $\rho = 880$ kg/m^3,

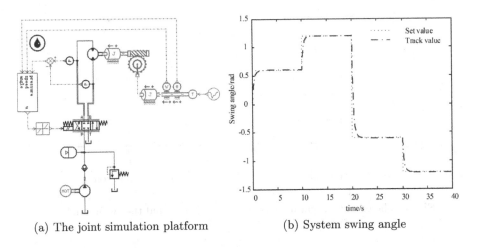

(a) The joint simulation platform   (b) System swing angle

Fig. 3. The AMEsim simulation platform and swing angle

4.1   Performances in Tracing the Pre-set Swing Angle

Taking the rock stratum in Jincheng Sihe Coal Mine as an example, the expected swing angles of an anchor-hole driller corresponds to the pre-determined installed locations of bolts are show in Fig. 3(b).

Under the designed EKF-based ASMC-DC controller, Fig. 3(b) depicts the response process of tracking the pre-set swing angle. No matter how large the step change of the expected swing angles, the controller designed in the paper has the good control performance in dynamic and steady indexes. Especially, the swing angles can track the expected values without the overshoot, which avoids the high-temperature of the emulsion resulted from the frequent forward and reversal of the hydraulic motor, and decreases the energy consumption.

4.2   The Role of the Dead-Zone Compensator

In this experiments, the performances of the EKF-based ASMC controller with or without the compensator are compared and analyzed experimentally, with the purpose of evaluating the effectiveness of the developed compensation strategy.

Figure 4(a) and Fig. 4(b) depict the dead-zone parameters estimated by the designed adaptive law. Obviously, both of them gradually coverage to their real values during the transient process. From the response process shown in Fig. 5, no matter the EKF-based ASMC controller with/without the dead-zone compensator, the overshoots are both zero and without chattering of output, which meets the requirement of the rotating control system. Moreover, the designed compensator overcome the nonlinearity and the uncertainty of the rotating system, which plays a positive role in the shorter settling time and the smaller tracking error of the swing angles.

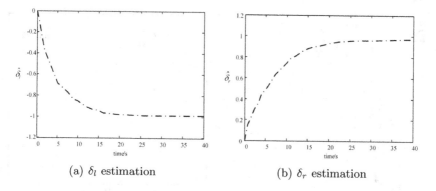

(a) δ_l estimation   (b) δ_r estimation

Fig. 4. The estimated parameters of the dead-zone and the swing angles

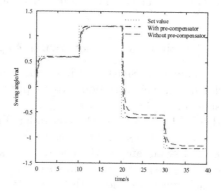

Fig. 5. System swing angle.

4.3 Comparison on the Control Performances of Different Controllers

To further evaluate the effectiveness of the designed EKF-based ASMC-DC (ASMC-DC + EKF), the experiments are carried out to compare EKF-based

PID (PID + EKF) and EKF-based ASMC (ASMC + EKF) with a step distur-
bance with $50\,\text{N·m}$ at $t = 9\,\text{s} - 10\,\text{s}$ simulating the real one is loaded in the
experiment. The core parameters are set as follows. PID + EKF: $K_p = 2.53$,
$K_i = 0.08$, $K_d = 0.003$; ASMC + EKF: $k_1 = 4.78$, $k_2 = 3.81$, $K = 85$, $\gamma_2 = 5$;
ASMC-DC + EKF: $k_1 = 4.78$, $k_2 = 3.81$, $K = 85$, $\gamma_1 = 8$, $\gamma_2 = 5$, $Q = R = 10$.

Figure 6(a) and Fig. 6(b) depict the swing angle and the tacking error
obtained by the three controllers. Two figures indicate that all controllers can
respond to the expected swing angle without overshoot. Moreover, the perturbation happened at 9th second can be significantly suppressed by EKF-based
ASMC with/without the dead-zone compensator, however, has an obvious oscil-
lation for EKF-based PID controller. To sum up, adaptively sliding mode controller designed in the paper provides an effective way to track the pre-set swing
angle with better dynamic and steady-state indexes, as well as fully overcome the
load disturbance. By integrating it with the dead-zone compensator, EKF-based
ASMC-DC shows better robustness.

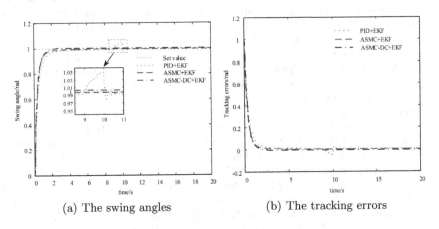

(a) The swing angles   (b) The tracking errors

Fig. 6. The control performances

5 Conclusions

To overcome the negative influence of the nonlinearity and the un-modeled factors on the control performance, an EKF-based ASMC-DC is proposed. Taking
the load disturbance and the dead-zone, the rotation system is firstly modeled.
Then, a dead-zone compensator is developed. Following that, an adaptively sliding mode controller is designed. Moreover, extended Kalman filter is employed to
predict the swing angle in the next control period. The experimental results show
that the designed controller can rapidly track the pre-set swing angle without
overshoot and oscillations.

Acknowledgment. This work was supported by the National Natural Science Foundation of China under Grant 61973305, and the Future Scientists Program of China
University of Mining and Technology under Grant 2020WLKXJ029.

References

1. Kang, H.P., Lin, J., Fan, M.J.: Investigation on support pattern of a coal mine roadway within soft rocks a case study. Int. J. Coal Geol. **140**, 31–40 (2015)
2. He, Y., Guo, Y.N., Gong, D.W.: Asynchronous active disturbance rejection balance control for hydraulic support platforms. Control Theor. Appl. **36**(1), 151–163 (2019)
3. Deng, W.X., Yao, J.Y., Ma, D.W.: Robust adaptive precision motion control of hydraulic actuators with valve dead-zone compensation. ISA Trans. **70**, 269–278 (2017)
4. He, S.P., Jun, S.: Finite-time sliding mode control design for a class of uncertain conic nonlinear systems. IEEE/CAA J. Automatica Sin. **4**(4), 809–816 (2017)
5. Haris, S.Z., Irfan, K.T., Ismail, L.: Design and adaptive sliding mode control of hybrid magnetic bearings. IEEE Trans. Ind. Electron. **65**(3), 2537–2547 (2018)
6. Liu, J.K., Sun, F.C.: Research and development on theory and algorithms of sliding-mode control. Control Theor. Appl. **24**(3), 407–418 (2007)
7. Kang, H.-S., Lee, Y., Hyun, C.-H., et al.: Design of sliding-mode control based on fuzzy disturbance observer for minimization of switching gain and chattering. Soft Comput. **19**(4), 851–858 (2014). https://doi.org/10.1007/s00500-014-1412-8
8. Palli, G., Strano, S., Terzo, M.: Sliding-mode observers for state and disturbance estimation in electro-hydraulic systems. Soft Comput. **74**, 58–70 (2018)
9. Hui, Y.X., Chen, Z.H., Yuan, Y.B., et al.: Design of fuzzy sliding-mode controller for hydraulic turbine regulating system via input state feedback linearization method. Energy **93**, 173–187 (2015)
10. Cerman, O., Hušek, P.: Adaptive fuzzy sliding-mode control for electro-hydraulic servo mechanism. Expert Syst. Appl. **39**(11), 10269–10277 (2012)
11. Deng, H., Luo, J.H., Duan, X.G., et al.: Adaptive inverse control for gripper rotating system in heavy-duty manipulators with unknown dead-zones. IEEE Trans. Ind. Electron. **64**(10), 7952–7961 (2017)
12. Hu, C.X., Yao, B., Wang, Q.F.: Performance-oriented adaptive robust control of a class of nonlinear systems preceded by unknown dead-zone with comparative experimental results. IEEE/ASME Trans. Mechatron. **18**(1), 178–189 (2013)
13. Zhou, J., Wen, C., Zhang, Y.: Adaptive output control of nonlinear systems with uncertain dead-zone nonlinearity. IEEE Trans. Autom. Control **51**(3), 504–511 (2006)
14. Lu, X., Wang, L.L., Wang, H.X., et al.: Kalman filtering for felayed singular systems with multiplicative noise. IEEE/CAA J. Autom. Sin. **3**(1), 51–58 (2016)
15. Zhang, Y.L., Xu, Q.S.: Adaptive sliding mode control with parameter estimation and Kalman filter for precision motion control of a piezo-driven microgripper. IEEE Trans. Control Syst. Technol. **25**(2), 728–735 (2017)
16. Rajabi, N., Abolmasoumi, A.H., Soleymani, M.: Sliding-mode trajectory tracking control of a ball-screw-driven shake table based on online state estimations using EKF/UKF. Struct. Control Health Monit. **25**(5), 1–13 (2017)
17. He, Y.D., Wang, J.Z., Hao, R.J.: Adaptive robust dead-zone compensation control of electro-hydraulic servo systems with load disturbance rejection. J. Syst. Sci. Complexity **28**(2), 341–359 (2014). https://doi.org/10.1007/s11424-014-2243-5
18. Merritt, H.E.: Hydraulic Control System. Wiley, New York (1967)
19. Guo, Y.N., Cheng, W.D., Gong, W., et al.: Adaptively robust rotary speed control of an anchor-hole driller under varied surrounding rock environments. Control Eng. Pract. **86**, 24–36 (2019)

Exploring Multi-scale Deep Encoder-Decoder and PatchGAN for Perceptual Ultrasound Image Super-Resolution

Jianyong Liu[1], Heng Liu[1(✉)], Xiaoyu Zheng[1], and Jungong Han[2]

[1] Anhui University of Technology, Maanshan, Anhui, China
jianyong-liu@qq.com, hengliusky@aliyun.com, zhengxiaoyu51888@163.com
[2] The University of Warwick, Conventry CV4 7AL, UK
jungonghan77@gmail.com

Abstract. Ultrasound visual imaging is currently one of three mainstream image diagnosis technologies in the medical industry, but due to the limitations of sensors, transmission media and ultrasound characteristics, the quality of ultrasound imaging may be poor, especially its low spatial resolution. We incorporate, in this paper, a new multi-scale deep encoder-decoder structure into a PatchGAN (patch generative adversarial network) based framework for fast perceptual ultrasound image super-resolution (SR). Specifically, the entire algorithm is carried out in two stages: ultrasound SR image generation and image refinement. In the first stage, a multi-scale deep encoder-decoder generator is employed to accurately super-resolve the LR ultrasound images. In the second stage, we advocate the confrontational characteristics of the discriminator to impel the generator such that more realistic high-resolution (HR) ultrasound images can be produced. The assessments in terms of PSNR/IFC/SSIM, inference efficiency and visual effects demonstrate its effectiveness and superiority, when compared to the most state-of-the-art methods.

Keywords: Ultrasound image super-resolution · Deep encoder-decoder · Multi-scale · PatchGAN

1 Introduction

Ultrasonograph [6] is an effective visual diagnosis technology in medical imaging industry and has unique advantages over modalities, such as magnetic resonance imaging (MRI), X-ray, and computed tomography (CT). In the actual diagnosis through ultrasound imaging, doctors usually judge pathological changes by visually perceiving the region of interest (ROI) in the ultrasound image, such as the shape contour or edge smoothness. This means that the higher the resolution of ultrasound image, the more conducive to visual perception for better medical diagnosis. However, due to the acoustic diffraction limit of the medical industry,

ⓒ Springer Nature Singapore Pte Ltd. 2020
H. Zhang et al. (Eds.): NCAA 2020, CCIS 1265, pp. 47–59, 2020.
https://doi.org/10.1007/978-981-15-7670-6_5

it is difficult to acquire HR ultrasound data. Therefore, for enhancing the resolution of ultrasound data, image SR may become a potential solution, which is of great significance for medical clinical diagnosis based on visual perception [7,16].

Over the past few years, the methods based on deep learning have emerged in various fields of natural image processing, ranging from image de-noising [2] to image SR [5,10], and video segmentation [21]. Recently, they have also been applied into diverse medical image processing tasks, ranging from CT image segmentation [24] to ultrasound image SR [3,14]. Umehara *et al.* [26] firstly apply SR convolutional neural network (SRCNN [5]) to enhance the image resolution of chest CT images. A recent work [9] demonstrates that the deeper and wider network can result in preferable image SR results and seems to have no problem of poor generalization performance. Actually, this experience may not always applicable to SR of medical data (including ultrasound images), because there may not be abundant medical image samples for training in practice. Thus, how to design an appropriate deep network structure becomes one key to improve the efficiency of medical image SR.

Other deep supervised models, especially 'U-net' convolutional networks [20,25], are proposed and explored recently for bio-medical image segmentation as well as ultrasound image SR. With no fully connected layers, U-net consists of only convolution and deconvolution operations, where the former is named as encoder and the latter is called decoder. However, in such a U-net, the pooling layers and the single-scale convolutional layers may fail to take advantage of various image details and multi-range context of SR. Note that the terms of encoder and decoder in the following sections mean the convolution operation and deconvolution operations respectively.

At the same time, Ledig *et al.* [10] presented a new deep structure, namely SR generative adversarial network (SRGAN), producing photo-realistic SR images. Instead of using CNNs, Choi *et al.* [3] have applied the SRGAN model for high-speed ultrasound SR imaging. Such GANs based works claimed to obtain better image reconstruction effect with good visual quality. Unfortunately, Yochai *et al.* in their recent work [1] analyze that the visual perception quality and the distortion decreasing of an image restoration algorithm are contradictory with each other.

Actually, almost all the above-mentioned deep methods are from the perspective of single-scale spatial domain reconstruction, without a consideration of multi-scale or even frequency domain behavioral analysis of network feature learning. Motivated by the frequency analysis of neural network learning [19,28] and the structure simulation of multi-resolution wavelet analysis [13], in this work, we present a novel deep multi-scale encoder-decoder based approach to super-resolve the LR ultrasound images. Moreover, inspired by the analysis of [1], our model integrates the PatchGAN [8] way to better tradeoff the reconstruction accuracy and the visual similarity to real ultrasound data. We perform extensive experiments on different ultrasonic data sets and the results demonstrate our approach not only achieves high objective quality evaluation value, but also holds rather good subjective visual effect.

As far as we know, the methods that deal with the resolution enhancement of a single ultrasound image are few, let alone a thorough exploration of multi-scale and adversarial learning to achieve accurate reconstruction with perception trade-off. The contributions of the work are generalized as follows:

- By simulating the structure of multi-resolution wavelet analysis, we propose a new end-to-end deep multi-scale encoder-decoder framework that can generate a HR ultrasound image, given a LR input for 4× up-scaling.
- We integrate the PatchGAN adversarial learning with the VGG feature loss and the ℓ_1 pixel-wise loss to jointly supervise the image SR process at different levels during training. The experimental results turn out that the integrated loss is good at recovering multiple levels of details of ultrasound images.
- We evaluate the proposed approach on several public ultrasound datasets. We also compare the variants of our model and analyze the performance and the differences to others, which might be useful for future ultrasound image SR research.

The rest of the paper is organized as follows. Related works are outlined in Sect. 2. Section 3 describes our proposed approach and its feasibility analysis. Lots of experimental results and analysis are shown in Sect. 4. Finally, the conclusion of this paper is summarized in Sect. 5.

2   Related Work

2.1   Natural Image SR

Given the powerful non-linear mapping, CNN based image SR methods can acquire better performance than the traditional methods. The pioneering SRCNN, proposed by Dong et al. [5], only utilized three convolutional layers to learn the mapping function between the LR images and the corresponding HR ones from numerous LR-HR pairs. However, the fact that SRCNN has only three convolutional layers indicates the model actually is not good at capturing image features. Considering that the prior knowledge may be helpful for the convergence, Liang et al. [11] used Sobel edge operator to extract gradient features to promote the training convergence. In spite of speeding up the training procedure, the improvement for reconstruction performance is very limited. Recently, based upon the structure simulation of multi-resolution wavelet analysis, Liu et al. [13] proposed a multi-scale deep network with phase congruency edge map guidance model (MSDEPC) to super-resolve single LR images.

Aiming to improve the reconstruction quality, Ledig et al. [10] applied such adversarial learning strategy to form a novel image SR model - SRGAN, of which the generator network is used to super-resolved the LR input efficiently while the discriminator network determines whether the super-resolved images approximate the real HR ones.

Recognizing that batch normalization may make the extracted features lose diversity and flexibility, Lim *et al.* [12] proposed an enhanced deep residual SR model (named as EDSR) by removing the batch normalization operation. In order to hold the flexibility of the path, they also adjusted the residual structure so that the sum of different paths will no longer pass through the ReLU layer.

Different from SRGAN [10] that only distinguishing the generated image itself, Park *et al.* [18] presented to add additional discrimination network which acts on the feature domain (called as SRFeat), so that the generator can generate high-frequency features related to the image structure. Moreover, long range jump connections were used in the generator to make information flow more easily in layers far away from each other.

2.2   Ultrasound Image SR

Compared with the flourishing situation of natural images, little attention has been paid to SR of medical images overall. Recently, Zhao *et al.* [29] explored the properties of the decimation matrix in the Fourier domain and managed to acquire an analytical solution with ℓ_2 norm regularizer for the problem of ultrasound image SR. Unlike many studies focusing on ultrasound lateral resolution enhancement, Diamantis *et al.* [4] pay their attention to axial imaging. Being conscious of the accuracy of ultrasound axial imaging mainly depends on image-based localization of single scatter, they use a sharpness based localization approach to identify the unique position of the scatter, and successfully translate SR axial imaging from optical microscopy into ultrasound imaging.

Having seen the power of deep learning in natural image processing, Umehara *et al.* [26] firstly applied the SRCNN to enhance the resolution of some chest CT images and the results demonstrated the CNN based SR model is also suitable for medical images. Moreover, in order to ease the problem of lacking numerous training samples in common medical datasets, Lu *et al.* [14] utilized dilated CNNs and presented a new unsupervised SR framework for medical ultrasound images. However, this is not a real unsupervised method in the sense that their model still needs LR patches as well as their corresponding HR labels.

Very recently, Van Sloun *et al.* [25] applied U-Net [20] deep model to improve upon standard ultrasound localization microscopy (Deep-ULM), and obtained SR vascular images from high-density contrast-enhanced ultrasound data. Their deep-ULM model is observed to be suitable for real time applications, resolving about 1250 HR patches per second. Aiming to improve the texture reconstruction of ultrasound image SR, Choi *et al.* [3] slightly modified the architecture of SRGAN [10] to improve the lateral resolution of ultrasound images. Despite its surprisingly good performance, some evidence [17] (including our corresponding observations in Fig. 3 and Fig. 4) showed that the produced super-resolution image is easy to contain some linear aliasing artifacts.

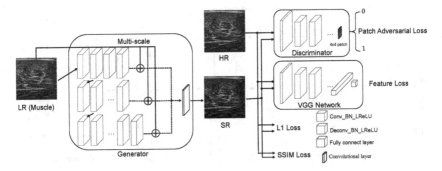

Fig. 1. The proposed ultrasound image SR model: multi-scale ultrasound super-resolved image generator (left) and patches discriminator based on four levels losses (right).

3 Methodology

According to the wavelet and multi-resolution analysis (MRA) [15], an image $f(x)$ is expressed as

$$f(x) = \sum_{k \in Z}^{N} a_k^{j_0} \phi_k^{j_0}(x) + \sum_{j=j_0}^{J} \sum_k b_k^j \psi_k^j(x), \tag{1}$$

in which j varies from j_0 to J, k indexes the basis function, and $\{a_k^{j_0}\}$, $\{b_k^j\}$ act as weighting parameters to associate to the scale function $\phi(x)$ and the wavelet function $\psi(x)$, respectively. Concretely, the image $f(x)$ consists of two components (see Eq. (1)), which are the approximation (the first item, low frequency component) and the details (the second item, high frequency components). From the point of view of deep learning, Eq. (1) may be looked on as a combination reconstruction of different scales branches and each scale reconstruction can be realized by network deconvolution (decoder). Moreover, in Eq. (1), the low frequency (approximation) coefficients a_k^j and the high frequency (detail) coefficients b_k^j can be calculated as:

$$a_k^j = \langle f(x), \phi_k^j(x) \rangle = \sum_i p_{ik}^j f_i$$
$$b_k^j = \langle f(x), \psi_k^j(x) \rangle = \sum_i q_{ik}^j f_i \tag{2}$$

Here, the image $f(x)$ can be represented as $f = \{f_1, f_2, \cdots, f_i, \cdots\}$, the scale function ϕ_k^j is relaxed to $\{p_{1k}^j, p_{2k}^j, \cdots, p_{ik}^j, \cdots\}$, and the detail function ψ_k^j can be loosened to $\{q_{1k}^j, q_{2k}^j, \cdots, q_{ik}^j, \cdots\}$. Obviously, if regarding the weights p_{ik}^j and q_{ik}^j as the convolution kernels at scale j and using the inner projection as the feature encoding, Eq. (2) can be easily implemented by one scale convolution (encoder) on the image $f(x)$. Thus, based on such structure simulation analysis,

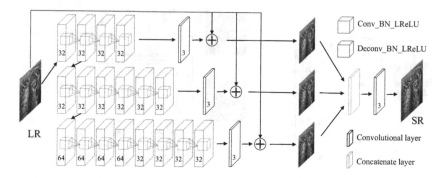

Fig. 2. The pipeline of multi-scale deep encoder-decoder SR network.

obviously we can construct a multi-scale deep encoder and decoder network to renew some lost high-frequency details for image SR.

The pipeline of our proposed ultrasound image SR approach is shown in Fig. 1. Our overall model can be seen as a GAN framework, which consists of two parts: one is a multi-scale encoder-decoder based SR generator to enhance the resolution of the ultrasound images with four levels losses; the other is patches discriminator, which is used to further recover the ultrasound images.

3.1   Multi-scale Encoder-Decoder SR

Inspired by above multi-resolution analysis and the structure simulation, in this work, we specially construct a multi-scale encoder-decoder deep network for the task of the ultrasound image SR and treat it as a generator. The detailed architecture of the proposed multi-scale encoder-decoder SR network is shown in Fig. 2. The specific configurations of the network can be found in Table 1. Here, according to Eq. (1), if regarding the LR image as the approximation component of the HR one, The optimization goal of multi-scale encoder-decoder learning can be treated as:

$$\tilde{f} = \arg\min_f(\|(y + \sum_j F_j(y, \Theta_j)) - f\|_1), \tag{3}$$

where y and f represent the LR image and the corresponding HR image, and $F(\cdot)$ indicates the reconstruction function. Θ is the learned parameter of the network and the symbol j denotes a specific scale.

In the multi-scale structure, the LR image I_{LR} is firstly sent to three scales encoder-decoder branches to obtain the image details of different scales. Then, these detail maps are directly added to LR image input to get three reconstruction images at different scale thanks to that the LR image can be regarded as the approximation (low frequency) component of the HR one. Finally, we concatenate three reconstruction images and warp them to acquire the super-resolved ultrasound image I_{SR}.

Table 1. The configuration of three scales encoder-decoder streams

Scale1	$(\text{conv3-32})^{\text{a}} \times 2$
	$(\text{deconv3-32}) \times 2$
Scale2	$(\text{conv3-32}) \times 2$
	$(\text{conv3-32}) \times 2$
	$(\text{deconv3-32}) \times 4$
Sclae3	$(\text{conv3-32}) \times 2$
	$(\text{conv3-32}) \times 2$
	$(\text{conv3-64}) \times 2$
	$(\text{deconv3-64}) \times 2$
	$(\text{deconv3-32}) \times 4$

[a]The convolution and deconvolution layers are denoted as conv/deconv (kernel size)-(number of filters)

3.2   Patches Discrimination and Loss Function

Recent works [8,10] show that only using MSE pixel wise loss to supervise SR generation tends to produce over-smooth results. Therefore, in our proposed model, we incorporate four levels loss functions (one pixels loss and three details loss) to supervise the generated SR images to approach the ground-truth HR ones at all levels of details. Moreover, to encourage high-frequency structure, our model takes the similar structure of PatchGAN [8] to discriminate the patches between the generated SR images and the true HR ones.

The MSE loss (ℓ_2 loss) between the generated version $G(y)$ and the real HR one f can be calculated and treated as an objective function for minimization during training. However, due to the energy average characteristics, MSE loss will lead to over-smooth phenomenon. Thus, we replace MSE (ℓ_2) loss with ℓ_1 loss. Here the ℓ_1 loss becomes a measure of the proximity of all the corresponding pixels between such two images. Actually, the ℓ_1 loss is used in the proposed multi-scale encoder and decoder structure (see Eq. (3)).

Given a set of LR and HR image pairs $\{f_i, y_i\}_{i=1}^N$ and assuming the components of network reconstruction at multiple scales can be obtained, then the ℓ_1 pixel-wise loss function for the proposed network can be denoted as:

$$\ell_{pixel} = \sum_{i=1}^{N} ||(y_i + \sum_{j} \lambda_j F_j(y_i, \Theta_j)) - f_i||_1, \tag{4}$$

where λ is regulation coefficient for different details reconstruction term (usually can be set as the reciprocal of the number of scales). Here and in the following, $G(y_i) = y_i + \sum_j \lambda_j F_j(y_i, \Theta_j)$.

We also use the feature loss to guarantee edge features when acquiring super-resolved ultrasound images. Different from the pixel-wise loss, we firstly transform y_i and f_i into certain common feature space or manifold with a mapping function $\phi(\cdot)$. Then we can calculate the distance between them in such feature space. Usually, the feature loss can be described as:

$$\ell_{feature} = \sum_{i=1}^{N} ||\phi(G(y_i)) - \phi(f_i)||_2, \tag{5}$$

For the mapping function $\phi(\cdot)$ in our proposed model, in practice we use the combination of the output of the 12th and the 13th convolution layers of the VGG [23] network to realize it. By this way, we can recover the clear edge feature details in super-resolved ultrasound images.

Since SSIM [27] measures the structure similarity in a neighborhood of certain pixel between the generated image and the ground truth one, we may directly apply this measure as one kind of loss:

$$\ell_{ssim} = \sum_{i=1}^{N} ||SSIM(G(y_i)) - SSIM(f_i)||_2, \tag{6}$$

Based on the adversarial mechanism of PatchGAN [8], the adversarial loss of our multi-scale SR generator can be defined as:

$$\ell_{adv} = \sum_{i=1}^{N} -log(D(G(y_i))) \tag{7}$$

where $D(\cdot)$ is the discriminator of PatchGAN.

The loss of the patches discriminator can be described as the following:

$$\mathcal{L}_{dis} = \sum_{i=1}^{N} log(D(f_i)) + \sum_{i=1}^{N} log(1 - D(G(y_i))), \tag{8}$$

Finally, the total generator loss utilized to supervise the training of the network can be expressed as:

$$\mathcal{L}_{gen} = \alpha\ell_{pixel} + \beta\ell_{feature} + \gamma\ell_{ssim} + \eta\ell_{adv}, \tag{9}$$

where the α, β, γ, η are the weighting coefficients, which in our experiments are set with 0.16, 2e−6, 0.84 and 1e−4, respectively.

4   Experimental Results and Analysis

4.1   Datasets and Training Details

The reconstruction experiments and the performance comparisons are performed on two ultrasound image datasets: CCA-US[1] and US-CASE[2]. The CCA-US

---

[1] http://splab.cz/en/download/databaze/ultrasound.
[2] http://www.ultrasoundcases.info/Cases-Home.aspx.

dataset contains 84 B-mode ultrasound images of common carotid artery (CCA) of ten volunteers (mean age 27.5 ± 3.5 years) with different body weight (mean weight 76.5 ± 9.7 kg). The US-CASE dataset contains 115 ultrasound images of liver, heart and mediastinum, etc. For the objective quality measurement on the super-resolved images, the well-known PSNR [dB], IFC [22], and SSIM [27] metrics are used. The running code of our work is publicly available at https:// github.com/hengliusky/UltraSound_Image_Super_Resolution.

For all ultrasound images in CCA-US and US-CASE, by flipping and rotating, each one is enhanced to 32 images, resulting in a training set of 5760 images and a test set of 640 images. Training images are cropped into small overlapped patches with a size of 64×64 pixels and a stride of 14. The cropped ground truth patches are treated as the HR patches. The corresponding LR ones are obtained by two bi-cubic interpolation of the ground truth.

Table 2. Performance comparisons for 4× SR on two ultrasound datasets. The best results are indicated in Bold.

DataSets	Bicubic	SRCNN [5]	SRGAN [10]	Our variant (w/o PatchGAN)	Our proposed
	PSNR/IFC	PSNR/IFC	PSNR/IFC	PSNR/IFC	PSNR/IFC
CCA-US	20.913/1.213	20.673/0.972	25.331/1.127	22.035/1.226	**25.711/2.290**
US-CASE	26.299/1.055	25.636/1.009	**29.069**/1.102	26.373/1.080	29.038/**2.127**

Table 3. Performance comparisons on average PSNR and SSIM for 4× SR. The best results are indicated in Bold.

DataSets	EDSR [12]	SRFeat [18]	Our variant (w/o PatchGAN)	Our proposed
	PSNR/SSIM	PSNR/SSIM	PSNR/SSIM	PSNR/SSIM
CCA-US	25.290/0.740	25.602/0.721	24.755/**0.834**	**25.714**/0.814
US-CASE	27.432/0.804	28.864/0.808	28.922/0.840	**29.038/0.881**

The comparative experiments are to perform different ultrasound image SR methods for fulfilling 4× SR task. In Table 2 and Table 3, we provide the quantitative evaluation comparisons, and in Fig. 3 and Fig. 4, some visual comparison examples with PSNR/IFC or PSNR/SSIM measures are also provided. In addition, to compare the efficiency of our approach with other methods, we also show the inference speed, the model capacity and the throughput of data processing of all methods in Table 4.

4.2 Experimental Comparisons and Analysis

From Table 2 and Table 3, we can see that our proposed model can achieve the best or the second best PSNR measure on different ultrasound image datasets. As for IFC and SSIM measures, our approach will always get the best evaluation value even both on such two ultrasound datasets. Moreover, even without the help of PatchGAN, the multi-scale deep encoder-decoder also demonstrates quite good SR performance (see Table 2), which indicates that it does recover multiple scales image details. While according to Table 3, we see that the feature loss makes sense in enhancing the SR performance of ultrasound data.

Table 4. The comparisons of model inference efficiency. Blue text indicates the best performance and green text indicate the second best performance

	SRCNN [5]	SRGAN [10]	SRFeat [18]	EDSR [12]	Our proposed
Platform	MATLAB	TensorFlow	TensorFlow	TensorFlow	TensorFlow
Test image size	600 * 448	150 * 112	150 * 112	150 * 112	600 * 448
Inference time	188 ms	53 ms	136 ms	49 ms	176 ms
Throughout (Kb/ms)	1.189	0.929	0.362	1.0	4.474
Model capacity	270 KB	9.1 MB	37.2 MB	9.1 MB	1.6 MB

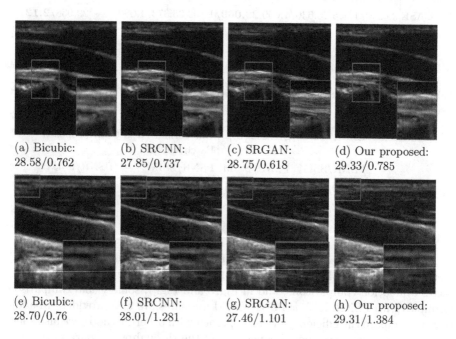

(a) Bicubic: 28.58/0.762

(b) SRCNN: 27.85/0.737

(c) SRGAN: 28.75/0.618

(d) Our proposed: 29.33/0.785

(e) Bicubic: 28.70/0.76

(f) SRCNN: 28.01/1.281

(g) SRGAN: 27.46/1.101

(h) Our proposed: 29.31/1.384

Fig. 3. The Visual and PSNR/IFC comparisons of super-resolved (4×) ultrasound images from CCA-US dataset by (a, e) Bicubic, (b, f) SRCNN, (c, g) SRGAN, and (d, h) Our proposed.

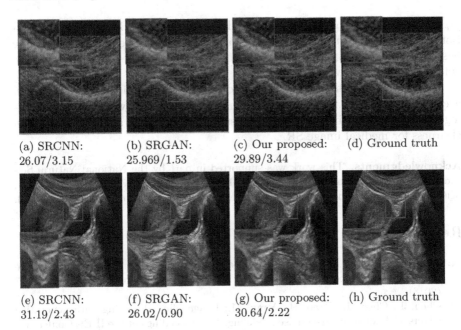

(a) SRCNN:
26.07/3.15

(b) SRGAN:
25.969/1.53

(c) Our proposed:
29.89/3.44

(d) Ground truth

(e) SRCNN:
31.19/2.43

(f) SRGAN:
26.02/0.90

(g) Our proposed:
30.64/2.22

(h) Ground truth

Fig. 4. The Visual and PSNR/IFC comparisons of super-resolved (4×) ultrasound images from US-CASE dataset by (a, e) SRCNN, (b, f) SRGAN, (c, g) Our proposed, and (d, h) Ground truth.

From Fig. 3 and Fig. 4, it is obvious that comparing with other approaches, our presented method can not only acquire the clearer SR visual results but also not to introduce the unwanted aliasing artifacts. In addition, if comparing the structure of our model with that of SRGAN [10], we can see that integrating four levels losses and adopting the local patches discrimination really play a crucial role on accurately super-resolve the LR ultrasound images. Therefore, all the quantitative comparisons and the visual results on different ultrasound image datasets illustrate the excellent performance and the wide effectiveness.

Furthermore, according to Table 4, this demonstrates that our proposed approach is most efficient and hereby to some extent is practically valuable for ultrasonic imaging visual diagnosis in medical industry.

5   Conclusion

In this work, different from the previous approaches, we propose a novel multi-scale deep encoder-decoder and PatchGAN based approach for ultrasound image SR of medical industry. Firstly, an end-to-end multi-scale deep encoder-decoder structure is employed to accurately super-resolve the LR ultrasound images with a comprehensive imaging loss, including the pixel-wise loss, the feature loss, the SSIM loss and the adversarial loss. And then we utilize the discriminator to urge the generator to get more realistic ultrasound images. Based on the evaluations on two different ultrasound image datasets, our approach demonstrates the best performance not only in objective qualitative measures and the inference efficiency but also in visual effects.

It is noted that the SR of ultrasound image requires higher reconstruction accuracy than that of natural image. Thus, our future work will focus on exploring the relationship between the reconstruction accuracy and the perception clearness for medical image SR.

Acknowledgments. This work was supported in part by the National Natural Science Foundation of China under Grant No. 61971004 and by the Key Project of Natural Science of Anhui Provincial Department of Education under Grant No. KJ2019A0083.

References

1. Blau, Y., Michaeli, T.: The perception-distortion tradeoff. In: Proceedings of the IEEE Conference on Computer Vision and Pattern Recognition, pp. 6228–6237 (2018)
2. Chen, J., Chen, J., Chao, H., Yang, M.: Image blind denoising with generative adversarial network based noise modeling. In: Proceedings of the IEEE Conference on Computer Vision and Pattern Recognition, pp. 3155–3164 (2018)
3. Choi, W., Kim, M., HakLee, J., Kim, J., BeomRa, J.: Deep CNN-based ultrasound super-resolution for high-speed high-resolution B-mode imaging. In: Proceedings of the IEEE International Ultrasonics Symposium, pp. 1–4, October 2018. https://doi.org/10.1109/ULTSYM.2018.8580032
4. Diamantis, K., Greenaway, A.H., Anderson, T., Jensen, J.A., Dalgarno, P.A., Sboros, V.: Super-resolution axial localization of ultrasound scatter using multi-focal imaging. IEEE Trans. Biomed. Eng. **65**(8), 1840–1851 (2018)
5. Dong, C., Loy, C.C., He, K., Tang, X.: Image super-resolution using deep convolutional networks. IEEE Trans. Pattern Anal. Mach. Intell. **38**(2), 295–307 (2016)
6. Hoskins, P.R., Martin, K., Thrush, A.: Diagnostic Ultrasound: Physics and Equipment. Cambridge University Press, Cambridge (2010)
7. Hudson, J.M.: Dynamic contrast enhanced ultrasound for therapy monitoring. Eur. J. Radiol. **84**(9), 1650–1657 (2015)
8. Isola, P., Zhu, J.Y., Zhou, T., Efros, A.A.: Image-to-image translation with conditional adversarial networks. In: Proceedings of the IEEE Conference on Computer Vision and Pattern Recognition, pp. 1125–1134 (2017)
9. Kim, J., Kwon Lee, J., Mu Lee, K.: Accurate image super-resolution using very deep convolutional networks. In: Proceedings of the IEEE Conference on Computer Vision and Pattern Recognition, pp. 1646–1654 (2016)
10. Ledig, C., et al.: Photo-realistic single image super-resolution using a generative adversarial network. In: Proceedings of the IEEE Conference on Computer Vision and Pattern Recognition, pp. 4681–4690 (2017)
11. Liang, Y., Wang, J., Zhou, S., Gong, Y., Zheng, N.: Incorporating image priors with deep convolutional neural networks for image super-resolution. Neurocomputing **194**, 340–347 (2016)
12. Lim, B., Son, S., Kim, H., Nah, S., Mu Lee, K.: Enhanced deep residual networks for single image super-resolution. In: Proceedings of the IEEE Conference on Computer Vision and Pattern Recognition Workshops, pp. 136–144 (2017)

13. Liu, H., Fu, Z., Han, J., Shao, L., Hou, S., Chu, Y.: Single image super-resolution using multi-scale deep encoder-decoder with phase congruency edge map guidance. Inf. Sci. **473**, 44–58 (2019)
14. Lu, J., Liu, W.: Unsupervised super-resolution framework for medical ultrasound images using dilated convolutional neural networks. In: Proceedings of the IEEE 3rd International Conference on Image, Vision and Computing, pp. 739–744. IEEE (2018)
15. Mallat, S.: A Wavelet Tour of Signal Processing. Academic Press, San Diego (1999)
16. Morin, R., Bidon, S., Basarab, A., Kouamé, D.: Semi-blind deconvolution for resolution enhancement in ultrasound imaging. In: Proceedings of the IEEE International Conference on Image Processing, pp. 1413–1417. IEEE (2013)
17. Odena, A., Dumoulin, V., Olah, C.: Deconvolution and checkerboard artifacts. Distill **1**(10), e3 (2016)
18. Park, S.J., Son, H., Cho, S., Hong, K.S., Lee, S.: SRFeat: single image super-resolution with feature discrimination. In: Proceedings of the European Conference on Computer Vision (ECCV), pp. 439–455 (2018)
19. Rahaman, N., et al.: On the spectral bias of neural networks. In: Proceedings of the 36th International Conference on Machine Learning, pp. 5301–5310. PMLR (2019)
20. Ronneberger, O., Fischer, P., Brox, T.: U-net: convolutional networks for biomedical image segmentation. In: Navab, N., Hornegger, J., Wells, W.M., Frangi, A.F. (eds.) MICCAI 2015. LNCS, vol. 9351, pp. 234–241. Springer, Cham (2015). https://doi.org/10.1007/978-3-319-24574-4_28
21. Sakkos, D., Liu, H., Han, J., Shao, L.: End-to-end video background subtraction with 3D convolutional neural networks. Multimedia Tools Appl. **77**(17), 23023–23041 (2017). https://doi.org/10.1007/s11042-017-5460-9
22. Sheikh, H.R., Bovik, A.C., De Veciana, G.: An information fidelity criterion for image quality assessment using natural scene statistics. IEEE Trans. Image Process. **14**(12), 2117–2128 (2005)
23. Simonyan, K., Zisserman, A.: Very deep convolutional networks for large-scale image recognition. arXiv preprint arXiv:1409.1556 (2014)
24. Skourt, B.A., El Hassani, A., Majda, A.: Lung CT image segmentation using deep neural networks. Procedia Comput. Sci. **127**, 109–113 (2018)
25. van Sloun, R.J., Solomon, O., Bruce, M., Khaing, Z.Z., Eldar, Y.C., Mischi, M.: Deep learning for super-resolution vascular ultrasound imaging. In: ICASSP 2019-2019 IEEE International Conference on Acoustics, Speech and Signal Processing (ICASSP), pp. 1055–1059. IEEE (2019)
26. Umehara, K., Ota, J., Ishida, T.: Application of super-resolution convolutional neural network for enhancing image resolution in chest CT. J. Digit. Imaging **31**(4), 441–450 (2018)
27. Wang, Z., Bovik, A.C., Sheikh, H.R., Simoncelli, E.P.: Image quality assessment: from error visibility to structural similarity. IEEE Trans. Image Process. **13**(4), 600–612 (2004)
28. Xu, Z.-Q.J., Zhang, Y., Xiao, Y.: Training behavior of deep neural network in frequency domain. In: Gedeon, T., Wong, K.W., Lee, M. (eds.) ICONIP 2019. LNCS, vol. 11953, pp. 264–274. Springer, Cham (2019). https://doi.org/10.1007/978-3-030-36708-4_22
29. Zhao, N., Wei, Q., Basarab, A., Kouamé, D., Tourneret, J.Y.: Single image super-resolution of medical ultrasound images using a fast algorithm. In: Proceedings of the IEEE 13th International Symposium on Biomedical Imaging, pp. 473–476. IEEE (2016)

Reliable Neighbors-Based Collaborative Filtering for Recommendation Systems

Li Zhang[1,2]($\boxtimes$) (iD), Xiaohan Sun[1], Jiangfei Cheng[1], and Zepeng Li[1]

[1] School of Computer Science and Technology, Joint International Research Laboratory of Machine Learning and Neuromorphic Computing, Soochow University, Suzhou 215006, Jiangsu, China
zhangliml@suda.edu.cn
[2] Provincial Key Laboratory for Computer Information Processing Technology, Soochow University, Suzhou 215006, Jiangsu, China

Abstract. The sparsity of data and the diversity of ratings have a great effect on the performance of recommendation systems. To deal with these two issues, this paper proposes a reliable neighbors-based collaborative filtering (RNCF) method for recommendation systems. RNCF first uses the item similarity to pre-estimated missing ratings to solve the issue of data sparsity, which enriches the sample data and reduces the impact of extremely sparse ratings on the recommendation effect. Then, RNCF finds a set of reliable users for each item by defining a reliable degree for each item. Next, RNCF adopts the K-means clustering algorithm to find the similar users for a target user. The set of reliable neighbors is the intersection of sets of reliable users and similar users. Finally, RNCF provides a predicted rating of a target user on some item using the pre-estimated ratings of corresponding reliable neighbors. Experimental analysis shows that RNCF can improve the performance at the extreme levels of sparsity and rating diversity in recommendation systems.

Keywords: Collaborative filtering · Recommendation system · Data sparsity · Reliable degree

1 Introduction

The development of Internet technology has brought convenience to our lives, and also caused a serious problem. Faced with the increasing information, people can hardly recognize the desired information out of billions of bytes. In this situation, recommendation systems were proposed to deal with these problems.

This work was supported in part by the Natural Science Foundation of the Jiangsu Higher Education Institutions of China under Grant No. 19KJA550002, by the Six Talent Peak Project of Jiangsu Province of China under Grant No. XYDXX-054, by the Priority Academic Program Development of Jiangsu Higher Education Institutions, and by the Collaborative Innovation Center of Novel Software Technology and Industrialization.

© Springer Nature Singapore Pte Ltd. 2020
H. Zhang et al. (Eds.): NCAA 2020, CCIS 1265, pp. 60–71, 2020.
https://doi.org/10.1007/978-981-15-7670-6_6

Since early 1990s, recommendation systems have shown their great efficiency in the issue of information overload [8,13,15]. Recommendation systems mainly help users to filter the information they do not need and provide the interesting information to them.

There are mainly two-category methods for recommendation systems [3]: content-based (CB) [4,11] and collaborative filtering (CF) [3,7]. CB methods generate recommendations by analyzing a certain content like texts, images and sound [1,12,17]. So if you ever made choice on items, CB methods would constructs user preferences based on items and recommend you items similar to this chosen items. Unlike CB methods, CF methods need item ratings of users. If we can collect ratings of a set of items such as movies, music and news, we can make recommendations using CF methods.

As we know, neighborhood-based collaborative filtering (NCF) and model-based collaborative filtering (MCF) are the most basic algorithms in CF methods. MCF approaches learn a predictive model to make predictions using these ratings [14]. In contrast to MCF, NCF usually utilizes similarity measures to find neighbors for a target user, and predicts ratings of the target user on items [6]. In [2], Ayub et al. proposed the Jaccard coefficient (JAC) method to improve the recommendation performance. However, JAC cannot perform well when the dataset is very sparse. In recommendation systems, data sparsity is a serious issue when finding neighbors [10]. To solve this issue, some techniques have been successfully applied to recommendation systems. In [19], Zhang et al. proposed a cluster-based method, called BiFu to cope with the sparse and cold-start problem. BiFu fuses both user and item information to make predictions. Besides, there is a phenomenon of rating diversity of users, which would result in biased ratings and have a bad influence in the recommendation performance. To avoid the rating diversity, Xie et al. presented a rating-based recommender algorithm (RBRA) which represents a user by constructing a vector consisting of mean, variance, and range of this user's ratings [18]. In this way, the found neighbor users could be more reliable.

To solve the issues of data sparsity and rating diversity, this paper presents a reliable neighbors-based collaborative filtering (RNCF) method for recommendation systems. In RNCF, we define a reliable degree for each item. Using reliable degree of each item, RNCF finds a set of reliable users for this item. Then by adopting the K-means clustering algorithm, RNCF finds the similar users for a target user. The neighbor users can be obtained by intersecting the reliable users and the similar users, which are further utilized to estimate the item ratings for the target user. The main contributions of this paper lie in two aspects. (1) RNCF could efficiently relieve the issue of sparse data by making pre-estimation for the missing values in the user-item rating matrix. In doing so, all the users in the database would have a rating for each item. (2) RNCF could solve the issue of rating diversity by evaluating the reliable degree for every item. According to the reliable degree of a given item, RNCF gets the reliable users for this item, which should be fairer when recommending it.

The remaining portion of this paper is organized as follows. Section 2 describes our proposed method RNCF in detail. Section 3 reports experimental settings and results. Section 4 concludes our work with future directions. The notations in this paper is given in Table 1.

2   Reliable Neighbors-Based Collaborative Filtering

2.1   Estimation of Missing Ratings

Recommendation quality could be very poor for the issue of data sparsity. To solve this, we estimate the missing ratings in the user-item rating matrix $\mathbf{R}$ in advance. First, we cluster items into k clusters using K-means with cosine similarity, where k is a positive integer. Each cluster M_q is an item subset and $M_q \subset \mathcal{I}$, $q = 1, \cdots, k$.

Table 1. Notations in this paper

Notation	Explanation
$\mathbf{R}$	User-item rating matrix
m	Number of users in the user-item rating matrix
n	Number of items in the user-item rating matrix
$\mathcal{U}$	User set, $u \in \mathcal{U}$, $|\mathcal{U}| = n$
$\mathcal{I}$	Item set, $i \in \mathcal{I}$, $|\mathcal{I}| = m$
r_{ui}	Rating of user u on item i
$\overline{r}_u$	Average rating of user u
$\widehat{\mathbf{R}}$	Estimated user-item rating matrix obtained by estimating the missing values in $\mathbf{R}$

In the original user-item rating matrix $\mathbf{R}$, r_{ui} is the rating of user u on item i, and the missing ratings are set to zero. If $r_{ui} = 0$, we can roughly estimate it by

$$\widehat{r}_{ui} = \overline{r}_u + \frac{\sum_{j \in M_q} J(i,j)(r_{uj} - \overline{r}_u)}{\lambda + \sqrt{\sum_{j \in M_q} J(i,j)^2 (r_{uj} - \overline{r}_u)^2}} \tag{1}$$

where $\widehat{r}_{ui}$ is the pre-estimated value of user u on item i, $\lambda > 0$ is a small number to avoid the denominator to be zero, $J(i,j)$ is the Jaccard similarity coefficient between item i and j and defined as:

$$J(i,j) = \frac{|U_i \cap U_j|}{|U_i \cup U_j|} \tag{2}$$

where U_i is the set of users rated item i, and $|\cdot|$ is the function of finding the cardinality of a set $\cdot$.

Algorithm 1 describes the estimation procedure of missing ratings. After estimating the unrated entries in the original rating matrix $\mathbf{R}$, we can obtain a new user-item rating matrix $\widehat{\mathbf{R}}$, whose u-th row and i-th column is

$$\widehat{r}_{ui} = \begin{cases} \widehat{r}_{ui}, & if \ r_{ui} = 0, \\ r_{ui}, & otherwise. \end{cases} \tag{3}$$

Algorithm 1. Algorithm of estimating missing ratings

Input: The original user-item rating matrix $\mathbf{R}$, the item cluster number k, and the parameter λ;

Output: The estimated user-item rating matrix $\widehat{\mathbf{R}}$;

1: Perform K-means on items based on $\mathbf{R}$ and generate k clusters M_q, $q = 1, \cdots, k$;
2: **for** $u \in \mathcal{U}$ **do**
3:     Find the unrated item subset N of user u;
4:     **for** $i \in N$ **do**
5:         Find the cluster M_q to which item i belongs;
6:         Estimate the rating of user u on item i by (1).
7:     **end for**
8: **end for**

2.2 Search of Neighbor Users

Here, by neighbor user we mean that a user is not only similar to the given user but also reliable when rating items. The term similar users is frequently used in recommendation systems. However, similar users may be unreliable. The main reason is the diversity of ratings, which is due to users with subjective emotions when rating items. Some users, for example, are likely to give a high rating when they were happy, or give a strict rating because of some personal reasons. All these makes the original data be biased and affects recommended performance. Thus, we need to find reliable users. To do so, we define a reliable degree based on truth discovery to measure the reliability between users.

Reliable Users. The problem of truth discovery has been received a great deal of attention in many fields [9]. In many scenarios, the sources may be inaccurate, conflicting or even biased from the beginning if they come from subjective evaluation [5]. In this case, truth discovery seeks the useful information from possibly conflicting data, and filters the data that we really need. On the basis of truth discovery, a reliable degree is defined as follows.

Definition 1. *Given the user set U, the user-item matrix $\mathbf{R}$ and item i, the reliable degree of item i can be calculated by:*

$$t_i = \frac{1}{|U_i|} \sum_{u \in U_i} r_{ui} \tag{4}$$

where $U_i \subseteq U$ is the set of users rated item i, and r_{ui} is the rating of user u on item i.

In fact, the reliable degree t_i of item i in Definition 1 is the average rating on item i. On the basis of statistical properties, the average value approaches the true value. Thus, it is reasonable for us to adopt Definition 1 without any prior knowledge on items. In the light of Definition 1, we define the set of reliable users for an item.

Definition 2. *Given the set user U, the user-item matrix $\mathbf{R}$, item i, and a tolerable deviation $\sigma > 0$, the set of reliable users for item i is defined as:*

$$T_i = \{u | u \in U, r_{ui} \in [t_i - \sigma, t_i + \sigma]\} \tag{5}$$

where $T_i \subseteq U$.

According to Definition 2, we could obtain sets of reliable users for items given the tolerable deviation σ. We thought that reliable users for an item are relatively fair when they rated the corresponding item.

Neighbor Users. Now, we can find neighbor users for a target user and item based on reliable users and similar users. In other words, neighbor users are reliable and similar to the target user when rating the target item.

Let the number of user clusters be k'. We partition users into k' clusters using the K-means algorithm on the estimated user-item rating matrix $\widehat{\mathbf{R}}$. Thus, we have clusters $C_1, \cdots, C_{k'}$, where $C_p \subset \mathcal{U}$. For a target user, it is easy to get its similar users by determining which cluster it belongs to. Without loss of generality, let v be the target user. Assume that C_p is the set of similar users of user v, or $v \in C_p$. Traditional methods would take users in C_p as the neighbors of v.

For the diversity of user rating styles, some users are used to give a high rating to an item while other users prefer a low rating. In that case, we need to find reliable users for each item. Consider the target item j. By Definition 2, we have its reliable user set T_j.

When recommending the unrated item j for the target user v, we need to find the set of neighbor users as follow:

$$S_j^v = \{u | u \in C_p \cap T_j\} \tag{6}$$

where S_j^v is the set of neighbor users of user v on item j

For the target user, we need to find a neighbor user set for every unrated item when calculating the final rating score. While the traditional methods find only one similar user set for the target user in despite of item information. In this way, the neighbor user sets for all unrated items are more representative compare to the only one similar user set, so RNCF can make the recommendation be more accurate.

2.3   Prediction of Final Ratings

Now, we predict the final rating of the target user v on item j and have

$$r'_{vj} = \overline{\widehat{r}}_v + \frac{\sum_{u \in S_j^v} sim(v,u)(\widehat{r}_{uj} - \overline{\widehat{r}}_u)}{\sum_{u \in S_j^v} sim(v,u)} \tag{7}$$

where $\widehat{r}_{vj}$ is the pre-estimated rating of the target user v on item j in $\widehat{\mathbf{R}}$, $\overline{\widehat{r}}_v$ is the average estimated rating of the target user v in $\widehat{\mathbf{R}}$, and $sim(v,u)$ is the similarity between the target user v and user u:

$$sim(v,u) = \frac{\sum_{i=1}^{n} (\widehat{r}_{vi} - \overline{\widehat{r}}_v)(\widehat{r}_{ui} - \overline{\widehat{r}}_u)}{\sqrt{\sum_{i=1}^{n} (\widehat{r}_{vi} - \overline{\widehat{r}}_v)^2} \sqrt{\sum_{i=1}^{n} (\widehat{r}_{ui} - \overline{\widehat{r}}_u)^2}} \tag{8}$$

Algorithm 2 shows the procedure of predicting final ratings for the recommended items.

Algorithm 2. Algorithm of predicting final ratings

Input: The original user-item rating matrix $\mathbf{R}$, the pre-estimated user-item rating matrix $\widehat{\mathbf{R}}$, the user cluster number k', and the tolerable deviation σ;
Output: The final user-item rating matrix $\mathbf{R}'$;
1: Perform K-means on users based on $\widehat{\mathbf{R}}$ and generate k' clusters C_p, $p = 1, \cdots, k'$;
2: **for** $v \in \mathcal{U}$ **do**
3:     Find the similar user set C_p for user v;
4:     Find the unrated item subset N of user v based on $\mathbf{R}$;
5:     **for** $j \in N$ **do**
6:         Find the reliable user set T_j for item j by (4) and (5);
7:         Let the neighbor user set of user v on item j be $S_j^v = T_j \cap C_p$;
8:         **for** $u \in S_j^v$ **do**
9:             Calculate $sim(v,u)$ according to (8);
10:        **end for**
11:        Calculate the final rating r'_{vj} of user v on item j according to (7).
12:    **end for**
13: **end for**

---

3   Experiments

To validate the efficiency of RNCF, we compare it with other methods. All these methods are implemented in MATLAB R2014b on the personal computer with an Intel Core i7 processor with 16 GB RAM.

3.1   Databases

In our experiments, three databases are used to show the performance of our proposed method RNCF. Table 2 shows the statistic of the three datasets, where the sparsity level of these datasets is calculated by the method in [16].

The MovieLens database is a publicly collaborative filtering benchmark database collected during the seven-month period from September 19th, 1997 through April 22nd, 1998. MovieLens-100K and MovieLens-1M contain 100,000 ratings from 943 users on 1,628 movies and 1,000,209 anonymous ratings of 3,952 movies made by 6,040 MovieLens users, respectively. In Both datasets, each user has rated at least 20 movies whose rating values were integers $\{1, 2, 3, 4, 5\}$.

FilmTrust is a commonly used publicly dataset, which contains 35,494 ratings from 1,508 users on 2,701 movies and 1,853 trust values. In this paper, we only utilize the ratings without considering the trust values.

Table 2. Description of datasets used in this paper

Dataset	User count	Item count	Rating count	Average rating	Sparsity
MovieLens-100K	943	1,682	100,000	3.5299	0.0630
MovieLens-1M	6,040	3,952	1,000,209	3.5816	0.0419
FilmTrust	1,508	2,701	35,494	3.0027	0.0114

3.2   Performance Measure

We take the whole original dataset as the training set and the real rating set of users as the test set. To measure the performance of algorithms, two indexes, the mean absolute error (MAE) and root mean square error (RMSE) are adopted here. MAE and RMSE are defined as:

$$MAE = \frac{1}{n'} \sum_{(u,i) \in Test} |r_{ui} - r'_{ui}| \tag{9}$$

and

$$RMSE = \frac{1}{n'} \sqrt{\sum_{(u,i) \in Test} (r_{ui} - r'_{ui})^2} \tag{10}$$

where $Test = \{(u,i)|r_{ui} \neq 0, u \in \mathcal{U}, i \in \mathcal{I}\}$ is the test set, $n' = |Test|$ is the number of test set and r'_{ui} is the final predicted rating of r_{ui}. The smaller the MAE and RMSE are, the better the recommendation performance is.

3.3   Experimental Results

Analysis of Parameter Sensitivity. In the phase of estimation of missing ratings, RNCF requires the number of item clusters k, and the parameter λ

used to avoid the denominator to be zero in (1). In the procedure of finding the neighbor users, RNCF needs the number of user clusters k' and the tolerable deviation σ. These parameters are critical for the stability and performance of RNCF. Now we design experiments on the three datasets to show the effect of these parameters on the performance of RNCF.

First, we observe the effect of k when fixing $k' = 150$, $\lambda = 0.01$ and $\sigma = 1$. The value of k varies in the set $\{5, 10, 25, \cdots, 50\}$. Figure 1 shows the curves of performance indexes vs. k. From Fig. 1, we can see that RNCF achieves the best results on both performance indexes of MAE and RMSE when partitioning items into 20 clusters. The performance of RNCF is worse with larger k. Thus, let $k = 20$ in the following experiments.

Then we observe the effect of λ when fixing $k = 20$, $k' = 150$, and $\sigma = 1$. The value of λ varies in the set $\{10^{-6}, 10^{-5}, \cdots, 1, 5, 10, 100\}$. Figure 2 shows the curves of performance indexes vs. λ. From the curves of MAE and RMSE, we can see that the best results on three datasets are obtained when $\lambda = 0.01$. In following experiments, $\lambda = 0.01$ is set for all datasets.

Next we observe the effect of k' when fixing $k = 20$, $\lambda = 0.02$, and $\sigma = 1$. The value of k' varies from 50 to 500. The curves of performance indexes vs. k' are shown in Fig. 3. Roughly speaking, RNCF gets worse and worse MAE as the increase of k' for three datasets, while RNCF has a turn point on RMSE for the MovieLens-100K dataset. According to both MAE and RMSE, we let $k' = 150$ in our experiments for all datasets.

Finally, we observe the effect of σ when fixing $k' = 150$, $\lambda = 0.01$ and $k = 20$. σ varies in the set $\{0.25, 0.5, \cdots, 2.5\}$. Figure 4 shows the curves of performance indexes vs. σ over all the datasets. With the increase of σ, the MAE performance of RNCF become worse and worse, the RMSE performance of RNCF has turn points. For all concerned, we set $\sigma = 1$ in our experiments.

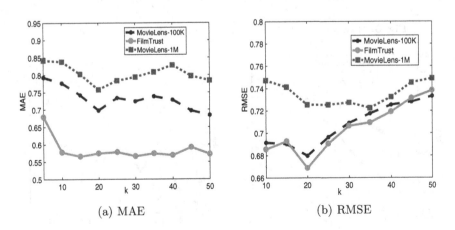

(a) MAE   (b) RMSE

Fig. 1. MAE (a) and RMSE (b) vs. item cluster number for three datasets.

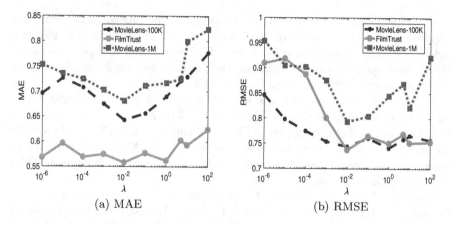

Fig. 2. MAE (a) and RMSE (b) vs. λ for three datasets.

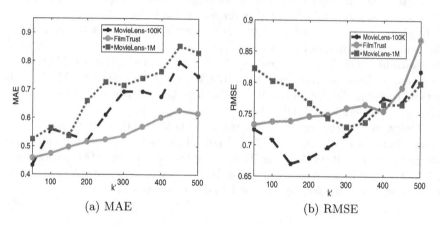

Fig. 3. MAE (a) and RMSE (b) vs. user cluster number for three datasets.

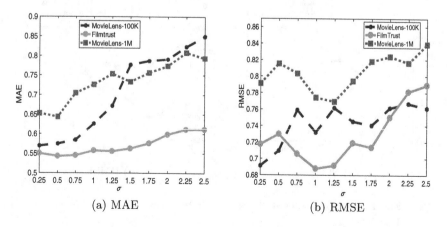

Fig. 4. MAE (a) and RMSE (b) vs. σ for three datasets.

Comparison of Methods. In this part, we compare our method with some other recommender methods, which are listed as follows:

– User-based method with Pearson similarity (UBPCC) [6] is one of the traditional NCF methods.
– User-based method with Cosine similarity (UBCos) [6] is one of the traditional NCF method.
– JAC [2] is a method with the Jaccard-based similarity to improve performance of recommendation systems. In [2], the authors only used the Jaccard coefficients without considering clustering methods.
– RBRA [18] is a rating-based algorithm which can solve the sparsity issue in recommendation systems. RBRA uses triples to represent users when calculating the similarity between users.
– BiFu [19] uses a bi-clustering method to partition both users and items, and also has a smooth phase for data sparsity.

Table 3 shows MAE and RMSE of all six methods on the FilmTrust dataset. We can see that RNCF can achieve the best performance among the six methods. RNCF has the lowest MAE (0.5253), followed by RBRA (0.8264). Compared to the second best RBRA, MAE and RMSE of RNCF increases by 36.44% and 31.00%, respectively.

Table 4 and 5 show the performance of all the six methods on two MovieLens datasets, respectively. Similar to the results in Table 3, RNCF has the best results among other methods. On MovieLens-100k, RNCF increases MAE and RMSE 39.20% and 30.87% compared to BiFu, respectively. On MovieLens-1M, the MAE of RNCF is increased by 26.80% compared to RBRA, and RMSE of RNCF 27.55% compared to BiFu.

Table 3. MAE and RMSE on FilmTrust.

Methods	MAE	RMSE
UBPCC	0.8911	1.0271
UBCos	0.8491	1.0729
BiFu	0.8354	0.9885
RBRA	0.8264	1.0340
JAC	0.8807	1.0223
RNCF	0.4581	0.6867

Table 4. MAE and RMSE on Movielens-100K.

Methods	MAE	RMSE
UBPCC	0.8823	1.0909
UBCos	0.8724	1.0860
BiFu	0.8359	1.0196
RBRA	0.8479	1.0387
JAC	0.8538	1.0734
RNCF	0.5199	0.6703

Table 5. MAE and RMSE on Movielens-1M.

Methods	MAE	RMSE
UBPCC	0.8917	1.0955
UBCos	0.9090	1.1318
BiFu	0.8835	1.0242
RBRA	0.8754	1.0704
JAC	0.8946	1.1683
RNCF	0.6525	0.7074

4   Conclusion

In this paper, we propose an efficient rating-based recommendation algorithm RNCF by introducing reliable users for items. To cope with the sparsity issue, RNCF estimates the missing ratings for sparsity data by partitioning items and utilizing the Jaccard coefficient. Specially, to remedy the rating diversity issue, we define the reliable users based on reliable degrees. The neighbor users for a target user on a target item can be generated by combining the reliable users and similar users obtained by K-means. We conduct experiments on three commonly used recommender datasets: MovieLens-100K, MovieLens-1M and FilmTrust. Experimentally, the recommendation quality in terms of RMSE and MAE shows that RNCF generates more accurate recommendations with less prediction error compare with other methods.

Although RNCF achieves good performance on three datasets. However, these datasets are in small scale. If there are too many users and items in datasets, RNCF would be time-consuming. In the future, we try to make RNCF apply to large scale datasets.

References

1. Alekseev, A., Nikolenko, S.: User profiling in text-based recommender systems based on distributed word representations, pp. 196–207 (2016)

2. Ayub, M., Ghazanfar, M.A., Maqsood, M., Saleem, A.: A jaccard base similarity measure to improve performance of CF based recommender systems. In: International Conference on Information Networking, pp. 1–6 (2018)

3. Bobadilla, J., Ortega, F., Hernando, A.: Recommender systems survey. Knowl. Based Syst. **46**(1), 109–132 (2013)

4. Deldjoo, Y., Elahi, M., Cremonesi, P., Garzotto, F., Piazzolla, P., Quadrana, M.: Content-based video recommendation system based on stylistic visual features. J. Data Semant. **5**(2), 1–15 (2016)

5. Dong, X.L., Berti-Equille, L., Srivastava, D.: Integrating conflicting data: the role of source dependence. Proc. VLDB Endowment **2**(1), 550–561 (2009)

6. Dou, Y., Yang, H., Deng, X.: A survey of collaborative filtering algorithms for social recommender systems. In: International Conference on Semantics, Knowledge and Grids, pp. 40–46 (2016)

7. Elahi, M., Ricci, F., Rubens, N.: A survey of active learning in collaborative filtering recommender systems. Comput. Sci. Rev. **20**(C), 29–50 (2016)

8. Goldberg, D., Nichols, D., Oki, B.M., Terry, D.: Using collaborative filtering to weave an information tapestry. Communi. ACM **35**(12), 61–70 (1992)

9. Li, S., Xu, J., Ye, M.: Approximating global optimum for probabilistic truth discovery (2018)

10. Li, X., She, J.: Collaborative variational autoencoder for recommender systems. In: The ACM SIGKDD International Conference, pp. 305–314 (2017)

11. Musto, C., Semeraro, G., Gemmis, M.D., Lops, P.: Learning word embeddings from wikipedia for content-based recommender systems (2016)

12. Oramas, S., Ostuni, V.C., Noia, T.D., Serra, X., Sciascio, E.D.: Sound and music recommendation with knowledge graphs. ACM Trans. Intell. Syst. Technol. **8**(2), 21 (2017)

13. Puglisi, S., Parra-Arnau, J., Forné, J., Rebollo-Monedero, D.: On content-based recommendation and user privacy in social-tagging systems. Comput. Stand. Interfaces **41**, 17–27 (2015)

14. Ricci, F., Rokach, L., Shapira, B., Kantor, P.B. (eds.): Recommender Systems Handbook. Springer, Boston (2011). https://doi.org/10.1007/978-0-387-85820-3

15. Rodríguez-Hernández, M.D.C., Ilarri, S.: Pull-based recommendations in mobile environments. Comput. Stand. Interfaces **44**, 185–204 (2016)

16. Sarwar, B., Karypis, G., Konstan, J., Riedl, J.: Analysis of recommendation algorithms for E-commerce. In: Proceedings of ACM on E-Commerce (EC-00) (2000)

17. Sejal, D., Rashmi, V., Venugopal, K.R., Iyengar, S.S., Patnaik, L.M.: Image recommendation based on keyword relevance using absorbing markov chain and image features. Int. J. Multimedia Inf. Retrieval **5**(3), 1–15 (2016)

18. Xie, F., Xu, M., Chen, Z.: RBRA: a simple and efficient rating-based recommender algorithm to cope with sparsity in recommender systems. In: International Conference on Advanced Information Networking and Applications Workshops, pp. 306–311 (2012)

19. Zhang, D., Hsu, C.H., Chen, M., Chen, Q., Xiong, N., Lloret, J.: Cold-start recommendation using BI-clustering and fusion for large-scale social recommender systems. IEEE Trans. Emerg. Topics Comput. **2**(2), 239–250 (2014)

Downhole Condition Identification for Geological Drilling Processes Based on Qualitative Trend Analysis and Expert Rules

Yupeng Li[1,2], Weihua Cao[1,2(✉)], Wenkai Hu[1,2], and Min Wu[1,2]

[1] School of Automation, China University of Geosciences, Wuhan 430074, China
{yupengli,weihuacao,wenkaihu,wumin}@cug.edu.cn
[2] Hubei key Laboratory of Advanced Control and Intelligent Automation for Complex Systems, Wuhan, China

Abstract. In geological drilling processes, downhole incidents usually pose a serious threat to drilling safety. In order to improve the performance of drilling safety monitoring, a systematic downhole condition identification method is proposed based on the Qualitative Trend Analysis (QTA) and expert rules. Qualitative trends directly related to downhole conditions are extracted by QTA. Then, the existing expert rules to identify downhole conditions are summarized to create a knowledge base. The qualitative trends extracted from historical drilling data serve as new rules to extend the knowledge base. For online application, with new data collected, the qualitative trends are calculated, and then the downhole condition can be identified by querying the knowledge base. The effectiveness and practicality of the proposed method are demonstrated by case studies involving real drilling processes.

Keywords: Geological drilling · Qualitative Trend Analysis · Condition identification · Knowledge bases

1 Introduction

Geological drilling is a main way for geological exploration and ore prospecting. However, the hostile downhole environment would pose challenges to the safety of drilling processes [1]. As reported, nearly 20% of the total drilling time was wasted in dealing with accidents [2]. If downhole conditions are correctly identified, it can help to prevent downhole accidents in the early stages and reduce the costs of drilling maintenance.

Supported by the National Natural Science Foundation of China under Grants 61733016 and 61903345, the National Key R&D Program of China under Grant 2018YFC0603405, the Hubei Provincial Technical Innovation Major Project under Grant 2018AAA035, the Hubei Provincial Natural Science Foundation under Grant 2019CFB251, the 111 project under Grant B17040, and the Fundamental Research Funds for the Central Universities under Grant CUGCJ1812.

© Springer Nature Singapore Pte Ltd. 2020
H. Zhang et al. (Eds.): NCAA 2020, CCIS 1265, pp. 72–82, 2020.
https://doi.org/10.1007/978-981-15-7670-6_7

At the present stage, safety monitoring of geological drilling processes relies mainly on the expertise of drilling operators [3]. But obtaining such expertise is not easy; even if obtained from one drilling project, it is difficult to use it for other projects. With the installation of computerized systems, abundant information related to downhole conditions is easily obtained and historized in historical drilling database. Therefore, exploring a data-based condition identification method to deal with downhole incidents, is vital to improve drilling safety while avoiding the bias and proneness to errors of humans.

Mechanism model-based condition identification methods were usually implemented under some ideal assumptions on drilling processes [4,5]. Good results could be achieved only when the downhole phenomena and interactions were modeled well. Recently, data driven approaches for downhole condition identification have received increasing studies, due to the easy storage and retrieval of data. Key parameters were developed from the multi-regression analysis in [6] to estimate the severity of the lost circulation incident. The Bayesian trees augmented native bayes algorithm was used for dynamic risk analysis [7]. Reference [8] defined two influx indicators based on the anomalous signals, to calculate the probability of influx faults. However, these studies were based on only process signals, but did not consider the time-dependent relationship of process signals. In this vein, reference [9] developed a real-time diagnosis method by combining the qualitative trend of process signals with a support vector machine model. Reference [10] presented a pattern recognition model considering the similarity of signal trends to calculate the probability of downhole incidents.

In addition, reference [11] proposed a fuzzy system to diagnose the downhole incident, where the change of plausibility of premises was considered as expert systems rules. To detect abnormalities, reference [12] presented a knowledge-based system using drilling variables that behave irregularly to detect faults. A major problem with this kind of application is that the knowledge base contains less rules, and thus many conditions can not be identified.

According to the analysis of existing studies, most downhole condition identification approaches are only based on either mechanism models or original signals; the trend based expert rules are rarely considered while such information is useful and commonly used in practice. However, the problem with the existing knowledge-based systems is that the expert rules might be not always up-to-date and usually do not incorporate the system changes. Therefore, it is necessary to distill such trend information from historical drilling data to complement the list of expert rules.

Motivated by above discussions, a downhole condition identification method based on qualitative trend analysis and expert rules is proposed. The method consists of the following four key steps: (i) The knowledge base is established based on expert rules for downhole condition identification. (ii) Trend information is extracted from historical drilling data through a qualitative trend analysis method. (iii) The trends of historical data are taken as new rules to complement the existing knowledge base. (iv) For online identification, after the trend of new acquired signals is extracted, it is compared with rules in the knowledge base to identify the downhole condition.

The remainder of this paper is organized as follows: Preliminaries about geological drilling processes and the condition identification problem are introduced in Sect. 2. The proposed method for downhole condition identification is systematically presented in Sect. 3. Then the effectiveness of the proposed method is demonstrated by case studies in Sect. 4, some conclusions are covered in Sect. 5.

2 Problem Description

The structural diagram of a geological drilling process is shown in Fig. 1. A drilling rig drives the bit at the end of the drillstring, to break underground rocks. While the mud pump pumps the drilling mud to the bottom of the well through the drillstring, to transport downhole cuttings from the annulus to the mud pit. Common abnormal conditions are the lost circulation and the kick. Correctly identifying these conditions is important to prevent accidents, such as the blowout. In addition, the normal switching conditions that occur after the adjustment of operating parameters are also considered, because such actions may lead to similar characteristics in drilling signals as those caused by abnormal conditions. In drilling processes, operators ensure the drilling safety by monitoring the real-time signals, including standpipe pressure (SPP), mud flow in (MFI), mud flow out (MFO), and mud pit volume (MPV). In summary, 4 common conditions, including the normal, normal switching, lost circulation and kick, are analyzed.

The objective of this paper is to identify the drilling downhole condition, based on the variational trends (obtained from expert rules and quantitative trend analysis) of m measured drilling process variables $\mathbf{x} = [x_1, x_2, \cdots, x_m]^T$.

The occurrence of downhole incidents would cause variational trend changes in signals of the m process variables. A knowledge base R contains both general expert rules and qualitative trend rules is created for downhole conditions identification. Considering limited samples associated with abnormal conditions during the drilling process, an instance based identification is adopted to extract new rules from historical data. Further, since the size of the knowledge base is small, the online calculated trend vector is compared with all rules each time. If it is consistent with any rule, the condition is identified and presented to the operator.

3 The Proposed Method for Downhole Condition Identification

In this section, the first part introduces the way of establishing the expert rules based knowledge base, the second part presents steps for extracting qualitative trends of process signals, and the third part proposes a method to extend the knowledge base.

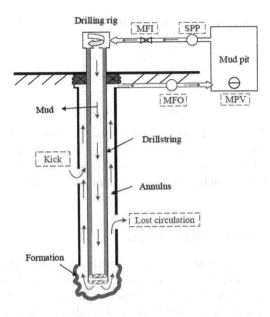

Fig. 1. A structural diagram of a geological drilling process. The blue arrows represent flow directions of the drilling fluid. Abnormal conditions considered in this study are highlighted by dashed red rectangles. Process variables are shown in dashed green rectangles. (Color figure online)

3.1   Establishment of Expert Knowledge Base

Some expert rules to identify downhole conditions are concluded based on experience, and these expert rules are utilized to create an expert knowledge base.

According to the engineering manual [14], three types of variational trends of drilling process signals under different conditions are summarized in Table 1, where ↗ denotes a slow increasing trend, ↘ stands for a slow decreasing trend, and − indicates no significant trend change [13].

Table 1. Variational trends of drilling process signals under the Lost Circulation (LC), KicK(KK), NorMal Switching (NMS), and NorMal (NM) conditions.

Process variables	LC	KK	NMS	NM
SSP	↘	↘	↗	-
MFI	-	-	↗	-
MFO	↘	↗	↗	-
MPV	↘	↗	-	-

Based on the rules in Table 1, the expert knowledge base R_g can be summarized as

$$R_g = \begin{bmatrix} \beta_{11} & \beta_{12} & \cdots & \beta_{1m} \\ \beta_{21} & \beta_{22} & \cdots & \beta_{2m} \\ \vdots & \vdots & \ddots & \vdots \\ \beta_{n1} & \beta_{n2} & \cdots & \beta_{nm} \end{bmatrix}^T, \tag{1}$$

where β denotes the trend of signals, the increasing, decreasing, and no significant change correspond to 1, -1, 0, respectively. For example, one rule related to the kick is $[1\ 0\ 1\ 1]^T$. The number of rules is n, each column denotes an expert rule corresponds to a downhole condition, where one condition may correspond to multiple rules.

3.2 Extraction of Qualitative Trends

In drilling processes, abnormal downhole conditions can be reflected by various drilling process variables. To extract trends of process variables, the Qualitative Trend Analysis (QTA) and sliding windows are adopted. The sliding window contains a signal segment, and it is to be updated when new samples are collected. Since the variational trend can be captured in a linear manner, the first-polynomial is utilized to extract trends of drilling process variables [9].

Because different original process variables x^o have different ranges, the normalized signal $x(k)$ is calculated as

$$x(k) = \frac{x^o(k) - x^o_{min}}{x^o_{max} - x^o_{min}}, \tag{2}$$

where x^o_{min} and x^o_{minmax} represent the minimum and maximum of $x^o(k)$, respectively. Then, the $x(k)$ is normalized to the range of [0,1].

Considering a time series $x(k)$, the trend index at $k = t$ is defined as $s(t)$. In the time interval $k \in [t - L + 1, t]$, the least square model is

$$\hat{x}(k) = s(t)k + b(t), \tag{3}$$

where t denotes the current sampling time, and L indicates the length of the sliding window. The optimal value of $\hat{s}(t)$ is obtained by minimizing the estimation errors, i.e.,

$$\min [x(k) - \hat{x}(k)]^2. \tag{4}$$

The sliding window moves in sample by sample. After obtaining the trend index $\hat{s}(t)$ at t, the window moves to the next time interval $k \in [t + 2 - L, t + 1]$ to compute $\hat{s}(t + 1)$.

Since the drilling process have multiple signals, the signals at t can be represented as a vector

$$\mathbf{x}(t) = [x_1(t), \cdots, x_m(t)]^T. \tag{5}$$

Multivariate drilling process signals are denoted as $X = [\mathbf{x}_1, \mathbf{x}_2, \ldots, \mathbf{x}_m]$, and the qualitative trend $\beta_p(t)$ for the pth variable is defined as

$$\beta_p(t) = \begin{cases} 1 & s_p(t) \in (h_p, +\infty) \\ 0 & s_p(t) \in [-h_p, h_p] \\ -1 & s_p(t) \in (-\infty, -h_p) \end{cases}, \tag{6}$$

where $p \in \{1, 2, \ldots, m\}$, $s_p(t)$ is the trend index of $\mathbf{x}_p$ at t, and h_p is a non-negative trend threshold for the pth variable. The thresholds are used to reduce the noise interference for trend extraction. Since the variation speed of different signal is different, the corresponding trend threshold is also various. The symbol 1, 0, -1 in Eq. (6) correspond to the three trends of increase, no significant change, and decrease, respectively.

Both the sliding window length L and trend threshold vector $\mathbf{h} = [h_1, \ldots, h_m]^T$ have great influence on the identification result. If the L is long, it will contain more historical information and weaken the trend information closer to the current moment. On the contrary, the long-term change features cannot be comprehensively captured. Meanwhile, for a single variable, a higher threshold h is more likely to miss some trends, while a lower h can be interfered by disturbances like measurement noises to produce false alarms. It should be emphasized that different variables correspond to different optimal values of L and h. Therefore, the grid search dynamic subspace model is introduced to determine L and h from historical data for each variable. A finite subset of parameter L selects an integer from 2 to 5 samples, and the trend threshold h is taken from 0 to 0.3 with a fixed step of 0.001. For example, based on the estimated qualitative trend $\tilde{\beta}_p$, the prediction accuracy of the pth variable A_p is defined as

$$A_p = 1 - \frac{\sum_{j=1}^{N} (T_p(j))}{N} \times 100\%, \tag{7}$$

$$T_p(j) = \begin{cases} 0 & \text{if } \tilde{\beta}_p(j) - \beta_p(j) = 0, \\ 1 & \text{otherwise}, \end{cases} \tag{8}$$

where N denotes the number of total training samples, and $\beta_p(j)$ represents the real historical trend.

The goal is to find m pairs of L and h, corresponding to the maximum prediction accuracy for each class of variable, respectively. Figure 2 presents the relationships between L, h and prediction accuracy of the 4 variables. For the SPP variable, the relationship between the threshold h_{SPP} and the prediction accuracy under different window lengths L_{SPP} is presented. Obviously, the accuracy was optimal when $L_{SPP} = 3$ and $h_{SPP} = 0.05$, and thus this pair parameters was exploited to extract the trend of SPP. Similarly, the most suitable pair of L and h was determined for the other three variables, respectively.

Using the optimal L and h, the trend vector $Q(t)$ is given by

$$Q(t) = [\beta_1(t), \beta_2(t), \ldots, \beta_m(t)]^T. \tag{9}$$

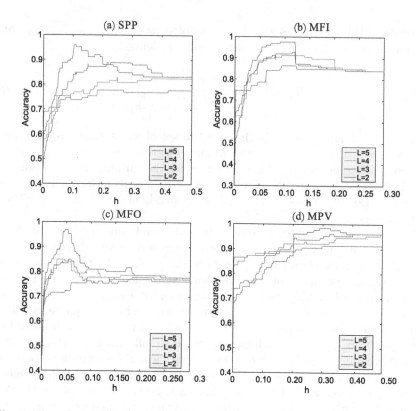

Fig. 2. The relationships between L, h and prediction accuracy of the 4 variables.

Then, the trend vectors extracted from historical data are summarized as new rules for establishing the data-based new knowledge base, to identify downhole conditions.

3.3 Identification of Downhole Conditions Based on Qualitative Trend Rules

The expert rules concluded in Sect. 3.1 are limited. A new rule base creation method is thus required to summarize the historical data into useful knowledge. Meanwhile, the variation trends of process signals have some similarities in the same region. Hence, for the purpose of extending the knowledge base, a new rules extraction method from historical data is proposed based on qualitative trend analysis.

For the historical drilling data in the same region, calculate the qualitative trend vectors under different conditions. Then, select new qualitative trend vectors that are not involved in the existing expert rules based knowledge base, to constitute the data-based new knowledge base R_n.

It should be noted that duplicated rules are removed when establishing the new rule base. Since a large amount of vectors belong to normal conditions, only rules of the normal switching, faulty conditions, and part of normal conditions are extracted. Suppose that downhole conditions are lost circulation at $k \in \{t_1, t_2, ..., t_n\}$, the qualitative trend matrix S corresponding to the lost circulation can be expressed as

$$S = \begin{bmatrix} \beta_1(t_1) & \beta_2(t_1) & \cdots & \beta_m(t_1) \\ \beta_1(t_2) & \beta_2(t_2) & \cdots & \beta_m(t_2) \\ \vdots & \vdots & \ddots & \vdots \\ \beta_1(t_n) & \beta_2(t_n) & \cdots & \beta_m(t_n) \end{bmatrix}^T, \tag{10}$$

where n is the number of extracted new rules. Then, each column of the matrix in Eq. (10) would be included as a rule belongs to the lost circulation condition. In the same way, if the condition's label is normal switching, the vector would be included into the knowledge base that belongs to the normal switching. Next, the new rule base R_n for a special region is established based on many matrices of S. Then, the comprehensive rule base

$$R = [R_g, R_n] \tag{11}$$

is created based on R_g and R_n.

In the matrix R, all trend vectors have a fixed relationship with downhole conditions, but each kind of downhole condition corresponds to a variety of rules. If the collected new trend vector satisfies any of the rules in the comprehensive rule base R, the downhole condition will be identified as the corresponding condition. Mathematically, the corresponding relation is represented by

$$[\beta_1(t), \beta_2(t), \cdots, \beta_m(t)]^T - R(:, i) = 0, \tag{12}$$

where $R(:, i)$ denotes a rule in R. On the contrary, if Eq. (12) is not equal to zero, which means the trend vector does not comply with this rule.

There are two main steps in the online downhole condition identification phase. First, the qualitative trend of drilling process signals is calculated. Then, the trend vector is compared with each rule in R to identify the downhole condition.

4   Industrial Case Study

This section presents an industrial case study to illustrate the effectiveness of the proposed method. The industrial data was collected from a geological well located in Heilongjiang Province, China. First, the shallow drilling process data was used to extract new rules. Then, a signal segment under the normal switching condition was taken as examples, where downhole conditions were identified based on the knowledge rule base.

Expert rules from existing process knowledge are shown in Table 2 [14], which contains the relationships between signal trends of SSP, MFI, MFO and MPV, and the four downhole conditions, i.e, Lost Circulation (LC), KicK (KK), Nor-Mal (NM), and NorMal Switching (NMS). For example, when a lost circulation incident occurs, the drilling signals of SSP, MFO and MPV may show decrease trends while MFI remains unchanged, these trends are denoted as $R_{LC}=$ $[-1\ -1\ 0\ -1]$. Then, the expert rules based knowledge base R_g was established. In addition, the new rule base R_n in Table 2 was created based on trend vectors of historical drilling data under abnormal conditions.

Table 2. List of the existing expert rules that compose the expert knowledge base R_g.

Variable	LC	KK	NMS	NM
SSP	−1	−1	1	0
MFI	0	0	1	0
MFO	−1	1	1	0
MPV	−1	1	0	0

Table 3. List of the data-based qualitative trend rules that compose the new knowledge base R_n.

Variable	LC	LC	KK	NMS	NMS	NMS	NM	NM	NM
SPP	0	1	0	−1	−1	1	0	0	0
MFI	0	0	0	−1	−1	1	0	0	0
MFO	−1	0	1	−1	−1	0	0	1	−1
MPV	−1	−1	1	1	0	0	−1	0	1

The expert rules matrix R_g corresponds to Table 2, and the new rules matrix R_n corresponds to Table 3. Then, the comprehensive knowledge base $R = [R_g : R_n]$ can be established as

$$R=\begin{bmatrix} -1 & 0 & 1 & -1 & 0 & -1 & -1 & 1 & 1 & 0 & 0 & 0 & 0 \\ 0 & 0 & 0 & 0 & 0 & -1 & -1 & 1 & 1 & 0 & 0 & 0 & 0 \\ -1 & -1 & 0 & 1 & 1 & -1 & -1 & 0 & 1 & 1 & 0 & -1 & 0 \\ -1 & -1 & -1 & 1 & 1 & 0 & 1 & 0 & 0 & 0 & -1 & 1 & 0 \end{bmatrix}. \tag{13}$$

The case is a normal switching condition. Figure 3 presents 46 samples of the 4 signals $x_{SPP}(t)$, $x_{MFI}(t)$, $x_{MFO}(t)$ and $x_{MPV}(t)$ at the drilling depth from 2825.15 m to 2825.71 m. The downhole condition was normal before $t = 17$. Since the operator changed MFI at $t = 17$, both SPP and MFO decreased from $t = 17$ to $t = 18$, while the MPV increased from $t = 17$ to $t = 18$. The qualitative trend signals of the 4 variables $\beta_{SPP}(t)$, $\beta_{MFI}(t)$, $\beta_{MFO}(t)$ and $\beta_{MPV}(t)$ are shown in Fig. 4. It can be seen that the trend vectors before $t = 17$ were in accordance with the rules belonging to the normal condition. But the trend vector changed to $[-1-1-1\ 1]$ at $t = 18$, which was not consistent with any expert rule in R_g. However, it can be found in the new rule base R_n. Hence, it was identified as a switching condition by the comprehensive rule base R.

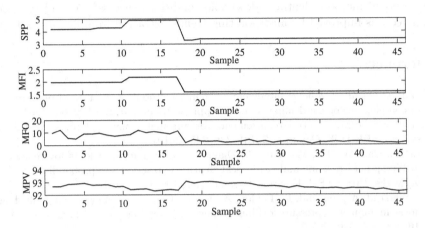

Fig. 3. The 4 signals of a normal switching downhole condition.

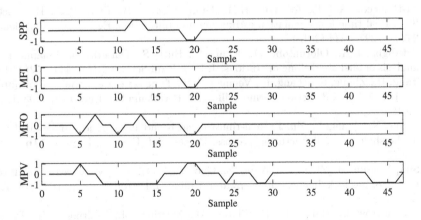

Fig. 4. Qualitative trend signals of a normal switching downhole condition.

5   Conclusion

Motivated by the fact that the historical drilling process data contains various information that is associated with downhole conditions. This paper proposes an expert rules knowledge base establishment method considering the rules obtained from historical data, to identify downhole conditions for geological drilling processes. Since downhole conditions can be reflected by the trend of process signals, the qualitative trend analysis is used to calculate trends of signals. The main challenge faced by the knowledge base is that the number of expert rules is too small. Therefore, the qualitative trends of historical data are extracted as new rules to complement the knowledge base. According to the industrial case study, the proposed method identified downhole conditions successfully, and the performance was improved by incorporating trends extracted from historical data.

References

1. Gan, C., et al.: Two-level intelligent modeling method for the rate of penetration in complex geological drilling process. Appl. Soft Comput. **25**, 592–602 (2019)
2. Godhavn, J.-M.: Control requirements for automatic managed pressure drilling system. SPE Drill. Complet. **25**(3), 336–345 (2010)
3. Li, Y., Cao, W., Hu, W., Wu, M.: Diagnosis of downhole incidents for geological drilling processes using multi-time scale feature extraction and probabilistic neural networks. Process Saf. Environ. Protect. **137**, 106–115 (2020)
4. Willersrud, A., Imsland, L., Pavlov, A., Kaasa, G.-O.: A framework for fault diagnosis in managed pressure drilling applied to flow-loop data. IFAC Proc. Vol. **46**(32), 625–630 (2013)
5. Nikoofard, A., Aarsnes, U.J.F., Johansen, T.A., Kaasa, G.-O.: State and parameter estimation of a drift-flux model for underbalanced drilling operations. IEEE Trans. Control Syst. Technol. **25**(6), 2000–2009 (2017)
6. Al-Hameedi, A.T.T., Alkinani, H.H., Dunn-Norman, S., Flori, R.E., Hilgedick, S.A.: Real-time lost circulation estimation and mitigation. Egyptian J. Petroleum **27**(4), 1227–1234 (2018)
7. Adedigba, S.A., Oloruntobi, O., Khan, F., Butt, S.: Data-driven dynamic risk analysis of offshore drilling operations. J. Petroleum Sci. Eng. **165**, 444–452 (2018)
8. Tang, H., Zhang, S., Zhang, F., Venugopal, S.: Time series data analysis for automatic flow influx detection during drilling. J. Petroleum Sci. Eng. **172**, 1103–1111 (2019)
9. Zhang, X., Zhang, L., Hu, J.: Real-time diagnosis and alarm of down-hole incidents in the shale-gas well fracturing process. Process Saf. Environ. Protect. **116**, 243–253 (2018)
10. Sun, X., Sun, B., Zhang, S., Wang, Z., Gao, Y., Li, H.: A new pattern recognition model for gas kick diagnosis in deepwater drilling. J. Petrol. Sci. Eng. **167**, 418–425 (2018)
11. Sheremetov, L., Batyrshin, I., Filatov, D., Martinez, J., Rodriguez, H.: Fuzzy expert system for solving lost circulation problem. Appl. Soft Comput. **8**(1), 14–29 (2008)
12. Skalle, P., Aamodt, A., Gundersen, O.E.: Detection of symptoms for revealing causes leading to drilling failures. SPE Drill. Complet. **28**(2), 182–193 (2013)
13. Willersrud, A., Blanke, M., Imsland, L.: Incident detection and isolation in drilling using analytical redundancy relations. Control Eng. Pract. **41**, 1–12 (2015)
14. Sun, Z.: Drilling Anomaly Prediction Technology. Petroleum Industry Press, Beijing (2006)

Adaptive Neural Network Control for Double-Pendulum Tower Crane Systems

Menghua Zhang[1,2] and Xingjian Jing[1(✉)]

[1] Department of Mechanical Engineering, The Hong Kong Polytechnic University, Hung Hom, Hong Kong, China
xingjian.jing@polyu.edu.hk
[2] School of Electrical Engineering, University of Jinan, Jinan 250022, China

Abstract. In practical applications, tower crane systems always exhibit double-pendulum effects, because of non-ignorable hook mass and large payload scale, which makes the model more complicated and most existing control methods inapplicable. Additionally, most available control methods for tower cranes need to linearize the original dynamics and require exact knowledge of system parameters, which may degrade the control performance significantly and make them sensitive to parametric uncertainties. To tackle these problems, a novel adaptive neural network controller is designed based on the original dynamics of double-pendulum tower crane systems without any linear processing. For this reason, the neural network structures are utilized to estimate the parametric uncertainties and external disturbances. Based on the estimated information, and adaptive controller is then designed. The stability of the overall closed-loop system is proved by Lyapunov techniques. Simulation results are illustrated to verify the superiority and effectiveness of the proposed control law.

Keywords: Tower cranes · Double-pendulum effects · Adaptive control · Neural network · Lyapunov techniques

1 Introduction

Researches in cranes have been going on for a long time to transport construction materials and cargoes. As a class of underactuated systems, the controller design for cranes is still open and challenging because of the highly underactuated nature, heavily nonlinearity, strong coupling, and inevitable disturbances [1, 2]. For example, the system parameters including the hook and payload mass, friction-related parameters, cable length, etc., are usually difficult to be measured accurately. In addition, the system stability may be influenced significantly by the external disturbance, such as the unavoidable wind. Therefore, it is significant to design controllers taking robustness requirements into full account.

In order to solve these problems, researchers have conducted a lot of work on robust controller design, mainly consisting of adaptive control [3], sliding mode control [4, 5], model predictive control [6], fuzzy logic control [7], observed-based control [8], neural network control [9], and so forth, for different types of cranes, including overhead cranes [10], gantry cranes [11], double-pendulum overhead cranes [12],

© Springer Nature Singapore Pte Ltd. 2020
H. Zhang et al. (Eds.): NCAA 2020, CCIS 1265, pp. 83–96, 2020.
https://doi.org/10.1007/978-981-15-7670-6_8

rotary cranes [13], offshore cranes [14], etc. As a class of essential transport tools in construction sites, tower crane systems have received a lot of interest in recent times. Compared with other cranes, control for tower cranes is more difficult due to their more complex dynamics and larger work space [15]. To obtain accurate jib and trolley positioning, rapid payload swing suppression and elimination, some tower crane-related controllers have been proposed, mainly including gain-scheduling feedback controller [16], path-following controller [17], adaptive controller [18], neural network controller [19], etc.

However, all the aforementioned tower crane controllers completely neglect the hook mass and the distance between the hook's CoG (center of gravity) and the payload's CoG. In reality, the hook mass cannot simply be omitted, and the payload size is usually very large. In this situation, the payload will rotate around the hook and present double-pendulum effect. Compared with the single-pendulum model, the double- pendulum model is much closer to the actual condition. Therefore, although the double-pendulum tower crane presents more complicated dynamics, study of their control problems is of both theoretical and applied significance. In our previous work [20], the dynamical model of double-pendulum tower crane systems is established by Eulerian-Lagrangian method, which provides a model for realizing trolley and jib positioning and hook and payload swing suppression control.

Aiming at the above problems, an adaptive neural network control is designed for double-pendulum tower crane system on the basis of [20]. In addition to accurate jib and trolley positioning performance, rapid hook and payload swing suppression and elimination can be guaranteed simultaneously. In addition, the stability of the closed-loop system and the convergence of the system states are proved by Lyapunov techniques. The superior control performance of the designed control method is further verified by some numerical simulation results. The main merits of this paper lie in the following aspects:

1. The adaptive neural network controller is designed and analyzed based on the original double-pendulum tower crane model without any linear treatment, and rigorous analyses of the stability and state convergence are provided.
2. In practical applications, the parametric uncertainties and external disturbances always exist, which may degrade the control performance. Hence, a radial basis function neural network is introduced into the designed control law to systematically solve the disturbances.

The rest of this paper consists of the following contents. In Sect. 2, the double-pendulum tower crane dynamics and the main control objectives are illustrated. The main results including the adaptive neural network controller design and stability analysis are given in Sect. 3. Some simulation results are given in Sect. 4 to verify the effectiveness and superiority of the proposed control scheme. The main work is concluded in Sect. 5.

2 Problem Statement

The dynamic model and control objectives of double-pendulum tower crane systems are given in this section.

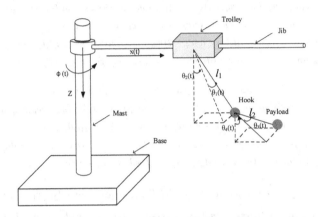

Fig. 1. Schematic diagram of double-pendulum tower crane systems.

2.1 Dynamic Model of Double-Pendulum Tower Crane Systems

Figure 1 shows the schematic diagram of double-pendulum tower crane systems. By using Lagrange's modeling method, the dynamic equations for such systems are provided as follows [20]:

$$
\begin{aligned}
&\left(\begin{array}{l}
(M_t + m_h + m_p)x^2 + (m_p l_1^2 + m_p l_1^2)(s_2^2 + s_1^2 c_2^2) + O + m_p l_2^2 (s_4^2 + s_3^2 c_4^2) \\
+ 2m_p l_1 l_2 (s_2 s_4 + s_1 c_2 s_3 c_4) + 2(m_h + m_p) l_1 x s_1 c_2 + 2m_p l_2 x s_3 c_4 + 2m_p l_2 x \dot\theta_4 c_4
\end{array}\right)\ddot\phi \\
&- (m_h l_1 s_2 + m_p l_1 s_2 + m_p l_2 s_4)\ddot x + 2\left(l_1(m_p + m_h)\left(\dot\theta_1 c_1 c_2 - \dot\theta_2 s_1 s_2\right) + m_p l_2\left(\dot\theta_3 c_3 c_4 - \dot\theta_4 s_3 s_4\right)\right)x\dot\phi \\
&+ 2\left((M_t + m_h + m_p)x + (m_p + m_h)l_1 s_1 c_2 + m_p\left(l_2 s_3 c_4 + l_2 \dot\theta_4 c_4\right)\right)\dot x\dot\phi - m_p l_2 c_3 c_4 (l_1 s_2 + l_2 s_4)\ddot\theta_3 \\
&+ 2\left((m_h + m_p)l_1^2\left(\dot\theta_2 c_1^2 s_2 c_2 + \dot\theta_1 s_1 c_1 c_2^2\right) + m_p l_2^2 c_4\left(\dot\theta_4 c_3^2 s_4 + \dot\theta_3 s_3 c_3 c_4\right)\right)\dot\phi + m_p l_2 s_3 c_4 (l_1 s_2 + l_2 s_4)\dot\theta_3^2 \\
&- l_1 c_1 c_2\left(m_p l_2 s_4 + (m_h + m_p)l_1 s_2\right)\ddot\theta_1 + l_1 s_1 c_2\left((m_h + m_p)l_1 s_2 + m_p l_2 s_4\right)\dot\theta_1^2 + 2m_p l_2 c_3 s_4 (l_1 s_2 - l_2 s_4)\dot\theta_3\dot\theta_4 \\
&+ l_1\left((m_h + m_p)(l_1 s_1 + xc_2) + m_p l_2 (s_1 s_2 s_4 + c_2 s_3 c_4)\right)\ddot\theta_2 + m_p l_2\left(l_1 (s_2 s_3 c_4 - s_1 c_2 s_4) - 2x\dot\phi s_4\right)\dot\theta_4^2 \\
&+ m_p l_2\left(l_2 s_3 + l_1 (s_2 s_3 s_4 + s_1 c_2 c_4) + 2x\dot\phi c_4\right)\ddot\theta_4 + l_1\left(m_p l_2 (s_1 c_2 s_4 - s_2 s_3 c_4) - (m_h + m_p)x s_2\right)\dot\theta_2^2 \\
&+ 2l_1 c_1\left((m_h + m_p)l_1 s_2^2 + m_p l_2 s_2 s_4\right)\dot\theta_1\dot\theta_2 - m_p l_2 \dot x\dot\theta_4 c_4 + 2m_p l_1 l_2 \dot\phi\dot\theta_1 c_1 c_2 s_3 c_4 \\
&+ 2m_p l_1 l_2 \dot\phi\dot\theta_2 (c_2 s_4 - s_1 s_2 s_3 c_4) + 2m_p l_1 l_2 \dot\phi\dot\theta_3 s_1 c_2 c_3 c_4 + 2m_p l_1 l_2 \dot\phi\dot\theta_4 (s_2 c_4 - s_1 c_2 s_3 s_4) = u_1 - M_f - D_1,
\end{aligned}
\tag{1}
$$

$$
\begin{aligned}
&- (m_h l_1 s_2 + m_p l_1 s_2 + m_p l_2 s_4)\ddot{\phi} + (M_t + m_h + m_p)\ddot{x} + (m_h + m_p)l_1\ddot{\theta}_1 c_1 c_2 + m_p l_2 \ddot{\theta}_3 c_3 c_4 \\
&- (m_h + m_p)l_1\dot{\theta}_2 s_1 s_2 - m_p l_2 \ddot{\theta}_4 s_3 s_4 - (m_h + m_p)l_1\dot{\theta}_1^2 s_1 c_2 - (m_h + m_p)l_1\dot{\theta}_2^2 s_1 c_2 - m_p l_2 \dot{\theta}_3^2 s_3 c_4 \\
&- 2(m_h + m_p)l_1\dot{\theta}_1\dot{\theta}_2 c_1 s_2 - \left((M_t + m_h + m_p)x + (m_h + m_p)l_1 s_1 c_2 + m_p l_2 s_3 c_4\right)\dot{\phi}^2 \\
&- m_p l_2 \dot{\theta}_4^2 s_3 c_4 - 2m_p l_2 \dot{\theta}_3\dot{\theta}_4 c_3 s_4 - 2(m_h + m_p)l_1\dot{\phi}\dot{\theta}_2 c_2 - 2m_p l_2 \dot{\phi}\dot{\theta}_4 c_4 = u_2 - F_f - D_2,
\end{aligned}
\tag{2}
$$

$$
\begin{aligned}
&- l_1 c_1 c_2 \left((m_h + m_p)l_1 s_2 + m_p l_2 s_4\right)\ddot{\phi} - l_1\left((m_h + m_p)(x + l_1 s_1 c_2) + m_p l_2 s_3 c_4\right)c_1 c_2 \dot{\phi}^2 \\
&+ (m_h + m_p)l_1^2\ddot{\theta}_1 c_1^2 + (m_h + m_p)l_1\ddot{x} c_1 c_2 + m_p l_1 l_2 \ddot{\theta}_3 c_{1-3} c_2 c_4 + m_p l_1 l_2 \ddot{\theta}_4 S_{1-3} c_2 s_4 \\
&+ m_p l_1 l_2 \dot{\theta}_3^2 s_{1-3} c_2 c_4 + m_p l_1 l_2 \dot{\theta}_4^2 s_{1-3} c_2 c_4 - 2(m_h + m_p)l_1^2\dot{\theta}_1\dot{\theta}_2 s_2 c_2 - 2m_p l_1 l_2 \dot{\theta}_3\dot{\theta}_4 c_{1-3} c_2 s_4 \\
&- 2(m_h + m_p)l_1^2\dot{\phi}\dot{\theta}_2 c_1 s_2^2 - 2m_p l_1 l_2 \dot{\phi}\dot{\theta}_4 c_1 c_2 c_4 + (m_h + m_p)g l_1 s_1 c_2 = -d_{q1} l_1^2 c_2^2 \dot{\theta}_1,
\end{aligned}
\tag{3}
$$

$$
\begin{aligned}
&l_1\left((m_h + m_p)(xc_2 + l_1 s_1) + m_p l_2 (s_1 s_2 s_4 + c_2 s_3 c_4)\right)\ddot{\phi} - (m_h + m_p)l_1\ddot{x} s_1 s_2 + (m_h + m_p)l_1^2\ddot{\theta}_2 \\
&- m_p l_1 l_2 \ddot{\theta}_3 s_{1-3} s_2 c_4 + m_p l_1 l_2 \ddot{\theta}_4 (c_{1-3} s_2 s_4 + c_2 c_4) + 2(m_h + m_p)l_1\dot{x}\dot{\phi} c_2 + (m_h + m_p)l_1^2\dot{\theta}_1^2 s_2 c_2 \\
&+ l_1\dot{\phi}^2\left((m_h + m_p)(xs_1 s_2 - l_1 c_1^2 s_2 c_2) - m_p l_2 (c_2 s_4 - s_1 s_2 s_3 c_4)\right) + m_p l_1 l_2 \dot{\theta}_3^2 c_{1-3} s_2 c_4 \\
&- m_p l_1 l_2 \dot{\theta}_1\dot{\theta}_3 c_{1-3} s_2 c_4 + m_p l_1 l_2 \dot{\theta}_4^2 (c_{1-3} s_2 c_4 - c_2 s_4) + 2m_p l_1 l_2 \dot{\theta}_3\dot{\theta}_4 s_{1-3} s_2 s_4 + 2m_p l_1 l_2 \dot{\phi}\dot{\theta}_3 c_2 c_3 c_4 \\
&+ 2(m_h + m_p)l_1^2\dot{\phi}\dot{\theta}_1 c_1 c_2^2 + (m_h + m_p)g l_1 c_1 s_2 + 2m_p l_1 l_2 \dot{\phi}\dot{\theta}_4 (s_1 s_2 c_4 - c_2 s_3 s_4) = -d_{q1} l_1^2\dot{\theta}_2,
\end{aligned}
\tag{4}
$$

$$
\begin{aligned}
&- m_p c_3 c_4 (l_1 s_2 + l_2 s_4)\ddot{\phi} + m_p l_2 c_3 c_4 \ddot{x} + m_p l_1 l_2 \ddot{\theta}_1 c_1 c_{1-3} c_2 c_4 - m_p l_1 l_2 \ddot{\theta}_2 s_{1-3} s_2 c_4 + m_p l_2^2\ddot{\theta}_3 c_4^2 \\
&- 2m_p l_2^2\dot{\theta}_3\dot{\theta}_4 s_4 c_4 + m_p l_2 \dot{x}\dot{\theta}_4 c_3 s_4 - m_p l_2 c_3 c_4 (l_2 s_3 c_4 + l_1 s_1 c_2 + x)\dot{\phi}^2 - m_p l_1 l_2 \dot{\theta}_1^2 s_{1-3} c_2 c_4 \\
&- m_p l_1 l_2 \dot{\theta}_2^2 s_{1-3} c_2 c_4 - 2m_p l_1 l_2 \dot{\theta}_1\dot{\theta}_2 c_{1-3} s_2 c_4 - 2m_p l_2 \dot{\phi}\dot{\theta}_2 c_2 c_3 c_4 - 2m_p l_2^2\dot{\phi}\dot{\theta}_4 c_3 s_4^2 \\
&+ m_p g l_2 s_3 c_4 = -d_{q2} l_2^2 c_4^2 \dot{\theta}_3,
\end{aligned}
\tag{5}
$$

$$
\begin{aligned}
&m_p l_2 \ddot{\phi}(xc_4 + l_1 s_2 s_3 s_4 + l_1 s_1 c_2 c_4 + l_2 s_3) - m_p l_2 \ddot{x} s_3 s_4 + m_p l_1 l_2 \ddot{\theta}_1 s_{1-3} c_2 s_4 + m_p l_1 l_2 \ddot{\theta}_2 (c_{1-3} s_2 s_4 + c_2 c_4) \\
&+ m_p l_2^2\ddot{\theta}_4 + m_p l_1 l_2 \dot{\theta}_1^2 c_{1-3} s_4 + m_p l_2 \dot{\phi}^2 (xs_3 s_4 - l_1 s_2 c_4 + l_1 s_1 c_2 s_3 s_4 - l_2 c_3^2 s_4 c_4) + m_p l_2^2\dot{\theta}_3^2 s_4 c_4 \\
&+ m_p l_1 l_2 \dot{\theta}_2^2 (c_{1-3} s_4 - s_2 c_4) + 2m_p l_2 \dot{x}\dot{\phi} c_4 - 2m_p l_1 l_2 \dot{\theta}_1\dot{\theta}_2 s_{1-3} s_2 s_4 + 2m_p l_1 l_2 \dot{\phi}\dot{\theta}_1 c_1 c_2 c_4 + m_p g l_2 c_3 s_4 \\
&+ 2m_p l_1 l_2 \dot{\phi}\dot{\theta}_2 (c_2 s_3 s_4 - s_1 s_2 c_4) + 2m_p l_2^2\dot{\phi}\dot{\theta}_3 c_3 s_4^2 = -d_{q2} l_2^2\dot{\theta}_4,
\end{aligned}
\tag{6}
$$

with s_1, s_2, s_3, s_4, c_1, c_2, c_3, c_4, s_{1-3}, and c_{1-3} standing for abbreviations of $\sin\theta_1$, $\sin\theta_2$, $\sin\theta_3$, $\sin\theta_4$, $\cos\theta_1$, $\cos\theta_2$, $\cos\theta_3$, $\cos\theta_4$, $\sin(\theta_1 - \theta_3)$, and $\cos(\theta_1 - \theta_3)$, respectively. The physical meaning of the variables and parameters in (1)–(6) are provided in Table 1.

Table 1. Variables and parameters of double-pendulum tower crane systems

Variables/Parameters	Physical meaning	Units
x	trolley translation displacement	m
ϕ	jib slew angle	rad
θ_1, θ_2	hook swing angles	rad
θ_3, θ_4	payload swing angles	rad
l_1	distance from the hook to the trolley	m
l_2	distance from the payload to the hook	m

(continued)

Table 1. (*continued*)

Variables/Parameters	Physical meaning	Units
d_{q1}, d_{q2}	air resistance-related coefficients	NA
M_f, F_f	friction force and torque	N·m, N
O	jib inertia moment	kg·m²
M_t, m_h, m_p	trolley, hook, and payload mass	kg
g	gravity constant	m/s²
u_1, u_2	control inputs	N·m, N
D_1, D_2	Uncertain/unknown disturbances	N·m, N

As can be seen from (1)–(6) that, there are 2 control inputs (u_1 and u_2) but 6 degrees-of-freedom (ϕ, x, θ_1, θ_2, θ_3, and θ_4) to be controlled for double-pendulum tower crane systems. As a consequence, it is difficult and challenging to design available controllers for such highly underactuated systems.

To facilitate the subsequent controller development and analysis, (1)–(6) can be rewritten into the following compact form:

$$M(q)\ddot{q} + C(q,\dot{q})\dot{q} + G(q) = U + F_d, \tag{7}$$

where the detailed expressions of these vectors and matrices are provided as follows:

$$q = [\phi\, x\, \theta_1\, \theta_2\, \theta_3\, \theta_4]^T,$$

$$M(q) = \begin{bmatrix} Q_{11} & Q_{12} & Q_{13} & Q_{14} & Q_{15} & Q_{16} \\ Q_{12} & M_t+m_h+m_p & Q_{23} & Q_{24} & Q_{25} & Q_{26} \\ Q_{13} & Q_{23} & Q_{33} & 0 & Q_{35} & Q_{36} \\ Q_{14} & Q_{24} & 0 & Q_{44} & Q_{45} & Q_{46} \\ Q_{15} & Q_{25} & Q_{35} & Q_{45} & m_2 l_2^2 C_4^2 & 0 \\ Q_{16} & Q_{26} & Q_{36} & Q_{46} & 0 & m_2 l_2^2 \end{bmatrix},$$

$$G(q) = \begin{bmatrix} 0 \\ 0 \\ (m_h+m_p)gl_1s_1c_2 \\ (m_h+m_p)gl_1c_1s_2 \\ m_pgl_2s_3c_4 \\ m_pgl_2c_3s_4 \end{bmatrix}, \quad U = \begin{bmatrix} M \\ F \\ 0 \\ 0 \\ 0 \\ 0 \end{bmatrix}, \quad F_d = \begin{bmatrix} -M_f - D_1 \\ -F_f - D_2 \\ -d_{q1}l_1^2c_2^2\dot{\theta}_1 \\ -d_{q1}l_1^2\dot{\theta}_2 \\ -d_{q2}l_2^2c_4^2\dot{\theta}_3 \\ -d_{q2}l_2^2\dot{\theta}_4 \end{bmatrix}$$

$$Q_{11} = \begin{pmatrix} (M_t + m_h + m_p)x^2 + (m_p l_1^2 + m_h l_1^2)(s_2^2 + s_1^2 c_2^2) + O + m_p l_2^2(s_4^2 + s_3^2 c_4^2) \\ + 2m_p l_1 l_2(s_2 s_4 + s_1 c_2 s_3 c_4) + 2(m_h + m_p)l_1 x s_1 c_2 + 2m_p l_2 x s_3 c_4 + 2m_p l_2 x \dot{\theta}_4 c_4 \end{pmatrix},$$

$$Q_{12} = -(m_h l_1 s_2 + m_p l_1 s_2 + m_p l_2 s_4), \ Q_{23} = (m_h + m_p)l_1 c_1 c_2, \ Q_{24} = -(m_h + m_p)l_1 s_1 s_2,$$

$$Q_{13} = -l_1 C_1 C_2(m_p l_2 s_4 + (m_h + m_p)l_1 s_2), \ Q_{15} = -m_p l_2 c_3 c_4(l_1 s_2 + l_2 s_4), \ Q_{25} = m_p l_2 c_3 c_4,$$

$$Q_{14} = l_1((m_h + m_p)(l_1 s_1 + x c_2) + m_p l_2(s_1 s_2 s_4 + c_2 s_3 c_4)), \ Q_{46} = m_p l_1 l_2(c_{1-3} s_2 s_4 + c_2 c_4)$$

$$Q_{16} = m_p l_2 \left(l_2 s_3 + l_1(s_2 s_3 s_4 + s_1 c_2 c_4) + 2x \dot{\phi} c_4 \right), \ Q_{26} = -m_p l_2 s_3 s_4, \ Q_{33} = (m_h + m_p)l_1^2 c_2^2,$$

$$Q_{44} = (m_h + m_p)l_1^2, \ Q_{36} = m_p l_1 l_2 s_{1-3} c_2 s_4, \ Q_{35} = m_p l_1 l_2 c_{1-3} c_2 c_4, \ Q_{45} = -m_p l_1 l_2 s_{1-3} s_2 c_4.$$

Remark 1 : For simplicity, the detailed expression of $C(q, \dot{q})$ is not given.

For double-pendulum tower crane systems, the following assumption, which is commonly adopted in [1, 2, 7, 20], is made.

Assumption 1 [1, 2, 7, 20]: The swing angles of the hook and payload are always kept within the following ranges:

$$\theta_1, \theta_2, \theta_3, \theta_4 \in \left(-\frac{\pi}{2}, \frac{\pi}{2} \right). \tag{8}$$

2.2   Control Objectives

The control objective of this paper is to design an adaptive neural network controller for double-pendulum tower crane systems, which could realize the following properties:

1) The jib slew angle ϕ, the trolley displacement x are driven to the target positions $p_{d\phi}$ and p_{dx}, respectively, in the presence of parametric uncertainties and external disturbances, in the sense that

$$\lim_{t \to \infty} \phi(t) = p_{d\phi}, \ \lim_{t \to \infty} x(t) = p_{dx}. \tag{9}$$

2) The hook and payload swings are suppressed and eliminated rapidly, that is

$$\lim_{t \to \infty} \theta_1(t) = 0, \ \lim_{t \to \infty} \theta_2(t) = 0, \ \lim_{t \to \infty} \theta_3(t) = 0, \ \lim_{t \to \infty} \theta_4(t) = 0. \tag{10}$$

3   Controller Development and Stability Analysis

In this section, the main results including the adaptive neural network controller and stability analysis are detailed derived.

3.1   Adaptive Neural Network Controller Development

To achieve the aforementioned objectives, we introduce the positioning errors of the jib and trolley as

$$e_\phi = \phi - p_{d\phi}, \tag{11}$$

$$e_x = x - p_{dx}. \tag{12}$$

Then, the error vector of the double-pendulum tower crane system is defined as

$$\boldsymbol{e} = \begin{bmatrix} e_\phi \, e_x \, \theta_1 \, \theta_2 \, \theta_3 \, \theta_4 \end{bmatrix}^T. \tag{13}$$

Based on the aforementioned error vector and control objectives, we construct the following non-negative function as

$$V = \frac{1}{2}\dot{\boldsymbol{e}}^T \boldsymbol{M}(\boldsymbol{q})\dot{\boldsymbol{e}} + (m_h + m_p)gl_1(1 - c_1 c_2) + m_p g l_2 (1 - c_3 c_4). \tag{14}$$

Taking the time derivative of (19), it is derived that

$$
\begin{aligned}
\dot{V} &= \dot{\boldsymbol{e}}^T \boldsymbol{M}(\boldsymbol{q})\ddot{\boldsymbol{e}} + \frac{1}{2}\dot{\boldsymbol{e}}^T \dot{\boldsymbol{M}}(\boldsymbol{q})\dot{\boldsymbol{e}} + (m_h + m_p)gl_1 s_1 c_2 \dot{\theta}_1 + (m_h + m_p)gl_1 c_1 s_2 \dot{\theta}_2 + m_p g l_2 s_3 c_4 \dot{\theta}_3 + m_p g l_2 c_3 s_4 \dot{\theta}_4 \\
&= \dot{\boldsymbol{e}}^T (\boldsymbol{U} - \boldsymbol{G}(\boldsymbol{q}) + \boldsymbol{F}_d) + (m_h + m_p)gl_1 s_1 c_2 \dot{\theta}_1 + (m_h + m_p)gl_1 c_1 s_2 \dot{\theta}_2 + m_p g l_2 s_3 c_4 \dot{\theta}_3 + m_p g l_2 c_3 s_4 \dot{\theta}_4 \\
&= \dot{\phi}(u_1 - M_f - D_1) + \dot{x}(u_2 - F_f - D_2) - d_{q1} l_1^2 c_2^2 \dot{\theta}_1^2 - d_{q1} l_1^2 \dot{\theta}_2^2 - d_{q2} l_2^2 c_4^2 \dot{\theta}_3^2 - d_{q2} l_2^2 \dot{\theta}_4^2
\end{aligned}
\tag{15}
$$

Here, the property $\boldsymbol{\xi}^T\left[(1/2)\dot{\boldsymbol{M}}(\boldsymbol{q}) - \boldsymbol{C}(\boldsymbol{q},\dot{\boldsymbol{q}})\right]\boldsymbol{\xi} = 0$, $\forall \boldsymbol{\xi} \in \mathfrak{R}^6$ is used for the deduction.

Define $\Delta_1 \triangleq -M_f - D_1$, $\Delta_2 \triangleq -F_f - D_2$. In practical applications, the system parameters, frictions, external disturbances, et al., are unknown/uncertain, therefore, Δ_1 and Δ_2 are unknown/uncertain disturbances.

The bounded and continuous function Δ_1 and Δ_2 can be approximated by the following double-layer neural network structures:

$$\Delta_1 = \boldsymbol{W}_1^T \sigma(\boldsymbol{T}_1^T \boldsymbol{X}) + \varepsilon_1, \quad \Delta_2 = \boldsymbol{W}_2^T \sigma(\boldsymbol{T}_2^T \boldsymbol{X}) + \varepsilon_2, \tag{16}$$

wherein $\boldsymbol{T}_1$ and $\boldsymbol{T}_2$ stand for the bounded input weight matrices, $\boldsymbol{W}_1$ and $\boldsymbol{W}_2$ refer to the bounded weight vector, ε_1 and ε_2 represent the approximation errors with $|\varepsilon_1| \le \bar{\varepsilon}_1$, $|\varepsilon_2| \le \bar{\varepsilon}_2$, respectively, $\boldsymbol{X}$ is denoted by

$$\boldsymbol{X} = \begin{bmatrix} \theta_1 \, \theta_2 \, \theta_3 \, \theta_4 \, \dot{\phi} \, \dot{x} \, \dot{\theta}_1 \, \dot{\theta}_2 \, \dot{\theta}_3 \, \dot{\theta}_4 \, \dot{\phi}_d \, \ddot{\phi}_d \, \dot{x}_d \, \ddot{x}_d \end{bmatrix}^T$$

and the activation function $\sigma(\boldsymbol{X})$ is selected as

$$\sigma(X) = \tanh(X), \tag{17}$$

Then, the estimated errors of W_1, W_2, T_1, and T_2 are constructed as

$$\tilde{W}_1 = W_1 - \hat{W}_1, \ \tilde{W}_2 = W_2 - \hat{W}_2, \ \tilde{T}_1 = T_1 - \hat{T}_1, \ \tilde{T}_2 = T_2 - \hat{T}_2. \tag{18}$$

wherein $\hat{W}_1$, $\hat{W}_2$, $\hat{T}_1$, and $\hat{T}_2$ stand for the estimates of W_1, W_2, T_1, and T_2, respectively, which are obtained by the subsequent update laws.

According to the structure of (15), the neural network controller and the adaptive law are designed as

$$u_1 = -k_{p\phi}e_\phi - k_{d\phi}\dot{\phi} - k_{s\phi}\mathrm{sgn}\left(\dot{\phi}\right) - \hat{W}_1^T\sigma\left(\hat{T}_1^T X\right), \tag{19}$$

$$u_2 = -k_{px}e_x - k_{dx}\dot{x} - k_{sx}\mathrm{sgn}(\dot{x}) - \hat{W}_x^T\sigma\left(\hat{T}_x^T X\right), \tag{20}$$

$$\dot{\hat{W}}_1 = \beta_1\dot{\phi}\Pi_1\left[\sigma\left(\hat{T}_1^T X\right) - \dot{\sigma}\left(\hat{T}_1^T z\right)\hat{T}_1^T X\right], \dot{\hat{T}}_1 = \alpha_1\dot{\phi}\Gamma_1\hat{W}_1^T\dot{\sigma}\left(\hat{T}_1^T X\right)X, \tag{21}$$

$$\dot{\hat{W}}_2 = \beta_2\dot{x}\Pi_2\left[\sigma\left(\hat{T}_2^T X\right) - \dot{\sigma}\left(\hat{T}_2^T X\right)\hat{T}_2^T X\right], \dot{\hat{T}}_2 = \alpha_2\dot{x}\Gamma_2\hat{W}_2^T\dot{\sigma}\left(\hat{T}_2^T X\right)X, \tag{22}$$

where Π_1, Γ_1, Π_2, Γ_2 denote positive definite symmetric matrices, α_1, β_1, α_2, β_2, $k_{p\phi}$, $k_{d\phi}$, k_{px}, k_{dx}, $k_{s\phi}$, $k_{sx} \in \Re^+$ refer to positive control gains, which are tuned to meet the following conditions:

$$\alpha_1, \beta_1, \alpha_2, \beta_2, k_{p\phi}, k_{d\phi}, k_{px}, k_{dx} > 0, \ k_{s\phi} > \Omega_1, \ k_{sx} > \Omega_2. \tag{23}$$

where $\bar{\Omega}_1$ and $\bar{\Omega}_2$ will be defined later.

In order to achieve the purpose of rapid hook and payload swing reduction, the hook and payload swing related information is introduced to the designed controller, and in so doing, (18) and (19) are improved by

$$u_1 = -k_{p\phi}e_\phi - k_{d\phi}\dot{\phi} - k_{s\phi}\mathrm{sgn}\left(\dot{\phi}\right) - \hat{W}_1^T\sigma\left(\hat{T}_1^T X\right) - k_{h\phi}\left(\dot{\theta}_1^2 + \dot{\theta}_2^2 + \dot{\theta}_3^2 + \dot{\theta}_4^2\right)\dot{\phi}, \tag{24}$$

$$u_2 = -k_{px}e_x - k_{dx}\dot{x} - k_{sx}\mathrm{sgn}(\dot{x}) - \hat{W}_x^T\sigma\left(\hat{T}_x^T X\right) - k_{hx}\left(\dot{\theta}_1^2 + \dot{\theta}_2^2 + \dot{\theta}_3^2 + \dot{\theta}_4^2\right)\dot{x}, \tag{25}$$

with k_{h_ϕ} and $k_{hx} \in \Re^+$ being positive control gains.

3.2   Stability/Convergence Analysis

To accelerate the following analysis, we introduce the compact set as

$$B = \left\{\tilde{W}_1 \in \mathbb{R}^N, \ \tilde{W}_2 \in \mathbb{R}^N : \|\tilde{W}_1\| \le \bar{\omega}_1, \|\tilde{W}_2\| \le \bar{\omega}_2\right\}, \tag{26}$$

with N standing for the number of neuros, $\bar{\omega}_1$ and $\bar{\omega}_2$ being the radii of the ellipsoid B.

For the given compact set B, it is derived that

$$\|\hat{W}_1\| \leq \kappa_1, \ \|\hat{W}_2\| \leq \kappa_2 \rightarrow \left|\hat{\Delta}_1\right| \leq \kappa_1, \ \left|\hat{\Delta}_2\right| \leq \kappa_2, \tag{27}$$

where κ_1 and κ_2 refer to upper bounds of $\hat{W}_1$ and $\hat{W}_2$, respectively. Then, the following error signal is introduced as

$$e_i = -\hat{W}_i^T \sigma\left(\hat{T}_i^T X\right) + W_i^T \sigma\left(T_i^T X\right) + \varepsilon_i, \ i = 1,\ 2. \tag{28}$$

To calculate e_i, the Taylor expansion of $\sigma\left(T_i^T z\right)$ is first provided as

$$\sigma\left(T_i^T X\right) = \sigma\left(\hat{T}_i^T X\right) + \dot{\sigma}\left(\hat{T}_i^T X\right)\tilde{T}_i^T z + \mathrm{O}\left(\tilde{T}_i^T X\right)^2, \tag{29}$$

with $\mathrm{O}\left(\tilde{T}^T z\right)^2$ referring to the high-order residual term.

It follows from (28) and (29) that

$$
\begin{aligned}
e_i &= \left(\hat{W}_i^T + W_i^T\right)\left[\sigma\left(\hat{T}_i^T X\right) + \dot{\sigma}\left(\hat{T}_i^T X\right)\tilde{T}_i^T X + \mathrm{O}\left(\tilde{T}_i^T X\right)^2\right] - \hat{W}_i^T \sigma\left(\hat{T}_i^T X\right) + \varepsilon_i \\
&= \tilde{W}_i^T \sigma\left(\hat{T}_i^T X\right) + \tilde{W}_i^T \dot{\sigma}\left(\hat{T}_i^T X\right)\tilde{T}_i^T X + \hat{W}_i^T \dot{\sigma}\left(\hat{T}_i^T X\right)\tilde{T}_i^T X + W_i^T \mathrm{O}\left(\tilde{T}_i^T X\right)^2 + \varepsilon_i \\
&= \tilde{W}_i^T \left[\sigma\left(\hat{T}_i^T X\right) - \dot{\sigma}\left(\hat{T}_i^T X\right)\tilde{T}_i^T X\right] + \hat{W}^T \dot{\sigma}\left(\hat{T}^T X\right)\tilde{T}_i^T X + \tilde{W}_i^T \dot{\sigma}\left(\hat{T}_i^T X\right)T_i^T X + W^T \mathrm{O}\left(\tilde{T}_i^T X\right)^2 + \varepsilon_i \\
&= \tilde{W}_i^T \left[\sigma\left(\hat{T}_i^T X\right) - \dot{\sigma}\left(\hat{T}_i^T X\right)\hat{T}_i^T X\right] + \hat{W}_i^T \dot{\sigma}\left(\hat{T}_i^T X\right)\tilde{T}_i^T X + \Omega_i,
\end{aligned} \tag{30}
$$

where Ω_i represents the following auxiliary function:

$$\Omega_i = \tilde{W}_i^T \dot{\sigma}\left(\hat{T}_i^T z\right)T_i^T z + W^T \mathrm{O}\left(\tilde{T}_i^T z\right)^2 + \varepsilon_i. \tag{31}$$

For the compact set B, it is easy to derive that

$$\Omega_i \leq \bar{\Omega}_i, \tag{32}$$

with $\bar{\Omega}_i$ standing for the upper bound of Ω_i.

Theorem 1: The designed adaptive neural network control laws(21)–(22), (24)–(25), can drive the jib and trolley to their desired positions, in the meantime, suppress and eliminate the hook and payload swing, in the sense that

$$
\begin{aligned}
&\lim_{t\to\infty}[\phi \, x \, \theta_1 \, \theta_2 \, \theta_3 \, \theta_4]^T = \left[p_{d\phi}\, p_{dx}\, 0\, 0\, 0\, 0\right]^T, \\
&\lim_{t\to\infty}\left[\dot{\phi}\, \dot{x}\, \dot{\theta}_1\, \dot{\theta}_2\, \dot{\theta}_3\, \dot{\theta}_4\right]^T = [0\, 0\, 0\, 0\, 0\, 0]^T.
\end{aligned} \tag{33}
$$

Proof : The Lyapunov candidate function is constructed as

$$V_{all} = V + \frac{1}{2}k_{p\phi}e_\phi^2 + \frac{1}{2}k_{px}e_x^2 + \frac{1}{2\alpha_1}\text{Tr}\left(\tilde{T}_1^T\Gamma_1^{-1}\tilde{T}_1\right)$$
$$+ \frac{1}{2\beta_1}\tilde{W}_1^T\Pi_1^{-1}\tilde{W}_1 + \frac{1}{2\alpha_2}\text{Tr}\left(\tilde{T}_2^T\Gamma_2^{-1}\tilde{T}_2\right) + \frac{1}{2\beta_2}\tilde{W}_2^T\Pi_2^{-1}\tilde{W}_2. \tag{34}$$

Differentiating (34) with respect to time, and substituting the results of (21)–(22), (24)–(25), and (30) into the obtained equation, we are led to

$$\dot{V}_{all} = \dot{V} + k_{p\phi}e_\phi\dot{\phi} + k_{px}e_x\dot{x} + \frac{1}{\alpha_1}\text{Tr}\left(\tilde{T}_1^T\Gamma_1^{-1}\dot{\tilde{T}}_1\right) + \frac{1}{\beta_1}\tilde{W}_1^T\Pi_1^{-1}\dot{\tilde{W}}_1 + \frac{1}{\alpha_2}\text{Tr}\left(\tilde{T}_2^T\Gamma_2^{-1}\dot{\tilde{T}}_2\right) + \frac{1}{\beta_2}\tilde{W}_2^T\Pi_2^{-1}\dot{\tilde{W}}_2$$
$$= \dot{\phi}\left(-k_{d\phi}\dot{e}_\phi - k_{s\phi}\text{sgn}\left(\dot{\phi}\right) - k_{h\phi}\left(\dot{\theta}_1^2 + \dot{\theta}_2^2 + \dot{\theta}_3^2 + \dot{\theta}_4^2\right)\dot{\phi} + e_1\right) - d_{q1}l_1^2c_2^2\dot{\theta}_1^2 - d_{q1}l_1^2\dot{\theta}_2^2$$
$$+ \dot{x}\left(-k_{dx}\dot{e}_x - k_{sx}\text{sgn}(\dot{x}) - k_{hx}\left(\dot{\theta}_1^2 + \dot{\theta}_2^2 + \dot{\theta}_3^2 + \dot{\theta}_4^2\right)\dot{x} + e_2\right) - d_{q2}l_2^2c_4^2\dot{\theta}_3^2 - d_{q2}l_2^2\dot{\theta}_4^2$$
$$- \frac{1}{\alpha_1}\text{Tr}\left(\tilde{T}_1^T\Gamma_1^{-1}\dot{\tilde{T}}_1\right) + \frac{1}{\beta_1}\tilde{W}_1^T\Pi_1^{-1}\dot{\tilde{W}}_1 + \frac{1}{\alpha_2}\text{Tr}\left(\tilde{T}_2^T\Gamma_2^{-1}\dot{\tilde{T}}_2\right) + \frac{1}{\beta_2}\tilde{W}_2^T\Pi_2^{-1}\dot{\tilde{W}}_2$$
$$= -k_{d\phi}\dot{\phi}^2 - k_{h\phi}\left(\dot{\theta}_1^2 + \dot{\theta}_2^2 + \dot{\theta}_3^2 + \dot{\theta}_4^2\right)\dot{\phi}^2 - k_{s\phi}|\dot{\phi}| + \Omega_1\dot{\phi} - k_{dx}\dot{x}^2 - k_{hx}\left(\dot{\theta}_1^2 + \dot{\theta}_2^2 + \dot{\theta}_3^2 + \dot{\theta}_4^2\right)\dot{x}^2$$
$$- k_{sx}|\dot{x}| + \Omega_2\dot{x} - d_{q1}l_1^2c_2^2\dot{\theta}_1^2 - d_{q1}l_1^2\dot{\theta}_2^2 - d_{q2}l_2^2c_4^2\dot{\theta}_3^2 - d_{q2}l_2^2\dot{\theta}_4^2$$
$$\leq -k_{d\phi}\dot{\phi}^2 - k_{h\phi}\left(\dot{\theta}_1^2 + \dot{\theta}_2^2 + \dot{\theta}_3^2 + \dot{\theta}_4^2\right)\dot{\phi}^2 - k_{dx}\dot{x}^2 - k_{hx}\left(\dot{\theta}_1^2 + \dot{\theta}_2^2 + \dot{\theta}_3^2 + \dot{\theta}_4^2\right)\dot{x}^2$$
$$- d_{q1}l_1^2c_2^2\dot{\theta}_1^2 - d_{q1}l_1^2\dot{\theta}_2^2 - d_{q2}l_2^2c_4^2\dot{\theta}_3^2 - d_{q2}l_2^2\dot{\theta}_4^2$$
$$\leq 0 \tag{35}$$

where (23) is used for the deduction.

Therefore, it can be concluded that the closed-loop system is Lyapunov stable [21], and the following results can be obtained:

$$\dot{\phi} = \dot{x} = \dot{\theta}_1 = \dot{\theta}_2 = \dot{\theta}_3 = \dot{\theta}_4 = 0. \tag{36}$$

From (36), it is obtained that

$$\ddot{\phi} = \ddot{x} = \ddot{\theta}_1 = \ddot{\theta}_2 = \ddot{\theta}_3 = \ddot{\theta}_4 = 0. \tag{37}$$

It follows from (36)–(37), and (3)–(6) that

$$\left(m_h + m_p\right)gl_1s_1c_2 = 0 \rightarrow \theta_1 = 0, \tag{38}$$

$$\left(m_h + m_p\right)gl_1c_1s_2 = 0 \rightarrow \theta_2 = 0, \tag{39}$$

$$m_pgl_2s_3c_4 = 0 \rightarrow \theta_3 = 0, \tag{40}$$

$$m_p g l_2 c_3 s_4 = 0 \rightarrow \theta_4 = 0, \tag{41}$$

where Assumption 1 is used.

Substituting (24), (36)–(41) into (1) yields

$$e_\phi = 0, \tag{42}$$

where the assumption $\lim\limits_{t \to \infty} \varepsilon_1 = 0$ is used in view of the fact that the neural network can approximate the disturbances very well.

In a similar way, one has

$$e_x = 0. \tag{43}$$

Based on the previous conclusions in (36), (38)–(41), (42)–(43), Theorem 1 is proven.

4 Simulation Results and Analysis

The performance of the designed controller is validated in this section, which are carried out by Matlab/Simulink. To better illustrate the control performance of the proposed control law, the PD method is selected for comparison. The nominal values of the double-pendulum tower crane system parameters are set as follows:

$M_t = 4.5$ kg, $m_h = 0.8$ kg, $m_p = 1$ kg, $l_1 = 0.7$ m, $l_2 = 0.5$ m, $O = 6.8$ kg $\cdot$ m^2, $g = 9.8$ m/s^2, $f_{ro\phi} = 5.2$, $k_{r\phi} = -1$, $\varepsilon_\phi = 0.01$, $f_{rox} = 5.4$, $k_{rx} = -1.5$, $\varepsilon_x = 0.01$, $d_{q1} = 0.1$, $d_{q2} = 0.1$.

The actual values of the system parameters are:

$M_t = 4.8$ kg, $m_h = 1$ kg, $m_p = 0.9$ kg, $l_1 = 0.68$ m, $l_2 = 0.52$ m, $O = 6.5$ kg $\cdot$ m^2, $g = 9.8$ m/s^2, $f_{ro\phi} = 4.8$, $k_{r\phi} = -1.4$, $\varepsilon_\phi = 0.01$, $f_{rox} = 5.8$, $k_{rx} = -1.8$, $\varepsilon_x = 0.01$, $d_{q1} = 0.2$, $d_{q2} = 0.2$.

The desired and initial positions of the jib and trolley are set as

$$p_{d\phi} = 50 \text{ deg}, \ p_{dx} = 1 \text{ m}, \ \phi(0) = 0°, \ x(0) = 0 \text{ m}.$$

Table 2. Control gains of the designed controller and PD method

Controllers	Control gains
PD controller	$k_{p\phi} = 10$, $k_{d\phi} = 18$, $k_{px} = 13$, $k_{dx} = 18$
Designed controller	$k_{p\phi} = 15$, $k_{d\phi} = 22$, $k_{s\phi} = 100$, $k_{px} = 6.3$, $k_{dx} = 12$, $k_{sx} = 100$, $k_{h\phi} = 100$, $k_{hx} = 100$, $\beta_1 = 15$, $\beta_2 = 15$, $\alpha_1 = 15$, $\alpha_2 = 15$, $\Pi_1 = 10$, $\Pi_2 = 10$, $\Gamma_1 = 10$, $\Gamma_2 = 10$

The control gains of the proposed control law and the comparative PD method are illustrated in Table 2, which are a result of careful tuning. The simulation time is set as 10 s.

The simulation results of the PD controller and the proposed adaptive neural network control method are shown in Figs. 2 and 3, respectively. One can see from these figures that with similar transportation time (both within 6 s), the designed controller exhibits superior control performance in terms of the anti-swing of the hook and payload, compared with the PD control law. More precisely, both the maximum and residual hook/payload swing amplitudes of the designed controller are much smaller than those of the PD control method.

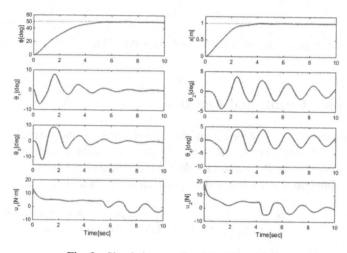

Fig. 2. Simulation results of the PD method.

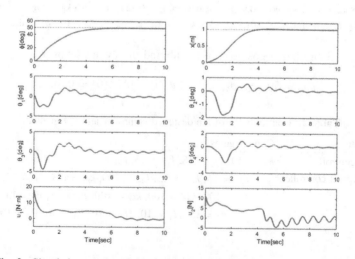

Fig. 3. Simulation results of the designed adaptive neural network controller.

5 Conclusion

To guarantee the jib and trolley accurately arrive at their desired positions and simultaneously ensure rapid hook and payload swing suppression and elimination, a novel adaptive neural network control scheme is designed for double-pendulum tower crane systems. The neural network structure is utilized to estimate the uncertain/unknown disturbance, and the adaptive controller is then proposed. Moreover, to guarantee the anti-swing performance of the hook and payload, the hook and payload swing information are introduced to the designed controller. In addition, the stability of the equilibrium point of the controlled system is demonstrated in light of Lyapunov techniques without any linear processing. Finally, the superiority and robustness of the designed control strategy are verified by a series of numerical simulation results. In the subsequent studies, we will try to apply the designed control method in industry practice.

Acknowledgements. This work is partially supported by the Innovation and Technology Fund (ITF) Project of HK ITC (Ref. ITP/020/19AP), the Strategic Research Fund of the Research Institute of Urban Sustainable Development, HK Polytechnic University (PolyU), and the Project of Strategic Importance of HK PolyU, the General Research Fund of HK RGC under Grant No. 15206717, the Key Research and Development (Special Public-Funded Projects) of Shandong Province under Grant No. 2019GGX104058, the National Natural Science Foundation for Young Scientists of China under Grant No. 61903155.

References

1. Chen, H., Sun, N.: Nonlinear control of underactuated systems subject to both actuated and unactuated state constraints with experimental verification. In: IEEE Transactions on Industrial Electronics, In Press. https://doi.org/10.1109/tie.2019.2946541
2. Sun, N., Liang, D., Wu, Y., Chen, Y., Qin, Y., Fang, Y.: Adaptive control for pneumatic artificial muscle systems with parametric uncertainties and unidirectional input constraints. IEEE Trans. Ind. Inf. **16**(2), 969–979 (2020)
3. Pan, H., Jing, X., Sun, W., Gao, H.: A bio-inspired dynamics-based adaptive tracking control for nonlinear suspension systems. IEEE Trans. Control Syst. Technol. **26**(3), 903–914 (2018)
4. Qian, D., Ding, H., Lee, S.G., Bae, H.: Suppression of chaotic behaviors in a complex biological system by disturbance observer-based derivative integral terminal sliding mode. IEEE/CAS J. Automatica Sinica **7**(1), 126–135 (2020)
5. Kim, G.H., Hong, K.S.: Adaptive sliding-mode control of an offshore container crane with unknown disturbances. IEEE/ASME Trans. Mechatron. **24**(6), 2850–2861 (2019)
6. Cahyono, R.T., Flonk, E.J., Jayawardhana, B.: Discrete-event systems modeling and the model predictive allocation algorithm for integrated berth and quay crane allocation. In: IEEE Transactions on Intelligent Transportation Systems, In Press. https://doi.org/10.1109/tits.2019.- 2910283
7. Zhang, M., Jing, X.: A bioinspired dynamics-based adaptive fuzzy SMC method for half-car active suspension systems with input dead zones and saturations. In: IEEE Transactions on Cybernetics, In Press. https://doi.org/10.1109/tcyb.2020.2972322

8. Zhang, Z., Li, L., Wu, Y.: Disturbance-observer-based antiswing control of underactuated crane systems via terminal sliding mode. IET Control Theor. Appl. **12**(18), 2588–2594 (2018)

9. Le, A.T.: Neural observer and adaptive fractional-order back stepping fast terminal sliding mode control of RTG cranes. In: IEEE Transactions on Industrial Electronics, In Press. https://doi.org/10.1109/tie.2019.2962450

10. Chen, H., Fang, Y., Sun, N.: A swing constraint guaranteed MPC algorithm for underactuated overhead cranes. IEEE/ASME Trans. Mechatron. **21**(5), 2543–2555 (2016)

11. Kolar, B., Rarms, H., Schlacher, K.: Time-optimal flatness based control of a gantry crane. Control Eng. Practice **60**, 18–27 (2017)

12. Sun, N., Wu, Y., Liang, X., Fang, Y.: Nonlinear stable transportation control for double-pendulum shipboard cranes with ship-motion-induced disturbances. IEEE Trans. Ind. Electron. **66**(12), 9467–9479 (2019)

13. Ouyang, H., Wu, X., Zhang, G.: Tracking and load sway reduction for double-pendulum rotary cranes using adaptive nonlinear control approach. International Journal of Robust and Nonlinear Control, In Press. https://doi.org/10.1002/rnc.4854

14. Ngo, Q.H., Nguyen, N.P., Nguyen, C.N., Tran, T.H., Ha, Q.P.: Fuzzy sliding mode control of an offshore container crane. Ocean Eng. **140**, 125–134 (2017)

15. Le, A.T., Lee, S.G.: 3D cooperative control of tower cranes using robust adaptive techniques. J. Franklin Institute **354**(18), 8333–8357 (2017)

16. Omar, H.M., Nayfeh, A.H.: Gain scheduling feedback control of tower cranes with friction compensation. J. Vibrat. Control **10**(2), 269–289 (2004)

17. Böck, M., Kugi, A.: Real-time nonlinear model predictive path-following control of a laboratory tower crane. IEEE Trans. Control Syst. Technol. **22**(4), 1461–1473 (2014)

18. Wu, T.S., Karkoub, M., Yu, W.S., Chen, C.T., Her, M.G., Wu, K.W.: Anti-sway tracking control of tower cranes with delayed uncertainty using a robust adaptive fuzzy control. Fuzzy Sets Syst. **290**, 118–137 (2016)

19. Duong, S.C., Uezato, E., Kinjo, H., Yamamoto, T.: A hybrid evolutionary algorithm for recurrent neural network control of a three-dimensional tower crane. Autom. Construct. **23**, 55–63 (2012)

20. Zhang, M., Zhang, Y., Ji, B., Ma, C., Cheng, X.: Modeling and energy-based sway reduction control for tower crane systems with double-pendulum and spherical-pendulum effects. Measure. Control **53**(1–2), 141–150 (2020)

21. Khalil, H.K.: Nonlinear Systems. 3rd edn. Prentice-Hall, New Jersey (2002)

Generalized Locally-Linear Embedding: A Neural Network Implementation

Xiao Lu[1], Zhao Kang[1,2(✉)], Jiachun Tang[3], Shuang Xie[4], and Yuanzhang Su[1]

[1] University of Electronic Science and Technology of China, Chengdu 611731, China
Zkang@uestc.edu.cn

[2] Trusted Cloud Computing and Big Data Key Laboratory of Sichuan Province,
Chengdu, China

[3] Fraklin and Marshall College, Lancaster, PA 17603, USA

[4] Chengdu University, Chengdu 610106, China

Abstract. Locally-linear embedding (LLE) is a prominent dimension reduction method by exploiting the local symmetries of linear reconstructions. Recently, auto-encoders have achieved great success in learning data representation via the deep neural networks (DNN). It is interesting to get the best of both worlds by implementing LLE with DNN. To this end, we introduce an extra fully-connected layer whose weight works as a reconstruction coefficient (i.e., relation among the samples). Consequently, the latent representation can well preserve the neighborhood structure. Experiments on dimension reduction and classification have validated the superiority of the proposed method.

Keywords: Dimensionality reduction and manifold learning · Auto-encoder · Deep neural networks · Similarity learning · Classification

1 Introduction

Real-world data, such as images and videos, usually have a high dimension. This leads to the well-known curse of dimensionality. However, the potentially meaningful structure of high dimensional data is of much lower dimensionality [15,16,19,24,30]. Dimension reduction (DR) has been studied to discover the underlying intrinsic structure [1,8,10,28,31], which plays an important role in many fields, such as data mining, machine learning, computer vision, etc., as a crucial preprocessing step.

Various dimension reduction methods have been proposed from different perspectives. Among these methods, variance-based methods reduce data dimension from the statistics of data, including principal component analysis (PCA) [12], linear discriminant analysis (LDA) [3], and marginal fisher analysis (MFA) [39]. Different from PCA, LDA is a supervised method and it seeks a linear subspace to discriminate data from different classes. MFA extends the Gaussian assumptions of LDA into a more non-parameteric setting. These methods only discover the Euclidean structure and ignore to explicitly model the data relations [20].

© Springer Nature Singapore Pte Ltd. 2020
H. Zhang et al. (Eds.): NCAA 2020, CCIS 1265, pp. 97–106, 2020.
https://doi.org/10.1007/978-981-15-7670-6_9

While embedding-based methods take geometry of data into consideration, including Isomap [35], Locally linear embedding (LLE) [32,43], and Locality preserving projection (LPP) [6,37]. In particular, Isomap retains the geodesic distance between pairwise data in the original data space; LLE pursues a low-dimensional manifold by representing each data point as a linear combination of its neighbors; LLP preserves the locality structure by minimizing the pairwise distance in the projected space weighted by the corresponding distance in the input space. These methods have been proved to be very effective in exploiting the underlying manifold structure [29]. Despite their wide applications, all these methods discover data structure in the input space, with less representation powers of data. In addition, they often rely on a pre-defined similarity matrix which specifies the relation between data points. However, the relation is fixed and defined on the original high-dimensional space. Such a relation may not be valid on the manifold.

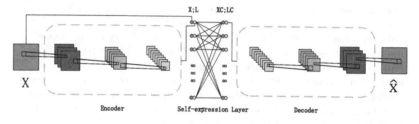

Fig. 1. Architecture of gLLE. The input X is mapped to L through an encoder, L is self-expressed by LC, and then reconstructed as $\hat{X}$ through a decoder. The squares in the dashed boxes represent the channels after convolution or deconvolution. For illustration, we just show three convolutional encoder layers, one self-expression layer, and three deconvolutional decoder layers.

With the capability of nonlinear mapping, auto-encoders (AE) [7] is also a popular dimension reduction technique. AE and its variants (including [23,27,36,38,40]) mainly focus on the primitive characteristics of data via reconstructing the data from a latent representation. However, most of them neglect the geometry of data, especially the relation among the samples. As a result, the rich information might get lost and the latent space becomes less representative [2].

To remedy above issues, in this paper, we implement LLE using a deep auto-encoder architecture. On the one hand, the proposed architecture encodes the neighborhood relation among the samples. Rather than just reconstructing each sample separately, the learned representation preserves well local neighborhood manifold. The LLE enjoys the powerful representation capability of neural network. Moreover, unlike traditional LLE, where the weights are fixed, the relation is iteratively adjusted so that the neighborhood structure is well maintained in the original and latent space. As displayed in Fig. 1, the architecture differs from the traditional auto-encoder network in that a fully connected self-expression layer appends after the encoder. The weights of the self-expression layer denotes the reconstruction coefficient.

The main contributions of this paper are listed as follows:

- We implement LLE in neural work, so that it can enjoy the powerful representation ability of deep neural network.
- The embedding process is supervised by the relation among the samples. As a result, the learned representation can preserve the local neighborhood structure on the data manifold.
- Experiments on dimension reduction and classification have demonstrated the effectiveness of the proposed method in the abilities of representation learning and dimension reduction.

The rest of this paper is organized as follows. Notations are given in Sect. 2 and LLE is revisited in Sect. 3. Our proposed method is presented in Sect. 4. We present dimension reduction and classification experiments in Sect. 5. Conclusion is given in Sect. 6.

2   Notations

We first introduce the notations that are used throughout this paper. Matrices are denoted by uppercase letters. For matrix X, its i-th column is written as X_i. The ℓ_2-norm of vector $h \in \mathcal{R}^{m \times 1}$ is defined as $\|h\| = \sqrt{\sum_{i=1}^{m} h_i^2}$. The Frobenius norm of the matrix $X \in \mathcal{R}^{m \times n}$ is defined as $\|X\|_F = \sqrt{\sum_{i=1}^{m} \sum_{j=1}^{n} X_{ij}^2}$.

3   Locally Linear Embedding Revisited

For a dataset $X \in \mathcal{R}^{m \times n}$ having n samples of m dimensions, we aim to find a compact representation $L \in \mathcal{R}^{k \times n}$, where $k \ll m$, that preserves local relation of each point. To satisfy this requirement, LLE assumes that each point is a linear combination of those in its neighborhood [32]. To find such a coefficient matrix C, for each X_i, LLE finds a set N_i containing all the indices of the neighbors of X_i according to some distance threshold and then solves the following problem:

$$\min_{C} \sum_{i=1}^{n} \|X_i - \sum_{j \in N_i} C_{ij} X_j\|^2 \quad s.t. \quad \sum_{j} C_{ij} = 1, \tag{1}$$

where $C_{ij} = 0$ if $X_j \notin N_i$. Then LLE finds latent representation L by solving

$$\min_{L} \sum_{i=1}^{n} \|L_i - \sum_{j \in N_i} C_{ij} L_j\|^2. \tag{2}$$

We can see that there are several limitations associated with current approach [17,18]. First, we need to specify N_i in advance based on some heuristic strategies, which may not be flexible enough to reflect the complicated structure of real data. Second, coefficient C, which has a direct impact on the learned representation L, is fixed in Eq. (2). This could cause inaccuracy since C might not reveal the underlying local manifold, which often occur when the data is contaminated with noise and outliers. Third, Eq. (1) is performed in the raw space. Along with the widespread adoption of deep neural networks, it is appealing to implement LLE based on it.

4   Proposed Methodology

Handcrafting a neighbors set N_i is challenging since the structure of the data space is often very complicated [14,33]. To obtain a good solution in general, a principled way is to learn it from data. Inspired by the self-expression property [4,21,25,34,41,42], we can express a point as a linear combination of all other points without specifying the neighborhood, i.e., $X \approx XC$ and $diag(C) = 0$. By minimizing certain norms of C, C is guaranteed to have a block-diagonal structure, i.e., $C_{ij} \neq 0$ iff point X_i and X_j are similar to each other to some extent. Furthermore, we can adopt a joint learning approach so that C can be iteratively updated.

Mathematically, our generalized LLE (gLLE) can be formulated as:

$$\min_{C,L} \frac{1}{2}||X - XC||_F^2 + \frac{\beta}{2}||L - LC||_F^2 + \alpha||C||_p \tag{3}$$
$$s.t. \quad diag(C) = 0,$$

where $\alpha, \beta > 0$ are trade-off parameters and $||\cdot||_p$ represents an arbitrary matrix norm.

Different from traditional LLE, Eq. (3) does not pre-define neighborhood for each point, which is automatically determined later through optimization. Consequently, this eliminates the uncertainty in LLE caused by heuristic approaches. Different types of norm for C can be applied, e.g., the nuclear norm to guarantee low-rank property [13], the $\ell1$-norm to obtain sparse solution [4], and the Frobenius norm [11]. These norms help make Eq. (3) robust to noise and outliers [9]. The first two terms keep the consistency of relation among data points between the original space and latent spaces. Different from LLE, Eq. (3) optimizes the similarity and latent representation simultaneously.

To take advantage of deep neural networks, we make use of deep auto-encoders (AE). As displayed in Fig. 1, a fully-connected layer is added to the traditional AE. Let Θ denote the network parameter, which include encoder parameters Θ_e, self-expression layer parameters C, and decoder parameters Θ_d. Then the latent representation can be written as L_{Θ_e}, which is a function of Θ_e. The weights of the fully-connected layer correspond to the elements of the coefficient matrix C. The decoder recovers data from the transformed latent representation $L_{\Theta_e}C$. Thus, the output $\hat{X}$ of the decoder is a function of $\{\Theta_e, C, \Theta_d\}$.

To combine the loss of auto-encoder, our objective function can be formulated as:

$$L(\Theta) = \frac{1}{2}||X - \hat{X}_\Theta||_F^2 + \alpha||C||_p + \frac{\beta}{2}||L_{\Theta_e} - L_{\Theta_e}C||_F^2$$
$$+ \frac{\gamma}{2}||X - XC||_F^2 \quad s.t. \quad diag(C) = 0,$$
(4)

where the first term is the reconstruction loss from auto-encoder, which helps the latent representation to keep enough information in the raw data. Note that all the unknowns in Eq. (4) are function of network parameters. This network can be trained by back-propagation. Once the network architecture is optimized, we obtain the lower-dimensional representation L and relation matrix C. Compared to the existing work in the literature, our proposed gLLE has the following advantages:

- The proposed LLE is powered by a deep auto-encoder. Hence, it enjoys the benefits of deep neural networks. Moreover, solving low-dimensional representation L converts to network optimization.
- From another perspective, the auto-encoder takes into account the data relation. It outputs relation preserving representation, i.e., local neighborhoods of data points are retained in the low-dimensional space.
- The relation matrix measures sample similarity, thus our method also solves similarity measure fundamental problem. Moreover, we convert the similarity learning into network parameters optimization problem.
- In traditional LLE, the relation C is fixed and defined on the original high-dimensional space. Here, the relation is adaptively learned and is consistent both in the input space and the low-dimensional space.
- Rather than following a two-step approach as in traditional LLE, our method implements it in a joint framework. The latent representation L and relation C can be iteratively boosted by using the result of the other towards an overall optimal solution.

5 Experiments

In this section, we show that our method does well in manifold learning and dimension reduction, which facilitates the subsequent classification.

5.1 Experimental Setup

We implement our dimension reduction experiments on the MNIST dataset. The MNIST dataset consists of handwritten digits of "0" through "9" , the size of each image is 28 * 28. We randomly select 500 images for each digit from the MNIST dataset, totally 5000 images. The encoder consists of two convolutional layers and one fully connected layer. The convolutional layers have 6 and 12 channels, their kernel sizes are 5 and 3, respectively. The ReLU activation function is applied. The fully connected layer maps data into the dimension we want to reduce to, which is set to 30 in this experiment. The Sigmoid activation function is used in this layer. No drop-out is applied in our experiments. The auto-encoder is pre-trained without the self-expression layer. Then we train the whole network, with the pre-trained weights of encoder and decoder as the initial state. We initialize the weight of self-expression layer with random numbers following the normal distribution. Adam is employed to do the optimization [22]. The full batch of the dataset is inputted to our network.

5.2 Dimension Reduction

To visualize the dimension reduction effect, we map the low-dimension digits into a two-dimensional manifold space with t-SNE [26]. We compare the performance of several classic methods, i.e., LDA [3], LLE [32], LPP [6], MFA [39], and auto-encoder (AE) [7]. For AE, we use the same architecture and parameters as our gLLE.

The visualization results are shown in Fig. 2. The data points of the 10 different digits are drawn in different colors. As we can see, low dimension points derived from LDA, LPP, LLE, and MFA crowd together and are not able to be distinguished. Whereas, the data points mapped by AE tend to form more distinctive clusters. For our gLLE, we implement both $\ell1$-norm (gLLE-L1) and $\ell2$-norm (gLLE-L2). We can observe that gaps between classes become more obvious and the clusters are more discriminative with gLLE. This owns to the consideration of neighborhood manifold in LLE.

Table 1. Classification error rate (%). The reduced dimensions are in parentheses.

Method	PCA	LDA	LPP	MFA	NPE	AE	gLLE-L2
MNIST	6.2 (55)	16.1 (9)	7.9 (55)	9.5 (45)	7.1 (30)	4.7 (30)	**4.2** (30)
PIE	20.6 (150)	4.5 (68)	7.9 (220)	4.5 (80)	12.9 (750)	5.5 (120)	**4.2** (120)

5.3 Classification

To quantitatively examine the advantage of low-dimensional representation, we perform classification experiments. To make the testing set for MNIST dataset, we randomly select another 500 images for each digit. We also implement on CMU PIE dataset. It contains 68 subjects, whose face images with size 32 * 32 are captured under varying poses, illumination, and expression. We randomly select 85 images for training and another 85 images for testing each subject. Similar to previous architecture, two convolutional layers and one fully connected layer compose the encoder. The channels of convolutional layers are 8 and 6, and their kernel sizes are 5 and 3. Output dimension of the fully connected layer is 120. After training the network, we forward pass our testing data with the encoder to obtain the low-dimensional representation L, then a nearest-neighbor classifier is applied.

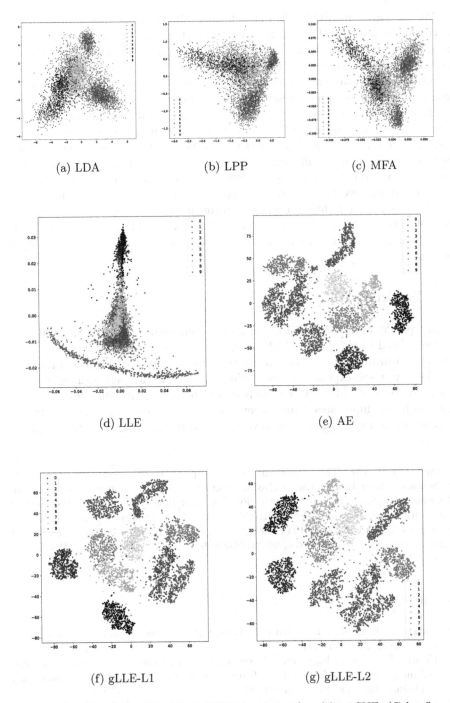

Fig. 2. 2D visualization of the learned digit image produced by t-SNE. (Color figure online)

We compare with PCA [12], LDA [3], LPP [6], MFA [39], neighborhood preserving embedding (NPE) [5]. Since LLE suffers from the out-of-sample problem, we use NPE to substitute it, which serves as a linear approximation to LLE to handle that issue. We select the dimension that each method reaches its best performance. To examine the effect of self-expression layer, we also compute the result of AE, i.e., without the self-expression layer.

Table 1 reports the error rate of each method. As a baseline, we apply the nearest-neighbor classifier on the original data and find the error rates are 6.5% and 12.5% on MNIST and PIE dataset respectively. As we can see, our method performs the best among all comparison approaches. LDA and its extension MFA give comparable performance since they used label information. In particular, we can still improve 1.3% upon auto-encoder on PIE dataset. This justifies the benefit of considering the data relation on the manifold when we perform dimension reduction. The traditional auto-encoder only considers the reconstruction of original data, and it does not guarantee the latent feature to lie in the linear subspaces.

6   Conclusion

In this paper, we implement LLE in a new deep auto-encoder network, named as generalized LLE (gLLE). gLLE integrates representation learning, LLE, and similarity learning into a unified framework. In particular, gLLE learns the data relation by a self-expression layer guided by the local neighborhood via the locally-linear embedding (LLE) method, which ensures the consistency of data relation in the input space and latent space. Extensive experiments on dimension reduction and classification have validated the superiority of the proposed method.

Acknowledgments. This paper was in part supported by Grants from the Natural Science Foundation of China (No. 61806045), the National Key R&D Program of China (No. 2018YFC0807500), the Fundamental Research Fund for the Central Universities under Project ZYGX2019Z015, the Sichuan Science and Techology Program (Nos. 2020YFS0057, 2019YFG0202), the Ministry of Science and Technology of Sichuan Province Program (Nos. 2018GZDZX0048, 20ZDYF0343, 2018GZDZX0014, 2018GZDZX0034).

References

1. Burges, C.J., et al.: Dimension reduction: a guided tour. Found. Trends® Mach. Learn. **2**(4), 275–365 (2010)
2. Charte, D., Charte, F., García, S., del Jesus, M.J., Herrera, F.: A practical tutorial on autoencoders for nonlinear feature fusion: taxonomy, models, software and guidelines. Inf. Fus. **44**, 78–96 (2018)
3. Duda, R.O., Hart, P.E., Stork, D.G.: Pattern Classification. Wiley, Hoboken (2012)
4. Elhamifar, E., Vidal, R.: Sparse subspace clustering: algorithm theory and applications. IEEE Trans. Pattern Anal. Mach. Intell. **35**(11), 2765–2781 (2013)

5. He, X., Cai, D., Yan, S., Zhang, H.J.: Neighborhood preserving embedding. In: Tenth IEEE International Conference on Computer Vision, 2005. ICCV 2005, vol. 2, pp. 1208–1213. IEEE (2005)
6. He, X., Niyogi, P.: Locality preserving projections. In: Advances in Neural Information Processing Systems, pp. 153–160 (2004)
7. Hinton, G.E., Salakhutdinov, R.R.: Reducing the dimensionality of data with neural networks. Science **313**(5786), 504–507 (2006)
8. Hou, C., Zhang, C., Wu, Y., Jiao, Y.: Stable local dimensionality reduction approaches. Pattern Recogn. **42**(9), 2054–2066 (2009)
9. Huang, J., Nie, F., Huang, H.: A new simplex sparse learning model to measure data similarity for clustering. In: IJCAI, pp. 3569–3575 (2015)
10. Huang, S., Kang, Z., Xu, Z.: Auto-weighted multi-view clustering via deep matrix decomposition. Pattern Recogn. **97**, 107015 (2020)
11. Ji, P., Salzmann, M., Li, H.: Efficient dense subspace clustering. In: 2014 IEEE Winter Conference on Applications of Computer Vision (WACV), pp. 461–468. IEEE (2014)
12. Jolliffe, I.: Principal component analysis. In: International Encyclopedia of Statistical Science, pp. 1094–1096. Springer, New York (2011). https://doi.org/10.1007/b98835
13. Kang, Z., Lu, X., Lu, Y., Peng, C., Chen, W., Xu, Z.: Structure learning with similarity preserving. Neural Netw. **129**, 138–148 (2020)
14. Kang, Z., Pan, H., Hoi, S.C., Xu, Z.: Robust graph learning from noisy data. IEEE Trans. Cybern. **50**(5), 1833–1843 (2020)
15. Kang, Z., Peng, C., Cheng, Q.: Robust PCA via nonconvex rank approximation. In: 2015 IEEE International Conference on Data Mining, pp. 211–220. IEEE (2015)
16. Kang, Z., Peng, C., Cheng, Q.: Top-n recommender system via matrix completion. In: Proceedings of the Thirtieth AAAI Conference on Artificial Intelligence, pp. 179–184. AAAI Press (2016)
17. Kang, Z., Peng, C., Cheng, Q.: Twin learning for similarity and clustering: a unified kernel approach. In: Thirty-First AAAI Conference on Artificial Intelligence (2017)
18. Kang, Z., Peng, C., Cheng, Q., Xu, Z.: Unified spectral clustering with optimal graph. In: Thirty-Second AAAI Conference on Artificial Intelligence (2018)
19. Kang, Z., Wen, L., Chen, W., Xu, Z.: Low-rank kernel learning for graph-based clustering. Knowl. Based Syst. **163**, 510–517 (2019)
20. Kang, Z., Xu, H., Wang, B., Zhu, H., Xu, Z.: Clustering with similarity preserving. Neurocomputing **365**, 211–218 (2019)
21. Kang, Z., et al.: Partition level multiview subspace clustering. Neural Networks **122**, 279–288 (2020)
22. Kingma, D.P., Ba, J.L.: Adam: a method for stochastic optimization. In: Proceedings 3rd International Conference Learning Representations (2014)
23. Lange, S., Riedmiller, M.: Deep auto-encoder neural networks in reinforcement learning. In: The 2010 International Joint Conference on Neural Networks (IJCNN), pp. 1–8. IEEE (2010)
24. Li, A., Chen, D., Wu, Z., Sun, G., Lin, K.: Self-supervised sparse coding scheme for image classification based on low rank representation. PLoS One **13**(6), 450–462 (2018)
25. Liu, G., Lin, Z., Yan, S., Sun, J., Yu, Y., Ma, Y.: Robust recovery of subspace structures by low-rank representation. IEEE Trans. Pattern Anal. Mach. Intell. **35**(1), 171–184 (2013)
26. Maaten, L.V.D., Hinton, G.: Visualizing data using t-SNE. J. Mach. Learn. Res. **9**, 2579–2605 (2008)

27. Masci, J., Meier, U., Cireşan, D., Schmidhuber, J.: Stacked convolutional auto-encoders for hierarchical feature extraction. In: Honkela, T., Duch, W., Girolami, M., Kaski, S. (eds.) ICANN 2011. LNCS, vol. 6791, pp. 52–59. Springer, Heidelberg (2011). https://doi.org/10.1007/978-3-642-21735-7_7

28. Nie, F., Xu, D., Tsang, I.W.H., Zhang, C.: Flexible manifold embedding: a framework for semi-supervised and unsupervised dimension reduction. IEEE Trans. Image Process. **19**(7), 1921–1932 (2010)

29. Peng, C., Chen, C., Kang, Z., Li, J., Cheng, Q.: Res-PCA: a scalable approach to recovering low-rank matrices. In: Proceedings of the IEEE Conference on Computer Vision and Pattern Recognition, pp. 7317–7325 (2019)

30. Peng, C., Chen, Y., Kang, Z., Chen, C., Cheng, Q.: Robust principal component analysis: a factorization-based approach with linear complexity. Inf. Sci. **513**, 581–599 (2020)

31. Peng, C., Kang, Z., Cheng, Q.: A fast factorization-based approach to robust PCA. In: 2016 IEEE 16th International Conference on Data Mining (ICDM), pp. 1137–1142. IEEE (2016)

32. Roweis, S.T., Saul, L.K.: Nonlinear dimensionality reduction by locally linear embedding. Science **290**(5500), 2323–2326 (2000)

33. Shakhnarovich, G.: Learning task-specific similarity. Ph.D. thesis, Massachusetts Institute of Technology (2005)

34. Tang, C., Zhu, X., Liu, X., Li, M., Wang, P., Zhang, C., Wang, L.: Learning joint affinity graph for multi-view subspace clustering. In: IEEE Transactions on Multimedia (2018)

35. Tenenbaum, J.B., De Silva, V., Langford, J.C.: A global geometric framework for nonlinear dimensionality reduction. Science **290**(5500), 2319–2323 (2000)

36. Vincent, P., Larochelle, H., Lajoie, I., Bengio, Y., Manzagol, P.A.: Stacked denoising autoencoders: learning useful representations in a deep network with a local denoising criterion. J. Mach. Learn. Res. **11**, 3371–3408 (2010)

37. Wang, R., Nie, F., Hong, R., Chang, X., Yang, X., Yu, W.: Fast and orthogonal locality preserving projections for dimensionality reduction. IEEE Trans. Image Process. **26**(10), 5019–5030 (2017)

38. Xie, J., Girshick, R., Farhadi, A.: Unsupervised deep embedding for clustering analysis. In: International Conference on Machine Learning, pp. 478–487 (2016)

39. Yan, S., Xu, D., Zhang, B., Zhang, H.J., Yang, Q., Lin, S.: Graph embedding and extensions: a general framework for dimensionality reduction. IEEE Trans. Pattern Anal. Mach. Intell. **29**(1), 40–51 (2007)

40. Zhang, C., Liu, Y., Fu, H.: Ae2-nets: autoencoder in autoencoder networks. In: Proceedings of the IEEE Conference on Computer Vision and Pattern Recognition, pp. 2577–2585 (2019)

41. Zhang, J., et al.: Self-supervised convolutional subspace clustering network. In: Proceedings of the IEEE Conference on Computer Vision and Pattern Recognition, pp. 5473–5482 (2019)

42. Zhou, P., Hou, Y., Feng, J.: Deep adversarial subspace clustering. In: Proceedings of the IEEE Conference on Computer Vision and Pattern Recognition, pp. 1596–1604 (2018)

43. Zhou, P., Du, L., Fan, M., Shen, Y.D.: An LLE based heterogeneous metric learning for cross-media retrieval. In: Proceedings of the 2015 SIAM International Conference on Data Mining, pp. 64–72. SIAM (2015)

Semi-supervised Feature Selection Using Sparse Laplacian Support Vector Machine

Li Zhang[1,2(✉)] , Xiaohan Zheng[1], and Zhiqiang Xu[1]

[1] School of Computer Science and Technology, Joint International Research Laboratory of Machine Learning and Neuromorphic Computing, Soochow University, Suzhou 215006, Jiangsu, China
zhangliml@suda.edu.cn

[2] Provincial Key Laboratory for Computer Information Processing Technology, Soochow University, Suzhou 215006, Jiangsu, China

Abstract. Semi-supervised feature selection is an active topic in machine learning and data mining. Laplacian support vector machine (LapSVM) has been successfully applied to semi-supervised learning. However, LapSVM cannot be directly applied to feature selection. To remedy it, we propose a sparse Laplacian support vector machine (SLapSVM) and apply it to semi-supervised feature selection. On the basis of LapSVM, SLapSVM introduces the ℓ_1-norm regularization, which means the solution of SLapSVM has sparsity. In addition, the training procedure of SLapSVM can be formulated as solving a quadratic programming problem, which indicates that the solution of SLapSVM is unique and global. SLapSVM can perform feature selection and classification at the same time. Experimental results on semi-supervised classification problems show the feasibility and effectiveness of the proposed semi-supervised learning algorithms.

Keywords: Support vector machine · Semi-supervised learning · Feature selection · ℓ_1-norm regularization · Quadratic programming

1 Introduction

Recently, semi-supervised feature selection has attracted substantial attention in machine learning and data mining [23,24]. There are two reasons. One reason is that data collected from real-world applications would have a lot of features. In this case, it is necessary to reduce dimension to achieve better learning performance. As a technique for dimension reduction, feature selection has always been

This work was supported in part by the Natural Science Foundation of the Jiangsu Higher Education Institutions of China under Grant No. 19KJA550002, by the Six Talent Peak Project of Jiangsu Province of China under Grant No. XYDXX-054, by the Priority Academic Program Development of Jiangsu Higher Education Institutions, and by the Collaborative Innovation Center of Novel Software Technology and Industrialization.

H. Zhang et al. (Eds.): NCAA 2020, CCIS 1265, pp. 107–118, 2020.
https://doi.org/10.1007/978-981-15-7670-6_10

of concern. The other reason is that labeling examples is expensive and time-consuming while there are large numbers of unlabeled examples available in many practical problems, which results in semi-supervised learning methods. In semi-supervised learning, algorithms construct their models from a few labeled examples together with a large collection of unlabeled data, including graph-based methods [1,4,11,13,30,31], methods based on support vector machine (SVM) [2,3,6,8], and others [5,14,16,19,20,25,26]. This paper focuses on SVM-based methods and discusses the issue of semi-supervised feature selection.

Famous semi-supervised methods based on SVM include transductive support vector machine (TSVM) [8], semi-supervised support vector machine (S3VM) [3], and Laplacian support vector machine (LapSVM) [2]. Bennett et al. proposed S3VM to construct an SVM using both the labeled and unlabeled data [3]. S3VM is iteratively tagging unlabeled data in the training procedure and is usually time consuming. Due to its way to utilize the unlabeled data, S3VM cannot directly classify unseen instances. To implement feature selection using S3VM, Hoai et al. proposed sparse semi-supervised SVM (S4VM) replacing ℓ_2-norm by ℓ_0-norm in S3VM. The objective of S4VM is solved by applying DC (difference of convex) programming [12]. Moreover, Lu et al. cast semi-supervised learning into an ℓ_1-norm linear reconstruction problem and presented an ℓ_1-norm semi-supervised learning method [14]. However, these methods cannot classify new instances directly due to their "closed" nature. LapSVM, an extension of SVM to the semi-supervised field, introduces an additional regularization term on the geometry of both labeled and unlabeled samples by using a graph Laplacian [2]. LapSVM follows a non-iterative optimization procedure and can be taken as a kind of graph-based methods. Gasso et al. proposed an ℓ_1-norm constraint Laplacian SVM (ℓ_1-NC LapSVM) by adding an extra ℓ_1-norm constraint to the optimization problem of LapSVM [9]. The sparseness of the solution to ℓ_1-NC LapSVM is determined by the size of regularization parameter. However, experimental results show that the sparseness of ℓ_1-NC LapSVM is limited for feature selection.

In fact, real data often contains noise, including redundant features, which would have a negative effect on the model performance. In order to eliminate the effect of noise or redundancy on data, it is necessary to generate a sparse decision model to implement feature selection. To implement it, this paper proposes a sparse Laplacian support vector machine (SLapSVM) to perform feature selection. To get a sparse decision model, we adopt the hinge loss and ℓ_1-norm regularization simultaneously. It is known that the hinge loss can lead to a sparse model representation for SVM [15,18]. In addition, the ℓ_1-norm regularization penalty as a substitution of the ℓ_2-norm regularization penalty can also induce a sparse solution [10,21,27,32]. Through the sparse decision model, of SLapSVM, we achieve feature selection. Similar to LapSVM, SLapSVM can be formulated as a quadratic programming problem, which indicates that its solution is unique and global.

The rest of the paper is outlined as follows. Section 2 presents SLapSVM. Section 3 shows experimental results on real-world datasets. Section 4 concludes and discusses further work.

2   SLapSVM

In this section, we propose the model of SLapSVM for semi-supervised learning. For a semi-supervised classification problem, suppose that we have a data set which consists of ℓ labeled and u unlabeled examples. Let $X_\ell = \{(\mathbf{x}_i, y_i)\}_{i=1}^{\ell}$ be the labeled set with $\mathbf{x}_i \in \mathbb{R}^d$ and $y_i \in \{+1, -1\}$, and $X_u = \{\mathbf{x}_i\}_{i=\ell+1}^{\ell+u}$ be the unlabeled set with $\mathbf{x}_i \in \mathbb{R}^d$, where d is the number of features. To integrate these two sets, let $X = \{\mathbf{x}_i\}_{i=1}^{\ell+u}$ be the instance set and $Y = \{y_i\}_{i=1}^{\ell}$ be the label set. Without loss of generalization, the first ℓ examples in the set X correspond to the labeled ones.

The goal of SLapSVM is to find an optimal decision function (model) f from a set of linear hypothesis functions

$$F = \{f(\mathbf{x}) | f(\mathbf{x}) = \boldsymbol{\alpha}^T \mathbf{x} + b, \boldsymbol{\alpha} \in \mathbb{R}^d, b \in \mathbb{R}\} \tag{1}$$

where $\boldsymbol{\alpha} = [\alpha_1, \cdots, \alpha_d]^T$ is the weight vector, and b is the bias.

To obtain the hypothesis function, we replace the ℓ_2-norm regularization in LapSVM by the ℓ_1-norm regularization, and propose LapSVM, which solves the following optimization problem:

$$\min_{\boldsymbol{\alpha}, b, \boldsymbol{\xi}} \quad \frac{1}{\ell} \sum_{i=1}^{\ell} \xi_i + \gamma_A(\|\boldsymbol{\alpha}\|_1 + \sigma\|b\|_1) + \frac{\gamma_I}{2} \sum_{i=1}^{\ell+u} \sum_{j=1}^{\ell+u} (f(\mathbf{x}_i) - f(\mathbf{x}_j))^2 W_{ij}$$

$$s.t. \quad y_i(\boldsymbol{\alpha}^T \mathbf{x}_i + b) \geq 1 - \xi_i, \ \xi_i \geq 0, \ i = 1, \cdots, \ell \tag{2}$$

where $\| \cdot \|_1$ is the ℓ_1-norm, $\boldsymbol{\xi} = [\xi_1, \cdots, \xi_\ell] \in \mathbb{R}^\ell$ is the slack vector for labeled samples, W_{ij} is the similarity between $\mathbf{x}_i$ and $\mathbf{x}_j$, σ is a small positive constant to ensure a unique solution, $\gamma_A \geq 0$ and $\gamma_I \geq 0$ are the regularization parameters. The first term in Eq. (2) is the hinge loss function that is very popular in SVM-like methods and can induce sparsity in theory. The second term $\|\boldsymbol{\alpha}\|_1$ is the ℓ_1-norm regularization term that can also induce sparsity in the ℓ_1-norm SVM [27,32] and sparse signal reconstruction methods [27]. The third term is the Laplacian regularization.

Next, we rewrite the formula Eq. (2) to solve it easily. Since there is no constrain on $\boldsymbol{\alpha}$, the absolute value sign would exist in the objective function Eq. (2) when calculating $\|\boldsymbol{\alpha}\|_1$. In this case, it is not easy to solve Eq. (2). We introduce two vectors $\boldsymbol{\alpha}^+$ and $\boldsymbol{\alpha}^-$ with positive elements, and let

$$\boldsymbol{\alpha} = \boldsymbol{\alpha}^+ - \boldsymbol{\alpha}^- \tag{3}$$

Similarly, we define

$$b = b^+ - b^- \tag{4}$$

where $b^+ > 0$ and $b^- > 0$. In addition, the third term in Eq. (2) can be expressed as:

$$
\frac{1}{2} \sum_{i,j=1}^{\ell+u} (f(\mathbf{x}_i) - f(\mathbf{x}_j))^2 W_{ij} = \begin{pmatrix} f(\mathbf{x}_1) \\ \vdots \\ f(\mathbf{x}_{\ell+u}) \end{pmatrix}^T (\mathbf{D} - \mathbf{W}) \begin{pmatrix} f(\mathbf{x}_1) \\ \vdots \\ f(\mathbf{x}_{\ell+u}) \end{pmatrix}
$$

$$
= \begin{pmatrix} f(\mathbf{x}_1) \\ \vdots \\ f(\mathbf{x}_{\ell+u}) \end{pmatrix}^T \mathbf{L} \begin{pmatrix} f(\mathbf{x}_1) \\ \vdots \\ f(\mathbf{x}_{\ell+u}) \end{pmatrix} \tag{5}
$$

$$
= \mathbf{f}^T \mathbf{L} \mathbf{f}
$$

$$
= \boldsymbol{\alpha}^T \mathbf{X}^T \mathbf{L} \mathbf{X} \boldsymbol{\alpha}
$$

where $\mathbf{X} \in \mathbb{R}^{(\ell+u) \times d}$ is the sample matrix in which $\mathbf{x}_i$ is the i-th row, the graph Laplacian matrix $\mathbf{L} = \mathbf{D} - \mathbf{W}$, and $\mathbf{D} \in \mathbb{R}^{(\ell+u) \times (\ell+u)}$ is the diagonal matrix given by $D_{ii} = \sum_j W_{ij}$. Substituting Eqs. (3) and (5) into Eq. (2), we have the following programming:

$$
\min_{\alpha \pm, b\pm, \xi} \quad \frac{1}{\ell} \sum_{i=1}^{\ell} \xi_i + \gamma_A \left(\sum_{j=1}^{d} \left(\alpha_j^+ + \alpha_j^- \right) + \sigma(b^+ + b^-) \right) + \gamma_I \boldsymbol{\alpha}^T \mathbf{X}^T \mathbf{L} \mathbf{X} \boldsymbol{\alpha} \tag{6}
$$

$$
s.t \quad y_i \left(\mathbf{x}_i^T (\boldsymbol{\alpha}^+ - \boldsymbol{\alpha}^-) + (b^+ - b^-) \right) \geq 1 - \xi_i
$$

$$
b^+, b^- \geq 0, \alpha_j^+, \alpha_j^- \geq 0, j = 1, \cdots, d
$$

$$
\xi_i \geq 0, i = 1, \cdots, \ell
$$

The programming Eq. (6) can be rewritten in matrix form:

$$
\min_{\mathbf{u}} \quad \mathbf{c}^T \mathbf{u} + \frac{1}{2} \mathbf{u}^T \mathbf{Q} \mathbf{u}
$$

$$
s.t. \quad \mathbf{A}^T \mathbf{u} \geq \mathbf{1} \tag{7}
$$

$$
\mathbf{u} \geq \mathbf{0}
$$

where $\mathbf{u} = [(\boldsymbol{\alpha}^+)^T, (\boldsymbol{\alpha}^-)^T, b^+, b^-, \boldsymbol{\xi}^T]^T \in \mathbb{R}^{2d+\ell+2}$, $\mathbf{0}$ is the column vector of all zeros, $\mathbf{c} = [\gamma_A \mathbf{1}^T, \gamma_A \mathbf{1}^T, \sigma, \sigma, \mathbf{1}^T/\ell]^T$, $\mathbf{A} = [\mathbf{Y} \mathbf{X}_\ell, -\mathbf{Y} \mathbf{X}_\ell, \mathbf{y}, -\mathbf{y}, \mathbf{I}] \in \mathbb{R}^{\ell \times (2d+\ell+2)}$, $\mathbf{y} = [y_1, y_2, \cdots, y_\ell]^T$, $\mathbf{Y}$ is the diagonal matrix with the diagonal line of $\mathbf{y}$, $\mathbf{1}$ is the column vector of all ones, $\mathbf{I}$ is the $\ell \times \ell$ identity matrix, $\mathbf{X}_\ell$ is the sample matrix of labeled examples, and

$$
\mathbf{Q} = \begin{pmatrix} \gamma_I \mathbf{X}^T \mathbf{L} \mathbf{X} & -\gamma_I \mathbf{X}^T \mathbf{L} \mathbf{X} & \mathbf{0} \\ -\gamma_I \mathbf{X}^T \mathbf{L} \mathbf{X} & \gamma_I \mathbf{X}^T \mathbf{L} \mathbf{X} & \mathbf{0} \\ \mathbf{0} & \mathbf{0} & \mathbf{0} \end{pmatrix}.
$$

Obviously, Eq. (7) is a constrained quadratic program problem that has $(2d + \ell + 2)$ variables and ℓ inequality constraints. Because the matrix $\mathbf{Q}$ is symmetric and positive semi-definite, this optimization problem could be solved

---

Algorithm 1: SLapSVM

Input: Instance set $X = \{\mathbf{x}_i\}_{i=1}^{\ell+u}$ and label set $Y = \{y_i\}_{i=1}^{\ell}$, where the first ℓ examples in X have labels corresponding to ones in Y, regularization parameters γ_A and γ_I.

Output: Sparse weight vector $\boldsymbol{\alpha}$ and bias b.

1 **begin**

2   Construct the similarity matrix $\mathbf{W}$:

$$W_{ij} = \begin{cases} \exp(-\gamma \|\mathbf{x}_i - \mathbf{x}_j\|_2^2), & \text{if } \mathbf{x}_i \text{ and } \mathbf{x}_j \text{ are neighbors} \\ 0, & \text{otherwise} \end{cases}$$

3   Represent matrices $\mathbf{Q}$ and $\mathbf{A}$, and the vector $\mathbf{c}$;

4   Solve the quadratic programming Eq. (7) to obtain the solution $\mathbf{u}$;

5   Get $\boldsymbol{\alpha}^+$, $\boldsymbol{\alpha}^-$, b^+ and b^- from $\mathbf{u}$;

6   Return $\boldsymbol{\alpha} = \boldsymbol{\alpha}^+ - \boldsymbol{\alpha}^-$ and $b = b^+ - b^-$.

7 **end**

---

efficiently through some standard techniques, such as the active set. The algorithm description of SLapSVM is given in Algorithm 1. Step 2 is to construct the similarity matrix, where the parameter γ could be determined by applying the median method used in [28,29].

Once we have $\boldsymbol{\alpha}$ and b, we can obtain the classification hyperplane. For an unseen sample $\mathbf{x}$, SLapSVM predicts its label by

$$\hat{f}(\mathbf{x}) = sign\left(\boldsymbol{\alpha}^T\mathbf{x} + b\right) \tag{8}$$

where $sign(\cdot)$ is the sign function, where $\hat{f}(\mathbf{x})$ is the estimated label for the unseen sample $\mathbf{x}$.

Let $NZ = \{\alpha_i | \alpha_i \neq 0, i = 1, \cdots, d\}$ be the set of non-zero coefficients for Eq. 8, where $|\cdot|$ is the number of elements in a set. Because both the hinge loss and the ℓ_1-norm can induce sparsity, we could get a sparse vector $\boldsymbol{\alpha}$ that corresponds to weights of features. Thus, the inequality $|NZ| < d$ holds true, and we can perform the operation of feature selection. The set NZ can actually reflect the selected feature subset and show the sparsity of the decision model. The smaller $|NZ|$ is, the more sparsity the decision model has.

3 Experimental Results

In this section, we validate the effectiveness of the proposed method in feature selection on synthetic and UCI [7] datasets. To demonstrate the capabilities of our algorithm, this paper compares SLapSVM with the state-of-art algorithms for feature selection, including S3VM-PiE [12], S3VM-PoDC [12], S3VM-SCAD [12], S3VM-Log [12], S3VM-ℓ_1 [12], and Lap-PPSVM [22]. All numerical experiments are performed on a personal computer with an Inter Core I5 processor with 4 GB RAM. This computer runs Windows 7, with Matlab R2013a.

3.1 Data Description and Experimental Setting

Datasets used here include a toy and seven UCI ones, which are described as follows:

- Gaussian data

  In the Gaussian dataset, two-class synthetic samples are drawn from two Gaussian distributions: $N((0,0)^T, \mathbf{I})$ and $N((3,0)^T, \mathbf{I})$, where $\mathbf{I} \in \mathbb{R}^{2 \times 2}$ is the identify matrix. There are 600 samples total and 300 samples for each class. For each class, 80% of data are selected as the training samples and the rest as the test ones.
- UCI data

  Seven UCI datasets are summarized in Table 1. These datasets represent a wide range of fields (including pathology, vehicle engineering, biological information, finance and so on), sizes (from 267 to 1473) and features (from 9 to 34). All datasets are normalized such that the features scale in the interval $[-1, 1]$ before training and test. Similar to [22], in our experiments, each UCI dataset is divided into two subsets randomly: 70% for training and 30% for test.

When we compare the different methods, some performance indexes would be considered, such as accuracy, F1-measure, and sparsity. These three performance indexes are described as follows.

- Accuracy is defined as:

$$Accuracy = \frac{TP + TN}{TP + TN + FP + FN} \tag{9}$$

where TP means the number of true positive samples, TN means the number of true negative samples, FP means the number of false positive samples and FN means the number of false negative samples.
- F1-measure can be defined as:

$$F1 - measure = \frac{2P \times R}{P + R} \tag{10}$$

where $P = TP/(TP + FP)$ is precision and $R = TP/(TP + FN)$ is recall.
- Sparsity is measured by $|NZ|$.

All regularization parameters in compared methods are selected from the set $\{10^{-6}, \cdots, 10^2\}$ using two-fold cross-validation [17]. Once the parameters are selected, they would be returned to the training subset to learn the final decision function. Each experiment is repeated 10 times and the average results on test subsets are reported.

Table 1. Description of seven UCI datasets

	Australian	CMC	German	Ionosphere	Hearts	Spect	WDBC
#Sample	700	1473	1000	351	270	267	569
#Attribute	14	9	24	34	13	22	14
#Class	2	2	2	2	3	2	2

3.2 Gaussian Data

Consider the random Gaussian dataset. In the training subset, we randomly take 10% as the labeled set and the rest as the unlabeled set. In order to verify the ability to select features, we append m-dimensional noise to the training subset, where m takes a value in the set $\{20, 40, 60, 80, 100\}$. The noise in each dimension is the white Gaussian noise and has a signal-noise ratio (SNR) of 3 dB. Note that the original features are the first two ones in the $(m + 2)$-dimensional dataset. Consider the variation of m, we choose the accuracy and the sum of the first two feature weights as the metrics to compare SLapSVM with other methods.

The average experimental results are given in Fig. 1. Basically, SLapSVM always achieves the best average accuracy when $m > 20$, as shown in Fig. 1(a). For visualization, we normalize the weight vector so that the sum of all weights is equal to one. From Fig. 1(b), we can see the good performance of SLapSVM in eliminating noise or the good ability to select useful features. Only can SLapSVM pick up the first two useful features. In other words, SLapSVM can accurately select those features that are helpful for classification.

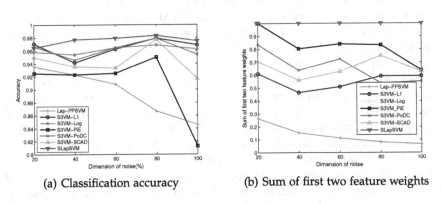

(a) Classification accuracy   (b) Sum of first two feature weights

Fig. 1. Performance vs. dimension of noise on Gaussian dataset, (a) classification accuracy and (b) sum of first two feature weights.

Further, we list the best performance and the corresponding weights of all methods in Table 2, where the best accuracy among these compared methods is in bold type, and "First weight" and "Second weight" mean that weights of the first and the second features, respectively. From Table 2, we can see that

SLapSVM has the best accuracy. Also, weights of the first and the second feature are evenly distributed. Note that the weight vector has been normalized, or the sum of all weights is equal to one.

Table 2. Comparison of seven methods on the Gaussian dataset

Methods	Accuracy(%)	First weight	Second weight
S3VM-ℓ_1	97.92 ± 2.26	0.3433	0.2519
S3VM-Log	96.71 ± 1.78	0.3643	0.2892
S3VM-PiE	95.00 ± 2.97	0.8175	0.0279
S3VM-PoDC	97.92 ± 3.77	0.2648	0.2721
S3VM-SCAD	97.71 ± 2.21	0.3342	0.4257
Lap-PPSVM	93.54 ± 1.67	0.1337	0.1461
SLapSVM	**98.33** ± 2.03	0.5298	0.4702

3.3   UCI Datasets

Seven UCI datasets are considered here. In the training subset, we randomly take 40% as the labeled set and the rest as unlabeled set, which follows the setting in [12]. We compare the effectiveness of seven algorithms and report the average results in Fig. 2, where Figs. 2(a), 2(b) and 2(c) show the average accuracy, F1-measure and the number of non-zero coefficients, respectively. Here, the number of non-zero coefficients reflects the ability of feature selection.

On the index of accuracy Fig. 2(a), SLapSVM performs the best among all seven methods on six out of seven datasets. On the Ionoshpere dataset, SLapSVM is slightly inferior to S3VM-Log. On both Australian and CMC datasets, SLapSVM has a great improvement in classification performance. On the other four datasets, SLapSVM is slightly superior to the compared methods.

On the index of F1-measure Fig. 2(b), SLapSVM also performs the best among all seven methods on six out of seven datasets. On the Heart dataset, SLapSVM is slightly inferior to Lap-PPSVM. On both Australian and CMC datasets, ℓ_1-norm has also a great improvement. On the other four datasets, SLapSVM is slightly superior to the compared methods.

From Figs. 2(a) and 2(b), we can conclude that SLapSVM performs very well compared to other six methods on the performance indexes of accuracy and F1-measure. Moreover, we focus on Fig. 2(c) that shows the ability to select features. We can see that SLapSVM has a significantly higher feature sparsity than other methods while maintaining the high classification performance. In other words, SLapSVM can achieve a better performance using less features.

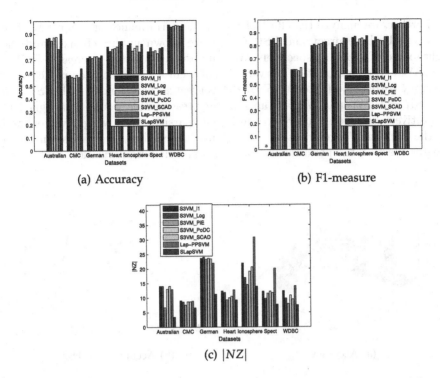

(a) Accuracy

(b) F1-measure

(c) $|NZ|$

Fig. 2. Different performance comparison on seven datasets, (a) accuracy, (b) F1-measure, and (c) $|NZ|$.

3.4   Parameter Sensitivity Analysis

As we can see above, SLapSVM has two parameters γ_A and γ_I. We are interested in the classification performance of our algorithm when the parameters γ_A and γ_I are changed and the sparsity of our algorithm when γ_A varies. In order to measure the sparsity of SLapSVM, we define the feature sparsity ratio (FSR) as follows:

$$FSR = 1 - \frac{|NZ|}{d}$$

where d is the number of features and $0 \leq FSR \leq 1$. $FSR = 0$ means that all features are selected and there has no sparsity, and $FSR = 1$ means that none of features are selected.

For this purpose, we choose the Ionosphere dataset. To observe the effect of regularization parameters on the algorithm performance, we change both γ_A and γ_I from 10^{-6} to 10^2. The resulted curve of accuracy vs. γ_A and γ_I obtained by SLapSVM on the Ionosphere dataset is shown in Fig. 3(a). From this figure, we can see that when γ_A is fixed, SLapSVM can achieve a better accuracy if γ_I is small. For a fixed γ_I, the performance of SLapSVM varies largely with changing γ_A. Thus, an appropriate γ_A would bring a good result.

Further, we analyze the effect of γ_A on the performance of SLapSVM. The curves of both accuracy and FSR vs. γ_A are shown in the left axes and the right axes of Fig. 3(b), respectively. We can observe that as γ_A increases, FSR of SLapSVM is getting greater. The variation of accuracy is slightly complexity. The accuracy corresponding to $0 < FSR < 1$ is greater than the one with $FSR = 0$ or $FSR = 1$. When $FSR = 1$, an arbitrary test sample would be assigned to a positive label. Note that γ_A controls the sparsity of the weight vector, and γ_I the Laplacian regularization term. Thus, the sparsity regularization has a greater effect on the performance than the Laplacian regularization does.

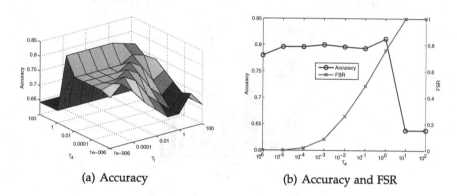

(a) Accuracy   (b) Accuracy and FSR

Fig. 3. Performance vs. regularization parameters on Ionoshpere, (a) accuracy, and (b) accuracy and FSR.

4 Conclusions

In this paper, we propose a novel sparse LapSVM for semi-supervised learning by replacing the ℓ_2-norm regularization with the ℓ_1-norm regularization, called SLapSVM. Extensive experiments are conducted to validate the feasibility and effectiveness of SLapSVM on feature selection. Among compared semi-supervised methods based on SVM, SLapSVM has the best ability of feature selection, which can be supported by experimental results on the Gaussian dataset. Furthermore, experimental results on seven UCI datasets also indicate the superiority of the SLapSVM in feature selection and classification.

References

1. Belkin, M., Niyogi, P.: Semi-supervised learning on riemannian manifolds. Mach. Learn. **56**(1–3), 209–239 (2004)
2. Belkin, M., Niyogi, P., Sindhwani, V.: Manifold regularization: a geometric framework for learning from labeled and unlabeled examples. J. Mach. Learn. Res. **7**(1), 2399–2434 (2006)

3. Bennett, K.P., Demiriz, A.: Semi-supervised support vector machines. In: Proceedings of International Conference on Neural Information Processing Systems, pp. 368–374 (1999)
4. Blum, A., Chawla, S.: Learning from labeled and unlabeled data using graph mincuts. In: Proceedings of Eighteenth International Conference on Machine Learning, pp. 19–26 (2001)
5. Blum, A., Mitchell, T.: Combining labeled and unlabeled data with co-training. In: Proceedings of the 1998 Conference on Computational Learning Theory, pp. 92–100 (1998)
6. Cheng, S.J., Huang, Q.C., Liu, J.F., Tang, X.L.: A novel inductive semi-supervised SVM with graph-based self-training. In: Yang, J., Fang, F., Sun, C. (eds.) IScIDE 2012. LNCS, vol. 7751, pp. 82–89. Springer, Heidelberg (2013). https://doi.org/10.1007/978-3-642-36669-7_11
7. Dheeru, D., Karra Taniskidou, E.: UCI machine learning repository (2017). http://archive.ics.uci.edu/ml
8. Gammerman, A., Vovk, V., Vapnik, V.: Learning by transduction. In: Proceedings of the Fourteenth Conference on Uncertainty in Artificial Intelligence, pp. 148–155. Morgan Kaufmann, San Francisco, CA (2013)
9. Gasso, G., Zapien, K., Canu, S.: Sparsity regularization path for semi-supervised SVM. In: Proceedings of International Conference on Machine Learning and Applications, pp. 25–30 (2007)
10. Jiang, J., Ma, J., Chen, C., Jiang, X., Wang, Z.: Noise robust face image superresolution through smooth sparse representation. IEEE Trans. Cybern. **47**(11), 3991–4002 (2017)
11. Kothari, R., Jain, V.: Learning from labeled and unlabeled data. In: Proceedings of International Joint Conference on Neural Networks, pp. 2803–2808 (2002)
12. Le, H.M., Thi, H.A.L., Nguyen, M.C.: Sparse semi-supervised support vector machines by DC programming and DCA. Neurocomputing **153**, 62–76 (2015)
13. Liu, W., He, J., Chang, S.F.: Large graph construction for scalable semi-supervised learning. In: Proceedings of International Conference on Machine Learning, pp. 679–686 (2010)
14. Lu, Z., Peng, Y.: Robust image analysis by l1-norm semi-supervised learning. IEEE Trans. Image Process. **24**(1), 176–188 (2015)
15. Poggio, T., Girosi, F.: A sparse representation for function approximation. Neural Comput. **10**(6), 1445–1454 (1998)
16. Rabiner, L.R.: A tutorial on hidden markov models and selected applications in speech recognition. Proc. IEEE **77**(2), 257–286 (1989)
17. Refaeilzadeh, P., Tang, L., Liu, H.: Cross-Validation. In: Lui, L., Özsu, M.T. (eds.) Encyclopedia of Database Systems, pp. 532–538. Springer, New York (2016)
18. Schölkopf, B.: Sparseness of support vector machines. Mach. Learn. **4**(6), 1071–1105 (2008)
19. Shahshahani, B.M., Landgrebe, D.A.: The effect of unlabeled samples in reducing the small sample size problem and mitigating the hughes phenomenon. IEEE Trans. Geosci. Remoto Sens. **32**(5), 1087–1095 (1994)
20. Shental, N., Bar-Hillel, A., Hertz, T., Weinshall, D.: Gaussian mixture models with equivalence constraints. In: Constrained Clustering: Advances in Algorithms, Theory, and Applications, pp. 33–58. Chapman & Hall, London (2009)
21. Sun, Y., et al.: Discriminative local sparse representation by robust adaptive dictionary pair learning. IEEE Trans. Neural Networks Learn. Syst. 1–15 (2020)

22. Tan, J., Zhen, L., Deng, N., Zhang, Z.: Laplacian p-norm proximal support vector machine for semi-supervised classification. Neurocomputing **144**(1), 151–158 (2014)
23. Tang, B., Zhang, L.: Semi-supervised feature selection based on logistic I-RELIEF for multi-classification. In: Geng, X., Kang, B.-H. (eds.) PRICAI 2018. LNCS (LNAI), vol. 11012, pp. 719–731. Springer, Cham (2018). https://doi.org/10.1007/978-3-319-97304-3_55
24. Tang, B., Zhang, L.: Multi-class semi-supervised logistic I-RELIEF feature selection based on nearest neighbor. In: Yang, Q., Zhou, Z.-H., Gong, Z., Zhang, M.-L., Huang, S.-J. (eds.) PAKDD 2019. LNCS (LNAI), vol. 11440, pp. 281–292. Springer, Cham (2019). https://doi.org/10.1007/978-3-030-16145-3_22
25. Yakowitz, S.J.: An introduction to Bayesian networks. Technometrics **39**(3), 336–337 (1997)
26. Yedidia, J.H.S., Freeman, W.T., Weiss, Y.: Generalized belief propagation. Adv. Neural Inf. Process. Syst. **13**(10), 689–695 (2000)
27. Zhang, L., Zhou, W.: On the sparseness of 1-norm support vector machines. Neural Networks **23**(3), 373–385 (2010)
28. Zhang, L., Zhou, W., Chang, P., Liu, J., Yan, Z., Wang, T., Li, F.: Kernel sparse representation-based classifier. IEEE Trans. Signal Process. **60**, 1684–1695 (2012)
29. Zhang, L., Zhou, W., Li, F.: Kernel sparse representation-based classifier ensemble for face recognition. Multimedia Tools Appl. **74**(1), 123–137 (2015)
30. Zhou, D., Schölkopf, B.: Learning from labeled and unlabeled data using random walks. In: Rasmussen, C.E., Bülthoff, H.H., Schölkopf, B., Giese, M.A. (eds.) DAGM 2004. LNCS, vol. 3175, pp. 237–244. Springer, Heidelberg (2004). https://doi.org/10.1007/978-3-540-28649-3_29
31. Zhou, D., Schölkopf, B.: A regularization framework for learning from graph data. In: ICML Workshop on Statistical Relational Learning and Its Connections to Other Fields, pp. 132–137 (2004)
32. Zhu, J., Rosset, S., Hastie, T., Tibshirani, R.: 1-norm support vector machines. In: Proceedings of the 16th International Conference on Neural Information Processing Systems, vol. 16, no. 1, pp. 49–56 (2003)

Tailored Pruning via Rollback Learning

Xin Xia[1], Wenrui Ding[1], Li'an Zhuo[1], and Baochang Zhang[1,2(✉)]

[1] Beihang University, Beijing, China
bczhang@buaa.edu.cn
[2] Shenzhen Academy of Aerospace Technology, Shenzhen, China

Abstract. Structured pruning of networks has received increased attention recently for compressing convolutional neural networks (CNNs). Existing methods are not yet optimal, however, because there is still an obvious performance loss after pruning. This conflicts with the learning theory's parsimony principle that prefers the simpler model, when they explain the same facts, and thus maintain a consistent level of performance. In this paper, we propose rollback learning for pruning (RLP)-an algorithm to efficiently prune networks while maintaining performance. RLP provides a generic and tailored pruning framework that is easily applied to existing CNNs. Unlike existing work, we conditionally prune the kernels, and only those reaching the performance limit (high filter sparsity) are ultimately removed. This guarantees an optimized pruning by backtracking and updating the dense filters to their full potential. Our implementation adds rollback pruning into generative adversarial learning to compute a soft mask to scale the output of the network structures by defining a new objective function. The soft mask is backtracked and duplicated before and after the batch normalization layer, leading to an optimized fit to the structure pruning. Extensive experiments demonstrate that the proposed method significantly outperforms state-of-the-art methods resulting in a $0.25\times$ FLOPs reduction and 0.04% accuracy increase over the full-precision ResNet-18 model on CIFAR10, and an $0.18\times$ FLOPs reduction and 0.54% accuracy increase over the full-precision ResNet-50 model on ImageNet.

Keywords: Convolutional Neural Network · Network Pruning · Classification

1 Introduction

Parsimony is fundamental in machine learning, providing a theoretical foundation for state-of-the-art approaches, such as support vector machine [21], sparse representation classifier [23], and convolutional neural networks (CNNs). CNNs have become larger and larger in attempt to solve more complicated problems, yet they are initialized by practicing the principle of simplicity in designing the models based on shared parameters and they demonstrate superior performance on many computer vision tasks such as object detection and image classification [10]. To meet the requirement of portability and real time performance for

© Springer Nature Singapore Pte Ltd. 2020
H. Zhang et al. (Eds.): NCAA 2020, CCIS 1265, pp. 119–131, 2020.
https://doi.org/10.1007/978-981-15-7670-6_11

embedded platforms, CNNs must go return to their roots and strive for model compactness, while at the same time preserving performance gains. To this end, extensive efforts have been made to perform CNNs compression and acceleration, including low-rank factorization [1], parameter quantization and binarization [19]. One promising direction to further reduce the redundancy of CNNs is network pruning [3,9,16]. Pruning is orthogonal to most other methods and can be applied in addition to them. For this reason, pruning has attracted a great deal of attention in both academia and industry.

Neural network pruning focuses on removing network connections in non-structured and/or structured manner. Early work in non-structured pruning [5] proposed a kind of saliency measurement to remove redundant weights determined by the second-order derivative matrix of the loss function w.r.t. the weights. Han et al. [4] proposed an iterative thresholding method to remove unimportant weights with small absolute values. Guo et al. [2] present a connection splicing to avoid incorrect weight pruning, which can reduce the accuracy loss of the pruned network. In contrast, structured pruning can reduce the network size and achieve rapid inference without specialized packages. Li et al. [12] propose a magnitude-based pruning to remove filters and their corresponding feature maps by calculating the ℓ_1-norm of filters in a layer-wise manner. A Taylor expansion-based criterion is proposed in [18] to iteratively prune one filter and then fine-tune the pruned network. Unlike these multistage and layer-wise pruning methods, [15] and [9] use a sparse but uncontrollable soft mask to prune the network end-to-end. Recently, kernel sparsity and entropy have been introduced as indicators to efficiently prune networks [13]. The method however is hardware-dependent and restricted for the general purpose applications.

Though a great deal of improvement has been gained through network pruning, the performance of the resulting network is still suboptimal. This is because 1) a guided parsimony is not considered in previous works to prune the network in a controllable way; and 2) the learning process is not directly related to the pruning task, which deteriorates the performance due to the entanglement with classification.

In this paper, we propose a rollback learning approach to prune existing networks in a generic and principled framework. Unlike existing work, in our framework the filters are first conditionally pruned, and only those whose full potential (high filter sparsity) are already achieved, but still with a low performance are ultimately removed. The full potential is evaluated based on the sparsity, which is theoretically founded in our paper. In our optimization, a soft mask is learned end-to-end to scale the output of the network structures. Through network backtracking, the corresponding filters will be updated to their full potential. We also employ generative adversarial learning to guide the network pruning. We simply regard the pruning network as the student model, the full-precision network as teacher model, and then build adversarial model by setting the student model as the generator and the output of the teacher model as the true value. The contributions of this work are summarized as follows:

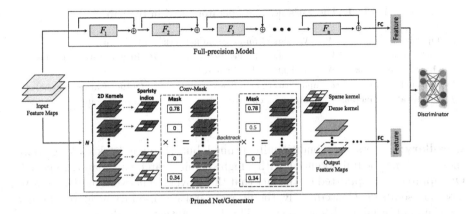

Fig. 1. The framework of rollback learning based on the generative adversarial learning (GAL). The rollback learning includes two steps: 1) a soft mask is learned but part of which will be backtracked based on the kernel sparsity; 2) rollback learning is implemented with generative adversarial learning based on the soft mask.

- We propose a principled framework to prune the network, based on a reasonable assumption that only the filters reaching their performance limit should really be pruned. The filter sparsity criterion is a practical and feasible evaluation of the filter performance limit.
- We introduce rollback pruning into generative adversarial learning to compute a soft mask, which will be subjected to backtracking to control the pruning process, leading to an optimized fit.

2 Rollback Learning

Our framework uses the sparsity of convolution filters to backtrack and tailor the pruning process through timely rollback. As shown in the experiments, the performance is improved even when the network is highly compressed. Figure 1 shows the two steps. In the first step, track through dense filters, and in the second step, the filters are updated to their full potential via rollback learning within the GAL framework. Table 1 summarizes the primary notations used in this paper.

2.1 Model Parsimony Based on Filter Sparsity

We address network pruning in a novel way by first defining which types of filters will be pruned. The obvious advantages of our framework comes from the well founded parsimony principle, in line with other state-of-the-art approaches.

Table 1. The variables and operators used in the paper.

x : Input image	W_B : Weights of ful-precision network	L : Loss function
m : Soft mask	W_G : Weights of pruned network	R : Regularization function
$\hat{m}$: backtracked mask	W_D : Weights of discriminator	
l : Layer index	i : Input feature map index	j : Output feature map index
$f(\cdot)$: Activation function	$*$: Convolution operator	η : Learning rate

Corollary: Filters are conditionally pruned but through backtracking, only those reaching the full potential, with a high sparsity, can be ultimately removed. Our approach is supported by the fact that the converged network model tends to have sparse filters. Consequently, the filters sparsity is used to evaluate whether their full potential has been achieved. This investigation provides us with a good performance evaluation measure for all filters based on the same parsimony criterion.

Here we show that whether the full potential of the kernel or filter has been achieved or not can be evaluated based on the kernel or filter sparsity. The proof is partially available but re-considered here for network pruning. While training a network, the update of W depends on the gradient $\frac{\partial L}{\partial W}$ and the weight decay $R(W)$:

$$W^{(t+1)} = W^{(t)} - \eta \frac{\partial L}{\partial W^{(t)}} - \eta_1 \frac{\partial R(W^{(t)})}{\partial W^{(t)}}, \tag{1}$$

where L represents the loss function and η is the learning rate. If the learning process converges, the term $\frac{\partial L}{\partial W^{(t)}}$ can be neglected.

$$W^{(t+1)} \approx W^{(t)} - \eta_1 \frac{\partial R(W^{(t)})}{\partial W^{(t)}}, \tag{2}$$

If $R(W^{(t)})$ is defined based on the ℓ_2-regularization, which makes the kernel sparse [11]. Thus, the kernels of a network tend to be sparse during training. Therefore, for a $4D$ convolution filter of size $N \times C \times K_h \times K_w$, we define the sparsity indices of the j-th filter to be used for our rollback learning as:

$$SI_j = \sum_{i=0}^{C-1} \left| W_{i,j}^l \right|_1, j \in 0, ..., N-1, \tag{3}$$

where C and N represent the number of input and output channels of current layer respectively. $W_{i,j}$ represents the 2D kernel of the i-th input channel of the j-th filter for the l-th layer with the size of $K_h \times K_w$. We note that the kernel or filter sparsity is widely used to evaluate the importance of the kernel or filter [13], which is re-investigated here and used to determine whether the performance limit is reached or not. A larger SI_j means the filter is less sparse, which also means that a performance potential needs to be further explored. Note that difference between kernel (2D) and filter (3D or 4D) lies in their dimensions.

2.2 Backtracking in Channel Selection

In this section, we introduce rollback learning to address the channel pruning problem. The channels are basic elements of any CNNs, and often have a significant redundancy. Due to simplicity and ease implementation, we use a mask m on output filters to guide channel pruning [9,15,22]. We have:

$$F_j^{l+1} = f(m_j \sum_i F_i^l * W_{i,j}^l), \qquad (4)$$

where F_j^l and F_j^{l+1} are the i-th input feature map and the j-th output feature map at the l-th layer. $*$ and $f(\cdot)$ refer to convolutional operator and activation respectively. The mask m can be learned end-to-end in the back propagation process. In our paper, we directly use the publicly available source code of [15] to compute m.

During training, we remove the feature map with a zero value in the soft mask that is associated with the corresponding channels in the current layer. However, there is a dilemma in the pruning-aware training in that the pruned filters are not evaluated according to their full performance before they are pruned. This can result in suboptimal pruning. To address this problem, we backtrack over the kernels with performance potential using the filter sparsity indices in Eq. 3 where we label each filter as:

$$\begin{aligned} S &= (S_0, ..., S_{N-1}) \\ &= (sign(SI_0 - \alpha), ..., sign(SI_{N-1} - \alpha)), \end{aligned} \qquad (5)$$

where $sign(x) = 1$ if $x > 0$, otherwise 0. α denotes the threshold, which will be elaborated in the experiment section below. We then define the backtracked mask $\hat{m}$ as:

$$\hat{m}_j = \begin{cases} max(m_j, 0.5) & S_j = 1 \\ m_j & S_j = 0 \end{cases}. \qquad (6)$$

This is the core concept of our rollback learning based on a tailored process. Based on Eq. 6, we can backtrack the less sparse filters ($S_j = 1$) by setting $\hat{m}_j$ to a non-zero value of at least 0.5. This means that the corresponding filters are still of a performance potential, which will be backtracked and updated again.

To achieve a higher pruning rate, we have to handle the batch normalization (BN) layer. This layer is widely used in a neural network with inputs that are zero mean/unit variance, and normalize the input layer by adjusting and scaling the activations. For the pruning task, the soft mask is applied after the convolutional layer but before the BN layer, which affects the structured pruning. To address this, we prune the output channel of the current layer and the input channel of next layer using the same mask simultaneously based on a clone process. It is implemented through the sign function, which is put after the activation function.

The advantages of backtracked operations are three-fold. First, the process is controllable, which optimizes the back propagation process by only pruning the filters that have reached their performance limit. Second, the rollback learning is generic and can be built on top of other state-of-the-art networks such as [15], for a better performance. Third, the evaluation criterion is theoretically founded.

2.3   Rollback Learning via GAL Based on $\hat{m}$

We backtrack the pruned filters during training, based on the backtracked mask $\hat{m}$. We use GAL as an example to describe our rollback learning algorithm. A pruned network obtained through GAL, with ℓ_1-regularization on the soft mask is used to mimic the full-precision baseline by aligning their outputs. The discriminator D with weights W_D is introduced to discriminate between the output of baseline and pruned network, and the pruned generator network G with weights W_G and soft mask m is learned together with D by using the knowledge from supervised features of baseline. Therefore, W_G, m, $\hat{m}$ and W_D are learned by solving the optimization problem as follows:

$$
\arg\min_{W_G, m} \max_{W_D, \hat{m}} \mathcal{L}_{Adv}(W_G, \hat{m}, W_D) + \mathcal{L}_{data}(W_G, \hat{m})
$$
$$
+ \mathcal{L}_{reg}(W_G, m, W_D), \tag{7}
$$

where $\mathcal{L}_{Adv}(W_G, \hat{m}, W_D)$ is the adversarial loss to train the two-player game between the baseline and the pruned network that compete with each other. This is defined as:

$$
\mathcal{L}_{Adv}(W_G, \hat{m}, W_D) = E_{f_b(x) \sim p_b(x)} \left[\log(D(f_b(x), W_D)) \right]
$$
$$
+ E_{f_g(x,z) \sim (p_g(x), p_z(z))} \left[\log(1 - D(f_g(x,z), W_D)) \right], \tag{8}
$$

where $p_b(x)$ and $p_g(x)$ represent the feature distributions of the baseline and the pruned network, respectively. $p_z(z)$ is the prior distribution of noise input z and the noise input z in the pruned network is used as the dropout. This dropout is active only while updating the pruned network. For notation simplicity, we omit z in the pruned network $f_g(x,z)$. In addition, $\mathcal{L}_{data}(W_G, \hat{m})$ is the data loss between output features from both the baseline and the pruned network, which is used to align the outputs of these two networks. Therefore, the data loss can be expressed using knowledge distillation loss [8]:

$$
\mathcal{L}_{data}(W_G, \hat{m}) = L_{KD}(W_G), \tag{9}
$$

where n is the number of mini-batch size. $f_b(.)$ and $f_g(.)$ are the full-precision and the pruned networks respectivly. Finally, $\mathcal{L}_{reg}(W_G, m, W_D)$ is a regularizer on W_G, m, and W_D, which can be spilt into three parts below:

$$
\mathcal{L}_{reg}(W_G, m, W_D) = R(W_G) + R_\lambda(m) + R(W_D), \tag{10}
$$

where $R(W_G)$ is the weight decay ℓ_2-regularization in the pruned network, which is defined as $\frac{1}{2}\|W_G\|_2^2$. $R_\lambda(m)$ is a sparsity regularizer for m with weight λ. If $m_i = 0$, we can remove the corresponding structure as its corresponding output has no significant contribution to the subsequent computation. In practice, we employ the widely used ℓ_1-regularization to constrain m, which is defined as $\lambda\|m\|_1$. $R(W_D)$ is a discriminator regularizer used to prevent the discriminator from dominating the training, while retaining the network. The RLP algorithm based on GAL is given in Algorithm 1.

Algorithm 1. Rollback learning for network pruning.

Require: The training dataset; the full-precision model with the pretrained weights W_B; sparisty factor λ; hyper-parameters such as initial learning rate η, weight decays, the maximum epoch T.

Ensure: A Pruned Model.

1: Initialize $W_G = W_B$, $m \sim \mathcal{G}aussian(0,1)$, and $t = 1$;
2: **repeat**
3:     **for** each batch **do**
4:         Use Eq. 3 to calculate kernel sparsity;
5:         The soft mask and corresponding filters are backtracked by using Eq. 6. The backtracked mask $\hat{m}$ is duplicated before and after the BN layer;
6:         Fix W_G and update W_D using Eq. 7. The loss is calculated based on the backtracked soft mask $\hat{m}$;
7:         Fix W_D and update W_G using Eq. 7. In this step, we calculate the soft mask m;
8:     **end for**
9: $t = t + 1$
10: **until** training error below threshold or the maximum epoch T.

---

3 Experiments

We have evaluated RLP on object classification using CIFAR10 and ILSVRC12 ImageNet datasets, using ResNets, MobileNetV2 as the backbone networks.

3.1 Datasets and Implementation Details

Datasets: CIFAR10 is a natural image classification dataset containing a training set of $50,000$ and a testing set of $10,000$ 32×32 color images distributed over 10 classes including: airplanes, automobiles, birds, cats, deer, dogs, frogs, horses, ships, and trucks.

The ImageNet object classification dataset is more challenging due to its large scale and greater diversity. There are $1,000$ classes, 1.2 million training images and 50k validation images.

Implementation Details: We use PyTorch to implement RLP with 4 Tesla V100 GPUs. The weight decay and the momentum are set to 0.0002 and 0.9 respectively. The hyper-parameter λ is selected through cross validation in the range $[0.01, 0.1]$ for ResNets and MobileNetv2. The drop rate in dropout is set to 0.1. The other training parameters are described on per experiment basis.

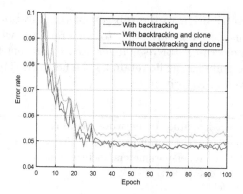

Fig. 2. Error rate on CIFAR10 with backtracking or not, with clone or not.

3.2   Experiments Setting on CIFAR10

We evaluated our method on CIFAR10 for two popular networks, ResNets and MobileNetV2. The stage kernels are set to 64-128-256-512 for ResNet-18 and 16-32-64 for ResNet-20/32/56/110. For all networks, we add the soft mask only after the first convolutional layer within each block to prune the output channel of current convolutional layer and input channel of next convolutional layer, simultaneously. The mini-batch size is set to be 128 for 100 epochs, and the initial learning rate is set to 0.01, scaled by 0.1 over 30 epochs. α is calculated based on sorting the sparsity indice of the filters, and used to determine the percentage to be backtracked. For example, if the percentage is 50%, α is set to the median of SI.

Fine-tuning: To achieve better performance, we fine-tune the pruned network with a smaller model size and faster forward process. We use the same batch size of 256 for 60 epochs as in training. The initial learning rate is changed to be 0.1 and scaled by 0.1 over 15 epochs. Note that a similar fine-tuning strategy was used in GAL.

3.3   Ablation Study

We use ResNet-18 on CIFAR10 for an ablation study to evaluate the effectiveness of RLP, including backtracking operation and the backtracking rate α.

Effect on Backtracking Operation: For a fair comparison, we train the pruning network with and without backtracking by using the same parameters. It is clearly shown in Fig. 2 that pruning with backtracking converge faster and more smoothly than that without backtracking, leading to a better performance on CIFAR10. As shown in Table 2, we obtain an error rate 4.70% and a 0.31× FLOPs reduction with backtracking, compared to the error rate is 5.15% and similar FLOPs reduction without backtracking. The performance is comparable with that of the full-precision model which validates the effectiveness of rollback

Table 2. Pruning results on CIFAR10 with backtracking or not, with clone or not. M means million (10^6).

Backtracking	Clone	Acc.(%)	'zero' rate(%)	FLOPs (M)
×	×	94.85	52.18	377.69
√	×	95.30	51.21	388.52
√	√	95.27	50.63	272.15

Table 3. Pruning results of ResNets and MobilenetV2 on CIFAR10. M means million (10^6). 'FT' means finetuning.

Model	ResNet-20[6]	RLP-0.50	RLP-0.75	RLP-0.95	
Accuracy/+FT (%)	91.73	89.48/90.13	90.97/91.28	91.64/**91.74**	
FLOPs (B)	40.55	13.78	22.10	28.69	
Model	ResNet-32[6]	RLP-0.50	RLP-0.75	RLP-0.95	
Accuracy/+FT (%)	92.63	91.50/91.85	91.84/92.27	92.49/**92.64**	
FLOPs (B)	68.86	28.40	35.91	49.93	
Model	ResNet-56[6]	L1[12]	CP[7]	NISP[22]	GAL-0.6[15]
Accuracy/+FT (%)	93.26	91.80	93.06	93.01	92.98/93.38
FLOPs (B)	125.49	90.90	62.00	81.00	78.30
Model	GAL-0.8[15]	RLP-0.50	RLP-0.75	RLP-0.95	
Accuracy/+FT (%)	90.36/91.58	92.45/93.01	93.01/93.12	93.25/**93.77**	
FLOPs (B)	49.99	47.87	62.34	88.16	
Model	ResNet-110[6]	GAL-0.1[15]	RLP-0.50	RLP-0.75	RLP-0.95
Accuracy/+FT (%)	93.68	92.65/93.59	93.19/93.37	93.37/93.46	93.65/**93.88**
FLOPs (B)	252.89	205.70	91.99	125.17	176.76
Model	MobileNet-V2[20]	RLP-0.50	RLP-0.75	RLP-0.95	
Accuracy/+FT (%)	94.43	94.21	**94.48**	**94.63**	
FLOPs (B)	91.15	47.40	66.54	83.86	

learning. Moreover, the experiment also demonstrates that the clone process leads to a higher FLOPs reduction ($388.52B$ v.s. $272.15B$, which has negligible impact on the accuracy (95.30% v.s. 95.27%) shown in Table 2.

Effect on α : α denotes the threshold corresponding to the backtrack rate[1]. Three backtrack rates are tested including 0.5, 0.75 and 0.95. We only backtrack once in the first batch of every epoch and the results with different α are shown in Tables 3 and 4. We note that RLP-a means that the backtrack rate is a% with an approximate pruning rate $1 - a$%.

3.4   Results on CIFAR10:

Two kinds of network are tested on the CIFAR10 database below.

ResNets: As is shown in Table 3, our method achieves new state-of-the-art results. Among other structured pruning methods for ResNet-56, our RLPs

---

[1] α and the backtrack rate are functionally equivalent.

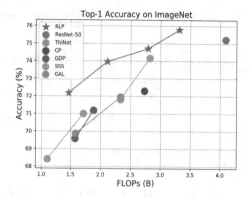

Fig. 3. Pruning results of ResNet-50 on ImageNet. Here, B means billion (10^9). The pretrained model of ResNet-50 our RLP use achieves the accuracy 75.24%[4](The model as our baseline downloads from https://download.pytorch.org/models/resnet50-19c8e357.pth.).

Table 4. Pruning results of ResNet-50 on ImageNet. B means billion (10^9).

Model	ResNet-50[6]	ThiNet-50[17]	ThiNet-30[17]	CP[7]	GDP-0.5[14]
Accuracy/+FT (%)	75.24	71.01	68.42	72.30	69.58
FLOPs (B)	4.09	1.71	1.10	2.73	1.57
Model	GDP-0.6[14]	SSS-26[9]	SSS-32[9]	GAL-0.1[15]	GAL-0.5[15]
Accuracy/+FT (%)	71.19	71.82	74.18	-/71.95	-/69.88
FLOPs (B)	1.88	2.33	2.82	2.33	1.58
Model	RLP-0.35	RLP-0.50	RLP-0.75	RLP-0.95	
Accuracy/+FT (%)	70.27/72.07	73.20/74.38	73.64/74.89	74.74/**75.78**	
FLOPs (B)	1.46	2.2	2.53	3.31	

achieves a smaller accuracy drop than GAL-0.6 (93.77% v.s. 93.38%) and GAL-0.8 (93.01% v.s. 91.58%) with a similar FLOPS reduction. Our RLP-0.5 has a larger FLOPs reduction than NISP (47.87M v.s. 81.00M), but with the same accuracy (93.01% v.s. 93.01%). These results demonstrate that RLPs are able to prune the network efficiently and generate a more compressed model with higher performance.

MobileNetV2: The pruning results of MobilnetV2 are summarized in Table 3. RLP-0.5 has a 0.48× FLOPs reduction, with a 0.22% accuracy drop compared to the full-precision model. With the pruning rate drop, RLP-0.75 achieves 0.05% accuracy increase, and RLP-0.95 achieves the highest performance 94.63% with 0.08× FLOPs reduction. These results mean that RLP is easily employed on depth-wise separable convolution, which is worth exploring in real applications.

3.5   Experiments ILSVRC12 ImageNet

For ILSVRC12 ImageNet, we test RLPs on ResNet-50 with different rates controlled by α. We train the network with the batch size of 256 for 60 epochs. The initial learning rate is set to 0.01 scaled by 0.1 over 15 epochs. Unlike the strategy on CIFAR10, we add the mask after both the first and second convolutional layers in each block. The fine-tuning process procedure follows the setting on CIFAR10 with the initial learning rate 0.00001.

Table 4 shows that RLPs achieve the state-of-the-art performance on the ILSVRC12 ImageNet. For ResNet-50, RLP-0.5 achieves a higher accuracy than SSS-32 with a higher FLOPs reduction (2.2B v.s. 2.82B). With the pruning rate decline, RLP-0.75 has a 0.71% greater improvement than SSS-32 with a similar FLOPs (2.53B v.s. 2.82B). Compared to GAL counterparts, our RLP-0.35 and RLP-0.5 achieve much better performance than GAL-0.5 (72.07% v.s. 69.88%) and GAL-0.1 (74.38% v.s. 71.95%) with similar FLOPs reduction. Furthermore, RLP-0.95 further shows a 0.18× FLOPs reduction while achieving a 0.54% improvement in the accuracy.

4   Conclusion

In this paper, we describe a rollback learning for pruning (RLP) algorithm to efficiently prune a network. Unlike previous work, we control the pruning process by only pruning the filters that reach their performance limit (high filter sparsity). Meanwhile, we update the dense filters to their full potential. By introducing rollback pruning into generative adversarial learning, we speed up the training and achieve an optimized fit to the structure pruning. Extensive experiments demonstrate that the proposed method significantly outperforms state-of-the-art methods with a 0.25× FLOPs reduction and 0.04% accuracy increase over its full ResNet-18 model on CIFAR10, and a 0.18× FLOPs reduction and 0.54% accuracy increase over its full ResNet-50 model on ImageNet.

Acknowledgements. Baochang Zhang is also with Shenzhen Academy of Aerospace Technology, Shenzhen, China, and he is the corresponding author. He is in part Supported by National Natural Science Foundation of China under Grant 61672079, Shenzhen Science and Technology Program (No.KQTD20-16112515134654). This study was supported by Grant NO. 2019JZZY011101 from the Key Research and Development Program of Shandong Province to Dianmin Sun.

References

1. Denton, E., Zaremba, W., Bruna, J., Lecun, Y., Fergus, R.: Exploiting linear structure within convolutional networks for efficient evaluation. arXiv preprint arXiv:1404.0736 (2014)
2. Guo, Y., Yao, A., Chen, Y.: Dynamic network surgery for efficient DNNs. In: Advances In Neural Information Processing Systems, pp. 1379–1387 (2016)

3. Han, S., Mao, H., Dally, W.J.: Deep compression: compressing deep neural networks with pruning, trained quantization and huffman coding. Fiber **56**(4), 3–7 (2015)
4. Han, S., Pool, J., Tran, J., Dally, W.: Learning both weights and connections for efficient neural network. In: Advances in Neural Information Processing Systems, pp. 1135–1143 (2015)
5. Hassibi, B., Stork, D.G.: Second order derivatives for network pruning: optimal brain surgeon. In: Advances in Neural Information Processing Systems, pp. 164–171 (1993)
6. He, K., Zhang, X., Ren, S., Sun, J.: Deep residual learning for image recognition. In: Proceedings of CVPR, pp. 770–778 (2016)
7. He, Y., Zhang, X., Sun, J.: Channel pruning for accelerating very deep neural networks. In: Proceedings of the IEEE International Conference on Computer Vision, pp. 1389–1397 (2017)
8. Hinton, G., Vinyals, O., Dean, J.: Distilling the knowledge in a neural network. Comput. Sci. **14**(7), 38–39 (2015)
9. Huang, Z., Wang, N.: Data-driven sparse structure selection for deep neural networks. In: Ferrari, V., Hebert, M., Sminchisescu, C., Weiss, Y. (eds.) ECCV 2018. LNCS, vol. 11220, pp. 317–334. Springer, Cham (2018). https://doi.org/10.1007/978-3-030-01270-0_19
10. Krizhevsky, A., Sutskever, I., Hinton, G.E.: ImageNet classification with deep convolutional neural networks. In: Advances in Neural Information Processing Systems, pp. 1097–1105 (2012)
11. Krogh, A., Hertz, J.A.: A simple weight decay can improve generalization. In: Advances in Neural Information Processing Systems, pp. 950–957 (1992)
12. Li, H., Kadav, A., Durdanovic, I., Samet, H., Graf, H.P.: Pruning filters for efficient convnets. arXiv preprint arXiv:1608.08710 (2016)
13. Li, Y., et al.: Exploiting kernel sparsity and entropy for interpretable CNN compression. In: Proceedings of CVPR, pp. 2800–2809 (2019)
14. Lin, S., Ji, R., Li, Y., Wu, Y., Huang, F., Zhang, B.: Accelerating convolutional networks via global & dynamic filter pruning. In: IJCAI, pp. 2425–2432 (2018)
15. Lin, S., et al.: Towards optimal structured CNN pruning via generative adversarial learning. In: Proceedings of CVPR, pp. 2790–2799 (2019)
16. Liu, Z., Li, J., Shen, Z., Huang, G., Yan, S., Zhang, C.: Learning efficient convolutional networks through network slimming. In: Proceedings of the IEEE International Conference on Computer Vision, pp. 2736–2744 (2017)
17. Luo, J.H., Wu, J., Lin, W.: ThiNet: a filter level pruning method for deep neural network compression. In: Proceedings of the IEEE International Conference on Computer Vision, pp. 5058–5066 (2017)
18. Molchanov, P., Tyree, S., Karras, T., Aila, T., Kautz, J.: Pruning convolutional neural networks for resource efficient inference. arXiv preprint arXiv:1611.06440 (2016)
19. Rastegari, M., Ordonez, V., Redmon, J., Farhadi, A.: XNOR-Net: ImageNet classification using binary convolutional neural networks. In: Leibe, B., Matas, J., Sebe, N., Welling, M. (eds.) ECCV 2016. LNCS, vol. 9908, pp. 525–542. Springer, Cham (2016). https://doi.org/10.1007/978-3-319-46493-0_32
20. Sandler, M., Howard, A., Zhu, M., Zhmoginov, A., Chen, L.C.: MobileNetV 2: inverted residuals and linear bottlenecks. In: Proceedings of CVPR, pp. 4510–4520 (2018)
21. Sch, C., Laptev, I., Caputo, B.: Recognizing human actions: a local SVM approach. In: International Conference on Pattern Recognition (2004)

22. Yu, R., et al.: NISP: pruning networks using neuron importance score propagation. In: Proceedings of CVPR, pp. 9194–9203 (2018)
23. Zhang, B., Perina, A., Murino, V., Del Bue, A.: Sparse representation classification with manifold constraints transfer. In: Proceedings of CVPR, June 2015

Coordinative Hyper-heuristic Resource Scheduling in Mobile Cellular Networks

Bei Dong[1,2(✉)], Yuping Su[1,2], Yun Zhou[3], and Xiaojun Wu[1,2]

[1] Key Laboratory of Modern Teaching Technology,
Ministry of Education, Xi'an 710062, China
[2] School of Computer Science, Shaanxi Normal University,
Xi'an 710119, China
dongbei@snnu.edu.cn
[3] School of Education, Shaanxi Normal University, Xi'an 710119, China

Abstract. In this work, a novel parallel-spacing coordinative hyper-heuristic algorithm is proposed for solving radio resource scheduling problem in mobile cellular networks. The task of this problem is to minimize the required bandwidth to satisfy diverse channel demand from all micro cellular, while without interference violation. Based on the undirected weighted graph generated by each network topology, six problem-related low-level heuristics are constructed. In the high-level heuristic space, a group of evolutionary strategies are implemented to manage the searching process in the low-level solution space. In classical hyper-heuristic framework, exploration ability might be partially decreased by non-single mapping from heuristic space to solution space. To that end, a group of problem distinctive local search mechanisms are developed and executed on elite population in the solution space parallelly and periodically. Effectiveness of parallel space coordinative searching technique is verified on a set of real-world problems, and the comparison results show that the proposed parallel-spacing coordinative hyper-heuristic algorithm works effectively on most problems.

Keywords: Hyper-heuristic · Resource scheduling problem · Mobile cellular network

1 Introduction

With the rapid developing of wireless communication technology, the number of wireless device and speed of wireless traffic data is increasing dramatically. Radio spectrum is the most important and rare resource, which is strictly managed and controlled by government. To meet the tremendous spectrum demand from mobile users and wireless services, it is extremely important and necessary to effectively schedule and utilize radio resource to alleviate the spectrum scarcity problem [1].

In modern wireless communication area, cellular network is still playing an important role on account of prominent coverage and stable performance. To guarantee maximum coverage and minimum overlapping, the entire service area of a network is divided into a set of non-overlapping hexagon cellular. Channel multiplexing is adopted

H. Zhang et al. (Eds.): NCAA 2020, CCIS 1265, pp. 132–143, 2020.
https://doi.org/10.1007/978-981-15-7670-6_12

to make groups of mobile users can use the same channels simultaneously if some interference constraints are satisfied, thus to improve spectrum utilization efficiency and mitigate interference [2]. The task of resource scheduling problem (RSP) in multi cellular networks is described as satisfying all the demand from each cell with minimum spectrum resource consumption while no interference constraints are violated.

Resource scheduling problem in wireless multi cellular networks is equivalent to the graph coloring problem which has proved to be NP-complete [3]. Research on solving RSP has attracted so much interesting, and a lot of methods have been presented to find global optimum or at least local optimal resource scheduling strategies [4–8]. These approaches can be categorized as transfer methods, heuristic algorithms and hyper heuristic approaches. Firstly, in transfer methods, resource scheduling problem was always transformed into a graph coloring problem, SAT problem and other classical optimization problems which already have been solved by a variety of deterministic methods [9–11]. Secondly, resource scheduling problem is modeled as a single objective optimization problem minimizing the number of required channels directly. To tackle this problem, simulated annealing [7], tabu search [12, 13, 16], genetic algorithm [14, 15], and other heuristic methods are applied. Thirdly, hyper-heuristic approaches are a novel kind of framework that is intended to solve RSP rapidly. In [7], a great deluge hyper heuristic algorithm is proposed to obtain near optimal solutions to the RSP. To further enhance searching ability, four problem dependent local search strategies are defined and served as low level heuristic in [8]. The aforementioned hyper heuristic algorithms show effectively solving the RSP, but have one thing in common, only searching in a single space (in heuristic space or solution space only). It is known that mapping from heuristic space to solution space is not one-to-one when applying hyper-heuristic algorithm, which may bring a decline in search and convergence in the later searching process. To bridge this gap, parallel collaborative searching of these two space must be regarded to improve efficiency and effectiveness.

In this paper, a novel parallel space coordinative hyper-heuristic algorithm (PSHHA) is introduced to tackle the resource scheduling problem in multi cellular networks. Different from previous hyper-heuristic based approaches, the proposed PSHHA searches in the heuristic and solution searching space simultaneously. In the early stage, searching is focused on the high level heuristic space to search for the permutations of low level heuristic sequences using evolutionary strategies. To facilitate the searching ability and speed up convergence, four local search operators are applied in the low-level solution space on the elitism individuals periodically in the later generations. Performance of the proposed algorithm is tested on a set of real benchmark problems, and the result demonstrates that with the parallel space coordinative technique, our method works well on most of the test problems especially on most difficult problems.

The rest of the paper is organized as follows: the mathematical formulation of the resource scheduling problem in multi cellular networks is described in Sect. 2. The proposed parallel-space coordinative hyper-heuristic algorithm is presented in Sect. 3. Simulation results on well-known benchmark problems are conducted in Sect. 4. Concluding remarks are given in Sect. 5.

2 Problem Formulation

Considering a multi cellular network which composed of N cellular is deployed in the service area and the total available frequencies are assumed to be divided into a series of channels. Each cellular has a different distribution of spectrum demands and the number of available mobile users. The objective of the resource scheduling problem is to determine optimal frequency schedule strategy which has minimum bandwidths required to fulfill the channel demand and no interference constraints are violated. The mathematical model of the resource scheduling problem is as follows:

$$\min_M F = \{f_{ij}\}_{M*N} \tag{1}$$

$$S.T. \quad \sum_{m=1}^{M} f_{mi} = d_i \tag{2}$$

$$|f_k - f_l| \geq c_{ij}, \text{ for all } i,j,k,l \text{ except}(i=j,k=l) \tag{3}$$

D is the resource demand vector, $D = \{d_1, d_2, \cdots, d_N\}(1 \leq i \leq N)$. d_i represents the bandwidth demand from ith cell.

C is a symmetric compatibility matrix $C = \{c_{ij}\}_{N*N}$, c_{ij} indicates the minimum channel separation distance to avoid interference between cell i and cell j.

F is channel allocation matrix (resource scheduling strategy), $F = \{f_{ij}\}_{M*N}$, and $f_{ij} = 1$ denotes channel i is assigned to cell j, and $f_{ij} = 0$, otherwise.

3 Parallel-Space Coordinative Hyper-heuristic Algorithm

In this section, we give a brief introduction of hyper-heuristic algorithm, then details of the proposed parallel-space coordinative hyper-heuristic algorithm is presented.

3.1 Hyper-heuristic Algorithm

Hyper-Heuristic algorithms are a kind of novel optimization methods and have been successfully applied in solving complex optimization problems [17–21]. Different from genetic algorithm, simulated annealing, tabu search and other classical heuristic approaches, there exists two searching space in hyper-heuristic framework: high level heuristic space and low-level solution space. A set of problem specific low-level heuristics (LLHs) are managed or generated by a high-level heuristic which is usually problem-independent during the optimization process. Concretely, feasible solutions of the original problem are coded by groups of LLHs sequences; then high-level heuristic are applied on the heuristic searching space (composed of heuristic sequences) to search for better LLH sequence in order to obtain optimal solution to the original problem.

For the sake of easier implementation and convenient transferring, hyper-heuristic has been widely used in complex combinatorial optimization problems, i.e., educational timetabling [22–24] and so on. The drawback of classical hyper-heuristic algorithm mainly generated by non-unique mapping from heuristic space to solution space. In the later searching process, it is easy to trap into local optimum due to degeneration of exploration ability.

3.2 Parallel-Spacing Coordinative Hyper-heuristic Algorithm

To improve convergence and searching capacity of classical hyper-heuristic algorithm, a novel parallel-space coordinative hyper-heuristic framework is proposed to solve the resource scheduling problem in multi cellular networks. In the novel hyper-heuristic framework, heuristic space and solution space searching are jointly worked: in the former stage, a set of evolutionary strategies is performed mainly in heuristic space to search for potential heuristic sequences to enhance exploration. To overcome the non-unique mapping problem in classical hyper-heuristic model, a group of local search strategies is generated and executed periodically on elitism population directly to escape from a local optimum and explore global optimal solutions in the solution space. Framework of the PSHHA algorithm is presented in Fig. 1.

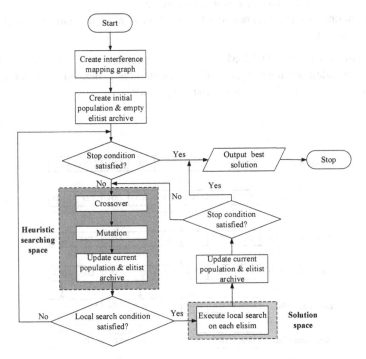

Fig. 1. Framework of PSHHA

Low-Level Heuristics

As mentioned above, resource scheduling problem in multi cellular networks can be modeled as a graph coloring problem. To effectively use knowledge of graph theory and problem-specific priori information, six low level heuristics are constructed as follows.

- LD (Largest Degree): in the interference mapping graph generated from the original multi cellular network, cells that have the largest degree means they has the most conflict constraints when assigning spectrum. This heuristic aims to consider the hardest condition firstly.
- LDM (Largest Demand): cells are considered in descending order with their channel demands, and the one that has the largest demand will be selected through this heuristic.
- LWD (Largest Weight Degree): weight along each link in the generated interference mapping graph representing the minimum spectrum separation between these two cells in order to avoid interference. This heuristic will choose the cell that has the largest spectrum constraints.
- LCD (Largest Conflict Degree): cells are dynamically sorting based on the conflict constraints with the assigned ones in this heuristic mechanism.
- LDs (Largest Distance): cells are sorting in descending order based on the physical distance between the former cell and themselves.
- RO (Random Ordering): generate a permutation of all cells randomly, and choose anyone from it.

Among this six LLHs, LD, LDM are static low-level heuristics which means the cell orderings generated are unchanged while cell orderings generated by LCD, LDs are dynamically changed in the spectrum assigning process.

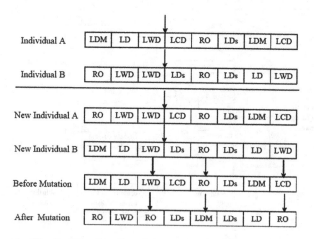

Fig. 2. Diagram of crossover and mutation operator

Evolutionary Strategies

In the proposed PSHHA, single-point crossover and multi-point mutation is adopted as evolutionary strategies in the heuristic searching space to generate new heuristic sequences. Figure 2 gives a diagram of the crossover and mutation mechanism.

It is worth mentioning that modification must be executed after crossover and mutation. When heuristic LDs or LCD is at the first position of a sequence, it would be replaced by one of other four LLHs; otherwise, an error will occur in cell selection.

Local Search Mechanism

In the proposed PSHHA, local search on elitist population in the solution space is triggered in the later searching stage when some conditions satisfied. As it mentioned above, non-unique mapping from heuristic space to solution space will decrease convergence and diversity. Local search mechanism should be performed in the solution space directly. Considering characteristics of the resource allocation problem, four problem-specific local search mechanisms are constructed.

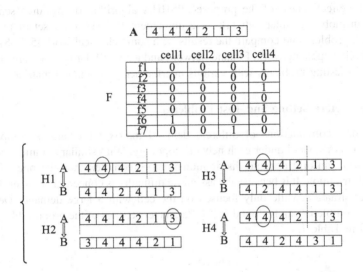

Fig. 3. Local search operators

In PSHHA, each high-level heuristic sequence L is corresponding to a communication request sequence A and channel allocation matrix F in solution space. Four local search strategies are given as follows:

- H1: find the communication request with the largest channel number, delete this communication request and insert it in a random position in A. Reassign the channels.
- H2: delete one communication request randomly, then insert it in a random position in A. Reassign the channels.
- H3: find the communication request with the largest channel number, delete this communication request and insert it in a random position in A. Different from H1, sequence that exist before the insertion point is reordered when reassigning channels.
- H4: find the communication request with the largest channel number, delete this communication request and insert it in a random position in A. Different from H3, sequence that exists after the insertion point is reordered when reassigning channels.

Details of the four local search operators are shown in Fig. 3.

It should be recognized that invalid searching will appear which means that the same individuals may produce after performing local search. Communication requests in the same cellular share a common encoding, and their orders have no effect on the assignment result. Ineffective searching phenomenon will arise when request ordering in same cellular is changed. To avoid this, prejudgment is used before executing local search. Once invalid phenomenon happened, select insertion point again.

4 Simulation Results

To test the effectiveness of the proposed PSHHA algorithm on resource scheduling problem in mobile cellular networks, we implement PSHHA on a set of real-world benchmark problem and compare the results with some classical work [5, 7, 8]. In this section, Philadelphia network, Finland network and a 55-cellular network are chosen to generate 20 testing problems with various interference and demand matrixes.

4.1 Parameters Setting and Test Problems

In this work, combinations of different demand matrix with various compatibility matrixes are considered under each network topology. With similar parameter settings, problem with uniform demand distribution is more complex than the one with non-uniform distribution. It is because the assigning difficulty is concentrated on each cell in the former situation while only focused on the cell with a large demand. Details of benchmark problems can be found in [5]. Parameter settings of the proposed PSHHA is suggested in Table 1.

Table 1. Parameter settings of PSHHA

Population size (P_n)	20
Elite population size (P_e)	$P_e/2$
Crossover probability (P_c)	0.95
Mutation probability (P_m)	0.2
Clonal size (q_c)	0.2
Maximum generation (G_{max})	5000

4.2 Effectiveness of Cooperative Searching Between Heuristic and Solution Space

To enhance exploitation ability of PSHHA, local search operators are performed on elite population during the optimization process periodically. We execute PSHHA and PSHHA without local search operator (NLS-PSHHA) on seven benchmark problems to verify the effectiveness of cooperation searching between heuristic and solution space. The parameters are setting as: population size $P_n = 5$, elite population size $P_e = 5$, clonal size $q_c = 2$, crossover probability $P_c = 0.95$, mutation probability $P_m = 0.2$, maximum generation number $G_{max} = 100$. The comparison results are presented in Fig. 4.

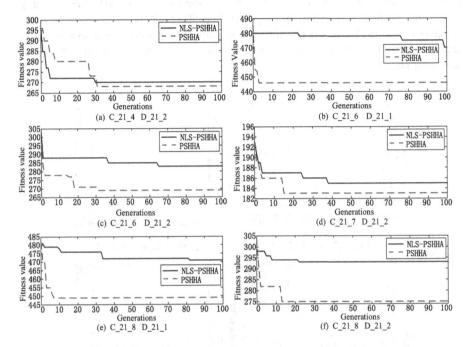

Fig. 4. Comparison results on cooperative searching strategy

From the comparison results shown in Fig. 4, PSHHA performs well than NLS-PSHHA, which means that the proposed cooperative searching strategy works well on convergence improvement. In Fig. 4(a), PSHHA can reach to a solution which is very close to the theoretic optimal solution (265) in 30 iterations while NLS-PSHHA still trap into local minima after 100 generations. Advantages are shown more obvious in Fig. 4(b) that a potential solution can be reached in 4 iterations while final optimal solution searched by NLS-PSHHA is much higher than that generated by PSHHA.

In the NLS-PSHHA, best solution found by global search in the later stage may far away from the true optimal solution due to the random mapping of low-level heuristic (RO)which will cause prematurity of the convergence. On the contrary, by introducing elite local search operator on the solution space periodically and directly, the above problem can be alleviated and searching ability can also be improved.

4.3   Performance of Local Search Mechanism

In the PSHHA, each elitism chooses one local search operator randomly other than a fixed one. To test performance of this integrated local search strategies, we execute

PSHHA with each local search mechanism independently (labeled as LS1, LS2, LS3, LS4) and integration mechanism group (labeled as LS5), the comparison results are given in Fig. 5.

Experimental results show that these five local search mechanisms perform the same on the test problem in Fig. 5(a). LS5 used in the proposed PSHHA has advantage results on benchmark problems in Fig. 5(b), Fig. 5(c) and Fig. 5(d) which are much complex problems. Based on the observation above, cooperative local search mechanism has better performance than adopt one local search operator independently in the searching process.

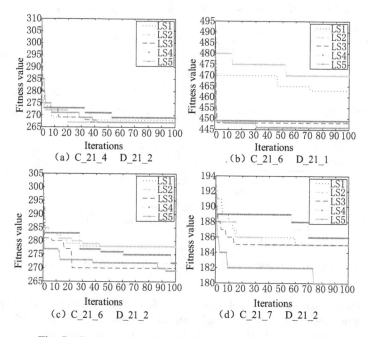

Fig. 5. Comparison results of different local search operators

4.4 Comparison Results on 20 Real-World Problems

The comparison results of PSHHA with other classical methods on 20 real-world benchmark problems are presented in Table 2. Each algorithm ran 10 times independently and the best solution is given.

The lower bound of each benchmark problem is given in Table 2 which is computed as $C_{ii} \times (d_i - 1)$, where d_i is the maximum channel demand in the demand matrix, C_{ii} is the minimum spectrum interval within each cellular.

It is clearly from Table 2 that global optima can be reached by PSHHA and other methods on most simple benchmark problems. On test problem 8, the theoretic optimal solution 265 can be obtained by PSHHA while the best value searched by other algorithms is larger than 274. On the complicated problems, advantages between PSHHA and other algorithms are much more obvious. Theoretic optimum of test problem 11 is 381, the optimal solution found by PSHHA is improved of 10.2 compared to the value given in [5], 4.19% compared to [7] and 3.07% compared to [8]. For the harder problem of 14, 15, 16, 20, the proposed PSHHA also has better performance than the related approaches, that means more spectrum resource can be saved which is very important in the wireless communication area.

Table 2. Comparison results on 20 problems

Index	Compatibility matrix	Demand vector	Lower bound	GOU	PSHHA	KENA	KENB	AUD
1	C_21_1	D_21_1	533	533	533	533	∅	534
2	C_21_1	D_21_2	309	309	309	309	∅	309
3	C_21_2	D_21_1	533	533	533	533	∅	535
4	C_21_2	D_21_2	309	309	309	309	∅	312
5	C_21_3	D_21_1	457	457	457	457	∅	∅
6	C_21_3	D_21_2	265	265	265	265	∅	∅
7	C_21_4	D_21_1	457	457	457	457	457	∅
8	C_21_4	D_21_2	265	280	265	274	270	∅
9	C_21_5	D_21_1	381	381	381	381	∅	382
10	C_21_5	D_21_2	221	221	221	221	∅	222
11	C_21_6	D_21_1	381	463	424	440	437	427
12	C_21_6	D_21_2	221	273	265	266	261	253
13	C_21_7	D_21_1	305	305	305	305	305	∅
14	C_21_7	D_21_2	177	197	180	187	182	∅
15	C_21_8	D_21_1	305	465	414	451	435	∅
16	C_21_8	D_21_2	177	278	248	275	268	∅
17	C_25_1	D_25_3	21	73	73	73	73	∅
18	C_25_1	D_25_4	89	121	200	200	200	∅
19	C_55_1	D_55_5	309	309	309	309	309	∅
20	C_55_1	D_55_6	71	79	71	72	71	∅

5 Concluding Remarks

In this paper, a novel coordinative hyper-heuristic framework is presented to solve resource scheduling problem in mobile cellular networks. Global search in high level heuristic space and local search in low level solution space is cooperated working during the whole process. To speed up convergence and escape from local optimum, four problem-dependent local search operators are designed and randomly executed on the elite individual when some conditions are satisfied. Experimental results demonstrate that the proposed parallel-spacing coordinative hyper-heuristic algorithm works effectively on the resource scheduling problem, and has better performance over some classical methods. Advantages are more distinct with the increase of problem difficulty.

Acknowledgement. This work was supported by the National Science Foundation of China (Grant No. 61703258, 61701291 and U1813205), the China Postdoctoral Science Foundation funded project (Grant No. 2017M613054, and 2017M613053) and the Shaanxi Postdoctoral Science Foundation funded project (Grant No. 2017BSHYDZZ33).

References

1. Teshome, A.K., Kibret, B., Lai, D.: A review of implant communication technology in WBAN: progresses and challenges. IEEE Rev. Biomed. Eng. **12**, 88–99 (2018)
2. Asadi, A., Wang, Q., Mancuso, V.: A survey on device-to-device communication in cellular networks. IEEE Commun. Surv. Tutorials **16**(4), 1801–1819 (2014)
3. Battiti, R., Bertossi, A., Cavallaro, D.: A randomized saturation degree heuristic for channel assignment in cellular radio networks. IEEE Trans. Veh. Technol. **50**(2), 364–374 (1999)
4. Castaneda, E., Silva, A., Gameiro, A., Kountouris, M.: An overview on resource allocation techniques for multi-user mimo systems. IEEE Commun. Surv. Tutor. **19**(1), 239–284 (2017)
5. Chakraborty, G.: An efficient heuristic algorithm for channel assignment problem in cellular radio networks. IEEE Trans. Veh. Technol. **50**(6), 1528–1539 (2001)
6. Coskun, C.C., Davalioglu, K., Ayanoglu, E.: Three-stage resource allocation algorithm for energy-efficient heterogeneous networks. IEEE Trans. Veh. Technol. **66**(8), 6942–6957 (2017)
7. Kendall, G., Mohamad, M.: Channel assignment in cellular communication using a great deluge hyper-heuristic. In: Proceedings of 12th IEEE International Conference on Networks (ICON 2004), vol. 2, pp. 769–773. IEEE (2004)
8. Kendall, G., Mohamad, M.: Channel assignment optimisation using a hyper heuristic. In: 2004 IEEE Conference on Cybernetics and Intelligent Systems, vol. 2, pp. 791–796. IEEE (2004)
9. Sharma, P.C., Chaudhari, N.S.: Channel assignment problem in cellular network and its reduction to satisfiability using graph k-colorability. In: 2012 7th IEEE Conference on Industrial Electronics and Applications (ICIEA), pp. 1734–1737. IEEE (2012)
10. Zhao, L., Wang, H., Zhong, X.: Interference graph based channel assignment algorithm for D2D cellular networks. IEEE Access **6**, 3270–3279 (2018)
11. Yu, J., Han, S., Li, X.: A robust game-based algorithm for downlink joint resource allocation in hierarchical OFDMA femtocell network system. IEEE Trans. Syst. Man Cybern. Syst. **50**(7), 2445–2455 (2020)
12. Peng, Y., Wang, L., Soong, B.H.: Optimal channel assignment in cellular systems using tabu search. In: 14th IEEE Proceedings on Personal, Indoor and Mobile Radio Communications 2003, PIMRC 2003, vol. 1, pp. 31–35. IEEE (2003)
13. Gözüpek, D., Genç, G., Ersoy, C.: Channel assignment problem in cellular networks: a reactive tabu search approach. In: ISCIS, pp. 298–303 (2009)
14. Khanbary, L.M.O., Vidyarthi, D.P.: A GA-based effective fault-tolerant model for channel allocation in mobile computing. IEEE Trans. Veh. Technol. **57**(3), 1823–1833 (2008)
15. Lima, M.A., Araujo, A.F., Cesar, A.C.: Adaptive genetic algorithms for dynamic channel assignment in mobile cellular communication systems. IEEE Trans. Veh. Technol. **56**(5), 2685–2696 (2007)
16. Audhya, G.K., Sinha, K.: A new approach to fast near-optimal channel assignment in cellular mobile networks. IEEE Trans. Mob. Comput. **12**, 1814–1827 (2013)

17. Burke, E.K., et al.: Hyper-heuristics: a survey of the state of the art. J. Oper. Res. Soc. **64**(12), 1695–1724 (2013). https://doi.org/10.1057/jors.2013.71
18. Sabar, N.R., Ayob, M., Kendall, G., Qu, R.: Grammatical evolution hyper-heuristic for combinatorial optimization problems. IEEE Trans. Evol. Comput. **17**(6), 840–861 (2013)
19. Amaya, I., et al.: Enhancing selection hyper-heuristics via feature transformations. IEEE Comput. Intell. Mag. **13**(2), 30–41 (2018)
20. Tyasnurita, R., Ozcan, E., John, R.: Learning heuristic selection using a time delay neural network for open vehicle routing. In: 2017 IEEE Congress on Evolutionary Computation (CEC), pp. 1474–1481. IEEE (2017)
21. Venkatesh, P., Singh, A.: A hyper-heuristic based artificial bee colony algorithm for k-interconnected multi-depot multi-traveling salesman problem. Inf. Sci. **463**, 261–281 (2018)
22. Burke, E.K., McCollum, B., Meisels, A., Petrovic, S., Qu, R.: A graph-based hyper heuristic for educational timetabling problems. Eur. J. Oper. Res. **176**(1), 177–192 (2007)
23. Liu, Y., Mei, Y., Zhang, M., Zhang, Z.: Automated heuristic design using genetic programming hyper-heuristic for uncertain capacitated arc routing problem. In: Proceedings of the Genetic and Evolutionary Computation Conference, pp. 290–297. ACM (2017)
24. Nguyen, S., Zhang, M.: A PSO-based hyper-heuristic for evolving dispatching rules in job shop scheduling. In: 2017 IEEE Congress on Evolutionary Computation (CEC), pp. 882–889. IEEE (2017)

Latent Sparse Discriminative Learning for Face Image Set Classification

Yuan Sun[1] , Zhenwen Ren[1,2(✉)] , Chao Yang[1(✉)], and Haoyun Lei[1]

[1] Department of National Defence Science and Technology,
Southwest University of Science and Technology, Mianyang 621010, China
sunyuan_work@163.com, ychao1983@126.com, swustlhy@163.com
[2] Department of Computer Science and Engineering, Nanjing University of Science and Technology, Nanjing 210094, China
rzw@njust.edu.cn

Abstract. Image set classification has drawn much attention due to its promising performance to overcome various variations. Recently, the point-to-point distance-based methods have achieved the state-of-the-art performance by leveraging the distance between the gallery set and the probe set. However, there are two drawbacks that need to be defeated: 1) they do not fully exploit the discrimination that exists between different gallery sets; 2) they face the great challenge of high computational complexity as well as multi-parameters, usually caused by some obvious sparse constraints. To address these problems, this paper proposes a novel method, namely *latent sparse discriminative learning* (LSDL), for face image set classification. Specifically, a new term is proposed to exploit the relations between different gallery sets, which can boost the set discrimination so as to improve performance. Moreover, we use a latent sparse constraint to reduce the trade-off parameters and computational cost. Furthermore, an efficient solver is proposed to solve our LSDL. Experimental results on three benchmark datasets demonstrate the advantages of our propose.

Keywords: Image set classification · Sparse representation · Face recognition · Video-based face classification · Point-to-point distance

1 Introduction

Different from traditional single image classification tasks [2, 8], image set classification models each image set by appropriate means and measure the dissimilarity between two sets to classify [2, 5, 13, 17, 22]. Overall, image set classification can not only effectively deal with a mount of appearance variations (*e.g.*, illumination, expression, pose, and disguise changes), but also enhance the discrimination power, accuracy and robustness [10, 22]. Hence, it is more promising in practical face classification applications.

Over the last decade, plentiful methods [1, 4, 5, 7, 17, 21–27] for image set classification have been proposed to solve two key problems: how to construct the

© Springer Nature Singapore Pte Ltd. 2020
H. Zhang et al. (Eds.): NCAA 2020, CCIS 1265, pp. 144–156, 2020.
https://doi.org/10.1007/978-981-15-7670-6_13

proper model for each image set and how to define a suitable similarity measure to obtain the between-set distance. Amongst all the existing methods, the point-to-point distance-based methods have gained the most widely attention [3,9]. However, these methods have some drawbacks: 1) they do not consider the relations between the whole gallery sets and each gallery set; and 2) they usually introduce the sparse representation to obtain a sparse solution. The former results in performance loss, and the latter brings out computational burden and multi-parameters.

To address the above problems, this paper presents an efficient and effective face image set classification method, termed *latent sparse discriminative learning* (LSDL). In summary, LSDL has the following contributions:

- To improve discrimination of different gallery sets, it simultaneously minimizes the nearest points distance between the whole gallery sets and the probe set, and the distances between the whole gallery set and each independent gallery set.
- It proposes the latent sparse normalization with capped simplex constraint to replace the sparse constraint for reducing the computational cost and the numbers of trade-off parameters, meanwhile, the robustness of sparse representation can be preserved.
- By evaluating LSDL on some benchmark datasets, the promising results have achieved.

2   Related Work

In recent years, a popular model is proposed by measuring distance of the between-set nearest points, namely point-to-point distance model. To date, some representative methods have been proposed [1,4,7,10,11,17,21–27]. For example, the distances of affine hull-based image set distance (AHISD) [1] and convex hull-based image set distance (CHISD) [1] can be used to compute the similarity between two affine/convex hulls of image sets. For solving the overlarge of affine/convex hulls, sparse approximated nearest points (SANP) [7] proposes to impose the sparsity constraint, and has achieved better performance. However, SANP model has high complexity; hence, regularized nearest points (RNP) [24] proposes to model image set as regularized affine hull. Nonetheless, these methods [12,19] only consider relation of pair sets in the training phase, and ignore relation of the set-to-set distance between a testing set and a large training set consisted of all the training sets. So far, there has been some more effective models based on collaborative representation, such as regularized hull based collaborative representation (RH-ISCRC) [27], collaborative regularized nearest points (CNRP) [25], and joint regularized nearest points (JRNP) [23]. Besides, prototype discriminative learning (PDL) [21] also gains the better effect by simultaneously learning the prototype set and a linear projection matrix, which keeps a prototype set from a same class closer and different classes farther. Similar ideas, pairwise linear regression model (PLRC) [4] considers the related subspace and the new unrelated subspace, simultaneously. Moreover, deep learning (*e.g.*, DRM-MV) [6] also performs image set classification tasks. Although these methods has achieved great success for classification tasks, there are some drawbacks that need to be defeated (see Sect. 1).

Overall, the relationship between the related works and the proposed method is that they are point-to-point distance model used for image set classification. The difference between them is that our method proposes to consider the relations between the whole gallery sets and each gallery for further boosting the set discrimination, and propose a latent sparse constraint to reduce the number of trade-off parameters and computational cost. Conversely, the existing point-to-point methods unwittingly ignore the important discrimination information, and usually introduce some sparse constraints to obtain sparse solution.

3 Proposed Method

In this section, we elaborate the details of the proposed method.

3.1 Problem Formulation

The goal of point-to-point image set classification method is to search a pair of virtual points, one from the whole gallery sets $X = [X_1, \cdots, X_m]$, and the other from the probe set Y, where the set size d is the numbers of samples in a gallery set or probe set. Inspired by [15,27], sparse constrain can lead to a sparse solution so as to resist noise and even outliers. What's more, we assume that the artificial virtual nearest point $X\alpha$ of whole gallery sets is not only close to the virtual nearest point $Y\beta$ of the probe set Y, but also close to that of each gallery set (*i.e.*, $\{X_i\gamma_i\}_{i=1}^m$). Regarding the mentioned above, our method is formulated as

$$\min_{\alpha,\beta,\gamma} \|X\alpha - Y\beta\|_2^2 + \lambda_1\|\alpha\|_1 + \lambda_2\|\beta\|_1 + \lambda\sum_{i=1}^m \|X\alpha - X_i\gamma_i\|_2^2 + \lambda_3\|\gamma_i\|_1 \quad (1)$$

where λ, λ_1, λ_2, and λ_3 are the trade-off parameters, α is the common coding coefficient of the whole gallery sets, and the i-th entry of $\gamma = [\gamma_1^T, \gamma_2^T, \cdots, \gamma_m^T]^T$ is the private coding coefficient of the i-th gallery set. Overall, the first term aims to obtain the minimization of the distance between the pair nearest points of the probe set and the whole gallery sets. The fourth term makes $X\alpha$ close to the point that produced by the unknown ground truth set, such that the discriminant ability of different gallery sets is improved. Moreover, the rest regularization are used to avoid trivial solution and obtain sparse solution (Fig. 1).

However, in (1), the parameters, λ_1, λ_2, and λ_3 need to be well tuned. Moreover, the introduced sparse terms lead to high computational complexity. Obviously, this is not user-friendly since the most suitable parameters for a specific dataset are usually challenging to decide. To address these problems, we normalize $\mathbf{1}^T\alpha = 1$, $\mathbf{1}^T\beta = 1$, and $\mathbf{1}^T\gamma_i = 1$, which make the second, the third, and the fifth terms constant, respectively. That is, the normalizations $\mathbf{1}^T\alpha = 1$,

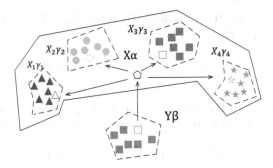

Fig. 1. The illustration of the proposed image set classification method. For example, there are four gallery sets $\{X_i\}_{i=1}^4$ and one probe set Y. The hollow icons and solid icons stand for the true samples and the artificial virtual nearest points, respectively. The arrows stand for the virtual point-to-point distance.

$\mathbf{1}^T\beta = 1$, and $\mathbf{1}^T\gamma_i = 1$ are equivalent to the latent sparse constraints on α, β, and γ_i, respectively. Therefore, problem (1) can be upgraded to

$$\min_{\alpha,\beta,\gamma} \|X\alpha - Y\beta\|_2^2 + \lambda \sum_{i=1}^m \|X\alpha - X_i\gamma_i\|_2^2 \tag{2}$$

$$\text{s.t. } \alpha \geq 0, \mathbf{1}^T\alpha = 1, \beta \geq 0, \mathbf{1}^T\beta = 1, \gamma_i \geq 0, \mathbf{1}^T\gamma_i = 1$$

Compared to other point-to-point methods, our method has latent sparse solutions and a concise formulation with only one parameter λ. Note that each set X_i could be replaced with a dictionary with k atoms, such as KSVD [27].

3.2   Optimization

Problem (2) consists of some ℓ_2-norm regularization terms, it obviously has good convergence. Correspondingly, we can very efficiently solve (2) by alternatively updating α, β, and γ.

Step 1: Update α, we fix β, and γ: First, by fixing the other variables, we solve α via the following problem.

$$\min_{\alpha} \|X\alpha - Y\beta\|_2^2 + \lambda \sum_{i=1}^m \|X\alpha - X_i\gamma_i\|_2^2 \quad \text{s.t. } \alpha \geq 0, \mathbf{1}^T\alpha = 1 \tag{3}$$

This problem can be expanded as follows.

$$\min_{\alpha} \sum_{i=1}^m (\|X\alpha\|_2^2 + \|X_i\gamma_i\|_2^2) - 2\operatorname{tr}(\alpha^T X^T (\frac{Y\beta + \lambda \sum_{i=1}^m x_i\gamma_i}{m\lambda + 1})) \tag{4}$$

For convenience, an auxiliary variable $P = \frac{1}{m\lambda+1}(Y\beta + \lambda \sum_{i=1}^m X_i\gamma_i)$ is introduced, problem (4) can be transformed to

$$\hat{\alpha} = \arg\min_{\alpha} \|X\alpha - P\|_2^2 \quad \text{s.t. } \alpha \geq 0, \mathbf{1}^T\alpha = 1 \tag{5}$$

Step 2: Update β, we fix γ, and α: By fixing the other variables, we solve β via the following problem.

$$\hat{\beta} = \arg\min_{\beta} \|X\alpha - Y\beta\|_2^2 \quad \text{s.t.} \quad \beta \geq 0, \mathbf{1}^T\beta = 1 \tag{6}$$

Step 3: Update γ, we fix β, and α: Similar to problem (6), we can directly obtain the following problem to update each γ_i while fixing the other variables.

$$\hat{\gamma}_i = \arg\min_{\gamma_i} \|X\alpha - X_i\gamma_i\|_2^2 \quad \text{s.t.} \quad \gamma_i \geq 0, \mathbf{1}^T\gamma_i = 1 \tag{7}$$

Interestingly, (5), (6), and (7) derive the same optimization manner. Considering that the non-negative and affined constraint is introduced to output a narrow solution, we call such a problem *capped simplex projection problem* (CSPP). For better flow of the paper, the details of solving CSPP are presented in the next section. Hereto, the pseudo-code is depicted as in Algorithm 1.

Algorithm 1. The algorithm of the proposed LSDL method

Input: Gallery sets $X = [X_1, \cdots, X_m]$ with set size d, probe set Y, and param λ.
1: Compress $\{X_i\}_{i=1}^m$ using KSVD [27] algorithm with k dictionary atoms;
2: Initialize $\alpha = \frac{1}{mk}$, $\beta = \frac{1}{n}$, $\gamma_i = \frac{1}{k}$, $\epsilon = 10^{-5}$, $iter = 1$, and $mit = 10^3$;
3: **repeat**
4:     Solve α via (5);
5:     Solve β via (6);
6:     Solve $\{\gamma_i\}_{i=1}^m$ via (7);
7:     Compute $iter++$, and compute the difference between two successive iterators;
8: **until** The difference is lower than ϵ or $iter < mit$;
Output: Perform image set classification via (18).

3.3 Solve Capped Simplex Projection Problem

In this paper, we introduce the accelerated projected gradient (APG) method to optimize the resultant CSPP. For the convenience of notations, we denote (6) as $\varphi(\beta)$ and Ω as the capped simplex constraint domain. Without loss of generality, we define CSPP as follows by taking the β-problem as an example.

$$\min_{\beta \in \Omega} \varphi(\beta) = \|\boldsymbol{X}\alpha - \boldsymbol{Y}\beta\|_2^2 \tag{8}$$

Then, we introduce an auxiliary variable η to convert such a problem into an easier paradigm, meanwhile, η is approximated to the solution β in an iterative manner. Then, an alternative optimization method is introduced to solve (8).

We first obtain β^0 by solving (1) with $\lambda_2 = 1$ and fixing α and γ_i without considering the capped simplex constraint, then initialize $\eta^0 = \beta^0$ to start the alternative optimization method. Now, η at iteration t is denoted as η^t, and the initial value of Newton acceleration coefficient is defined as c, which will be updated at each iteration.

Then, in the t-th iteration, β is approximated by using Taylor expansion up to second order, $i.e.$,

$$\beta^t = \underset{\beta}{\arg\min}\ f(\eta^{t-1}) + (\beta - \eta^{t-1})^T \varphi'(\beta^{t-1}) + \frac{L}{2}\|\beta - \eta^{t-1}\|_F^2 \qquad (9)$$

By dropping the irrelevant terms, problem (9) can be rewritten as a more compact form. Therefore, we have

$$\beta^t = \underset{\beta \in \Omega}{\arg\min}\frac{L}{2}\|\beta - (\eta^{t-1} - \frac{1}{L}\varphi'(\eta^{t-1}))\|_2^2 \qquad (10)$$

Let $e = (\eta^{t-1} - \frac{1}{L}\varphi'(\eta^{t-1}))$. Problem (10) can then be predigested to

$$\beta^t = \arg\min \frac{1}{2}\|\beta - e\|_2^2 \quad s.t. \ \ \beta \geq 0, 1^T\beta = 1 \qquad (11)$$

Analogously to [14,20], we write the Lagrangian function of problem (11) as

$$\mathcal{L}(z,\omega,\pi) = \frac{1}{2}\|z - e\|_2^2 - \pi(1^Tz - 1) - \omega^Tz \qquad (12)$$

where π is a scalar and ω is a Lagrangian coefficient vector.

Suppose the optimal solution of (11) is z^*, and the Lagrange coefficient vectors are ω^* and π^*, respectively. According to the Karush–Kuhn–Tucker (KKT) conditions, for $\forall j$, we have $\sum_{m=1}^r z_j^* - \sum_{m=1}^r e_j^m - \omega_j^* - \pi^* = 0$, $z_j^* \geq 0$, $\omega_j^* \geq 0$ and $z_j^*\omega_j^* = 0$. Evidently, we also have $rz^* - \sum_{m=1}^r e^m - \omega^* - \pi^*1 = 0$.

According to the constraint $1^Tz^* = 1$, we gain $\pi^* = (r - \sum_{m=1}^r 1^Te^m - 1^T\omega^*)/n$. So the optimal solution z^* is formulated as

$$\frac{\sum_{m=1}^r e^m}{r} + \frac{1}{n} - \frac{\sum_{m=1}^r 1^Te^m 1}{rn} - \frac{1^T\omega^* 1}{rn} + \frac{\omega^*}{r} \qquad (13)$$

Defining $g = \sum_{m=1}^r e^m/r + 1/n - \sum_{m=1}^r 1^Te^m 1/(rn)$ and $\hat{\omega}^* = 1^T\omega^*/(rn)$, (13) can then be simplified to $z^* = g - \hat{\omega}^*1 + \frac{\omega^*}{r}$. As a result, for $\forall j$, we have

$$z_j^* = g_j - \hat{\omega}^* + \frac{\omega_j^*}{r} = \lfloor g_j - \hat{\omega}^* \rfloor_+ \qquad (14)$$

Similarly, we can derive $\omega_j^* = r\lfloor \hat{\omega}^* - g_j \rfloor_+$. Due to $\hat{\omega}^* = 1^T\omega^*/(rn)$, the optimal solution $\hat{\omega}^*$ is represented as $\hat{\omega}^* = \frac{1}{n}\sum_{j=1}^n \lfloor \hat{\omega}^* - g_j \rfloor_+$. To solve the self-dependent $\hat{\omega}^*$, we define an auxiliary function as

$$\Theta(\hat{\omega}) = \frac{1}{n}\sum_{j=1}^n \lfloor \hat{\omega}^* - g_j \rfloor_+ - \hat{\omega} \qquad (15)$$

Note that $\hat{\omega} \geq 0$ and $\Theta'(\hat{\omega}) \leq 0$, and $\Theta'(\hat{\omega})$ is a piecewise linear and convex function, we can use Newton method to find the root $\hat{\omega}^*$ when $\Theta(\hat{\omega}) = 0$, i.e.,

$$\hat{\omega}_{t+1} = \hat{\omega}_t - \frac{\Theta(\hat{\omega}_t)}{\Theta'(\hat{\omega}_t)} \qquad (16)$$

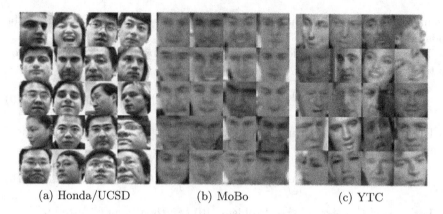

| (a) Honda/UCSD | (b) MoBo | (c) YTC |

Fig. 2. Sample images of the three datasets. These faces contain high intra-class variations in the form of different poses, illumination variations, expression deformations, and occlusions.

3.4   Classification

After obtaining the optimal coding coefficient vectors $\hat{\alpha}$, $\hat{\beta}$ and $\{\hat{\gamma_i}\}_{i=1}^{m}$, we can compute the point-to-point distance between the artificial virtual points of the whole gallery set and the probe set, $i.e.$, $X\hat{\alpha}$ and $Y\hat{\beta}$.

$$r_i = (\|X_i\|_* + \|Y\|_*) * (\|X_i\hat{\alpha}_i - Y\hat{\beta}\|_2^2 + \|X\hat{\alpha} - X_i\hat{\gamma}_i\|_2^2)/\|\hat{\alpha}_i\|_2^2 \qquad (17)$$

where $\|X_i\|_*$ and $\|Y\|_*$ are the nuclear norm ($i.e.$, the sum of the singular values). $\|X_i\hat{\alpha}_i - Y\hat{\beta}\|_2^2$ represents the contribution of i-th gallery set to generate the virtual point ($i.e.$, $Y\hat{\beta}$) of the probe set. $\|X_i\|_* + \|Y\|_*$ can eliminate the interference unrelated to the class information of image set. Finally, the label assignment of the probe image set Y is determined by

$$\text{label}\ (Y) = \arg\min_i\{r_i\} \qquad (18)$$

4   Experiments

In this section, we perform experiments on three benchmark image set datasets to make the comparison with some baseline methods. All experiments are performed on the PC with an Intel Core i5 (2.5 GHz) CPU and 8 GB RAM. The programming software used is MATLAB 2016b.

4.1   Experimental Setup

Three benchmark face set databases are adopted, including the Honda/UCSD [24,27], CMU Mobo (Mobo) [18,27] and YouTube Celebrities (YTC) [16,18,27] datasets, as shown in Fig. 2. In our method, we set the trade-off parameter λ and the numbers of dictionary atoms as $\lambda = 0.1$ and $k = 10$, respectively.

Our LSDL method is compared with several representative image set classification methods, including AHISD [1], CHISD [1], SANP [7], ISCRC-ℓ_1 [27], ISCRC-ℓ_2 [27], RNP [24], DRM-WV [6], PLRC [4] and PDL [21]. Note that the set size is varied from [50, 100, 200], and the average classification accuracies and standard deviations are reported in all experiments.

4.2   Results on the Honda/UCSD Dataset

In Honda/UCSD dataset, it contains 59 video sequences involving 20 different subjects. Each sequence consists about 12–645 frames, which covers large variations on different head poses and facial expressions. We resize each face image to size 20×20 after histogram equalization and divide all sequences into two groups. Randomly, choosing 20 video sequences for training and the remaining 39 ones for testing.

The experimental results under different number of training frames are presented in Table 1. It clearly shows that the proposed LSDL method is better than other comparison methods in all cases and obtain the best classification rates up to 100%, delightfully. The comparison methods achieve lower classification rates when the frame is 50 and 100.

4.3   Results on the CMU Mobo Dataset

In Mobo dataset, there are 96 video sequences of 25 subjects walking on a treadmill, with different patterns, including inclined, slow, fast, and carrying a ball. We resize each face images to size 30×30 after histogram equalization. For each subject, one image set is randomly selected for training, and the rest three ones for testing, and ten-fold cross validation experiments are conducted.

The experimental results are shown in Table 1. Obviously, the average classification accuracies of our LSDL are same as ISCRC, which is the state-of-the-art method. The average classification accuracies on Mobo dataset have obtained satisfying results when set size are 50, 100 and 200.

4.4   Results on the YouTube Celebrities Dataset

The YTC dataset contains 1910 videos of 47 celebrities (actors and politicians) from YouTube web site. The face tracking and classification task becomes relatively more difficult than Honda and Mobo since most of the face images are low-quality, highly compressed, and collected under unconstrained real-life condition with large pose/expression variation, and motion blur. We resize all face images to size 30×30 grayscale images, and then extract their LBP features for reducing the illumination variations. We perform experiments with randomly choosing three sets for training and six sets for testing, and five-fold cross validation experiments are conducted.

Table 1. Performance of different methods on Honda/UCSD and Mobo (%).

Method	Honda/UCSD			CMU Mobo		
	Set size 50	Set size 100	Set size 200	Set size 50	Set size 100	Set size 200
AHISD [1]	82.0	84.6	89.4	91.6 ± 2.8	94.1 ± 2.0	91.9 ± 2.6
CHISD [1]	82.0	87.2	92.3	91.2 ± 3.1	93.8 ± 2.5	96.0 ± 1.3
SANP [7]	84.6	92.3	94.9	91.9 ± 2.7	94.2 ± 2.1	97.3 ± 1.3
RNP [24]	87.2	94.9	97.4	91.9 ± 2.5	94.7 ± 1.2	97.4 ± 1.5
ISCRC-ℓ_1 [27]	89.7	97.4	100.0	93.5 ± 2.8	96.5 ± 1.9	**98.7 ± 1.7**
ISCRC-ℓ_2 [27]	89.7	94.8	100.0	93.5 ± 2.8	96.4 ± 1.9	98.4 ± 1.7
PLRC [4]	87.2	97.4	100.0	92.1 ± 1.6	94.6 ± 1.9	97.5 ± 1.8
PDL [21]	87.2	94.9	97.4	92.5 ± 2.3	94.8 ± 1.9	96.6 ± 2.6
DRM-MV [6]	96.9	99.3	100.0	92.9 ± 1.7	96.2 ± 0.9	98.1 ± 0.8
Our LSDL	**100.0**	**100.0**	**100.0**	**93.5 ± 2.2**	**96.5 ± 1.8**	98.5 ± 0.6

Table 2. Performance (%) and running time of different methods on YTC.

Method	Set size 50	Set size 100	Set size 200	Runing time (Set size 50)
AHISD [1]	57.1 ± 8.1	59.7 ± 6.4	57.1 ± 8.1	0.92 ± 0.01
CHISD [1]	57.9 ± 6.8	62.7 ± 7.2	64.2 ± 7.5	1.64 ± 0.02
SANP [7]	56.7 ± 5.5	61.9 ± 8.1	65.4 ± 6.8	8.55 ± 0.94
RNP [24]	58.4 ± 6.9	63.2 ± 8.4	65.4 ± 7.2	0.22 ± 0.00
ISCRC-ℓ_1 [27]	62.3 ± 6.2	65.6 ± 6.7	66.7 ± 6.4	0.08 ± 0.00
ISCRC-ℓ_2 [27]	57.4 ± 7.2	60.7 ± 6.5	61.4 ± 6.4	0.04 ± 0.00
PLRC [4]	61.7 ± 8.2	65.6 ± 7.9	66.8 ± 7.5	6.21 ± 0.88
PDL [21]	63.9 ± 6.8	65.7 ± 7.7	67.1 ± 7.6	62.54 ± 5.26
DRM-MV [6]	62.3 ± 5.5	68.2 ± 6.2	70.3 ± 4.8	376.44 ± 15.79
Our LSDL	**72.2 ± 8.7**	**77.2 ± 9.0**	**75.7 ± 10.5**	0.54 ± 0.01

The experimental results are shown in Table 2. We can find that our LSDL method achieves the maximum accuracy of 72.2%, 77.2% and 75.7% when set size is 50, 100 and 200, respectively, which is the highest performance reported so far. It is worth mentioning that our LSDL with 50 image frames has better performance than all the comparative methods with varying image frames. This phenomenon further proves that our LSDL is superior to all the comparative methods.

4.5   Running Time Comparison

Time cost is a matter of great concern in practical applications. We mainly compare different methods on YTC dataset with 50 images per set. As shown in the last column of Table 2, our LSDL method has very lower running time than other comparison methods. Moreover, the running time within 1 s has same order of magnitude as ISCRC and RNP; however, the classification accuracies of our LSDL are much higher than that of other methods.

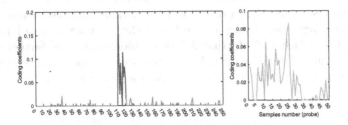

Fig. 3. Coding coefficients of the gallery sets (left part) and the probe set (right part), *i.e.*, α and β. Mobo dataset contains 25 classes, and the number of dictionary atoms m is set to 10. Red coefficients correspond to the correct class. (Color figure online)

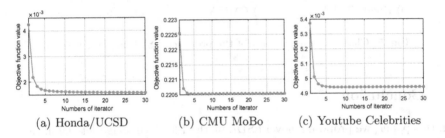

(a) Honda/UCSD   (b) CMU MoBo   (c) Youtube Celebrities

Fig. 4. Convergence of the proposed LSDL method.

4.6   Latent Sparse Property Analysis

As we mentioned in Sect. 3.1, our LSDL has latent sparse property. In order to demonstrate this, we perform experiment on the Mobo dataset with 50 frames and 10 dictionary atoms. The coding coefficients α and β are shown in the left part and right part of Fig. 3, respectively. As we seen, our method can obtain the sparse solutions of α and β. This demonstrates that our LSDL can not only obtain sparse solutions to resist noise but also can decreases the computing cost.

4.7   Convergence Analysis

It is easy to prove that our model is convex and has closed-form solutions for all variables. Experimentally, the convergence curves are showed in Fig. 4 on Honda, Mobo and YTC. We can see that the proposed algorithm converges faster and the objective value becomes stable within 5 iterations in all experiment.

4.8   Parameter Sensitivity

In our method, the one parameter λ needs to be tuned. To analyze its influence on classification accuracy, we tune λ to vary from $\{10^{-7}, \cdots, 10^4\}$ while setting $d = 50$. The results are shown in Fig. 5. It is obvious that our method obtains the best accuracy when fixing $\lambda = 0.1$ and the accuracy is relatively stable when $\lambda \leq 0.1$, which indicate that λ has a wide range. This further suggests that explicit minimizing the virtual point-to-point distance between the whole training set and each gallery set can improve the discrimination.

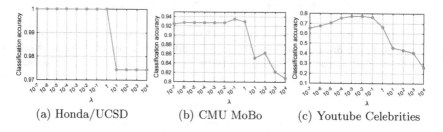

(a) Honda/UCSD   (b) CMU MoBo   (c) Youtube Celebrities

Fig. 5. Classification accuracy of the proposed LSDL method *w.r.t.* λ.

5   Conclusion

In this paper, we propose a novel LSDL method for face image set classification. Different from the existing point-to-point distance-based methods, LSDL proposes to consider the relations between the whole gallery sets and each gallery set; moreover, it proposes a latent sparse normalization with capped simplex constraint to avoid high computational cost as well as improving robustness. Numerous experiments on three benchmark face set databases showed that the proposed method significantly outperforms the competitors in terms of classification accuracy and computational speed.

Acknowledgements. This research was supported by the Intelligent Manufacturing and Robot Special Project of Major Science and Technology Project in Sichuan (Grant no. 2020ZDZX0014), the Science and Technology Special Project of Major Scientific Instruments and Equipment Project in Sichuan (Grant no. 19ZDZX0119), the Undergraduate Innovation and Entrepreneurship Project of Sichuan Province of Sichuan (Grant no. 19xcy099).

References

1. Cevikalp, H., Triggs, B.: Face recognition based on image sets. In: 2010 IEEE Computer Society Conference on Computer Vision and Pattern Recognition, pp. 2567–2573. IEEE (2010)

2. Cevikalp, H., Yavuz, H.S., Triggs, B.: Face recognition based on videos by using convex hulls. In: IEEE Trans. Circuits Syst. Video Technol. (2019)
3. Chen, L., Hassanpour, N.: Survey: how good are the current advances in imageset based face identification?–experiments on three popular benchmarks with a naïve approach. Comput. Vis. Image Underst. **160**, 1–23 (2017)
4. Feng, Q., Zhou, Y., Lan, R.: Pairwise linear regression classification for image set retrieval. In: Proceedings of the IEEE Conference on Computer Vision and Pattern Recognition, pp. 4865–4872 (2016)
5. Gao, X., Sun, Q., Xu, H., Wei, D., Gao, J.: Multi-model fusion metric learning for image set classification. Knowl.-Based Syst. **164**, 253–264 (2019)
6. Hayat, M., Bennamoun, M., An, S.: Deep reconstruction models for image set classification. IEEE Trans. Pattern Anal. Mach. Intell. **37**(4), 713–727 (2014)
7. Hu, Y., Mian, A.S., Owens, R.: Face recognition using sparse approximated nearest points between image sets. IEEE Trans. Pattern Anal. Mach. Intell. **34**(10), 1992–2004 (2012)
8. Huang, C., Li, Y., Chen, C.L., Tang, X.: Deep imbalanced learning for face recognition and attribute prediction. IEEE Trans. Pattern Anal. Mach. Intell. (2019)
9. Huang, Z., et al.: A benchmark and comparative study of video-based face recognition on cox face database. IEEE Trans. Image Process. **24**(12), 5967–5981 (2015)
10. Liu, B., Jing, L., Li, J., Yu, J., Gittens, A., Mahoney, M.W.: Group collaborative representation for image set classification. Int. J. Comput. Vis. **127**(2), 181–206 (2019)
11. Liu, X., Guo, Z., You, J., Kumar, B.V.: Dependency-aware attention control for image set-based face recognition. IEEE Trans. Inf. Forensics Secur. **15**, 1501–1512 (2019)
12. Mian, A., Hu, Y., Hartley, R., Owens, R.: Image set based face recognition using self-regularized non-negative coding and adaptive distance metric learning. IEEE Trans. Image Process. **22**(12), 5252–5262 (2013)
13. Moon, H.-M., Seo, C.H., Pan, S.B.: A face recognition system based on convolution neural network using multiple distance face. Soft. Comput. **21**(17), 4995–5002 (2016). https://doi.org/10.1007/s00500-016-2095-0
14. Nie, F., Wang, X., Jordan, M.I., Huang, H.: The constrained Laplacian rank algorithm for graph-based clustering. In: Thirtieth AAAI Conference on Artificial Intelligence (2016)
15. Ren, Z., Sun, Q.: Simultaneous global and local graph structure preserving for multiple kernel clustering. IEEE Trans. Neural Networks Learn. Syst. (2020)
16. Ren, Z., Sun, Q., Wu, B., Zhang, X., Yan, W.: Learning latent low-rank and sparse embedding for robust image feature extraction. IEEE Trans. Image Process. **29**(1), 2094–2107 (2019)
17. Ren, Z., Sun, Q., Yang, C.: Nonnegative discriminative encoded nearest points for image set classification. Neural Comput. Appl. **32**(13), 9081–9092 (2019). https://doi.org/10.1007/s00521-019-04419-y
18. Ren, Z., Wu, B., Zhang, X., Sun, Q.: Image set classification using candidate sets selection and improved reverse training. Neurocomputing **341**, 60–69 (2019)
19. Song, Z., Cui, K., Cheng, G.: Image set face recognition based on extended low rank recovery and collaborative representation. Int. J. Mach. Learn. Cybernet. **11**(1), 71–80 (2019). https://doi.org/10.1007/s13042-019-00941-6
20. Wang, H., Yang, Y., Liu, B., Fujita, H.: A study of graph-based system for multi-view clustering. Knowl.-Based Syst. **163**, 1009–1019 (2019)

21. Wang, W., Wang, R., Shan, S., Chen, X.: Prototype discriminative learning for face image set classification. In: Lai, S.-H., Lepetit, V., Nishino, K., Sato, Y. (eds.) ACCV 2016. LNCS, vol. 10113, pp. 344–360. Springer, Cham (2017). https://doi.org/10.1007/978-3-319-54187-7_23

22. Wei, D., Shen, X., Sun, Q., Gao, X., Yan, W.: Prototype learning and collaborative representation using grassmann manifolds for image set classification. Pattern Recogn. **100**, 107123 (2020)

23. Yang, M., Wang, X., Liu, W., Shen, L.: Joint regularized nearest points for image set based face recognition. Image Vis. Comput. **58**, 47–60 (2017)

24. Yang, M., Zhu, P., Van Gool, L., Zhang, L.: Face recognition based on regularized nearest points between image sets. In: 2013 10th IEEE International Conference and Workshops on Automatic Face and Gesture Recognition (FG), pp. 1–7. IEEE (2013)

25. Zhang, L., Yang, M., Feng, X., Ma, Y., Zhang, D.: Collaborative representation based classification for face recognition. arXiv preprint arXiv:1204.2358 (2012)

26. Zheng, P., Zhao, Z.Q., Gao, J., Wu, X.: Image set classification based on cooperative sparse representation. Pattern Recogn. **63**, 206–217 (2017)

27. Zhu, P., Zuo, W., Zhang, L., Shiu, S.C.K., Zhang, D.: Image set-based collaborative representation for face recognition. IEEE Trans. Inf. Forensics Secur. **9**(7), 1120–1132 (2014)

Sparse Multi-task Least-Squares Support Vector Machine

Xiaoou Zhang and Zexuan Zhu[✉]

College of Computer Science and Software Engineering, Shenzhen University,
Shenzhen 518060, China
zhuzx@szu.edu.cn

Abstract. Multi-task learning (MTL) can improve the learning performance by sharing information among different tasks. Multi-task least-squares support vector machine (MTLS-SVM) represents one of most widely used MTL methods to solve convex linear problems by extending the conventional least-squares support vector machine (LS-SVM). Nevertheless, MTLS-SVM suffers from an inherent shortage for being sensitive to noise features. In this paper, we propose a novel algorithm to learn a sparse feature subset for classification based on MTLS-SVM. In particular, a 0–1 multi-task feature vector is introduced into MTLS-SVM to control whether a feature is selected or not. The resultant algorithm namely sparse MTLS-SVM (SMTLS-SVM) iteratively selects the most informative feature through an efficient alternating minimization method. Comprehensive experimental results on both synthetic and real-world datasets show that SMTLS-SVM can obtain better or competitive performance compared with the original MTLS-SVM and other existing MTL methods in terms of classification performance.

Keywords: Multi-task learning · Support vector machine · Feature selection

1 Introduction

Multi-task learning (MTL) developed rapidly in the last decade and has been widely used in various fields such as computer vision [3,7], bioinformatics [17,28], natural language processing [32], web applications [1], and ubiquitous computing [20,27]. Single-task learning only focuses on one task and ignores the potential correlation between different tasks. In contrast, MTL leverages similar structures and information existing between related tasks to improve the overall learning performance of the tasks [6,30].

Evgenious and Pontil [8,9] first introduced MTL to support vector machine (SVM), and they assumed that all tasks share an average hyperplane. On this basis, MTL has been explored in literature for different SVMs, such as MTL using one-class v-SVM (OC-V-SVM) [31], multi-task least-squares SVMs (MTLS-SVM) [29], multi-task proximal SVM (MTPSVM) [16], multi-task asymmetric

© Springer Nature Singapore Pte Ltd. 2020
H. Zhang et al. (Eds.): NCAA 2020, CCIS 1265, pp. 157–167, 2020.
https://doi.org/10.1007/978-981-15-7670-6_14

least squares SVM (MT-aLS-SVM) [19], and multi-task least squares twin SVM (MTLS-TWSVM) [21]. Clustered MTL [10] and Graph-regularized [26] MTL have also been introduced to SVMs to improve the learning performance.

However, most of the existing MTL based SVMs do not consider the feature sparsity. Sparsity is an important structure for MTL, especially in some high dimensional real world applications [18]. It is necessary to introduce feature selection into multi-task SVM to reduce the computational complexity and improve classification accuracy. Wang et al. [25] proposed a feature selection method with shared information discovery model based on multi-task SVM. The method uses a $l_{2,1}$-norm regularization term to learn a sparse feature selection matrix for each learning task. Yet, on tackling high dimensional data, it is hard to set the parameters of $l_{2,1}$-norm regularization term [23].

In this paper, we proposed a novel MTL feature selection method by extending the MTLS-SVM [29]. The resultant algorithm namely sparse MTLS-SVM (SMTLS-SVM) introduces a feature vector for each task, and then iteratively selects the most influential features for all tasks. In every iteration, we just need to solve primal MTLS-SVM problem after dimensionality reduction. Comprehensive experimental results on both synthetic and real-world datasets demonstrate the efficiency of SMTLS-SVM in comparison with the primal MTLS-SVM and existing MTL methods.

The rest of this paper is organized as follows. Section 2 introduces the basics of MTLS-SVM. Section 3 details the proposed method. Section 4 presents the numerical experiments on synthetic and real-world data. Finally, Section 5 concludes this paper.

2   Multi-task Least-Squares Support Vector Machine

Before introducing the specific algorithms, the basic notations are defined as follows. Let $\mathbb{R}$ be the set of real numbers and $\mathbb{N}_n = \{1, 2, \cdots, n\}$ be the set of positive integers. A vector is denoted in lower-case letters $\mathbf{x} \in \mathbb{R}^d$ with x_i being an element. The transpose of $\mathbf{x}$ is written as $\mathbf{x}^{\mathbf{T}}$. A vector or matrix with all entries equal to one is represented as $\mathbf{1}_d \in \mathbb{R}^d$ or $\mathbf{1}_{m \times n} \in \mathbb{R}^{m \times n}$, respectively. The l_p-norm of a vector v is denoted as $\|v\|_p$. $A \odot B$ represents the element-wise product between two matrices A and B.

The MTLS-SVM was proposed in [29], which directly introduces regularized MTL (RMTL) into LS-SVM [22]. Similar to LS-SVM, this algorithm only sovles a convex linear system instead of convex quadratic programming (QP) problem. Suppose we have m learning tasks and $\forall i \in \mathbb{N}_m$, we have n_i training data $\{\mathbf{x}_{i,j}, y_{i,j}\}_{j=1}^{n_i}$, where $\mathbf{x}_{i,j} \in \mathbb{R}^d$ and $y_{i,j} \in \{-1, +1\}$. Thus, we have $n = \sum_{i=1}^{m} n_i$. For convenience, let $\mathbf{y} = (\mathbf{y}_1^{\mathbf{T}}, \mathbf{y}_2^{\mathbf{T}}, \cdots, \mathbf{y}_m^{\mathbf{T}})^{\mathbf{T}}$ with $\mathbf{y}_i = (y_{i,1}, y_{i,2}, \cdots, y_{i,n_i})^{\mathbf{T}}$ for $\forall i \in \mathbb{N}_m$

Inspired by the intuition of Hierarchical Bayes [2,5,12], MTLS-SVM assume all $\mathbf{w}_i \in \mathbb{R}^d (\forall i \in \mathbb{N}_m)$ can be written as $\mathbf{w}_i = \mathbf{w}_0 + \mathbf{v}_i$, where the vector $\mathbf{v}_i \in \mathbb{R}^d$ represents the bias between task i and the common mean vectors $\mathbf{w}_0$. MTLS-SVM solves the classification problem by finding $\mathbf{w}_0 \in \mathbb{R}^d$, $\{\mathbf{v}_i\}_{i=1}^{m} \in \mathbb{R}^{d \times m}$, and

$\mathbf{b} = (b_1, b_2, \cdots, b_m)^{\mathrm{T}} \in \mathbb{R}^m$ simultaneously to solve the following optimization problem:

$$\min \mathcal{J}(\mathbf{w}_0, \mathbf{v}_i, \boldsymbol{\xi}_i) = \frac{1}{2}\|\mathbf{w}_0\|_2^2 + \frac{1}{2}\frac{\lambda}{m}\sum_{i=1}^{m}\|\mathbf{v}_i\|_2^2 + \frac{\gamma}{2}\sum_{i=1}^{m}\boldsymbol{\xi}_i^{\mathrm{T}}\boldsymbol{\xi}_i \tag{1}$$

$$\text{s.t.} \quad \mathbf{A}_i^{\mathrm{T}}(\mathbf{w}_0 + \mathbf{v}_i) + b_i\mathbf{y}_i = \mathbf{1}_{n_i} - \boldsymbol{\xi}_i, i \in \mathbb{N}_m$$

where for $\forall i \in \mathbb{N}_m$, $\boldsymbol{\xi}_i = (\xi_{i,1}, \xi_{i,2}, \cdots, \xi_{i,n_i})^{\mathrm{T}} \in \mathbb{R}^{n_i}$, $\mathbf{A}_i = (y_{i,1}\mathbf{x}_{i,1}, y_{i,2}\mathbf{x}_{i,2}, ..., y_{i,n_i}\mathbf{x}_{i,n_i}) \in \mathbb{R}^{d \times n_i}$. λ, γ are two positive real regularized parameters and $\mathbf{A} = (\mathbf{A}_1, \mathbf{A}_2, \cdots, \mathbf{A}_m) \in \mathbb{R}^{d \times n}$.

3 The Proposed SMTLS-SVM

In this section, we propose an alternating minimization feature selection algorithm based on MTLS-SVM resulting in SMTLS-SVM.

3.1 Problem Formulation

MTLS-SVM is based on regularized MTL and assumes that all tasks share an average hyperplane, which, however, cannot induce sparse solutions. In order to obtain sparse solution of MTLS-SVM, we introduce a feature vector $\boldsymbol{\delta} = [\delta_1, \delta_2, \cdots, \delta_d]^{\mathrm{T}} \in \{0, 1\}^d$ to indicate whether the corresponding feature is selected ($\delta_j = 1$) or not ($\delta_j = 0$) for all tasks. Then, the objective of SMTLS-SVM can be simplified as:

$$\min \mathcal{J}(\boldsymbol{\delta}, \mathbf{w}_0, \mathbf{v}_i, \boldsymbol{\xi}_i) = \frac{1}{2}\|\mathbf{w}_0\|_2^2 + \frac{1}{2}\frac{\lambda}{m}\sum_{i=1}^{m}\|\mathbf{v}_i\|_2^2 + \frac{\gamma}{2}\sum_{i=1}^{m}\boldsymbol{\xi}_i^{\mathrm{T}}\boldsymbol{\xi}_i \tag{2}$$

$$\text{s.t.} \quad \widehat{\mathbf{A}}_i^{\mathrm{T}}(\mathbf{w}_0 + \mathbf{v}_i) + b_i\mathbf{y}_i = \mathbf{1}_{n_i} - \boldsymbol{\xi}_i, i \in \mathbb{N}_m$$

For $\forall i \in \mathbb{N}_m$, we have $\widehat{\mathbf{A}}_i = (y_{i,1}(\mathbf{x}_{i,1} \odot \boldsymbol{\delta}), y_{i,2}(\mathbf{x}_{i,2} \odot \boldsymbol{\delta}), ..., y_{i,n_i}(\mathbf{x}_{i,n_i} \odot \boldsymbol{\delta})) \in \mathbb{R}^{d \times n_i}$. The definition of $\boldsymbol{\xi}_i, \lambda, \gamma$ are the same as that used in Sect. 2. The Lagrangian function for the problem (2) is

$$\min \mathcal{L}(\boldsymbol{\delta}, \mathbf{w}_0, \mathbf{v}_i, \mathbf{b}, \boldsymbol{\xi}_i, \boldsymbol{\alpha}_i) = \frac{1}{2}\|\mathbf{w}_0\|_2^2 + \frac{1}{2}\frac{\lambda}{m}\sum_{i=1}^{m}\|\mathbf{v}_i\|_2^2 + \frac{\gamma}{2}\sum_{i=1}^{m}\boldsymbol{\xi}_i^{\mathrm{T}}\boldsymbol{\xi}_i$$

$$- \sum_{i=1}^{m}\boldsymbol{\alpha}_i^{\mathrm{T}}(\widehat{\mathbf{A}}_i^{\mathrm{T}}(\mathbf{w}_0 + \mathbf{v}_i) + b_i\mathbf{y}_i - \mathbf{1}_{n_i} + \boldsymbol{\xi}_i) \tag{3}$$

where $\forall i \in \mathbb{N}_m$, $\boldsymbol{\alpha}_i = (\alpha_{i,1}, \alpha_{i,2}, \cdots, \alpha_{i,n_i})^{\mathrm{T}}$ consists of Lagrange multipliers, and we have $\boldsymbol{\alpha} = (\boldsymbol{\alpha}_1^{\mathrm{T}}, \boldsymbol{\alpha}_2^{\mathrm{T}}, \cdots, \boldsymbol{\alpha}_m^{\mathrm{T}})^{\mathrm{T}} \in \mathbb{R}^n$. This problem is non-convex and difficult to solve.

3.2 Alternating Minimization Algorithm

To solve Eq. (3), in the section, we design an efficient alternating minimization algorithm which iteratively selects the features, and then solves a subproblem with the selected features only. The framework of the algorithm is described in Algorithm 1. Algorithm 1 involves two steps, i.e., the greedy feature selection step (also known as the best-case analysis described in Sect. 3.3) and the subproblem optimization step. First, we initialize the Lagrange multipliers α to $\mathbf{1}_n$ and find the most important features which limit the number to less than B, then fixed variables δ and solve a subproblem of Lagrange function (3). The whole process is repeated until the termination criterion is met.

Algorithm 1. Alternating minimization algorithm for problem (3)

1: Initialize $\alpha^0 = \mathbf{1}_n$ and iteration index $t = 1$;
2: **while** *stopping condition is not satisfied* **do**
3:    Perform Greedy Feature Selection: do best-case analysis to select the informative features based on α^{t-1};
4:    Subproblem Optimization: solve a subproblem of Lagrange function (3), obtaining the optimal solution α^t;
5:    $t = t + 1$;
6: **end while**

Obviously, the subproblem is a standard MTLS-SVM as problem (1) after we fixed variables δ. The primal problem with equal constraints for MTLS-SVM is a convex problem, the KKT conditions for optimality yield the following set of linear equations:

$$
\begin{cases}
\dfrac{\partial \mathcal{L}}{\partial \mathbf{w}_0} = 0 \Rightarrow \mathbf{w}_0 = \widehat{\mathbf{A}}\alpha \\[2mm]
\dfrac{\partial \mathcal{L}}{\partial \mathbf{v}_i} = 0 \Rightarrow \mathbf{v}_i = \frac{m}{\lambda}\widehat{\mathbf{A}}_i\alpha_i, & \forall i \in \mathbb{N}_m \\[2mm]
\dfrac{\partial \mathcal{L}}{\partial b_i} = 0 \Rightarrow \alpha_i^{\mathrm{T}}\mathbf{y}_i = 0, & \forall i \in \mathbb{N}_m \\[2mm]
\dfrac{\partial \mathcal{L}}{\partial \xi_i} = 0 \Rightarrow \alpha_i = \gamma\xi_i, & \forall i \in \mathbb{N}_m \\[2mm]
\dfrac{\partial \mathcal{L}}{\partial \alpha_i} = 0 \Rightarrow \widehat{\mathbf{A}}_i^{\mathrm{T}}(\mathbf{w}_0 + \mathbf{v}_i) + b_i\mathbf{y}_i - \mathbf{1}_{n_i} + \xi_i = \mathbf{0}_{n_i}, \forall i \in \mathbb{N}_m
\end{cases}
\tag{4}
$$

This problem can be solved by Krylow methods [11] efficiently and more details can be found in [29]. Let the corresponding solutions be $\alpha^* = (\alpha_1^{*\mathrm{T}}, \alpha_2^{*\mathrm{T}}, \cdots, \alpha_m^{*\mathrm{T}})^{\mathrm{T}}$ with $\alpha_i^* = (\alpha_{i,1}^*, \alpha_{i,2}^*, \cdots, \alpha_{i,n_i}^*)^{\mathrm{T}}$ and $\mathbf{b}^* = (\mathbf{b}_1^*, \mathbf{b}_2^*, \cdots, \mathbf{b}_m^*)^{\mathrm{T}}$. Then, we can fix solution α^* and $\mathbf{b}^*$, and substitute (4) into (3), then we get a minimization problem as follows:

$$
\min \mathcal{L}(\delta) = -\frac{1}{2}(\widehat{\mathbf{A}}\alpha^*)^{\mathrm{T}}(\widehat{\mathbf{A}}\alpha^*) - \frac{1}{2}\frac{m}{\lambda}\sum_{i=1}^{m}((\widehat{\mathbf{A}}_i\alpha_i^*)^{\mathrm{T}}(\widehat{\mathbf{A}}_i\alpha_i^*))
\tag{5}
$$

For convenience, let $\mathbf{c}_i(\boldsymbol{\alpha}_i) = \sum_{j=1}^{n_i} \alpha_{i,j}^* y_{i,j} \mathbf{x}_{i,j} \in \mathbb{R}^d$, for $\forall i \in \mathbb{N}_m$, we have

$$(\widehat{\mathbf{A}}_i \boldsymbol{\alpha}_i^*)^{\mathrm{T}} (\widehat{\mathbf{A}}_i \boldsymbol{\alpha}_i^*) = \| \sum_{j=1}^{n_i} \alpha_{i,j}^* y_{i,j} (\mathbf{x}_{i,j} \odot \boldsymbol{\delta}) \|^2 = \sum_{k=1}^{d} c_{i,k}^2 (\boldsymbol{\alpha}_i^*) \delta_k \qquad (6)$$

$$(\widehat{\mathbf{A}} \boldsymbol{\alpha}^*)^{\mathrm{T}} (\widehat{\mathbf{A}} \boldsymbol{\alpha}^*) = \| \sum_{i=1}^{m} \sum_{j=1}^{n_i} \alpha_{i,j}^* y_{i,j} (\mathbf{x}_{i,j} \odot \boldsymbol{\delta}) \|^2 = \sum_{k=1}^{d} (\sum_{i=1}^{m} c_{i,k} (\boldsymbol{\alpha}_i^*) \delta_k)^2 \qquad (7)$$

Substituting (6) and (7) into the problem (5), we have

$$\max \mathcal{L}(\boldsymbol{\delta}) = \frac{1}{2} \sum_{k=1}^{d} (\sum_{i=1}^{m} c_{i,k} (\boldsymbol{\alpha}_i^*) \delta_k)^2 + \frac{1}{2} \frac{m}{\lambda} \sum_{i=1}^{m} \sum_{k=1}^{d} c_{i,k}^2 (\boldsymbol{\alpha}_i^*) \delta_k \qquad (8)$$

This problem is still a challenging problem. Note that if we only have one task ($m = 1$), problem (8) can be further formulated as a linear programming problem. We can construct a feasible solution by first finding the informative features with the largest feature score, but in MTL this can be complicated. Accordingly, we designed a greedy feature selection method to select a common feature subset for all tasks.

3.3 Best-Case Analysis

For each i task, we define a feature score vector $\mathbf{s}_i$ to measure the importance of the features.

$$\mathbf{s}_i = [\mathbf{c}_i(\boldsymbol{\alpha}_i^*)]^2$$

After solving a subproblem with fixed features, we can get the feature score $\mathbf{s}_i$ for each task. However, for MTL, each task has its own feature score rank and we can select a common set of features among all task. In this way, we propose a *Score-based feature score fusion method* as below that fuses all tasks scores. The method selects the largest scored features greedily and sets the relative δ_j to 1.

The selection procedure repeats until the maximum number of iterations is reached or no more new features are selected. We can simply control the sparsity by setting the number of features B added in each iteration and the maximum number of iterations. Although our algorithm cannot obtain a globally optimal solution, the experimental results show that the selected features achieve good performance.

Score-Based Feature Score Fusion we first normalized each task score to the range of $[0, 1]$ as follows:

$$\mathbf{s}_i' = \frac{\mathbf{s}_i - \mathbf{s}_{i\,\min}}{\mathbf{s}_{i\,\max} - \mathbf{s}_{i\,\min}}$$

where $\mathbf{s}_{i\,\min}$ and $\mathbf{s}_{i\,\max}$ are the minimum and maximum values in vector $\mathbf{s}_i$, respectively. The average of the normalized scores is defined as

$$\mathbf{s} = \frac{1}{m}\sum_{i=1}^{m}\mathbf{s}_i'$$

where m is the number of tasks, we get a final score for all tasks. The larger the final score, the better the feature.

4 Experimental Study

In order to examine the generalization performance of the proposed algorithm, we conduct experiments on a number of synthetic and real-word multi-task dataset. This work focuses on multi-task binary classification problems. The detailed description of the datasets is provided in Table 1. The proposed SMTLS-SVM is compared with the single-task SVM (LSSVM), $l_{1,1}$ norm (Lasso) multi-task model [24], MTLS-SVM, $l_{2,1}$ norm (L21) multi-task mode [4], and the Trace MTL (Trace) [13]. Five-fold cross-validation is applied to determine the parameters in the range $\{10^{-6}, 10^{-5}, \cdots, 10^{3}\}$ for all the compared methods. The linear kernel is adopted in the experiments. To evaluate the performance of the MTL methods, ACC and Youden index (YI) is employed in this experimental study.

Table 1. The description of the datasets.

Data	No. tasks	No. samples	No. features
Mnist	10	1000	87
Usps	10	1000	64
Brain	4	50	10367
Lung	5	203	12600
Leukemia	3	72	7129
thyroid	4	168	2000
MLL	3	72	12582

4.1 Experiment on Synthetic Data

To test the robustness of the compared algorithms to noise features, in this subsection, we synthesize datasets with different number of feature dimensions. We consider feature dimensions with $d = 250, 500, 750, 1000$. The first 100 elements of weight vector for the ith task $\mathbf{w}_i$ are generated from normal distribution $N(1,1)$ and the rest elements are exactly zero. The input features for the ith task $\mathbf{X}_i$ are generated from $N(0,1)$ and each task owns 500 examples. The corresponding labels $\mathbf{Y}_i$ are computed by $sign(\mathbf{X}_i^{\mathbf{T}}\mathbf{w}_i + b_i)$, where b_i is a vector of noise in which each element is generated from $N(0, 0.1)$. We consider 50% of the data for training and the experiments are performed 30 times independently. As each task has 100 common beneficial features, we set B $=10$ and the maximum number iterations to 10 in our method. The experimental results on synthetic data are shown in Fig. 1, where SMTLS-SVM distinctly outperforms the other methods. As the number of noise features increases, only SMTLS-SVM and L21 can maintain good performance. L21 can select beneficial elements, but it also contains more noise entries than SMTLS-SVM. LS-SVM, MTLS-SVM, and Trace do not consider the sparse structure in the model. Therefore, they cannot exclude the noise features.

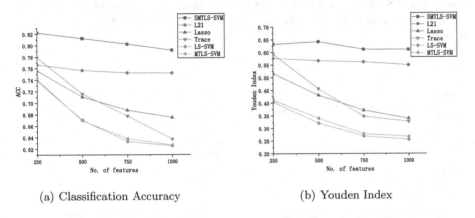

(a) Classification Accuracy   (b) Youden Index

Fig. 1. Classification performance vs. feature size on synthetic data

4.2   Experiment on Handwritten Data

In this subsection, we evaluate the proposed method on two handwritten digit recognition datasets: Mnist [14] and Usps [15]. Since the handwritten data is a multi-class data, so we neet to transform it into a multi-task data, for this we can treat each one-vs-rest classification problem as a single task and use undersampling to balance datasets. In this experiment, 30% of the data is used for training and the rest for testing. We simply set $B = 3$ and the maximum number of iterations to 12. The results averaged over 30 times are shown in Table 2. The results indicate that the multi-task methods are better than the single-task methods, but not the difference among the multi-task methods is not significant. SMTLS-SVM selects no more than 30 features, which means SMTL-SVM can achieve the similar or superior results to other algorithms while maintaining sparsity.

Table 2. The learning performances (±standard deviation) for the handwritten digit recognition

		Lasso	L21	Trace	LS-SVM	MTLS-SVM	SMTLS-SVM
Mnist	ACC	0.8783 ± 0.01	**0.8927 ± 0.01**	0.8821 ± 0.01	0.8808 ± 0.01	0.8873 ± 0.01	0.8918 ± 0.01
	YI	0.7567 ± 0.02	**0.7855 ± 0.01**	0.7642 ± 0.01	0.7616 ± 0.02	0.7848 ± 0.02	0.7836 ± 0.02
Usps	ACC	0.8893 ± 0.01	0.8940 ± 0.01	0.8873 ± 0.01	0.8917 ± 0.01	0.8820 ± 0.01	**0.8956 ± 0.01**
	YI	0.7787 ± 0.02	0.7921 ± 0.01	0.7747 ± 0.02	0.7834 ± 0.02	0.7865 ± 0.02	**0.7953 ± 0.02**

Similar to the synthetic data, we also add noise feature to the Mnist data to test the robustness of the algorithms. We add extra noise generated from N(1, 1) to the Mnist dataset in a range of $\{1000, \cdots, 5000\}$. The results of classification accuracy and Youden index against the number of features are plotted in Fig. 2, where SMTLS-SVM shows better performance than the other methods. SMTLS-SVM is more robust to the noise features thanks to the introduction of feature selection.

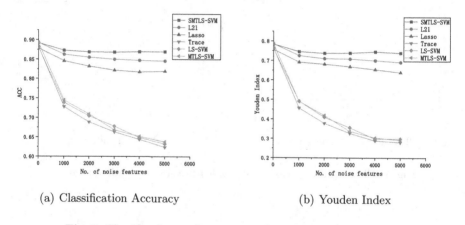

(a) Classification Accuracy   (b) Youden Index

Fig. 2. Classification performance vs. feature size on Mnist data

4.3 Experiment on Micorarray Data

In multi-class classification of Micorarray data, the classification of cancer may only be affected by a small number of genes (features). We evaluate the effectiveness of the proposed algorithm on cancer classification using five multi-class Micorarray datasets namely Brain, Lung, Leukemia, thyroid and MLL. For fair comparison, we randomly select 60% of the whole data set for training, and the rest for testing. SMTLS-SVM is configured with $B = 30$ and maximum number of iterations 12. The experimental setup is similar to the handwritten data. The experiments were performed 30 times and the average learning performance of the six algorithms on the Micorarray datasets is reported in Table 3. The experimental results consistently indicate that SMTLS-SVM outperforms the other competing methods thanks to the use of feature selection.

Table 3. The learning performances (±standard deviation) for the microarray datasets

		Lasso	L21	Trace	LS-SVM	MTLS-SVM	SMTLS-SVM
Brain	ACC	0.8262 ± 0.02	0.8107 ± 0.03	0.7464 ± 0.04	0.8119 ± 0.03	0.8298 ± 0.05	**0.8440 ± 0.04**
	YI	0.3975 ± 0.08	0.4725 ± 0.09	0.4903 ± 0.05	0.4467 ± 0.07	0.4333 ± 0.07	**0.5192 ± 0.06**
Lung	ACC	0.9746 ± 0.01	0.9735 ± 0.01	0.9708 ± 0.01	0.9674 ± 0.01	0.9578 ± 0.01	**0.9832 ± 0.01**
	YI	0.7815 ± 0.05	0.7845 ± 0.08	0.7037 ± 0.04	0.7690 ± 0.05	0.7928 ± 0.07	**0.8467 ± 0.05**
Leukemia	ACC	0.9655 ± 0.02	0.9621 ± 0.02	0.9138 ± 0.04	0.9298 ± 0.03	0.9368 ± 0.02	**0.9862 ± 0.02**
	YI	0.9003 ± 0.04	0.9216 ± 0.05	0.7358 ± 0.03	0.7934 ± 0.06	0.7979 ± 0.06	**0.9523 ± 0.03**
thyroid	ACC	0.8746 ± 0.02	0.8769 ± 0.02	0.8146 ± 0.03	0.8530 ± 0.02	0.8742 ± 0.01	**0.8929 ± 0.02**
	YI	0.5931 ± 0.05	0.6144 ± 0.05	0.6634 ± 0.05	0.5196 ± 0.03	0.5905 ± 0.05	**0.6782 ± 0.04**
MLL	ACC	0.9632 ± 0.02	0.9644 ± 0.02	0.8977 ± 0.03	0.9390 ± 0.02	0.9437 ± 0.03	**0.9839 ± 0.01**
	YI	0.9171 ± 0.03	0.9396 ± 0.03	0.8362 ± 0.05	0.8168 ± 0.04	0.8421 ± 0.05	**0.9501 ± 0.03**

5   Conclusions

In this paper, we propose a novel multi-task feature selection algorithm to learn a sparse feature subset for classification based on MTLS-SVM. The resultant algorithm SMTLS-SVM introduces a 0–1 multi-task feature vector into MTLS-SVM to control whether or not the features are selected. SMTLS-SVM then iteratively selects the most informative feature through an efficient alternating minimization method. Comprehensive experiments on both synthetic and real-world datasets demonstrate the effectiveness of SMTLS-SVM in terms of classification performance and feature sparsity.

Acknowledgement. This work was supported in part by the National Natural Science Foundation of China under Grant 61871272 and in part by the Shenzhen Scientific Research and Development Funding Program under Grant JCYJ20190808173617147.

References

1. Ahmed, A., Das, A., Smola, A.J.: Scalable hierarchical multitask learning algorithms for conversion optimization in display advertising. In: Proceedings of the 7th ACM International Conference on Web Search and Data Mining, pp. 153–162 (2014)
2. Allenby, G.M., Rossi, P.E.: Marketing models of consumer heterogeneity. J. Econometrics **89**(1–2), 57–78 (1998)
3. Almaev, T., Martinez, B., Valstar, M.: Learning to transfer: transferring latent task structures and its application to person-specific facial action unit detection. In: Proceedings of the IEEE International Conference on Computer Vision, pp. 3774–3782 (2015)
4. Argyriou, A., Evgeniou, T., Pontil, M.: Multi-task feature learning. In: Advances in Neural Information Processing Systems, pp. 41–48 (2007)
5. Arora, N., Allenby, G.M., Ginter, J.L.: A hierarchical Bayes model of primary and secondary demand. Mark. Sci. **17**(1), 29–44 (1998)
6. Caruana, R.: Multitask learning. Mach. Learn. **28**(1), 41–75 (1997)
7. Chu, X., Ouyang, W., Yang, W., Wang, X.: Multi-task recurrent neural network for immediacy prediction. In: Proceedings of the IEEE International Conference on Computer Vision, pp. 3352–3360 (2015)

8. Evgeniou, T., Micchelli, C.A., Pontil, M.: Learning multiple tasks with kernel methods. J. Mach. Learn. Res. **6**, 615–637 (2005)
9. Evgeniou, T., Pontil, M.: Regularized multi-task learning. In: Proceedings of the 10th ACM SIGKDD International Conference on Knowledge Discovery and Data Mining, pp. 109–117 (2004)
10. Gao, P.X.: Facial age estimation using clustered multi-task support vector regression machine. In: Proceedings of the 21st International Conference on Pattern Recognition, pp. 541–544 (2012)
11. Hamers, B., Suykens, J., De Moor, B.: A comparison of iterative methods for least squares support vector machine classifiers. Technical report, Internal Report 01–110, ESAT-SISTA, KU Leuven (Leuven, Belgium) (2001)
12. Heskes, T.: Empirical bayes for learning to learn. In: Proceedings of the 17th International Conference on Machine Learning, pp. 367–374. Morgan Kaufmann, San Francisco (2000)
13. Ji, S., Ye, J.: An accelerated gradient method for trace norm minimization. In: Proceedings of the 26th Annual International Conference on Machine Learning, pp. 457–464 (2009)
14. Kang, Z., Grauman, K., Sha, F.: Learning with whom to share in multi-task feature learning. In: Proceedings of the 28th International Conference on Machine Learning, pp. 521–528 (2011)
15. Lee, G., Yang, E., Hwang, S.: Asymmetric multi-task learning based on task relatedness and loss. In: International Conference on Machine Learning, pp. 230–238 (2016)
16. Li, Y., Tian, X., Song, M., Tao, D.: Multi-task proximal support vector machine. Pattern Recogn. **48**(10), 3249–3257 (2015)
17. Li, Y., Wang, J., Ye, J., Reddy, C.K.: A multi-task learning formulation for survival analysis. In: Proceedings of the 22nd ACM SIGKDD International Conference on Knowledge Discovery and Data Mining, pp. 1715–1724 (2016)
18. Liu, C., Zheng, C.T., Qian, S., Wu, S., Wong, H.S.: Encoding sparse and competitive structures among tasks in multi-task learning. Pattern Recogn. **88**, 689–701 (2019)
19. Lu, L., Lin, Q., Pei, H., Zhong, P.: The als-svm based multi-task learning classifiers. Appl. Intell. **48**(8), 2393–2407 (2018)
20. Lu, X., Wang, Y., Zhou, X., Zhang, Z., Ling, Z.: Traffic sign recognition via multi-modal tree-structure embedded multi-task learning. IEEE Trans. Intell. Transp. Syst. **18**(4), 960–972 (2016)
21. Mei, B., Xu, Y.: Multi-task least squares twin support vector machine for classification. Neurocomputing **338**, 26–33 (2019)
22. Suykens, J.A.K., Gestel, T.V., Brabanter, J.D., Moor, B.D., Vandewalle, J.: Least squares support vector machines. World Scientific (2002)
23. Tan, M., Tsang, I.W., Wang, L.: Towards ultrahigh dimensional feature selection for big data. J. Mach. Learn. Res. **15**, 1371–1429 (2014)
24. Tibshirani, R.: Regression shrinkage and selection via the lasso. J. Roy. Stat. Soc.: Ser. B (Methodol.) **58**(1), 267–288 (1996)
25. Wang, S., Chang, X., Li, X., Sheng, Q.Z., Chen, W.: Multi-task support vector machines for feature selection with shared knowledge discovery. Sig. Process. **120**, 746–753 (2016)
26. Widmer, C., Kloft, M., Görnitz, N., Rätsch, G.: Efficient training of graph-regularized multitask SVMs. In: Flach, P.A., De Bie, T., Cristianini, N. (eds.) ECML PKDD 2012. LNCS (LNAI), vol. 7523, pp. 633–647. Springer, Heidelberg (2012). https://doi.org/10.1007/978-3-642-33460-3_46

27. Xu, J., Tan, P.N., Zhou, J., Luo, L.: Online multi-task learning framework for ensemble forecasting. IEEE Trans. Knowl. Data Eng. **29**(6), 1268–1280 (2017)
28. Xu, J., Zhou, J., Tan, P.N.: Formula: factorized multi-task learning for task discovery in personalized medical models. In: Proceedings of the 2015 SIAM International Conference on Data Mining, pp. 496–504 (2015)
29. Xu, S., An, X., Qiao, X., Zhu, L.: Multi-task least-squares support vector machines. Multimedia Tools Appl. **71**(2), 699–715 (2013). https://doi.org/10.1007/s11042-013-1526-5
30. Xu, Z., Kersting, K.: Multi-task learning with task relations. In: IEEE 11th International Conference on Data Mining pp. 884–893. IEEE (2011)
31. Yang, H., King, I., Lyu, M.R.: Multi-task learning for one-class classification. In: The 2010 International Joint Conference on Neural Networks, pp. 1–8 (2010)
32. Zhao, L., Sun, Q., Ye, J., Chen, F., Lu, C.T., Ramakrishnan, N.: Multi-task learning for spatio-temporal event forecasting. In: Proceedings of the 21th ACM SIGKDD International Conference on Knowledge Discovery and Data Mining, pp. 1503–1512 (2015)

Discriminative Subspace Learning for Cross-view Classification with Simultaneous Local and Global Alignment

Ao Li[1]([✉]), Yu Ding[1], Deyun Chen[1], Guanglu Sun[1], and Hailong Jiang[2]

[1] School of Computer Science and Technology,
Harbin University of Science and Technology, Harbin 150080, China
dargonboy@126.com
[2] Department of Computer Science, Kent State University, Kent, USA

Abstract. With the wide applications of cross-view data, cross-view Classification tasks draw much attention in recent years. Nevertheless, an intrinsic imperfection existed in cross-view data is that the data of the different views from the same semantic space are further than that within the same view but from different semantic spaces. To solve this special phenomenon, we design a novel discriminative subspace learning model via low-rank representation. The model maps cross-view data into a low-dimensional subspace. The main contributions of the proposed model include three points. 1) A self-representation model based on dual low-rank models is adopted, which can capture the class and view structures, respectively. 2) Two local graphs are designed to enforce the view-specific discriminative constraint for instances in a pair-wise way. 3) The global constraint on the mean vector of different classes is developed for further cross-view alignment. Experimental results on classification tasks with several public datasets prove that our proposed method outperforms other feature learning methods.

Keywords: Cross-view classification · Subspace learning · Discriminative analysis · Low-rank constraint.

1 Introduction

Cross-view classification is a noteworthy technology in the area of pattern recognition due to the large distribution discrepancy of data from different views. So, a well designed cross-view feature extraction plays a significant role in improving the classification performance. Nevertheless, it is worth noting that the samples in same view space are closer than that in different view spaces but from the same class space. Therefore, how to deal with the large view-variance between data from two views is a meaningful topic for subspace learning.

The aim of subspace learning is to find a feature space that can represent the high-dimensional data more effectively, which has been studied in many literatures [4–13]. Principal component analysis (PCA) [1], as a typical method, seeks

© Springer Nature Singapore Pte Ltd. 2020
H. Zhang et al. (Eds.): NCAA 2020, CCIS 1265, pp. 168–179, 2020.
https://doi.org/10.1007/978-981-15-7670-6_15

a projection subspace by maximizing variance. In order to make the learned subspace discriminative, linear discriminant analysis (LDA) was proposed with the Fisher criterion [2]. The feature subspace learned by LDA makes the intra-class samples more similar, and the inter-class samples less similar. Unfortunately, LDA may lead to overfitting for corrupted data. Recently, low-rank models to solve overfitting problems are very popular. Robust PCA (RPCA) [3] was designed to recover noisy data through the rank-minimization technique. Inspired by RPCA, low-rank representation (LRR) was designed to explore the intrinsic latent manifold structures of data from multiple subspaces. Liu et al. proposed the latent LRR (LatLRR) by considering the latent feature of the data [4]. After that, the supervised regularization-based robust subspace (SRRS) framework built in [5] provides a discriminative feature learning method that unifies the low-rank constraint and discriminant to learning low-dimensional subspace. In [6], an unsupervised robust linear feature extraction method was designed, namely low-rank embedding (LRE). LRE can reduce the negative impact of samples being occluded and corrupted. Ren et al. extended the LRE by introducing the $l_{2,1}$-norm term to make it more robust and effective [7].

Recently, a large number of cross-view feature learning algorithms also have achieved satisfied achievement [14–17]. However, some of methods ignore that samples from same view also have valuable discriminative information. To address this problem, Kan et al. designed a multi-view discriminant analysis (MvDA) [14]. MvDA can learn a multi-view feature subspace by optimizing multiple linear transforms from different views. Next, a most recent method was proposed, namely robust cross-view learning (RCVL) [15], which aims to learn an effective discriminative subspace by two discriminative graphs. Nevertheless, the global discriminative information is lost in RCVL.

Inspired by the above mentioned subspace learning methods, we propose a novel feature subspace learning model by exploring the local and global discriminative constraints simultaneously to realize the cross-view alignment. The main values of our methods are presented as follows: (1) The dual low-rank constraints framework is built to describe the two latent structures in cross-view data, namely the view structure and the class structure. Our method reveals the potential manifold of cross-view data so that the learned subspace contains more valuable feature information. (2) A local alignment mechanism based on two local graph constraints is adopted to constrain the neighbor relationships of the samples in the feature subspace. This mechanism can make the two structures in (1) to be separated effectively. (3) We set up a global alignment constraint as a complement to further reduce the effect of view-variance within the classes between views. The illustration of our framework is shown in Fig. 1, which learns a invariant subspace by maximizing the distance of the inter-class within each view and minimizing intra-class between views from both of class and view perspective.

The structure for the remainder of this paper is as follows. Section 2 simply introduces some of the related methods. Section 3 presents the proposed model

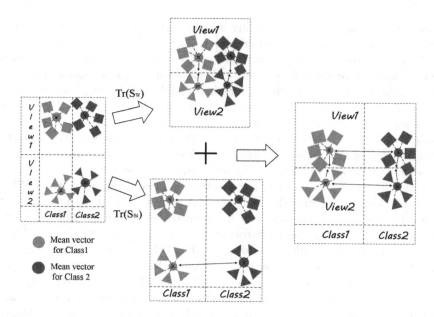

Fig. 1. Conceptual illustration of the proposed discriminative subspace learning framework

and its solution process. Section 4 shows the results of the comparison experiment and parameter experiment. At last, Section 5 summarizes this paper.

2   Related Works

There are two related methods involved in our framework: 1) low-rank representation, 2) linear discriminant analysis. Then, we discuss the role of two related works as follows.

2.1   Low-Rank Representation

Low-rank representation can cope well with data from multiple subspaces. Assuming $X = [X_1, X_2, ..., X_k]$ is a matrix of natural data from k categories. LRR can be expressed by

$$\min_{Z,E} rank(Z) + \lambda \|E\|_1 , s.t. X = XZ + E \tag{1}$$

in which Z is a low-rank linear combination coefficient matrix of data X. Generally, the samples data contains a lot of random noise. In function (1), matrix E denotes the noisy data, and we use l_1-norm to get randomness. In this way, XZ can recover the true data from noise. $\lambda > 0$ is the balanced parameter. LRR can dig and utilize the self-similar information hidden in data. Therefore, LRR can not only learn the original subspace of the data in noisy environments, but also have the ability to uncover the latent manifold structure of data, which is belief to be feasible and potential for representing the cross-view data.

2.2   Linear Discriminant Analysis

The principle of LDA is to find a discriminative subspace with the largest inter-class variance and the smallest intra-class variance. Assuming training data $\{X, y\} = \{(x_1, y_1), ..., (x_n, y_n)\}$ is from m classes, where X denotes samples and y means label. In addition, $\hat{x}$ represents the center of all samples, and $\hat{x}_i$ represents the center of the ith class samples. Hence, between-class and within-class scatter matrices are expressed by:

$$S_b = \sum_{i=1}^{m} n_i(\hat{x}_i - \hat{x})(\hat{x}_i - \hat{x})^T$$
$$S_w = \sum_{i=1}^{m} \sum_{x \in X_i} (x - \hat{x}_i)(x - \hat{x}_i)^T$$

(2)

where n_i is the number of samples from the ith class, and X_i is the ith class samples set. Therefore, LDA finds a projection by maximizing the generalized Rayleigh quotient as follows:

$$\max_{w} \frac{Tr(w^T S_b w)}{Tr(w^T S_w w)}$$

(3)

where $Tr(\cdot)$ denotes the trace operator and w is the projection matrix. In addition, the trace-ratio problem is not conducive to the solution of subsequent problems, so we transform function (3) into the trace-difference problem as follows:

$$\max_{w} Tr(w^T S_b w) - Tr(w^T S_w w)$$

(4)

LDA can reflect differences between samples based on a supervised discriminative constraint. However, its performance is not satisfied for cross-view analysis due to the large discrepancy of data distribution from different views.

3   The Proposed Algorithm

This section gives detailed discussions on our framework and develops a numerical scheme to obtain the approximate solutions iteratively.

3.1   Notations

Assuming $X = [X1, X2] \in R^{d \times n}$ is a set of cross-view samples with two views from c classes, where n means the number of the training samples and d denotes the dimensionality of natural data. We design two local graph-based constraints to seek two latent view-invariant structures, which are composed of class structure matrix $Z_c \in R^{n \times n}$ and view structure matrix $Z_v \in R^{n \times n}$, respectively. $E \in R^{d \times n}$ is a matrix of error data designed to obtain a robust subspace from noise. $P \in R^{d \times p}$ is the low-dimensional projection matrix. In addition, $V_1, V_2, \hat{V}_1$ and $\hat{V}_2 \in R^{n \times c}$ are constant coefficient matrices used to construct the discriminative global alignment constraint.

3.2   Objective Function

To address the discriminative cross-view analysis, we proposed a novel subspace learning with simultaneous local and global alignments, of which the objective function is as follows:

$$
\min_{Z_c, Z_v, E, P} \overbrace{\|Z_c\|_* + \|Z_v\|_* + \lambda_1 \|E\|_{2,1}}^{D(Z_c, Z_v, E)}
$$

$$
+ \overbrace{\alpha(Tr(P^T X Z_c L_c (P^T X Z_c)^T) - Tr(P^T X Z_v L_v (P^T X Z_v)^T))}^{U(P, Z_c, Z_v)} \qquad (5)
$$

$$
+ \overbrace{\lambda_2(Tr(S_W(P^T X Z)) - Tr(S_{B1}(P^T X Z)) - Tr(S_{B2}(P^T X Z)))}^{G(P, Z)}
$$

$$
s.t. X = X(Z_c + Z_v) + E, \ P^T P = I
$$

where $D(Z_c, Z_v, E)$ represents class and view structures of the cross-view space by dual low-rank representations. $U(P, Z_c, Z_v)$ enforces the view-specific discriminative local neighbor relationship among instances. $G(P, Z)(Z = Z_c + Z_v)$ presents the global discriminative constraint by the mean instance of each class cross different views. In short, we combine the local alignment constraint and the global alignment constraint on a dual low-rank framework to learn cross-view subspaces. In the following, the above terms are illustrated in detail.

Dual Low-Rank Representations: In general, we adopt a single rank minimum constraint to learn the latent information of data. But cross-view data contains class information and view information simultaneously. Even data from the same class have a large divergence. Hence, we use two structure matrices Z_c and Z_v to solve the specific problem that is the between-view samples from the same class are far away and the within-view samples from the different classes are closer. Thus, we define the first term with dual low-rank representations to strip down the class and view structures as follows:

$$
D(Z_c, Z_v, E) = \|Z_c\|_* + \|Z_v\|_* + \lambda_1 \|E\|_{2,1}, \ s.t. X = X(Z_c + Z_v) + E \qquad (6)
$$

where $\|\cdot\|_*$ is the symbol of the nuclear norm, which is an approximate representation of the rank minimum problem, and its solution relatively convenient. We adopt the $l_{2,1}$-norm to make matrix E have the sparsity as the noisy data. λ_1 is a positive balance parameter, which can be tuned in experiments.

Graph-Based Discriminative Local Alignment: To introduce the local discriminative constraint, two graph-based constraints are constructed on the each pair of synthetic samples with Z_c and Z_v from class and view subspaces respectively as follows, which can better cluster intra-class samples and decentralize inter-class ones.

$$
U_c = \sum_{i,j} (Y_{c,i} - Y_{c,j})^2 W_{i,j}^c
$$

$$
U_v = \sum_{i,j} (Y_{v,i} - Y_{v,j})^2 W_{i,j}^v \qquad (7)
$$

where $Y_{c,i}, Y_{v,i}$ denote the ith projected sample of cross-view data in the class space $Y_c = P^T X Z_c$ and view space $Y_v = P^T X Z_v$, respectively. Correspondingly, $Y_{c,j}, Y_{v,j}$ denote the jth projected sample. $W_{i,j}^c$ and $W_{i,j}^v$ denote graph weight matrices that are defined as follows:

$$W_{i,j}^c = \begin{cases} 1, & \text{if } x_i \in N_{k_1}^c(x_j), \text{and } l_i = l_j, \\ 0, & \text{otherwise} \end{cases}$$

$$W_{i,j}^v = \begin{cases} 1, & \text{if } x_i \in N_{k_2}^v(x_j), \text{but } l_i \neq l_j, \\ 0, & \text{otherwise} \end{cases} \tag{8}$$

where l_i and l_j are the labels of sample x_i, x_j, respectively. $x_i \in N_{k_1}^c(x_j)$ denotes that x_i belongs to the k_1 adjacent data sets of the same sample x_j. $x_i \in N_{k_2}^v(x_j)$ means that x_i belongs to the k_2 adjacent data sets of the same view sample x_j. With the help of trace operator, the pair-wise local discriminative constraint $U(P, Z_c, Z_v)$ can be rewritten with U_c and U_v based on Fisher criterion as follows.

$$U(P, Z_c, Z_v) = \alpha(Tr(P^T X Z_c L_c (P^T X Z_c)^T) - Tr(P^T X Z_v L_v (P^T X Z_v)^T)) \tag{9}$$

where L_c and L_v mean the Laplacian operators of W^c and W^v. α is a balance parameter, which can be tuned in experiments.

Discriminative Global Alignment: It is noteworthy that the $U(P, Z_c, Z_v)$ preserves the discriminant in a local way by focusing on each pair of samples, which is not powerful enough. So, to further improve the proposed our model, we design a global discriminative constraint for cross-view analysis as the third term $G(P, Z)$, which is denoted as $G(P, Z) = Tr(S_W(P^T X Z)) - Tr(S_{B1}(P^T X Z)) - Tr(S_{B2}(P^T X Z))$. In $G(P, Z)$, $S_W(P^T X Z)$ is within-class scatter matrix of two views, defined by $S_W(P^T X Z) = \sum_{j=1}^{c}(\mu_j^1 - \mu_j^2)(\mu_j^1 - \mu_j^2)^T$. $Tr(S_{Bi}(P^T X Z))(i = 1, 2)$ is between-class scatter matrices of the ith view, defined by $Tr(S_{Bi}(P^T X Z)) = \sum_{j=1}^{c}(\mu_j^i - \mu^i)(\mu_j^i - \mu^i)^T$, where μ_j^i denotes the mean projected sample of the jth class from the ith view, and μ^i denotes the overall mean projected sample from the ith view. To be computed efficiently, the third term can be designed as:

$$G(P, Z) = \lambda_2(Tr(S_W(P^T X Z)) - Tr(S_{B1}(P^T X Z)) - Tr(S_{B2}(P^T X Z)))$$

$$= \lambda_2(\|P^T X Z(V_1 - V_2)\|_F^2 - \|P^T X Z(V_1 - \hat{V}_1)\|_F^2 - \|P^T X Z(V_2 - \hat{V}_2)\|_F^2) \tag{10}$$

where $Z = Z_c + Z_v$ denotes the global representation. λ_2 is a balance parameter, which can be tuned in experiments. V_i and $\hat{V}_i(i = 1, 2)$ are the coefficient matrices of the within-class mean sample of each view and the global mean sample of each view, respectively. In detail, $V_i(k, m) = (1/n_i^m)$ only if x_k belongs to the mth class from the ith view, where n_i^m means the number of samples of the mth class from the ith view; otherwise, $V_i(k, m) = 0$. $\hat{V}_i(k, m) = (1/n_i)$ only if x_k belongs to the ith view, where n_i denotes the number of samples from the ith view; otherwise, $\hat{V}_i(k, m) = 0$. Equation (10) achieves global alignment by the mean vectors of joint synthetic samples from global representation and further enforces the view-invariant constraint on the same class.

3.3 Optimization Scheme

To facilitate the solution of Z_c and Z_v in Eq. (5), we add two auxiliary variables J_c and J_v. Then, Eq. (5) can be transformed into the following form:

$$\min_{Z_c, Z_v, E, P} \|J_c\|_* + \|J_v\|_* + \lambda_1 \|E\|_{2,1} + U(P, Z_c, Z_v) + G(P, Z)$$

$$s.t. X = X(Z_c + Z_v) + E, P^T P = I, J_c = Z_c, J_v = Z_v \tag{11}$$

Then, we transform the function (11) to the Augmented Lagrangian form, and the result is as follows:

$$\min_{Z_c, Z_v, E, P} \|J_c\|_* + \|J_v\|_* + \lambda_1 \|E\|_{2,1} + U(P, Z_c, Z_v) + G(P, Z)$$

$$+ Tr(Y_1^T (X - X(Z_c + Z_v) - E)) + Tr(Y_2^T (J_c - Z_c)) + Tr(Y_3^T (J_v - Z_v))$$

$$+ \frac{\eta}{2} (\|X - X(Z_c + Z_v) - E\|_F^2 + \|J_c - Z_c\|_F^2 + \|J_v - Z_v\|_F^2)$$

$$s.t. P^T P = I$$

$$\tag{12}$$

where Y_1, Y_2, Y_3 are the Lagrange multipliers and $\eta > 0$ is the penalty parameter. We use an alternating solution to iteratively optimize all variables. We define the left-bottom of the variable plus t as the t-th solution.

First, we solve the projection matrix P_t one by one, because P_t is an orthogonal matrix. Hence, enforcing the derived function to be zero, the objective function with respect to P is:

$$(XZ_t((V_1 - \hat{V}_1)(V_1 - \hat{V}_1)^T + (V_2 - \hat{V}_2)(V_2 - \hat{V}_2)^T - (V_1 - V_2)(V_1 - V_2)^T)Z_t^T X^T$$

$$- \alpha(X(Z_{c,t} L_c Z_{c,t} - Z_{v,t} L_v Z_{v,t}) X^T)) P_{i,t} = \epsilon_{i,t} P_{i,t}$$

$$\tag{13}$$

Updating J_c and J_c:

$$J_{c,t+1} = \min_{J_{c,t}} \frac{1}{\eta_t} \|J_{c,t}\|_* + \frac{1}{2} \|J_{c,t} - (Z_{c,t} + (Y_{2,t}/\eta_t))\|_F^2 \tag{14}$$

$$J_{v,t+1} = \min_{J_{v,t}} \frac{1}{\eta_t} \|J_{v,t}\|_* + \frac{1}{2} \|J_{v,t} - (Z_{v,t} + (Y_{3,t}/\eta_t))\|_F^2 \tag{15}$$

The singular value thresholding is an approximate method to solve the above two kernel norm minimization equations[18].

Then, we ignore the other variables except Z_c or Z_v and make the function to be zero.

$$Z_{c,t+1}((\lambda_2 H + \alpha L_c)/\eta_t) + X_N(I + X^T X) Z_{c,t+1}$$

$$= X_N(X^T (X - XZ_{v,t} - E_t)) + J_{c,t} + ((X^T Y_{1,t} - Y_{2,t})/\eta_t)) - \frac{\lambda_2}{\eta_t} Z_{v,t} H \tag{16}$$

$$Z_{v,t+1}((\lambda_2 H - \alpha L_v)/\eta_t) + X_N(I + X^T X)Z_{v,t+1}$$

$$=X_N(X^T(X - XZ_{c,t} - E_t)) + J_{v,t} + ((X^T Y_{1,t} - Y_{3,t})/\eta_t)) - \frac{\lambda_2}{\eta_t}Z_{c,t}H \quad (17)$$

where $H = V_1\hat{V}_1^T + V_2\hat{V}_2^T + \hat{V}_1V_1^T + \hat{V}_2V_2^T - V_1V_2^T - V_2V_1^T - \hat{V}_1\hat{V}_1^T - \hat{V}_2\hat{V}_2^T$ and $X_N = (X^T P_t P_t^T X)^{-1}$. It is obvious that Eq. (16) and Eq. (17) are two standard Sylvester equations, which can be easily solved.

Updating E:

$$E_{t+1} = \min_{E_t} \frac{\lambda_1}{\eta_t} \|E_t\|_{2,1} + \frac{1}{2} \|E_t - (X - X(Z_{c,t} + Z_{v,t}) + Y_1/\eta_t)\|_F^2 \quad (18)$$

The above equation is a $l_{2,1}$-norm minimization problem whose solution is shown in [19].

Algorithm 1

Input: data matrix X, parameters $\lambda_1, \lambda_2, \alpha$
Initialize: $\rho = 1.3$, $\theta = 10^{-9}$, $t = 0$, $t_{max} = 200$, $\eta_0 = 0.1$, $\eta_{max} = 10^{10}$;
while not converged or $t \leq t_{max}$ **do**
1. Calculate P_{t+1} using Eq.(13);
2. Calculate $J_{c,t+1}$ using Eq.(14);
3. Calculate $J_{v,t+1}$ using Eq.(15);
4. Calculate $Z_{c,t+1}$ using Eq.(16);
5. Calculate $Z_{v,t+1}$ using Eq.(17);
6. Calculate E_{t+1} using Eq.(18);
7. Calculate $Y_{1,t+1}$ by
   $Y_{1,t+1} = Y_{1,t} + \eta_t(X - X(Z_{c,t+1} + Z_{v,t+1}) - E_{t+1})$
8. Calculate $Y_{2,t+1}$ by $Y_{2,t+1} = Y_{2,t} + \eta_t(J_{c,t+1} - Z_{c,t+1})$
9. Calculate $Y_{3,t+1}$ by $Y_{3,t+1} = Y_{3,t} + \eta_t(J_{v,t+1} - Z_{v,t+1})$
10. Calculate the parameter η_{t+1} by $\eta_{t+1} = \min(\eta_{max}, \rho\eta_t)$;
11. Check convergence by
    $max(\|X - X(Z_{c,t+1} + Z_{v,t+1}) - E_{t+1}\|_\infty,$
    $\qquad\qquad \|J_{c,t+1} - Z_{c,t+1}\|_\infty, \|J_{v,t+1} - Z_{v,t+1}\|_\infty) < \theta;$
12. $t = t + 1$.
end while

Output: J_c, J_v, Z_c, Z_v, E, P

The solution of all variables in objective function (12) is shown in Algorithm 1. Where these parameters $\rho, \theta, t_{max}, \eta, \eta_{max}$ are set by experience. Moreover, the trade-off parameters $\alpha, \lambda_1, \lambda_2$ are tuned by the experiments and we initialize these matrices $Z_c, Z_v, E, Y_1, Y_2, Y_3$ as 0.

4 Experiments

In this section, we compare the proposed algorithm with excellent feature subspace learning algorithms. The data are mapped by the low-dimensional subspace as the features, on which the kNN classifier is implemented to valuate the performance.

4.1 Experimental Datasets

CMU-PIE Face dataset is composed of face pictures of 68 different people. Everyone has 21 different illumination conditions and 9 different poses. We adopt 4 poses Pose05, Pose09, Pose27 and Pose29. We divide the dataset equally to set different cross-view training and testing subsets. **Wikipedia dataset** is an image-text bimodal data set, consisting of 2866 image-text samples from 10 classes. Due to the inconsistency of dimensionality of the two feature, we use PCA to adjust the image dimensions for the next experiment. **COIL-20 object dataset** is composed of 20 objects from a level 360-degree view. There are 5 degrees between every two adjacent images, so each category has 72 samples. We divide the 72 images into two groups G1 and G2. In addition, G1 is composed of samples from $[0°, 85°]$ and V2 $[185°, 265°]$. Similarly, G2 is composed of samples from $[90°, 175°]$ and $[270°, 355°]$. **COIL-100 object dataset** is an extension of the COIL-20. The only difference is that the COIL-100 is composed of 20 objects from a level 360-degree view. Therefore, the set of the COIL-100 database is similar to the COIL-20 dataset.

4.2 Experimental Results and Analysis

In experiments, we select PCA, LDA, locality preserving projections (LPP) [20], LatLRR, SRRS, and RCVL as several comparison methods. To CMU-PIE dataset, we randomly select two poses to form a cross-view experiment set, where C1:{ Pose05, Pose09}, C2:{Pose05, Pose27}, C3:{Pose05, Pose29}, C4:{Pose09, Pose 27}, C5:{Pose09, Pose29}, C6: {Pose27, Pose29}. We show the classification results of all experimental algorithms in Table 1. For COIL-20 and COIL-100 object databases, we select two sets of samples from G1 and G2 as a cross-view training set, and the others set as a test set. Figure 2 displays the classification results of four experimental groups from COIL-20 and COIL-100 datasets. Moreover, Fig. 3 displays the results of comparison experiments on Wikipedia.

Table 1. Classification results (%) of all methods on CMU-PIE dataset.

Methods	C1	C2	C3	C4	C5	C6
LDA	62.96	66.76	62.16	61.50	56.54	61.83
PCA	48.28	50.50	49.07	48.43	45.51	49.68
LatLRR	65.10	66.61	62.47	63.09	61.04	60.42
LPP	62.40	60.17	61.97	62.13	58.34	60.72
SRRS	95.35	91.66	95.82	90.22	96.04	87.16
RCVL	97.14	93.70	97.26	92.99	97.55	88.60
Ours	98.43	93.47	98.23	91.52	98.53	89.42

The results of experiments prove that our method achieves the persistent higher classification results than other methods. Another result also can be found that the classification results of most methods based on LRR are better than that of other comparisons. It is due to that low-rank constraint can retain more valid latent structural information hidden in cross-view data. Nevertheless, among the comparison methods on different datasets, our proposed method is prominent and competitive consistently by considering more cross-view requirements on both local and global alignments.

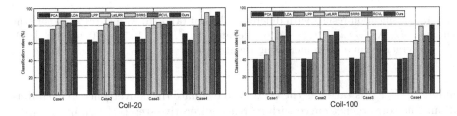

Fig. 2. Recognition results of different experiments on COIL-20 and COIL-100.

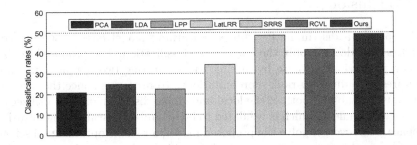

Fig. 3. Classification results of comparison experiments on Wikipedia dataset.

4.3   Performance Evaluations

In this section, we first point out how the parameters affect the classification performance. Then, the convergence analysis of the numerical scheme is also reported.

Our framework has three main parameters $\alpha, \lambda_1, \lambda_2$. We select CMU-PIE C1 to test their influence. The classification accuracy with variational parameters is shown in Fig. 4. When the parameters λ_1 and λ_2 changes within a certain range, the performance fluctuation on classification is slight. For α, the best performance happens at $\alpha = 10^2$. The results show that the parameters are not quite sensitive to the performance. More stable classification results can be obtained with the learned feature subspace of our proposed algorithm. In addition, the maximum value of $\|X - X(Z_{c,t+1} + Z_{v,t+1}) - E_{t+1}\|_\infty$, $\|J_c - Z_c\|_\infty$

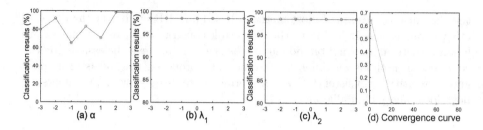

Fig. 4. Classification results with different values of (a) α,(b) λ_1 and (c) λ_2, where the parameters $\alpha, \lambda_1, \lambda_2$ value from -3 to 3 denotes $[10^{-3}, 10^{-2}, 10^{-1}, 1, 10, 10^2, 10^3,]$. (d) The convergence curve with increasing iterative step.

and $\|J_v - Z_v\|_\infty$ is taken as the criteria to evaluate the convergence in each iteration. The convergence curve with increasing iterative step is shown in Fig. 4(d), from which it is confirmed that our method converges fast within a few of iterations.

5   Conclusion

In this paper, a discriminative subspace learning model based low-rank constraint is provided to apply the cross-view classification. The proposed method learns a projection subspace by finding and separating two potential manifold structures with dual low-rank representations. To further enhance the discriminant and adaptability, both of the local and global discriminants are utilized for cross-view alignment, which is validated to be helpful for learning view-invariant subspace. Moreover, to solve the proposed model, a reliable optimization scheme is also designed to ensure convergence. Extensive results prove that our method is superior to existing conventional methods.

Acknowledgement. This work was supported in part by National Natural Science Foundation of China under Grant 61501147, University Nursing Program for Young Scholars with Creative Talents in Heilongjiang Province under Grant UNPYSCT-2018203, Natural Science Foundation of Heilongjiang Province under Grant YQ2019F011, Fundamental Research Foundation for University of Heilongjiang Province under Grant LGYC2018JQ013, and Postdoctoral Foundation of Heilongjiang Province under Grant LBH-Q19112.

References

1. Turk, M., Pentland, A.: Eigenfaces for recognition. J. Cogn. Neurosci. **3**(1), 71–86 (1991)
2. Belhumeur, P.-N., Hespanha, J.-P., Kriegman, D.-J.: Eigenfaces vs. fisherfaces: recognition using class specific linear projection. IEEE Trans. Pattern Anal. Mach. Intell. **19**(7), 711–720 (1997)

3. Cands, E.-J., Li, X., Ma, Y., et al.: Robust principal component analysis? J. ACM **58**(3), 1–37 (2011)
4. Liu, G., Yan, S.: Latent low-rank representation for subspace segmentation and feature extraction. In: IEEE International Conference on Computer Vision (ICCV), pp. 1615–1622. (2011)
5. Li, S., Fu, Y.: Learning robust and discriminative subspace with low-rank constraints. IEEE Trans. Neural Networks Learn. Syst. **27**(11), 2160–2173 (2016)
6. Wong, W.-K., Lai, Z., Wen, J., et al.: Low-rank embedding for robust image feature extraction. IEEE Trans. Image Process. **26**(6), 2905–2917 (2017)
7. Ren, Z., Sun, Q., Wu, B., et al.: Learning latent low-rank and sparse embedding for robust image feature extraction. IEEE Trans. Image Process. **29**, 2094–2107 (2020)
8. Li, A., Chen, D., Wu, Z., et al.: Self-supervised sparse coding scheme for image classification based on low-rank representation. PLoS ONE **13**(6), 1–15 (2018)
9. Li, A., Liu, X., Wang, Y., et al.: Subspace structural constraint-based discriminative feature learning via nonnegative low-rank representation. PLoS ONE **14**(5), 1–19 (2019)
10. Zhang, Z., Li, F., Zhao, M., et al.: Joint low-rank and sparse principal feature coding for enhanced robust representation and visual classification. IEEE Trans. Image Process. **25**(6), 2429–2443 (2016)
11. Zhang, Z., Ren, J., Li, S., et al.: Robust subspace discovery by block-diagonal adaptive locality-constrained representation. In: ACM International Conference on Multimedia, pp. 1569–1577 (2019)
12. Zhang, Z., Wang, L., Li, S., et al.: Adaptive structure-constrained robust latent low-rank coding for image recovery. In: IEEE International Conference on Data Mining (ICDM), pp. 846–855 (2019)
13. Zhang, Z., Zhang, Y., Liu, G., et al.: Joint label prediction based semi-supervised adaptive concept factorization for robust data Representation. IEEE Trans. Knowl. Data Eng. **32**(5), 952–970 (2020)
14. Kan, M., Shan, S., Zhang, H., et al.: Multi-view discriminant analysis. IEEE Trans. Pattern Anal. Mach. Intell. **38**(1), 188–194 (2016)
15. Ding, Z., Fu, Y.: Dual low-rank decompositions for robust cross-view learning. IEEE Trans. Image Process. **28**(1), 194–204 (2019)
16. Li, A., Wu, Z., Lu, H., et al.: Collaborative self-regression method with nonlinear feature based on multi-task learning for image classification. IEEE Access **6**, 43513–43525 (2018)
17. Zhang, Y., Zhang, Z., Qin, J., et al.: Semi-supervised local multi-manifold isomap by linear embedding for feature extraction. Pattern Recogn. **76**, 662–678 (2018)
18. Cai, J.-F., Cands, E.-J., Shen, Z.: A singular value thresholding algorithm for matrix completion. SIAM J. Optim. **20**(4), 1956–1982 (2010)
19. Yang, J., Yin, W., Zhang, Y., et al.: A fast algorithm for edge-preserving variational multichannel image restoration. SIAM J. Imaging Sci. **2**(2), 569–592 (2009)
20. He, X., Niyogi, P.: Locality preserving projections. In: Advances in Neural Information Processing Systems, pp. 153–160 (2004)

A Recognition Method of Hand Gesture Based on Dual-SDAE

Miao Ma[✉], Qingqing Zhu, Xiaoxiao Huang, and Jie Wu

School of Computer Science, Shaanxi Normal University, Xi'an, China
mmthp@snnu.edu.cn

Abstract. A gesture recognition method based on Dual-SDAE was proposed in this paper. In this method, one SDAE was constructed for the intensity information of gesture image, the other one was designed for the depth information of the gesture image. Then, with the conjunction on the outputs of the two networks, soft-max classifier was adopted to form the Dual-SDAE and used for gesture recognition. Besides, for a better understanding of the proposed method, the parameters' influence were also analyzed, including the number of hidden layers, the number of nodes in each hidden layer, and the intensity of noise and the regularization term. In the experiments, by comparing with the traditional SAE network, DBN and CNN, the best performance was achieved by the proposed method whose accuracy on the ASL dataset is 96.51%.

Keywords: Gesture recognition · Auto-encoder · Depth image · Neural network

1 Introduction

With the rapid development of wearable devices, intelligent home, IoT and other new technologies, creating a comprehensive and intelligent life has become one of the main trends of social development, for which human-computer interaction has become the key point to achieve this life. As a widely-used meaning for the communication in daily life, gesture is a natural way for the human-computer interaction. However, due to the ambiguity, diversity and similarity of gesture, the interference of background and different illumination environment, it is still difficult for computer to recognize gestures accurately and quickly.

For traditional gesture recognition methods [1,2], two processes are usually contained, namely the preprocessing (such as gesture region detection and segmentation, etc.) and the recognition. Actually, in these methods, preprocessing has a great impact on the effect of subsequent gesture recognition. However, the

Supported by National Science Foundation of China (No. 61877038, 61601274), Natural Science Basic Research Plan in Shaanxi Province of China (No. 2018JM6068, 2019JQ-093) and the Fundamental Research Funds for the Central Universities (No. GK202007029).

H. Zhang et al. (Eds.): NCAA 2020, CCIS 1265, pp. 180–192, 2020.
https://doi.org/10.1007/978-981-15-7670-6_16

usage of a more complex preprocessing can cause a heavy time-cost. So, an end-to-end based recognition method attracts many researchers' interests [3,4].

In 1986, Rumelhart proposed the concept of Autoencoder (AE) [5]. The automatic encoder learns the representation of the hidden layer from the input vector through the encoding process, and then reconstructs the input vector from the representation of the hidden layer through the decoding process. In 2007, Benjio et al. proposed Sparse Autoencoder (sparseAE) [6], in which the number of neurons in the hidden layer was reduced for the sparsity. In 2008, Denoising Autoencoder (DAE) was proposed by Vincent et al. [7], in which random noise was added in the first layer to improve the robustness of the network. Actually, the function of random noise is very similar to the sparsity adopted in sparseAE. In 2010, Vincent et al. proposed Stacked Denoising Autoencoder (SDAE) network [8]. In this method, with the stacking of several DAEs, SDAE network was formed to extract features. In the experiments, SDAE shows a better performance than that of Deep Belief Network (DBN) in many aspects.

For the capability of AE on the adaptive representation, AE has been widely used in the image processing. In 2014, two SDAE networks (one traditional SDAE and one marginalized SDAE) were build with global tuning strategy for pose action recognition in Chalearn 2013 dataset [9]. In [10], Sang et al. combined SDAE and SVM for the 3-D human motion recognition on MSR Action 3D dataset. In 2015, to distinguish the main object from other objects in multi-overlapping objects in an image of PASCAL VOC 2012 dataset, Liang et al. established an instance-level object segmentation subnetwork with DAE to generate a foreground mask of a main object [11]. In 2016, Santara et al. proposed a synchronous parallel algorithm for the pre-training of the deep network, whose performance on the MNIST dataset showed that the designed algorithm had the same reconstruction accuracy as the layer by layer greedy pre-training, and the speed was improved by 26% [12]. In the same year, Wang et al. proposed Collaborative Recurrent Autoencoder (CRAE), in which the Denoising Recurrent Autoencoder (DRAE) and Collaborative Filtering (CF) enable CRAE to make accurate recommendation and to fill gaps in content sequences [13]. By using the unsupervised learning property of the AE, Hasan et al. established the fully convolutional feedforward autoencoder network, and discovered the activity rule of the interested target from a long video sequence containing crowd activities on the CUHK-Avenue dataset [14].

Recently, with the development of the imaging technology, the depth information can be obtained in accompanied with the RGB image. Physically, in the depth image, the distance between the object and the camera is computed. It means, by using the depth information, the spatial relationship between different objects in 3-D space can be easily discriminated. This is very important for solving the interference of background in object recognition. Thus, in this paper, considering the effects of the depth and the intensity information for recognition, a Dual-SDAE network was designed for the gesture recognition. Moreover, the influences of the parameter-settings of the network (e.g., the structure of network, the noisy intensity and the regularity) are analyzed.

The rest of the paper is organized as: in Sect. 2, the principle of SDAE is given in detail. In Sect. 3, the method used for the construction of Dual-SDAE is discussed. In Sect. 4, by using the method discussed in Sect. 3, two SDAEs are built respectively for the intensity and the depth information. Then, by combining the two SDAEs, a Dual-SDAE is formed for gesture recognition and compared with some classical methods. Some conclusions are given in Sect. 5.

2   Stacked Denoising Autoencoder

For visual, linguistic or other artificial intelligence tasks, in order to learn more effective and abstract features, it is very necessary to build a model with multi-layer nonlinear mapping characteristics, such as the neural network, in which multiple hidden layers are introduced and each hidden layer is a non-linear transformation of the output of the upper layer for the extraction of more complex features. Specifically, in order to extract some useful information from natural images, the original pixels should be gradually transformed into more abstract expressions. The first hidden layer of the network can combine the pixels to detect lines or boundaries, and the second hidden layer can combine lines or boundaries to detect contours or simple "targets". This means, at a deeper level, these contours can be further combined to detect more abstract and advanced features [15]. Thus, many AEs are often cascaded to form a stacked autoencoder (SAE) network to extract more complex and abstract features in the form of deep network structure. Note that, the coding and the decoding processes are defined as Eq. (1) and Eq. (2) for the construction of a hidden layer.

$$Y = \mathbf{S}(WX + b) \tag{1}$$

$$\widehat{X} = \mathbf{S}(W^T Y + b') \tag{2}$$

Here, W is the connecting weight, X is the input vector, $\widehat{X}$ is the reconstructed vector, $\mathbf{S}(\cdot)$ indicates the S-type function, b and b' are the shifting values.

In Fig. 1, a four-layer SDAE network (including one input layer (I), two hidden layers (H_1, H_2) and one output layer (O)) is given in detail. The number of nodes contained in each layer (from top to bottom) is $n_i (i \in [1, 4])$. For the procedure of SDAE, random noise is firstly added to the input signal X to form a noisy input signal $\widetilde{X}$. Then, using Eq. (3) as the object function, the parameters in the rest layers are learned layer-by-layer.

$$\widehat{W} = \arg \max_W \sum_i \| X_i - \widehat{X}_i \|_2^2 \tag{3}$$

where X_i is the i-th input vector and $\widehat{X}_i$ is the corresponding decoded vector via Eq. (2).

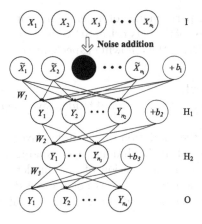

Fig. 1. The network structure of stacked denoising autoencoder

3   Structure of Dual-SDAE Network

Since the information contained in the depth image and the intensity image are different in physics, two SDAE networks are constructed respectively for these images in this paper. Moreover, due to the high level information extracted from the last several layers of neural network, the output of the last layers of the two SDAEs are stacked directly and used as the input of soft-max classifier for classification. The pipeline of Dual-SDAE is given in Fig. 2.

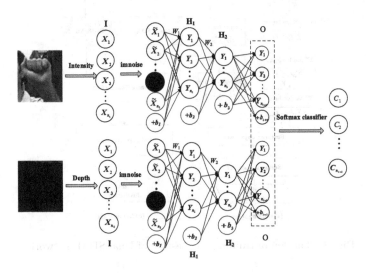

Fig. 2. Network structure diagram of Dual-SDAE

Note that, since the performance of a neural network depends largely on its structure, the designation of the network of SDAE is very important. In this paper, given the property of SDAE, the following aspects are considered in the designation of SDAE. They are:

a) Structure of network

It includes determination of the number of hidden layers and the number of nodes in each hidden layer.

In this paper, using the validation set, a hidden layer is firstly added and the number of nodes contained in this hidden layer is adjusted for a better performance. So, with the sequential addition of new hidden layer, the structure of SDAE is formed until the performance is not improved. In this procedure, the accuracy of the recognition is adopted as the metric to evaluate the network's performance.

b) Noise intensity on the input layer

In our method, pepper-like noise is adopted to enhance the robustness of SDAE. This means, the values inputted into the first layer are set to 0 with a certain probability. To obtain a proper intensity of noise, 200 iterations are used and the optimal noise intensity is selected according to the metric used in the determination of the structure of network.

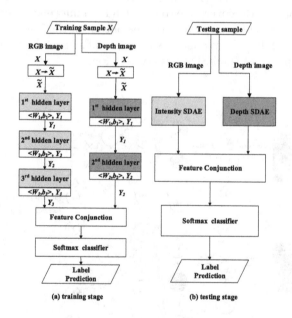

Fig. 3. Diagram of training and testing of Dual-SDAE networks.

c) **Regularity of object function**

It is well-known that too many parameters will lead to a higher complexity of the model, and prone to cause the over-fitting. Therefore, regularity is often introduced into the object function to reduce the complexity of the model. The commonly used regularities are L_1-norm and L_2-norm. Thus, the object function defined as Eq. (3) is modified as

$$\widehat{W} = \arg\max_{W} \sum_i \parallel X_i - \widehat{X}_i \parallel_2^2 + \lambda \parallel W \parallel_p, p \in \{1, 2\} \tag{4}$$

As indicated in [16], the usage of L_1-norm or L_2-norm can enhance the sparsity of the model. Actually, the function of the regularity is very similar to that of the adding of noise. The main difference is that the noise is only added in the first layer, while the regularity is used on each hidden layer. The training and the testing stages of the Dual-SDAE are shown in Fig. 3.

4 Experiments

In this section, American Sign Language (ASL) Dataset is firstly introduced. And then, using ASL Dataset, the parameter-settings of SDAE used respectively for the depth image and the intensity image are analyzed. Note that, all results are obtained with MATLAB 2014a running on Windows 7 (64 bit) OS (3.3 GHz, 32 GB RAM).

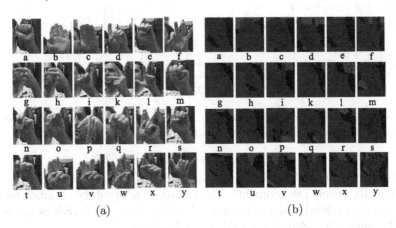

Fig. 4. Hand gesture images of 24 letters sampled from one person. (a) RGB images, (b) depth images corresponding to (a).

4.1 American Sign Language Dataset

American Sign Language (ASL) Dataset is a static image dataset published by Pugeault et al. in 2011. As shown in Fig. 4, it contains gestures expressing 24 different English letters (excluding the English letters j and z). The ASL dataset consists of the hand gestures coming from five people under different lighting conditions and backgrounds using Kinect. Each gesture is recorded via one RGB image and one depth image.

In the experiments, 56400 paired images are chosen and 50400 images are selected randomly as training samples while the rest is used for testing. It means 470 paired images are selected for each person and each letter. Note that, only the intensity information of the RGB image is resized to 32×32 in our method to train the intensity SDAE.

4.2 Influence of Parameters of SDAE

In this subsection, for a better analysis of the structure and the parameter settings of SDAE used respectively for intensity image and depth image, the structure of SAE is firstly analyzed. Then, with the addition of the noisy operator, the performance of SDAE is analyzed. At last, the regularity used in the training of each hidden layer is analyzed.

A. Structure of SAE Designed for Intensity and Depth Images

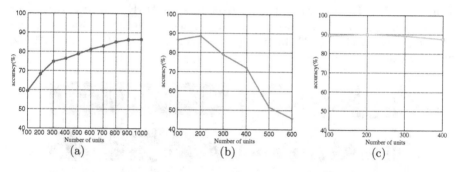

Fig. 5. Curves of the accuracy obtained with I-SAE by sequentially adding (a) the first hidden layer, (b) the second hidden layer and (c) the third hidden layer.

In this part, the number of the iteration is fixed as 200 and all results are the average of 10 repeated experiments. The variations of accuracy obtained for the SDAE build for intensity image are shown in Fig. 5. Since the SAE is designed for intensity image, the network is shortly named as I-SAE.

From Fig. 5(a), it can be seen that in the I-SAE with only one hidden layer, the recognition accuracy is improved with the increasing of the number of hidden layer nodes. When the number of hidden layer nodes is 900, the accuracy achieves 86.24%. Since only less than 0.2% is improved for the accuracy when the number nodes is increased in further, the number of nodes in the first hidden layer is fixed as 900. Then, by using the fixed first hidden layer, a new hidden layer is added and the variation of the accuracy with the increasing of the number of node is given in Fig. 5(b). It is found that a higher accuracy is achieved when the number of nodes in the second hidden layer is fixed as 200. Sequentially, by using this kind of greedy method, the number of the third hidden layer is fixed as 100. The variation of accuracy with the increasing of number of nodes is shown in Fig. 5(c). Note that, as no significant improvement is achieved when the forth hidden layer is added, the number of hidden layer of I-SAE is set as 3 and the corresponding number of node is set as 900, 200, 100.

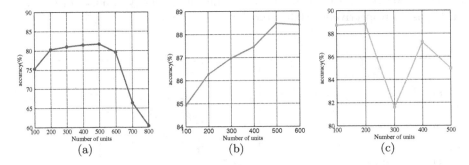

Fig. 6. Curves of the accuracy obtained with D-SAE by sequentially adding (a) the first hidden layer, (b) the second hidden layer and (c) the third hidden layer.

Using the same method described as the above, the structure of the depth-based SAE (D-SAE) network is determined. The variations of the accuracy are shown in Fig. 6. From Fig. 6, we can see that the highest accuracy is achieved by using two hidden layers whose nodes are respectively 500 and 500. Moreover, as the third hidden layer is added, the performance of D-SAE is not improved. Thus, the D-SAE with two hidden layers is adopted in latter experiments.

B. Intensity of Noisy Operation

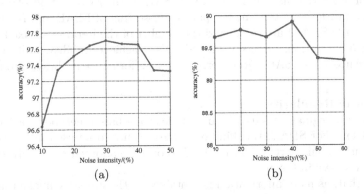

Fig. 7. Influence of different noisy intensity on the performance of (a) I-SDAE and (b) D-SDAE.

As aforementioned, the usage of noisy operator is good for the improvement on the robustness of SAE. So, in this subsection, using the same settings in Sect. 3, the influence of the intensity of noise on the two SAEs is analyzed. Note that, due to the usage of noisy operation, the results are the average of ten repeated experiments and the iteration number is fixed as 200. Thus, I-SAE and D-SAE are renamed as I-SDAE and D-SDAE, whose results are shown in Fig. 7. From Fig. 7, we can see that the intensity of noise is of significant influence on the performance of the two SDAEs. With a careful comparison, the intensity of noise for the two SDAEs is set respectively as 0.3 and 0.4.

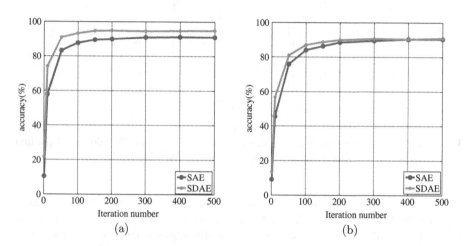

Fig. 8. The impact of adding noise for (a) intensity information and (b) depth information under different iterations.

To make a further analysis on the usage of the noisy operation, the performances of two SAEs (namely I-SAE and D-SAE) with or without noisy operation are shown in Fig. 8. Note that, the parameter-setting of I-SAE and I-SDAE (or D-SAE and D-SDAE) are same, and the number of the iteration is fixed as 500. Visually, 200 iterations are needed for SDAEs to achieve stability, while 400 iterations are needed for SAE. Moreover, a better performance is always obtained by SDAE.

C. Usage of Regularity

In literatures, regularity is usually used in learning based method to achieve the improvement. For SDAE, this kind of strategy can also be used in the training stage. Thus, in this part, L_1-norm and L_2-norm are adopted and analyzed. The results are shown in Fig. 9.

Note that, as more information is contained in the intensity image than that of the depth image, only the intensity images are used in this part to analyze the regularity. The weight of the regularity is searched in the grid $[10^{-7}, 10^{-6}, 10^{-5}, \ldots, 1, 10, 100]$, and the results are shown in Fig. 9(a). With a carefully observation, the best coefficients for L_1 and L_2 regularity are 10^{-5} and 10^{-6}, respectively. Note that, by using the regularity, the performance of SDAE is improved and a higher improvement is achieved by using L_2-norm.

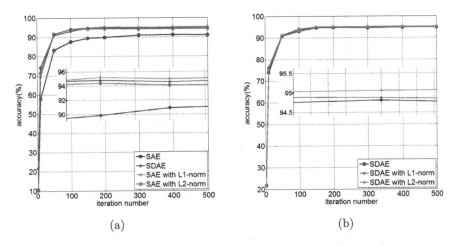

(a)   (b)

Fig. 9. The influence of adding noise, L_1-norm and L_2-norm under different iterations.

Since the influences of the noisy operator and the regularity are very similar, the performances of SAE with and without regularity are compared with that of SDAE in Fig. 9(b). Visually, the usages of the regularity and the noisy operation both improve the performance of SAE. However, the improvement obtained by using the regularity and the noisy operation are really different. This can be obtained from Figs. 9(a) and (b).

4.3   Dual-SDAE

Given above analysis, a Dual-SDAE is designed in this sub-section, in which I-SDAE and D-SDAE are combined together for the recognition. Specifically, the output of I-SDAE and D-SDAE are stacked together in Dual-SDAE as the input of Soft-max classifier (as shown in Fig. 3).

The performance of Dual-SDAE, I-SDAE and D-SDAE are given in Fig. 10 under different iterations. Visually, after 150 iterations, the curve of Dual-SDAE tends to be stable, which is quicker than that of I-SDAE and D-SDAE. Moreover, the accuracy obtained by Dual-SDAE is 96.51% which is the highest among the three methods.

To make a further analysis, the accuracy of each letter obtained by Dual-SDAE, I-SDAE and D-SDAE is given in Fig. 11. Obviously, for the recognition of most letters, a significant improvement is obtained by Dual-SDAE than that of I-SDAE and D-SDAE. This means the fusion of the intensity information and the depth information is very necessary for a better recognition of hand gesture.

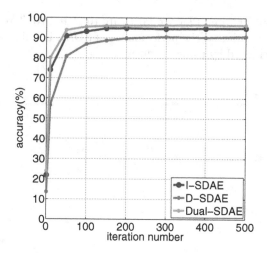

Fig. 10. The performance of Dual-SDAE under different iterations.

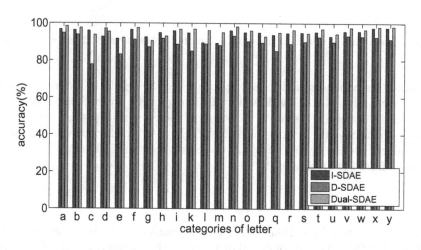

Fig. 11. The accuracy of 24 letters obtained via I-SDAE, D-SDAE and Dual-SDAE.

4.4 Comparison with Existing Gesture Recognition Methods

To verify the effectiveness of our method, HSF-RDF method [17], SIFT-PLS method [18], MPC method [19], DBN and CNN are adopted for comparison. Note that, HSF-RDF method uses random forest for classification. SIFT-PLS method extracts SIFT features and uses Partial Least Squares (PLS) classifier to recognize gesture. MPC method uses the center and the area of the hand gesture for recognition. For DBN, the number of hidden layer is fixed as 3 and corresponding number of nodes are set sequentially as 900, 200 and 100 respectively. For CNN, it consists of two convolution layers, two pooling layers and one full connection layer, and the size of convolution kernel is fixed as 5×5. The results are shown in Table 1.

Table 1. Recognition rate of multiple recognition methods

Recognition methods	HSF+RDF	SIFT+PLS	MPC	DBN	CNN	Dual-SDAE
Recognition rate	75%	71.51%	90.19%	84.65%	88.22%	**96.51%**

Obviously, the highest accuracy is achieved by Dual-SDAE. Besides, in our experiments, only 1 h is needed for Dual-SDAE in the training stage, while it is 1.8 h for DBN and 7 h for CNN.

5   Conclusions

In this paper, a Dual-SDAE is proposed for the gesture recognition, where two SDAEs designed respectively for the intensity image and the depth image are combined together. Using ASL Dataset, the performance of Dual-SDAE is analyzed in details. In the experiments, by comparing with some traditional methods and two deep neural networks, the proposed method shows the best performance. This means the combination of depth information and intensity information is good for the recognition.

References

1. Pavlovic, V.I.: Visual interpretation of hand gestures for human-computer interaction a review. IEEE Trans. Pattern Anal. Mach. Intell. **19**(7), 677–695 (1997)
2. Pavlovic, V.I.: Hand modeling, analysis and recognition. IEEE Signal Process. Mag. **18**(3), 51–60 (2001)
3. Murthy, G.R.S., Jadon, R.S.: Hand gesture recognition using neural networks. In: 2010 IEEE 2nd International on Advance Computing Conference (IACC), pp. 134–138 (2010)
4. Hasan, H., Abdul-Kareem, S.: Static hand gesture recognition using neural networks. Artif. Intell. Rev. **41**(2), 147–181 (2012). https://doi.org/10.1007/s10462-011-9303-1
5. Rumelhart, D.E.: Learning representations by back-propagating errors. Nature **323**(6088), 533–536 (1986)
6. Bengio, Y., Lamblin, P.: Greedy layer-wise training of deep networks. In: Advances in Neural Information Processing Systems 19 (NIPS 2006), pp. 153–160. MIT Press, Vancouver, British Columbia, Canada (2006)
7. Vincent, P., Larochelle, H.: Extracting and composing robust features with denoising auto-encoders. In: International Conference on Machine Learning, pp. 1096–1103 (2008)

8. Vincent, P.: Stacked denoising auto-encoders: learning useful representations in a deep network with a local denoising criterion. J. Mach. Learn. Res. **11**(6), 3371–3408 (2010)
9. Budiman, A., Fanany, M.I.: Stacked denoising auto-encoder for feature representation learning in pose-based action recognition. In: IEEE 3rd Global Conference on Consumer Electronics (GCCE), pp. 684–688 (2014)
10. Sang, R.X.: Discriminative feature learning for action recognition using a stacked denoising auto-encoder. Intell. Data Anal. Appl. **297**, 521–531 (2014)
11. Liang, X.D., Wei, Y.C.: Reversible recursive instance-level object segmentation. In: Proceedings of IEEE Conference on Computer Vision and Pattern Recognition, pp. 633–641 (2015)
12. Santara, A., Maji, D.: Faster learning of deep stacked autoencoders on multi-core systems using synchronized layer-wise pre-training. In: the Workshop on PDCKDD 2015 as a part of ECML/PKDD 2015 (2016)
13. Wang, H., Shi, X.J.: Collaborative recurrent auto-encoder: recommend while learning to fill in the blanks. In: 30th Conference on Neural Information Processing Systems (NIPS 2016) (2016)
14. Hasan, M., Choi, J.: Learning temporal regularity in video sequences. In: Proceedings of IEEE Conference on Computer Vision and Pattern Recognition, pp. 733–742 (2015)
15. Bengio, Y.: Learning deep architectures for AI. Found. Trends Mach. Learn. **2**(1), 1–127 (2009)
16. Zhu, Q.: Combining L1-norm and L2-norm based sparse representations for face recognition. Optik Int. J. Light Electron Opt. **126**(7–8), 719–724 (2015)
17. Pugeault, N., Bowden, N.: Spelling it out: real-time ASL fingerspelling recognition. In: 2011 IEEE International Conference on Computer Vision Workshops (ICCV Workshops), pp. 1114–1119 (2011)
18. Estrela, B.S.N., Campos, M.F.M.: Sign language recognition using partial least squares and RGB-D information. In: Proceedings of the 9th Computational Workshop (2013)
19. Pansare, J.R.: Real-time static hand gesture recognition for American Sign Language (ASL) in complex background. J. Sig. Inf. Process. **03**(3), 364–367 (2015)

Image Generation from Layout via Pair-Wise RaGAN

Ting Xu[ID], Kai Liu[ID], Yi Ji[(✉)], and Chunping Liu

College of Computer Science and Technology, Soochow University, Suzhou, China
{txu7,kliu0923}@stu.suda.edu.cn, {jiyi,cpliu}@suda.edu.cn

Abstract. Despite recent remarkable progress in image generation from layout, synthesizing vivid images with recognizable objects remains a challenging problem, object distortion and color imbalance occasionally happened in the generated images. To overcome these limitations, we propose a novel approach called Pair-wise Relativistic average Generative Adversarial Network (P-RaGAN) which includes a pair-wise relativistic average discriminator for enhancing the generative ability of network. We also introduce a consistency loss into our model to keep the consistency of original latent code and reconstructed or generated latent code for reducing the scope of solution space. A series of ablation experiments demonstrate the capability of our model in the task from layout to image on the complicated COCO-stuff and Visual Genome datasets. Extensive experimental results show that our model outperforms the state-of-the-art methods.

Keywords: Image generation · Layout · Relativistic average Discriminator · Consistency loss

1 Introduction

Image generation from layout is a novel hot topic in computer vision, which requires the function of dealing with incomplete information in the layout and the ability of learning how to handle multi-modal information(vision and language). Existing powerful algorithms can conveniently assist technicians to comprehend visual patterns. Therefore, there exists a number of visual applications, *e.g.* self-driving technology, it is a help for art creation and a novel aid in scene graph generation [25].

Here layout includes semantic layout (bounding boxes and object shapes) and spatial layout (bounding boxes and object category). It can be used not only as

Supported by National Natural Science Foundation of China (61972059, 61773272), The Natural Science Foundation of the Jiangsu Higher Education Institutions of China (19KJA230001), Key Laboratory of Symbolic Computation and Knowledge Engineering of Ministry of Education, Jilin University (93K172016K08), Suzhou Key Industry Technology Innovation-Prospective Application Research Project SYG201807), the Priority Academic Program Development of Jiangsu Higher Education Institutions.

initial domain in the task from layout to image but also as intermediate representation in other task such as text-to-image (known as T2I). Most promising results in deep learning are based on generative model like Variational Auto-Encoders (VAEs) and Generative Adversarial Networks (GANs). Hong *et al.* [9] proposed a two-step image synthesizing method with constructing a semantic layout from the text, which can generate an image conditioned on the layout and text description. But their methods did not follow the end-to-end training manner. Hinz *et al.* [8] introduced a novel method which added an object pathway to both the generator and discriminator for controlling the location of the objects. Compared with semantic layout, the spatial layout is a coarse-grained description. Image generation from the spatial layout does not need annotate the segmentation masks. There are two main challenges in this task. The first challenge is how to synthesize images which must correspond to layout. Second, from the perspective of human vision, the generated samples need to be real enough. Zhao *et al.* [22] proposed an approach called layout2im based on VAE-GAN network for synthesizing images from the spatial layout. Most their samples generated by layout2im did correspond to the given layout. However, object distortion and color imbalance has occasionally arisen in the generated samples.

Most existing image generation from layout to image tasks are GAN-based methods, but the training of vanilla GANs is notorious due to its uncertainty and instability. Therefore, many researchers are devoted to the method study of how to stabilize the training of GANs. The Wasserstein GAN (WGAN) [4] is a good illustration of stability training, which adopts Earth Mover Distance (EMD) instead of Jensen-Shannon (JS) divergence as an objective. Mao *et al.* [23] propose another method by replacing the cross entropy loss function with the least square loss function. The method also attempted to use different distance measures to build a more stable and converging adversarial network. Although these methods have partially solved the stability problem, some of them such as[3] are required more powerful computing performance than Standard GANs.

Our work aims to improve the generative ability of GAN for the high-quality of generated images. The overall framework of the proposed method is shown in Fig. 1. Consequently, we propose a novel method which contains two new components. One is to introduce a pair-wise relativistic average discriminator[12] for increasing generative capability, the other is to propose a consistency loss function for reducing the scope of solution space. The major contributions of our method are summarized here:

1) We propose a new model for generating realistic image from coarse layout.
2) We introduce a pair-wise relativistic average discriminator, which includes an image-wise discriminator that can generate more recognizable and vivid images and an object-wise discriminator that can address the problem of object distortion.
3) We adopt a consistency loss for narrowing the scope of solution space.

In the remainder of this paper is described below. We shortly present related work about our study in Sect. 2, and in Sect. 3, we illustrate in detail the components of P-RaGAN. Then we focus on the experimental results and analysis in Sect. 4.

2 Related Work

Conditional GANs. The introduction of GAN[6] has achieved promising results in image generation tasks. It has been an increasing attention on conditional image generation or manifold-valued Image Generation [24]. Conditions can be text [17,20,21], image [10], scene graph [11] or layout[22]. However, model collapse does exist with these GAN-based methods. In the Standard Generative Adversarial Networks (SGANs) methods, the discriminator is only responsible for obtaining the probability that the input sample is real data, and the generator is trained to synthesize samples that can "cheat" discriminator. But this way isn't appropriate in the sight of a priori knowledge that half of a mini-batch of data comes from the generated samples [12]. Moreover, from the perspective of human cognition, the discriminator should also output higher scores to generated sample which is more realistic than the real one. [12] proposed a relativistic discriminator to solve these problems and achieved the satisfied results. But it only judges whether the input image is relatively realistic at image level. For this, we discriminate at object level by comparing the authenticity between the input object patches and the opposite type.

Image Generation from Layout. The image generation from layout is formulated as:

$$I_g = G\left(L_S\right). \tag{1}$$

where L_S denotes the original layout which is composed of an appearance latent code z^r, the position of objects in images $B_{1:T}$ and the class label of object c. G is a generator network. I_g refers to the generated image. Since $z^r \in \mathbb{R}^{64}$ and $I_g \in \mathbb{R}^{C \times H \times W}$ (C refers to the channel, H and W are the height and the width of input image, respectively.) are essentially a low dimensional vector and a high dimensional vector, respectively, the layout-image pair can't be perfectly aligned and it prevents the generator from synthesizing more realistic and detailed images.

In most recent studies [8,9,11,14,22], layout-based image generation methods have been explored. These methods [8,9,11,14] divided T2I task into the following two steps: language → layout and layout → image. These approaches reduce the difficulty of generation because they employ layout as a bridge across a huge semantic gap between images and natural language descriptions. Despite that layout2im [22] takes layout as input, the approach can generate image by combining variational auto-encoders and generative adversarial network. Because of the standard generative adversarial network in Layout2im, the quality of the generated images is relatively poor.

One major challenge in layout to image task is how to generate images with more details and realistic objects without shape abnormality. To this end, we propose a novel Pair-wise Relativistic average Generative Adversarial Network (P-RaGAN). P-RaGAN consists of a pair-wise relativistic average discriminator. We also introduce a consistency loss for promising the consistency of the object latent code pairs (z^r, z_r') and (z^o, z_o').

3   Generating Realistic Image from Coarse Layout

It is proved that the training of traditional GANs network such as SGANs or DCGANs is troublesome from existing powerful empirical and theoretical evidence [15]. Therefore the task of generating high-quality images is difficult due to the training instability of traditional GANs. Thus, for generating more high-quality images, we construct a realistic image generation framework. The detailed framework is illustrated in Fig. 1. In order to generate images without object distortion by enhancing the power of GAN, we introduce the pair-wise relativistic average discriminator (PwRaD). And to reduce the scope of solution space, we also introduce the consistency loss function, as revealed in Fig. 1.

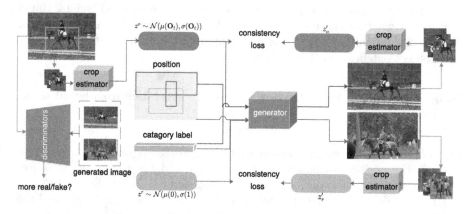

Fig. 1. Overview pipeline of the proposed model. The model first cuts out the objects in the real samples and employs the crop estimator to fit the distribution of these objects patches, then samples two latent codes from the distribution and the standard normal distribution, respectively. The layout consists of the code pair, together with the object's category label and location. Given the layout, it is fed into our generator to synthesize images. We also introduce a pair-wise relativistic average discriminator to generate reasonable and vivid images. In addition, consistency loss is designed to ensure the consistency between latent code pairs for constraining the range of solution space.

3.1   Relativistic Average Generative Adversarial Network

In [12], Jolicoeur-Martineau and Alexia pointed out that GANs based on non-Integral Probability Metrics (IPM) [16] couldn't generate the samples with high-quality, while IPM-based GANs only partially solved this problem. Hence, Jolicoeur-Martineau *et al.* proposed a relativistic discriminator network for stabilizing the training of GANs.

$$D(I_r, I_g) = \text{act}\left(F\left(I_r\right) - F\left(I_g\right)\right) \tag{2}$$

Here, *act* is activation function, $F(\cdot)$ represents the output of the last convolutional layer, I_r means real samples from dataset, I_g describes the false examples from generator.

$$D^{rev}(I_r, I_g) = \text{act}\left(F\left(I_g\right) - F\left(I_r\right)\right) \tag{3}$$

The model (formula (3)) not only increases the possibility of fake sample being more realistic, but also reduces the possibility of real data. To some extent, this makes the JS divergence between the two distributions ($\mathbb{P}_{data}$ and $\mathbb{P}_{fake}$) is minimized [12], which leads to the resolution of gradient disappearance in the GAN.

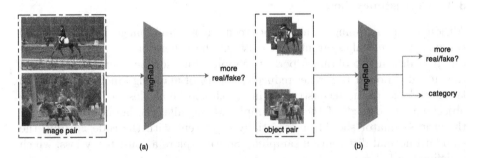

Fig. 2. Image-wise relativistic average discriminator (imgRaD) and object-wise relativistic average discriminator (objRaD). Both of imgRaD and objRaD need to compute the probability that the input is real. The objRaD also needs to output the probability that the input object belongs to each category, which is performed by a fully connected layer.

3.2   Pair-Wise Relativistic Average Discriminator

Inspired by the fact that the relativistic average discriminator (RaD) [12] can offer the promise of generalization ability, we propose a pair-wise relativistic average discriminator (Fig. 2) to estimate the probability of the given input data that is more realistic than the opposite type of input data on average. The discriminator includes an image-wise discriminator and an object-wise discriminator. Different from the standard GANs, the object-wise discriminator requires the similarity of input data and its opposite type in object level. The discriminators we proposed can be defined as:

$$\bar{D}_{img/obj}(\cdot) = \begin{cases} \text{act}\left(F(t) - \mathbb{E}_{I_g \sim \mathbb{P}_{fake}} F(z)\right) & \text{if } t \sim \mathbb{P}_{data} \\ \text{act}\left(F(t) - \mathbb{E}_{I_r \sim \mathbb{P}_{data}} F(I_r)\right) & \text{if } t \sim \mathbb{P}_{fake} \end{cases} \tag{4}$$

And the relativistic loss function of discriminator is:

$$L_{D_{img/obj}} = -\mathbb{E}_{I_r \sim \mathbb{P}_{data}} \left[\log\left(\bar{D}_{img/obj}(I_r)\right)\right] \\ -\mathbb{E}_{I_g \sim \mathbb{P}_{fake}} \left[\log\left(1 - \bar{D}_{img/obj}(I_g)\right)\right] \tag{5}$$

In the image-wise discriminator (D_{img}), I_r and z represent the ground-truth and the synthesized samples respectively. In the object-wise discriminator (D_{obj}), I_r and z refer to ground-truth and synthesized images patches containing objects, respectively. Neither of the two discriminators is added to the batch normalization layer. The proposed discriminator improves the corresponding loss function and ensures that the input data is more real than its opposite type on average, not only more real than a small part. Experiments show that the proposed model can generate pseudo-real-world image.

3.3   Consistency Loss

Although the mapping function between layout and image can be learned through the mutual confrontation between the generator and the discriminator. It is difficult to obtain a perfect mapping function between the input and the output. In order to further reduce the space of mapping function, the object latent code z'_o of the reconstructed image should be as close as possible to the object latent code z^o of the real sample. Meanwhile, the latent code z'_r from the crop estimator should be essentially consistent with the code z_r from the standard normal distribution sampling. So we explore a consistency loss, which is defined as follows:

$$\mathcal{L}_{con} = \mathbb{E}_{(z^o, z'_o) \sim (\mathbb{P}_{O_r}, \mathbb{P}_{O_g})} \left[\|z^o - z'_o\|_1\right] \\ + \mathbb{E}_{(z^r, z'_r) \sim (\mathbb{P}_{identity}, \mathbb{P}_{O_g})} \left[\|z^r - z'_r\|_1\right] \tag{6}$$

where $\mathbb{P}_{O_r}$ is the distribution of objects in the ground-truth images, $\mathbb{P}_{O_g}$ represents the distribution of objects in the generated.

3.4   Total Loss Function

Our final loss function is a weighted sum of these loss functions, which is defined in detail:

$$\mathcal{L}_{total} = \lambda_0 \mathcal{L}_{KL} + \lambda_1 \mathcal{L}_{con} + \lambda_2 \mathcal{L}_{img} + \lambda_3 \mathcal{L}_{obj} + \lambda_4 \mathcal{L}_{AC}^{obj} \tag{7}$$

where $\mathcal{L}_{KL}$ is the KL-divergence between the normal distribution and the distribution $\mathbb{P}_{O_t}$, $\mathcal{L}_{con}$ relates to the consistency loss, $\mathcal{L}_{img}$ and $\mathcal{L}_{obj}$ describe the relativistic loss from discriminators, $\mathcal{L}_{AC}^{obj}$ is the same as the classifier loss in layout2im. The λ_i is the hyper-parameter for balancing the proportion of various losses. We set λ_1 to 0.1, 0.5, 1 and 5, respectively, in the experiments. We find that when we set $\lambda_0 = 0.1$, $\lambda_1 = 0.5$, $\lambda_2 = 1$, $\lambda_3 = 1$ and $\lambda_4 = 1$, we get the best results.

4   Experimental Results and Analysis

In this section, we have carried out extensive experiments on the datasets with multi-objects and complex scene images, e.g. COCO-Stuff [5] and Visual Genome [13] datasets. At the same time, we also evaluate the performance of the proposed method from three aspects: the recognizability and stability of the generated image, the consistency with the image distribution in the datasets and the structural similarity of human visual perception. Finally, we further analyze the role of the pair-wise relativistic average discriminator and the consistency loss function for improving the quality of generated image through ablation experiments.

4.1   Evaluation Metrics

We adopt three evaluation indicators for evaluating the performance of our layout-conditional image generation method: Inception Score (IS) [18], Fre'chet Inception Distance (FID) [7], and structural similarity index (SSIM) [19]. The evaluation of inception scores on visual quality is considered to be related to human perspective [2]. For measuring the recognizability and diversity of synthesized samples, we apply the pre-trained classifier (VGG-net [1]) to all images generated by our model and baseline to study the statistical characteristics of its score distribution for computing the inception scores. The higher the score, the better. Unlike Inception Score, the FID score can measure whether the generated image is in the same distribution as the image in the dataset. We prefer getting lower scores for better performance. Evaluation based on Fre'chet Inception Distance (FID) is beneficial for reality evaluation. But when an image has its own feature (for example, as long as it has the feature of human face, even though the position of eyes and mouth is changed), IS and FID will still give it a high evaluation, which is obviously contrary to human cognition. We choose SSIM as another evaluation metric because SSIM is considered to be correlated with the image quality perception of the human visual system (HVS) [19]. The higher the quality of the image, the greater the SSIM value. Here we show it in percentage form.

Fig. 3. The partial samples of generated image using given layout. The results of the COCO-Stuff and Visual Genome datasets are shown in rows 1 to 4 and 5 to 8, respectively. For sg2im and layout2im, we use the pre-trained model to generate images.

Table 1. Quantitative evaluation results. GT indicates ground-truth layout. layout2im-p refers to the pre-trained model. layout2im-o: we train the layout2im from scratch.

Method	IS		FID		SSIM	
	COCO	VG	COCO	VG	COCO	VG
real image (64 × 64)	16.3 ± 0.4	13.9 ± 0.5	-	-	-	-
pix2pix [10]	3.5 ± 0.1	2.7 ± 0.02	121.97	142.86	-	-
sg2im(GT) [11]	7.3 ± 0.1	6.3 ± 0.2	67.96	74.61	-	-
layout2im [22]	9.1 ± 0.1	8.1 ± 0.1	**38.14**	**31.25**	-	-
layout2im-p	9.1 ± 0.1	8.1 ± 0.1	42.80	40.07	24.2	36.1
layout2im-o	9.1 ± 0.1	8.1 ± 0.1	43.80	39.39	24.1	36.1
ours	**10.1 ± 0.1**	**8.5 ± 0.1**	38.70	34.13	**25.7**	**38.2**

4.2 Comparing Analysis

According to the three evaluation indexes, we compare with the experimental results of three state-of-the-art image generation methods based on layout, and the quantitative evaluation results are summarized in Table 1. We present three evaluation results based on IS, FID and SSIM. Pix2pix [10] has achieved promising results in image translation. Here, we set the original domain as the layout and the real image as the target domain. Sg2im [11] adopts graph convolution to handle scene graph in the T2I. It can be applied to generate image from layout only by using ground-truth layout. Layout2im [22] is originally trained for taking layout as input to generate images.

From the IS index of two experimental datasets, our method is improved from 9.1 to 10.1 and from 8.1 to 8.5 respectively. Since this index represents the recognizability and diversity of object in the generated image, this demonstrates that our method can generate more recognizable and realistic objects. FID is more suitable to describe the diversity of GAN, and has better robustness to noise. The FID of our model is slightly lower than that reported by layout2im, because the images generated by our model are more the same. Our approach increases SSIM from 24.2 to 25.7 and from 36.1 to 38.2. SSIM is a full reference image quality evaluation index. It measures the similarity of two images from three aspects of brightness, contrast and structure. Its design takes the visual characteristics of the human visual system (HVS) into account, which is more consistent with the visual perception of the human eye than the traditional way. This also explains that from a human point of view, our model produces images that are more strongly correlated with the real samples in the dataset. In other words, it indicates that our method is easier to generate images that are better aligned with real samples. And our proposed approach significantly outperforms these existing approaches [10, 11, 22].

In addition, we respectively illustrate the visual effect of the generated image in Fig. 3 and Fig. 4. Given the coarse layout, Fig. 3 shows the partial generated image results of sg2im, layout2im and the proposed methods on two experimental datasets. Due to the combination of variational auto-encoders and generative adversarial network, layout2im method does synthesize images that correspond with layout. However, the object shape is distorted in Fig. 3 (a), (c), (g) and (j). Furthermore, compared with the object location of given layout, some objects' location in the generated image is wrong in Fig. 3 (k), (l) and (p), which makes the generated images look fake. Since their model ignores the fact that half of the input data is fake, this inevitably leads to the distortion in the generated images. In Fig. 3 (e) and (n), the reason why the images generated by sg2im look chaotic is that the information that scene graphs can express is very limited, and only six relationships are defined in the experiment.

In contrast, our model applies the consistency loss penalty model to ensure that the generated image is close enough to the ground-truth, and on this basis, we make full use of the fact that half of the data fed into the discriminator is fake, so the generated image with high recognition is more close to the ground-truth image.

Fig. 4. Partial samples of generated images on the COCO-Stuff (top) and Visual Genome (bottom) datasets. The ground-truth images are shown in first row. For sg2im (the second row) and layout2im (the third row), we use the pre-trained model to generate samples. (Color figure online)

Table 2. Ablation Study. We trained baseline from scratch. CL is the consistency loss, imgRaD and objRaD refer to image-wise and object-wise relativistic average discriminator, respectively, the PwRaD is the pair-wise relativistic average discriminators.

	IS		FID		SSIM	
Method	COCO	VG	COCO	VG	COCO	VG
baseline-o	9.1 ± 0.1	8.1 ± 0.1	43.80	39.39	24.1	36.1
baseline+CL	9.47 ± 0.2	8.2 ± 0.1	39.72	35.46	24.9	36.9
baseline+imgRaD	8.9 ± 0.2	7.8 ± 0.3	40.13	38.51	24.1	36.5
baseline+objRaD	9.4 ± 0.1	8.1 ± 0.1	39.83	37.65	24.2	37.1
baseline+PwRaD	9.7 ± 0.1	8.3 ± 0.1	39.14	35.38	25.0	37.7
baseline+PwRaD+CL	$\mathbf{10.1 \pm 0.1}$	$\mathbf{8.5 \pm 0.1}$	**38.70**	**34.13**	**25.7**	**38.2**

To further prove that our proposed model doesn't memorize the images, but has ability to generate images. We present the experimental results based on given layout and given the ground-truth images in Fig. 4. As shown in Fig. 4 (b), our model changes the appearance and direction of the bus, the image generated by layout2im is not so real, although it looks like a car. The shape of the car in sg2im is excessively distorted. As depicted in Fig. 4 (c), without changing the background of the sea, our model "upgrades" the ship in the image. In Fig. 4 (n), our model has changed the color of the sky and the river, and layout2im has changed the color of both, but it has turned the river yellow, which is obviously not in line with the cognition of the human visual system. Sg2im produces incomprehensible images.

From Fig. 4, we can also see that our model can better overcome the problems of the object shape distortion and color imbalance in sg2im and layout2im. For shape distortion, layout2im composes a tree in the wrong place in Fig. 4 (g) and (k). Sg2im directly ignores the existence of trees. Figure 4 (a), the car from layout2im looks more like a wall. The car synthesized by sg2im can't be recognized by human eyes. For color imbalance, as illustrated in Fig. 4 (o), the image generated by layout2im shows an unknown red object, sg2im struggles with the object shape and color. From these examples, we can see that our model can generate images with correct colors and shapes: Fig. 4 (g) shows two buses, (o) doesn't contains strange stuff.

Based on the analysis of the three evaluation indexes and the corresponding visualization results, we can see that our proposed method is effective to overcome the problems of shape distortion and color distortion of objects, and our model has the ability to generate images containing complex objects. It is undeniable that our model is not very good for some slender and tiny objects. We consider that our input is set to 64 × 64 low resolution images. How to generate high resolution image with clear tiny objects, obviously, is an area of special interest in our future work.

4.3   Ablation Study

Compared with the framework of baseline model, our proposed framework of image generation from coarse layout includes two new components: pair-wise relativistic average discriminators and a consistency loss function. In order to further fully demonstrate that the role of each component in improving the quality of the generated images, we conduct ablation experiments on two experimental datasets. The ablation results are summarized in Table 2.

Pair-Wise Relativistic Average Discriminators. Our pair-wise relativistic average discriminator includes image-wise relativistic average discriminator (imgRaD) and object-wise relativistic average discriminator(objRaD). Through the combination of baseline model and the different components, the evaluation results of two experimental datasets under three evaluation indexes are given in Table 2. We test the compact of removing the relativistic average discriminators.

Specifically, the model trained without objRaD which decreases inception score obtains poor performance, since the model cannot generate recognizable objects. The SSIM score is still high because of the image-wise discriminator. Without the constraint of image relativistic adversarial loss, the model leads to lower score as it ignores the priori knowledge that only half of the input data is real, and the model does not reduce the possibility of discriminating the real data as real. The pair-wise relativistic average discriminator-based GAN achieves higher scores, showing that the synthesized images include more vivid structures. For example, Fig. 3 (b) shows the skis, (p) sets tree in correct position.

Consistency Loss. A consistency loss term in total loss function promises the alignment between the two pairs of latent codes (z^r, z_r') and (z^o, z_o'). Does this result in better performance? Table 2 reports the scores of this variant. The model with consistency loss increases IS score from 9.1 to 9.4, decreases FID score to 39.7, and improves SSIM value to 24.9. In Fig. 4 (e), compared with sg2im and layout2im, the generative locomotive can be easily identified. Figure 4 (r) displays the mountain and sky which are very similar to those in ground-truth image. This demonstrates that the consistency loss penalizes the distance between the original latent codes z^r, z^o and reconstructed z_r' or generated z_o' latent code. Because L1 norm is more capable of handing the exception value, we use L1 norm to constrain the final loss function in the process of the image generation. And we also try L2 norm in the experiment, but the experimental results are not ideal.

5   Conclusion

In this paper, we propose an approach for image generation from layout which can synthesize more recognizable objects and vivid images. We improve the baseline model layout2im by introducing a pair-wise relativistic average discriminator and a consistency loss function. The quality of image generation is improved using the proposed image-wise and object-wise discriminators. We also design a consistency loss for constraining the scope of solution space. The proposed method Pair-wise Relativistic average Generative Adversarial Network (P-RaGAN) can achieve the better performance based on the objective evaluation metrics (IS, FID and SSIM) and the visual perception evaluation of generated image. Our future work is further enhance the tiny object of image generation.

References

1. Simonyan, K., Zisserman, A.: Very deep convolutional networks for large-scale image recognition. In: ICLR, Proceedings of the International Conference Learning Representation, pp. 1–14 (2015)
2. Odena, A., Olah, C., Shlens, J.: Conditional image synthesis with auxiliary classifier GAN. In: Proceedings of the 34th International Conference on Machine Learning-Volume 70. JMLR. org, pp. 2642–2651 (2017)

3. Gulrajani, I., Ahmed, F., Arjovsky, M., Dumoulin, V., Courville, A.C.: Improved training of Wasserstein GANs. In: Advances in Neural Information Processing Systems, pp. 5767–5777 (2017)
4. Arjovsky, M., Chintala, S., Bottou, L.: Wasserstein generative adversarial networks. In: Proceedings of the 34th International Conference on Machine Learning, ICML 2017, Sydney, NSW, Australia, 6–11 August 2017, Proceedings of Machine Learning Research, vol. 70 pp. 214–223, PMLR (2017)
5. Caesar, H., Uijlings, J., Ferrari, V.: Coco-stuff: Thing and stuff classes in context. In: Proceedings of the IEEE Conference on Computer Vision and Pattern Recognition, pp. 1209–1218 (2018)
6. Goodfellow, I., et al.: Generative adversarial nets. In: Advances in Neural Information Processing Systems, pp. 2672–2680 (2014)
7. Heusel, M., Ramsauer, H., Unterthiner, T., Nessler, B., Hochreiter, S.: GANs trained by a two time-scale update rule converge to a local NASH equilibrium. In: Advances in Neural Information Processing Systems, pp. 6626–6637 (2017)
8. Hinz, T., Heinrich, S., Wermter, S.: Generating multiple objects at spatially distinct locations. In: ICLR (Poster). OpenReview.net (2019)
9. Hong, S., Yang, D., Choi, J., Lee, H.: Inferring semantic layout for hierarchical text-to-image synthesis. In: Proceedings of the IEEE Conference on Computer Vision and Pattern Recognition, pp. 7986–7994 (2018)
10. Isola, P., Zhu, J.-Y., Zhou, T., Efros, A.A.: Image-to-image translation with conditional adversarial networks. In: Proceedings of the IEEE Conference on Computer Vision and Pattern Recognition, pp. 1125–1134 (2017)
11. Johnson, J., Gupta, A., Fei-Fei, L.: Image generation from scene graphs. In: Proceedings of the IEEE Conference on Computer Vision and Pattern Recognition, pp. 1219–1228 (2018)
12. Jolicoeur-Martineau, A.: The relativistic discriminator: a key element missing from standard GAN. In: 7th International Conference on Learning Representations, ICLR 2019, New Orleans, LA, USA, May 6–9, 2019. OpenReview.net (2019)
13. Krishna, R., et al.: Visual genome: connecting language and vision using crowd-sourced dense image annotations. Int. J. Comput. Vis. **123**(1), 32–73 (2017)
14. Li, W., et al.: Object-driven text-to-image synthesis via adversarial training. In: Proceedings of the IEEE Conference on Computer Vision and Pattern Recognition, pp. 12174–12182 (2019)
15. Martin, A., Lon, B.: Towards principled methods for training generative adversarial networks. In: NIPS 2016 Workshop on Adversarial Training. In Review for ICLR, vol. 2016 (2017)
16. Müller, A.: Integral probability metrics and their generating classes of functions. Adv. Appl. Probab. **29**(2), 429–443 (1997)
17. Reed, S., Akata, Z., Yan, X., Logeswaran, L., Schiele, B., Lee, H.: Generative adversarial text to image synthesis. In: International Conference on Machine Learning, pp. 1060–1069 (2016)
18. Salimans, T., Goodfellow, I., Zaremba, W., Cheung, V., Radford, A., Chen, X.: Improved techniques for training GANs. In: Advances in Neural Information Processing Systems, pp. 2234–2242 (2016)
19. Wang, Z., Bovik, A.C., Sheikh, H.R., Simoncelli, E.P., et al.: Image quality assessment: from error visibility to structural similarity. IEEE Trans. Image Process. **13**(4), 600–612 (2004)
20. Xu, T., et al.: ATTNGAN: fine-grained text to image generation with attentional generative adversarial networks. In: Proceedings of the IEEE Conference on Computer Vision and Pattern Recognition, pp. 1316–1324 (2018)

21. Zhang, H., et al.: StackGAN: Text to photo-realistic image synthesis with stacked generative adversarial networks. In: Proceedings of the IEEE International Conference on Computer Vision, pp. 5907–5915 (2017)
22. Zhao, B., Meng, L., Yin, W., Sigal, L.: Image generation from layout. In: Proceedings of the IEEE Conference on Computer Vision and Pattern Recognition, pp. 8584–8593 (2019)
23. Mao, X., Li, Q., Xie, H., Lau, R.Y.K., Wang, Z., Smolley, S.P.: Least squares generative adversarial networks. In: Proceedings of the IEEE International Conference on Computer Vision, pp. 2794–2802 (2017)
24. Huang, Z., Jiqing, W., Van Gool, L.: Manifold-valued image generation with Wasserstein generative adversarial nets. In: Proceedings of the AAAI Conference on Artificial Intelligence, vol. 33, pp. 3886–3893 (2019)
25. Gu, J., Zhao, H., Lin, Z., Li, S., Cai, J., Ling, M.: Scene graph generation with external knowledge and image reconstruction. In: Proceedings of the IEEE Conference on Computer Vision and Pattern Recognition, pp. 1969–1978 (2019)

Learning Unsupervised Video Summarization with Semantic-Consistent Network

Ye Zhao, Xiaobin Hu, Xueliang Liu$^{(\boxtimes)}$, and Chunxiao Fan

Hefei University of Technology, Hefei, China
huxb1108@gmail.com, liuxueliang1982@gmail.com
https://www.hfut.edu.cn/

Abstract. The aim of video summarization is to refine a video into concise form without losing its gist. In general, a summary with similar semantics to the original video can represent the original well. Unfortunately, most existing methods focus more on the diversity and representation content of the video, and few of them take video's semantic into consideration. In addition, most methods related to semantic pursue their own description of the video and this way will learn a biased mode. In order to solve these issues, we propose a novel *semantic-consistent unsupervised framework* termed ScSUM which is able to extract the essence of the video via obtaining the greatest semantic similarity, and requires no manual description. In particular, ScSUM consists of a frame selector, and a video descriptor which not only predict the description of summary, but also produce the description of original video used as targets. The main goal of our propose is to minimize the distance between the summary and original video in the semantic space. Finally, experiments on two benchmark datasets validate the effectiveness of the proposed methods and demonstrate that our method achieves competitive performance.

Keywords: Video summarization · Semantic · LSTM

1 Introduction

Every day a lot of videos are uploaded to the Internet, and it's time-consuming for people to browse these videos. It also requires a large capacity device to save. Thus, the mission of video summarization is to preserve the information of the original video to the greatest extent possible in a short form.

An ideal video summarization is to provide users the most important information of the target video without redundancy. And in order to reduce these redundancy, many video summarization approaches have been proposed which mainly focus on excluding the repeated content [2,26,27], but do not notice the

© Springer Nature Singapore Pte Ltd. 2020
H. Zhang et al. (Eds.): NCAA 2020, CCIS 1265, pp. 207–219, 2020.
https://doi.org/10.1007/978-981-15-7670-6_18

deep meaning of the video. Generally speaking, a summary with more semantic information is more in line with the information that humans obtain from the original video.

Recently, more researchers pay their attention into semantic analysis. For example, some latest works provided a query-conditioned video summarization solution to keep the semantic consistence [13,24]. In this way, the problem of video summarization is formulated as selecting a sparse subset of video frames according to user's query. With some words or sentences provided by users, it can generate a summary that matches the user query and the video content jointly. Lately, video caption [17], a technique that combines vision and language processing, also be applied into video summarization [19]. Similarly, they also use manually annotated description as prior knowledge to learn a selector.

However, we notice that the video summarization dataset such as SumMe [3] and TVSum [14] do not have manually labeled description (only a short video title provided), but the training of descriptor depends heavily on the description of the video content. To address this issue, we not only make descriptor to generate of the description of summary video but also work for raw video, so that we can get the target caption of video. In addition, we design a consistent-loss to measure the mutual information between summary video and original video by computing their distance in semantic space. By this way, even if we don't have annotations, we can also do as pseudo-supervised learning and avoid the bias of the model.

To this end, we propose a novel *semantic-consistent unsupervised video summarization (ScSUM)*, motivated by maximizing the mutual semantic information between summary video and original video. Structurally, our model consists of a frame selector and a semantic-consistent descriptor. While the selector takes input features extracted via pre-trained CNN, and predicts an importance score for each frame. In training process, summary features is weighted original video features with importance scores. In evaluation, a summary video is formed by the frames with high importance scores. And semantic-consistent descriptor aims to maximize semantic information preserving of the summary video. To achieve effective information preserving, descriptor employs attention mechanism to effectively generate the caption of the video. Besides, the objective function not only measures the cost of video captioning, but also penalizes the distances of the summary video and original video in deep semantic feature space to improve summary completeness.

The contributions of this paper are summarized as follows.

1. We proposes a unsupervised-based video summarization framework named *Unsupervised Video Summarization with Semantic-consistent Network*, which takes semantic of video into consideration.
2. We explore semantic analysis in the ScSUM framework without additional description of the target video, which can avoid the cost of description annotation by human.

3. The experiments on two benchmark datasets show that the our method outperforms the state-of-the-art ones in both supervised and unsupervised. These promising results verify the effectiveness of our ScSUM framework.

2   Related Works

Video summarization has been widely studied in recent years. And a mass of approaches have been proposed in the literature, and they mainly fall in several categories. There are unsupervised and supervised video summarization approaches. In the following, we will introduce them in detail, and in particular, our work is a unsupervised approach. Additionally, since our work explores semantic analysis, we will further review the existing content-based and semantic-based video summarization, respectively.

2.1   Supervised vs. Unsupervised Video Summarization

Supervised video summarization approaches has received much attention in recent years. It utilize videos and their manual annotations of key frames as training data to imitate human summarizing behavior and minimize the gap with ground-truth [2,23]. For example, Gong et al. [2] formulate video summarization as a supervised subset selection problem and propose a sequential determinantal point processing (seqDPP) based model to sample a representative and diverse subset from training data. Potapov et al. [10] explore a set od SCM classifiers to attach each segment in a video with an importance score. However, it's hard to provide sufficient and reliable human-annotated videos. Thus, several works try to exploit auxiliary information to assist video summarization, i.e.,video titles [14] and web-images [5].

Different with supervised ones, unsupervised approaches are generally designed to make the selector meet the expected objective, such as, representativeness, conciseness, and highlight. According to these different criterion, many methods have been developed. For instance, clustering-based method is a common means [1]. It clusters the similar frames into groups based on visual analysis, and consider these groups as the representative component of the video. Besides, RNN-based methods also are the most popular ones recently. Zhou et al. [27] treats a RNN layer as their selector and designs a DR objective function to evaluate the degree of diversity and representativeness of generated summaries for optimizing selector via reinforcement learning. Additionally, Mahasseni et al. [8] explore generative adversarial networks as their framework to minimize distance between training videos in an unsupervised way.

2.2   Content-Based Video Summarization

Most of the methods so far mainly analyze the visual content of the video. Early, workers pay their more attention into low-level visual features [1,3,6,7,9, 21], such as motion track, object and content frequency. In [6], they formulate video summarization as prediction of important people and objects and train a regression model from labeled training videos that scores any region's likelihood of belonging to an important person or object. And in [7], Lin et al. propose a context-based highlight detection method and predicts the contexts of each video segment.

Recently, the rise of deep learning has enabled people to analyze in depth, they mainly design a model to narrow the gap between the summary and the original video [4,8,22,23,25,27]. For example, Ji et al. [4] introduce attention mechanism into their encoder-decoder LSTM structure to mimic the way of selecting the keyshots of human. Similarly, Zhao et al. [25] consider video summarization as a structured prediction problem on sequential data, and two LSTM layers are designed to model the variable-range dependency in the video. The former layer works as a shot detection with sliding operation and latter one is to predict the importance of each detected segment. Besides, Cycle-SUM proposed by Yuan et al. [22] is designed to maximize the information preserving and compactness of the summary video. Above methods achieve the state-of-the art performance in the domain of video summarization, but they lack the analysis of the deep semantic information of the video.

2.3   Semantic-Based Video Summarization

Semantic aware summarization [12–14,16,19,24] is gradually gaining attention. Most of the semantic-related methods now focus on query-condition which takes user queries in the form of texts into consideration in order to learn more user-oriented summaries. Such as [16], Vasudevan et al. adopt a quality-aware relevance model and submodular mixtures to pick relevant and representative frames. In [24], Zhang et al. explore a query-conditioned three-player generative adversarial network to handle this task. The generator learns the joint representation of the query and content, and the discriminator judges three pairs of query-conditioned summaries to avoid the generation of random trivial summaries. Although these query-conditioned methods add semantic conditions, they are personalized and not extensive.

To the best of our knowledge, SASUM [19] is the only existing semantic-based video summarization which takes semantic information into consideration without queries. They explore video descriptor into framework to assist selector extract the most semantically relevant video segments. However, they learn the selector with manual annotated captions. On the one hand, these annotations of the target videos are not readily available and reliable, and on the other hand, using these annotations will make the model biased, this is not what we want. Therefore, our work is based on these issues and we develop a novel semantic model which does not rely on external annotations.

3   The Proposed Approach

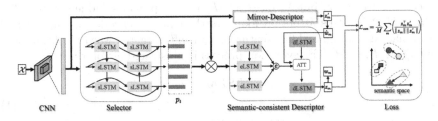

Fig. 1. Demonstration of proposed ScSUM architecture. There are two components in ScSUM: a selector for weighting each frame by an important score and the descriptor to preserve semantic information of summarized video. Firstly the deep feature is extracted from each frame χ by CNN; the selector then takes these features as input to predict the import scores p_t. During training, descriptor takes summary features as input and predicts caption w and original semantic features $z_m.$; the mirror-descriptor is the deep-copy of descriptor to generate original caption $\hat{w}$ and semantic features z_m'.

In this section, we introduce our ScSUM framework for video summarization. Figure 1 illustrates the whole framework of our approach. It main consists of two components: a frame selector and a semantic-consistent descriptor. Particularly, the frame selector learns to attach each frame with a score which indicate the importance of a frame to the summary; the semantic-consistent descriptor aims to consistent the summary with the semantics of the original video.

3.1   Keyframe Selector

Given an input video $\chi = \{x_1, x_2, ..., x_T\}$, where x_i is a frame and T denotes the total frame number, we first extract its deep features $F = \{f_1, f_2, ..., f_T\}$ via a deep CNN model. Then we adopt a LSTM as our selector to select a subset of keyframes, termed $sLSTM$. The frame selector predicts a sequence of important scores $s = \{s_t : s_t \in [0,1] | t = 1, 2, 3, ..., T\}$ indicating the semantic relevance level to the video. In training, the features of summary video are weighted features $S = \{f_t \cdot s_t | t = 1, 2, 3, ..., T\}$. And during testing, the important scores will be normalized into $\{0,1\}$, which are considered as indicator, while frame with big score will be selected to form summary.

3.2   Semantic-Consistent Descriptor

Given the summary features, $eLSTM$ will produce a T-length output feature vector $Y_e = \{y_1, y_2, ..., y_T\}$ and its corresponding hidden states h_t. Then the encoded reconstructed feature e will be fed into $dSLTM$ and converted to a sequence of semantic representation $\mathcal{Z} = \{z_1, z_2, ..., z_M\}$, where M donates the number of generated words.

In particular, we adopt attention mechanism into our descriptor. When we send encoded features e into $dLSTM$, we can obtain M-length feature $\mathcal{Z}$. When calculating the semantic information z_m, we first calculate the attention map α according to the output Y_e of $eLSTM$ and the previous hidden state h_d^{m-1} of $dLSTM$. With this attention map, we then weight encoded features e and concatenate it with embedding feature $\hat{w}_m$ of mth targeted word to obtain next input i_d^m of $dLSTM$. The evolution process is illustrated as follow:

$$y_t, h_t = eLSTM(\mathcal{S}, h_{t-1})$$
$$i_d^m = [\alpha \odot Y_e, \; \hat{w}_m] \tag{1}$$
$$z_m, h_d^m = dLSTM\left(i_d^m, h_d^{m-1}\right)$$

where $\odot$ donates element-wise multiplication. And the attention map α is calculated by

$$\alpha = \frac{\exp{(a_t)}}{\sum \exp{(a_t)}}, \; where \; a_t = W\left(\left[y_t, h_d^{m-1}\right]\right) + B \tag{2}$$

where W and B is the matrix and bias of attention block. After obtaining semantic feature $\mathcal{Z}$, we then convert it to a sequence of words $\omega = \{w_1, w_2, ..., w_M\}$, where w_i indicates a word.

As mentioned before, we do no have the annotated caption of each video in video summarization. The only relevant semantic description is the video title, which is obviously not suitable as target description. Even if we have these annotated caption, using them will make our model biased. Therefore, we exploit our descriptor to generating caption of original video as annotation. This is the only way to preserve the semantic information of the original video.

As shown in Fig. 1, we perform a mirror-descriptor to generate a description for original video, where mirror-descriptor is a deep-copy model from semantic-consistent descriptor. During the training phase, we treat descriptor obtained in each training as mirror-descriptor which will not be trained. Feeding original features F into it, we can get the description $\hat{\omega} = \{\hat{w}_1, \hat{w}_2, ..., \hat{w}_M\}$ of the original video. After obtaining these descriptions of the original videos, we can do "supervised" learning and avoid the problem that the model is biased when training with manual annotation.

3.3 Model Training

Captioning Loss. Inspired by the successes of probabilistic sequence models leveraged in captioning [18] and considering that semantic-consistent descriptor generates each word in order, we can define our captioning loss as

$$\mathcal{L}_{caption} = -\log P\left(\omega \,|z\right). \tag{3}$$

Consistent Loss. Since we expect summary to maintain as much semantic information as possible, the summary should be consistent with original video in semantic space. Based on such idea, we introduce the following consistent loss. We treat the output of $dLSTM$ $\mathcal{Z} = \{z_1, z_2, ..., z_M\}$ as the semantic representation of video, and compute the distance between summary and original in semantic space to make them closer and guarantee the semantic completeness of the summary video. So the consistent loss is given as

$$\mathcal{L}_{consistent} = \frac{1}{M} \sum_m \left(\frac{z_m^\top z_m'}{\|z_m\| \|z_m'\|} \right) \tag{4}$$

where z_m and z_m' denote the semantic representative feature of mth word of original video and summary in $dLSTM$.

Sparsity Loss. To avoid too many frames being selected to form the summary video, and inspired by [8], we implement a regularization term on the probability distributions s_t predicted by selector, and this penalty is defined as

$$\mathcal{L}_{sparsity} = \| \frac{1}{T} \sum_{t=1}^{T} s_t - \rho \| \tag{5}$$

where ρ is the percentage of frames to be selected.

Supervised Sparsity Loss. In the two benchmark datasets (TVSum and SumMe), ground-truth annotations of key frames are provided. Therefore, we can use cross-entropy loss to maximize the log-probability of expected keyframes when annotated labels are available. This objective is formalized as

$$\mathcal{L}_{sparsity}^{sup} = \frac{1}{T} \sum_t CE(s_t, \hat{s}_t) \tag{6}$$

4 Results

4.1 Experiment Setup

Datasets. We evaluate the proposed ScSUM framework on two publicly available benchmark datasets, SumMe [3] and TVSum [14]. SumMe consists of 25 user videos with several events such as cooking and sports. Similarly, TVSum contain 50 videos from YouTube covering various topics such as news, documents, etc. These videos are 1 to 5 min long. Besides, MSR-VTT [20], a dataset for general video captioning which has 10, 000 video clips, is used to pre-train the descriptor in our experiment (Fig. 2).

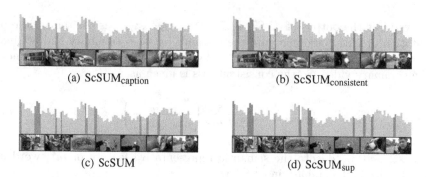

(a) ScSUM$_{caption}$ (b) ScSUM$_{consistent}$

(c) ScSUM   (d) ScSUM$_{sup}$

Fig. 2. Example summaries generated by our different variants from a sample video in TVSum [14]. The blue background represents the annotated importance scores, and the colored bars are the selected subset of frames by different variants (Color figure online).

Evaluation Metric. For fair comparison with the state-of-the-art approaches, we use the F-score [23] metric to evaluate our method. In all datasets (SumMe and TVSum), generated summaries are changed from frame-level scores to shot-level using the KTS method [10]. Particularly, let A donate the generated key-shots, and B donate the user-annotated key-shots, the precision(P) and recall(R) are defined based on the temporal overlaps between A and B, as follow,

$$
\begin{aligned}
P &= \frac{duration\ of\ overlap\ between\ A\ and\ B}{duration\ of\ A} \\
R &= \frac{duration\ of\ overlap\ between\ A\ and\ B}{duration\ of\ B}
\end{aligned}
\tag{7}
$$

Finally, F-score is defined as:

$$
F = \frac{2 \times P \times R}{(P + R)} \times 100\%
\tag{8}
$$

Implementation Details. We use the output of avgpool layer of the ResNet152 network [15], trained on ImageNet [11], as the feature for each frame by 2 fps. For applied LSTM, we use a BiLSTM with 1024 hidden units for $sLSTM$ and $eLSTM$. The $dLSTM$ of descriptor is an one-directional LSTM with 512 hidden units. Specially, we initialize the hidden state of $dLSTM$ with the final hidden state of $eLSTM$. The ratio ρ of frames to be selected is set to 0.3 as previous works [8]. For fair comparison, we randomly divide these two datasets five times, of which 80% are the training set and 20% are the test set.

Table 1. F-score results(%) of different variants of our method on benchmark datasets.

Method	SumMe	TVSum
ScSUM$_{caption}$	40.9	55.7
ScSUM$_{sparsity}$	41.3	57.4
ScSUM$_{consistent}$	41.7	57.9
ScSUM	42.2	58.1
ScSUM$_{sup}$	45.8	59.5

4.2   Quantitative Evaluation

Ablation Analysis. Since we design several loss function, it is important to show the performance of different variants of our method. Thus, we conduct the following ablation variants of ScSUM:

- **ScSUM$_{caption}$.** This variant discards all loss functions except caption loss $\mathcal{L}_{caption}$ to verify the effects of descriptor. And $\mathcal{L}_{caption}$ exists in each of the following variants as a basic model.
- **ScSUM$_{sparsity}$.** The sparsity loss $\mathcal{L}_{sparsity}$ is added into basic model to train our ScSUM with $\mathcal{L}_{caption}$.
- **ScSUM$_{consistent}$.** The consistent loss $\mathcal{L}_{consistent}$ and $\mathcal{L}_{caption}$ is jointly used to optimize our model. This variant is to prove that mutual semantic information between summary video and raw video is useful in our framework.
- **ScSUM$_{sup}$.** Our proposed ScSUM extend to supervised version by human annotated $\mathcal{L}_{sparsity}^{sup}$.

Table 1 shows the compared results of these variants of our framework. Comparing the F-scores of ScSUM$_{caption}$ and ScSUM$_{sparsity}$, we can see that the sparsity penalty has improved performance, proving the positive effects of sparsity penalty in video summarization. We also can find that the better performance of ScSUM$_{consistent}$ than ScSUM$_{caption}$ (with 0.8% and 2.2% surplus, on SumMe and TVSum respectively). It is because the designed consistent penalty successfully retain relevant semantic information which makes summary closer to semantics of the original video. Finally, we combine all objectives to learn a complete ScSUM, and it achieves better performance than each variant. By adding the supervision penalty to ScSUM, the performance of selector are further improved (7.9% improvement in SumMe and 2.4% in TVSum). This is because introduced labels have more manual information which can improve the algorithm.

Comparison with Unsupervised Approaches. Furthermore, to verify the performance of our method, we compare with other unsupervised approaches, as shown in Table 2. It can be seen that ScSUM outperforms the state-of-the-art unsupervised methods on both datasets. In particular, compared with

Table 2. Comparison of our proposed video summarization approach with **unsupervised** state-of-the-art.

Method	SumMe	TVSum
SASUM [19]	41.0	54.6
DR-DSN [27]	41.4	57.6
SUM-GAN [8]	39.1	51.7
Cycle-SUM [22]	41.9	57.6
ScSUM	**42.2**	**58.1**

SASUM [19] which also implement descriptor into model for considering semantic information, our method obtain better performance than it by a large margin. That's because SASUM learns the model by using manual annotations which is made by workers, but this may be cause model narrow. Our solution is different from them because we do not use annotated caption, but instead used the caption generated by semantic-consistent descriptor, thus avoiding prejudice and making summary close to original video.

Table 3. Results(%) of **supervised** methods on SumMe and TVSum.

Method	SumMe	TVSum
Bi-LSTM [23]	37.6	54.2
dpp-LSTM [23]	38.6	54.7
SASUM$_{sup}$ [19]	45.3	58.2
DR-DSN$_{sup}$ [27]	42.1	58.1
SUM-GAN$_{sup}$ [8]	41.7	56.3
ScSUM	42.2	58.1
ScSUM$_{sup}$	**45.8**	**59.5**

Comparison with Supervised Approaches. Table 3 reports the results of our method in supervised manner, ScSUM$_{sup}$, and other supervised approaches. It is easy to see that our ScSUM$_{sup}$ beats the others on the two datasets. And it is interesting that our ScSUM without manual annotations also achieves extraordinary performance and equal to some methods, even beyond. These all results demonstrate that our proposed semantic-consistent video summarization has competitive performance than the state-of-the-art unsupervised methods.

4.3 Qualitative Results

In order to clearly show the effect of the results, Fig. 3 illustrates the summarization examples from a sample video (#15) in TVSum. We compare the selected

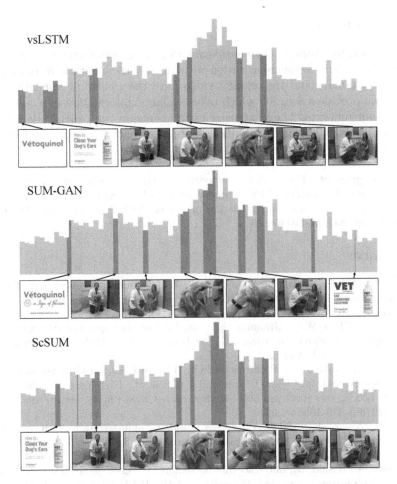

Fig. 3. Comparison of selected frames by ScSUM and other state-of-the-arts (vLSTM and SUM-GAN). The gray background represents ground-truth frame-level important scores; the red bars are selected subset shots of all frames (Color figure online).

frames of ScSUM with other representative state-of-the-arts (vsLSTM [23] and SUM-GAN [8]). As show in Fig. 3, our ScSUM picks out shorter, more key, and more representative shots than other two models. The important information of this example is basically distributed in the middle of the video, and our results have successfully covered this main region. It is interesting that the beginning and end of the video is very similar in content. VsLSTM and our method successfully avoid their recurrence, unlike SUM-GAN. In contrast to vsLSTM we have a more brief form without more content redundancy. This result clearly shows the advantages of our approach.

5   Conclusion

In this paper, we propose a *semantic-consistent video summarization (ScSUM)* method to tackle video summarization without manual annotations. In order to make the summary and the original video have the consistent semantic information, we provide a solution which takes semantic information into consideration without unwanted bias. Extensive experiments on two benchmark datasets demonstrates that our model produces more representative results and outperforms other state-of-the-art alternatives.

Acknowledgement. This work was supported in part by the National Key Research and Development Program of China under grant 2018AAA0102002, and in part by the National Natural Science Foundation of China (NSFC) under grants 61976076, 61632007, 61932009 and 61806066.

References

1. De Avila, S.E.F., Lopes, A.P.B., da Luz Jr., A., de Albuquerque Araújo, A.: Vsumm a mechanism designed to produce static video summaries and a novel evaluation method. Patt. Recogn. Lett. **32**(1), 56–68 (2011)
2. Gong, B., Chao, W.L., Grauman, K., Sha, F.: Diverse sequential subset selection for supervised video summarization. In: Advances in Neural Information Processing Systems, pp. 2069–2077 (2014)
3. Gygli, M., Grabner, H., Riemenschneider, H., Van Gool, L.: Creating summaries from user videos. In: Fleet, D., Pajdla, T., Schiele, B., Tuytelaars, T. (eds.) ECCV 2014. LNCS, vol. 8695, pp. 505–520. Springer, Cham (2014). https://doi.org/10.1007/978-3-319-10584-0_33
4. Ji, Z., Xiong, K., Pang, Y., Li, X.: Video summarization with attention-basedencoder-decoder networks. IEEE Trans. Circ. Syst. Video Technol. **30**(6), 1709–1717 (2019)
5. Khosla, A., Hamid, R., Lin, C.J., Sundaresan, N.: Large-scale video summarization using web-image priors. In: Proceedings of the IEEE Conference on Computer Vision and Pattern Recognition, pp. 2698–2705 (2013)
6. Lee, Y.J., Ghosh, J., Grauman, K.: Discovering important people and objects for egocentric video summarization. In: 2012 IEEE Conference on Computer Vision and Pattern Recognition, pp. 1346–1353. IEEE (2012)
7. Lin, Y.L., Morariu, V.I., Hsu, W.: Summarizing while recording: context-based highlight detection for egocentric videos. In: Proceedings of the IEEE International Conference on Computer Vision Workshops, pp. 51–59 (2015)
8. Mahasseni, B., Lam, M., Todorovic, S.: Unsupervised video summarization with adversarial LSTM networks. In: Proceedings of the IEEE conference on Computer Vision and Pattern Recognition, pp. 202–211 (2017)
9. Meng, J., Wang, H., Yuan, J., Tan, Y.P.: From keyframes to key objects: video summarization by representative object proposal selection. In: Proceedings of the IEEE Conference on Computer Vision and Pattern Recognition, pp. 1039–1048 (2016)
10. Potapov, D., Douze, M., Harchaoui, Z., Schmid, C.: Category-specific video summarization. In: Fleet, D., Pajdla, T., Schiele, B., Tuytelaars, T. (eds.) ECCV 2014. LNCS, vol. 8694, pp. 540–555. Springer, Cham (2014). https://doi.org/10.1007/978-3-319-10599-4_35

11. Russakovsky, O., et al.: Imagenet large scale visual recognition challenge. Int. J. Comput. Vis. **115**(3), 211–252 (2015)
12. Sharghi, A., Gong, B., Shah, M.: Query-focused extractive video summarization. In: Leibe, B., Matas, J., Sebe, N., Welling, M. (eds.) ECCV 2016. LNCS, vol. 9912, pp. 3–19. Springer, Cham (2016). https://doi.org/10.1007/978-3-319-46484-8_1
13. Sharghi, A., Laurel, J.S., Gong, B.: Query-focused video summarization: dataset, evaluation, and a memory network based approach. In: Proceedings of the IEEE Conference on Computer Vision and Pattern Recognition, pp. 4788–4797 (2017)
14. Song, Y., Vallmitjana, J., Stent, A., Jaimes, A.: Tvsum: Summarizing web videos using titles. In: Proceedings of the IEEE Conference on Computer Vision and Pattern Recognition, pp. 5179–5187 (2015)
15. Szegedy, C., et al.: Going deeper with convolutions. In: Proceedings of the IEEE Conference on Computer Vision and Pattern Recognition, pp. 1–9 (2015)
16. Vasudevan, A.B., Gygli, M., Volokitin, A., Van Gool, L.: Query-adaptive video summarization via quality-aware relevance estimation. In: Proceedings of the 25th ACM International Conference on Multimedia, pp. 582–590. ACM (2017)
17. Venugopalan, S., Rohrbach, M., Donahue, J., Mooney, R., Darrell, T., Saenko, K.: Sequence to sequence-video to text. In: Proceedings of the IEEE International Conference on Computer Vision, pp. 4534–4542 (2015)
18. Vinyals, O., Toshev, A., Bengio, S., Erhan, D.: Show and tell: a neural image caption generator. In: Proceedings of the IEEE Conference on Computer Vision and Pattern Recognition, pp. 3156–3164 (2015)
19. Wei, H., Ni, B., Yan, Y., Yu, H., Yang, X., Yao, C.: Video summarization via semantic attended networks. In: Thirty-Second AAAI Conference on Artificial Intelligence (2018)
20. Xu, J., Mei, T., Yao, T., Rui, Y.: Msr-vtt: a large video description dataset for bridging video and language. In: Proceedings of the IEEE Conference on Computer Vision and Pattern Recognition, pp. 5288–5296 (2016)
21. Yao, T., Mei, T., Rui, Y.: Highlight detection with pairwise deep ranking for first-person video summarization. In: Proceedings of the IEEE Conference on Computer Vision and Pattern Recognition, pp. 982–990 (2016)
22. Yuan, L., Tay, F.E., Li, P., Zhou, L., Feng, J.: Cycle-sum: cycle-consistent adversarial lstm networks for unsupervised video summarization (2019). arXiv preprint arXiv:1904.08265
23. Zhang, K., Chao, W.-L., Sha, F., Grauman, K.: Video summarization with long short-term memory. In: Leibe, B., Matas, J., Sebe, N., Welling, M. (eds.) ECCV 2016. LNCS, vol. 9911, pp. 766–782. Springer, Cham (2016). https://doi.org/10.1007/978-3-319-46478-7_47
24. Zhang, Y., Kampffmeyer, M., Liang, X., Tan, M., Xing, E.P.: Query-conditioned three-player adversarial network for video summarization (2018). arXiv preprint arXiv:1807.06677
25. Zhao, B., Li, X., Lu, X.: Hsa-rnn: hierarchical structure-adaptive rnn for video summarization. In: Proceedings of the IEEE Conference on Computer Vision and Pattern Recognition, pp. 7405–7414 (2018)
26. Zhao, B., Xing, E.P.: Quasi real-time summarization for consumer videos. In: Proceedings of the IEEE Conference on Computer Vision and Pattern Recognition, pp. 2513–2520 (2014)
27. Zhou, K., Qiao, Y., Xiang, T.: Deep reinforcement learning for unsupervised video summarization with diversity-representativeness reward. In: Thirty-Second AAAI Conference on Artificial Intelligence (2018)

Sustainable Competitiveness Evaluation for Container Liners Using a Novel Hybrid Method with Intuitionistic Fuzzy Linguistic Variables

Junzhong Bao[1]($\boxtimes$), Yuanzi Zhou[1], Peixin Shi[2], and Xizhao Wang[3]

[1] Navigation College, Dalian Maritime University, Dalian 116026, China
baojunzhong@hotmail.com, 1004581213@qq.com
[2] College of Transportation Engineering,
Dalian Maritime University, Dalian 116026, China
shipeixin9873@163.com
[3] China Ship Scientific Research Center, Wuxi 214082, China
wangxizhao@cssrc.com.cn

Abstract. In order to evaluate competitiveness of sustainability for major liners, multiple indexes are collected for reflecting liners' achievements based on publicly available sources. In consideration of existence of subjective factors of such kind of assessment, this study applies intuitionistic linguistic variables to depict the information of decision making with multiple attributes, and combination coefficients are introduced to fuse the objective entropy weight and the subjective weight of various attributes. An optimal combination weighting method based on square sums of distance is adopted to calculate the integrated evaluation values of liners. As for case studies, 5 larger container liners are selected as an illustrative example and they are ranked based on the integrated linguistic evaluation values and membership degrees as the basic measures of competitiveness.

Keywords: Group decision · Intuitionistic linguistic variables · Competitiveness

1 Introduction

The shipping industry acts a vital role for cargo flows in the global supply chain. It is believed that shipping is the most efficient, economical means for transporting large quantities of goods over significant distances (Pérez Lespier et al. 2018). However, the industry also has negative impacts for environment (Yang et al. 2013; Yuen et al. 2017). In 2015, 17 sustainable development goals have been adopted by the United Nations (UN's SDGs) for stimulation of actions for sustainable development of the international community over the coming 15 years. In light of the UN's 2030 Agenda for Sustainable Development, leading shipping liners have adopted sustainable development policies that are closely related to the UN's SDGs.

© Springer Nature Singapore Pte Ltd. 2020
H. Zhang et al. (Eds.): NCAA 2020, CCIS 1265, pp. 220–233, 2020.
https://doi.org/10.1007/978-981-15-7670-6_19

In liner shipping, sustainable competitiveness can be understood as performance of a liner company in implementing environment friendly policies, social well-being initiatives or economic growth strategy in compliance with the SDGs. In terms of research, several scholars have focused their studies on the influential factors of shipping firms' competitiveness. Fanam et al. (2016) elicited five critical factors influential to the liner shipping companies' competitiveness, including the freight rate, service quality, scheduling, handling equipment and information technology. Magnus et al. (2012) stated that advantages of liners depends on cost reduction, coordinated networks, economies of scale, regulation, pricing and shipper relationships. In the literature reviewed, factors involved in the competitiveness of shipping companies mainly focused on economic aspects rather than sustainable ones.

As for evaluation models, scholars have utilized various models to determine the competitiveness of ports or shipping liners. Singh et al. (2012) reviewed methodologies for sustainability assessment and highlighted that Principal Component Analysis (PCA) and Analytic Hierarchy Process (AHP) are widely applied in evaluation. In addition, Abbes (2015) evaluated the North and West African seaport competitiveness by using PCA, and found that effective planning and reform will improve the competitiveness of African seaports.

The major contributes of this paper are as follows. First, five liner companies are collected from the global top 30 ranked by the capacity of total 20-toot equivalent units (TEU) (UNCTAD 2018). The collected liners have published their respective corporate social responsibility reports (CSR) annually. The authors develop a hierarchical structure of factors influential to sustainable competitiveness of shipping liners which include intuitionistic initiatives or policies. Second, the authors apply a novel model of Intuitionistic Fuzzy Set (IFS) to evaluate the competitiveness of liner shipping companies, aiming at prioritizing the companies for sustainable development. An IFS is rarely tried for analyzing performance of liners in sustainability.

The remaining parts of the paper are as follows. In Sect. 2, the theoretical basis of intuitionistic fuzzy linguistic variables is elaborated; In Sect. 3, the decision making process for evaluation of sustainable competiveness of container liners is introduced; In Sect. 4, assessment attributes are selected and assessed in respect of five companies; In Sect. 5, the sorting results are analyzed; Sect. 6 presents concluding remarks.

2 Theories and Methods

This section recalls the basic theories of linguistic scale set, intuitionistic linguistic variables in details.

2.1 Linguistic Scale Set

When a decision maker carries out qualitative evaluation, he/she shall preset a proper linguistic value and linguistic assessment scale.

Definition 1 (Liu and Wang 2014; Xu 2006). Linguistic assessment scale set $S = \{s_i | i = 0, 1, 2, \ldots, l - 1\}$, where s_i is the linguistic variable, generally the number of terms in S takes odd. In general, linguistic scale set S, s_i and s_j satisfy the following conditions:

① If $i > j$, then $s_i > s_j$;
② Negative operator $neg(s_i) = s_{l-1-i}$ exists;
③ If $s_i \geq s_j$, then $\max(s_i, s_j) = s_i$;
④ If $s_i \leq s_j$, then $\min(s_i, s_j) = s_i$.

2.2 Intuitionistic Linguistic Variable

Definition 2 (Atanassov 1986; Liu and Wang 2014). An intuitionistic linguistic set (ILS) A in X is defined as:

$$A = \{ <x\,[h_{\theta(x)}, (u_A(x), v_A(x))] > \,| x \in X\} \tag{1}$$

where $h_{\theta(x)} \in \bar{S}$, $u_A : X \to [0, 1]$, $v_A : X \to [0, 1]$, and satisfy the condition:
$0 \leq u_A(x) + v_A(x) \leq 1, \forall x \in X$, $u_A(x)$ and $v_A(x)$ represent the membership degree and non-membership degree of element x to linguistic index $h_{\theta(x)}$, respectively. $\bar{S}$ is the expanded linguistic scale set, $\bar{S} = \{s_\alpha | \alpha = [0, q], q \geq l - 1\}$. When $u_A(x) = 1$, $v_A(x) = 0$, intuitionistic linguistic set degenerates to linguistic scale set.

Let $\pi(x) = 1 - v_A(x) - u_A(x), \forall x \in X$, then $\pi(x)$ is referred to as the hesitancy degree of x to linguistic index $h_{\theta(x)}$, and obviously, $0 \leq \pi(x) \leq 1, \forall x \in X$.

Definition 3 (Liu and Wang 2014; Wang and Li 2009). The ternary group $<h_{\theta(x)}, (u_A(x), v_A(x)) >$ is an intuitionistic linguistic number (ILN). $h_{\theta(x)}$ is the linguistic unit of the ILN. A can also be regarded as the set of the ILN. Generally it is recorded as $\tilde{a} = <s_{\theta(a)}, (u(a), v(a)) >$, where $u(a), v(a) \geq 0$ and $u(a) + v(a) \leq 1$.

Relative to the intuitionistic fuzzy number, the ILN increases the linguistic scale value $h_{\theta(x)}$, to cause the membership degree and non-membership degree corresponding to a more specific linguistic scale value; the ILN is more accurate and reasonable to reflect the decision makers' information (Wang and Li 2010).

Let $\tilde{a}_1 = <s_{\theta(a_1)}, (u(a_1), v(a_1)) >$ and $\tilde{a}_2 = <s_{\theta(a_2)}, (u(a_2), v(a_2)) >$ be two intuitionistic linguistic variables and $\lambda \geq 0$, then the following operation rules were defined as follows (Liu and Wang 2014; Wang and Li 2009):

$$\tilde{a}_1 + \tilde{a}_2 = \left\langle s_{\theta(a_1) + \theta(a_2)}, (1 - (1 - u(a_1))(1 - u(a_2)), v(a_1)v(a_2)) \right\rangle \tag{2}$$

$$\tilde{a}_1 \otimes \tilde{a}_2 = \left\langle s_{\theta(a_1) * \theta(a_2)}, (u(a_1)u(a_2), v(a_1) + v(a_2) - v(a_1)v(a_2)) \right\rangle \tag{3}$$

$$\lambda\tilde{a}_1 = \left\langle s_{\lambda \times \theta(a_1)}, \left(1 - (1 - u(a_1))^\lambda, (v(a_1))^\lambda\right) \right\rangle \tag{4}$$

$$\tilde{a}_1^{\lambda} = \left\langle s_{(\theta(a_1))^{\lambda}}, \left((u(a_1))^{\lambda}, 1 - (1 - v(a_1))^{\lambda} \right) \right\rangle \tag{5}$$

Definition 4 (Liu and Wang 2014; Wang and Li 2009). For any two ILNs, $\tilde{a}_1$ and $\tilde{a}_2$, rules of comparison of intuitionistic linguistic variables are determined as follows:
① If $E(\tilde{a}_1) > E(\tilde{a}_2)$, then $\tilde{a}_1 \succ \tilde{a}_2$
② If $E(\tilde{a}_1) = E(\tilde{a}_2)$, then
  If $S(\tilde{a}_1) > S(\tilde{a}_2)$, then $\tilde{a}_1 \succ \tilde{a}_2$
③ If $E(\tilde{a}_1) = E(\tilde{a}_2)$ and $S(\tilde{a}_1) = S(\tilde{a}_2)$, then

  If $H(\tilde{a}_1) > H(\tilde{a}_2)$, then $\tilde{a}_1 \succ \tilde{a}_2$;
  If $H(\tilde{a}_1) = H(\tilde{a}_2)$, then $\tilde{a}_1 = \tilde{a}_2$.

where,

$$E(\tilde{a}_1) = \frac{1}{2} \times (u(a_1) + 1 - v(a_1)) \times s_{\theta(a_1)} = s_{(\theta(a_1) \times (u(a_1) + 1 - v(a_1)))/2} \tag{6}$$

$$S(\tilde{a}_1) = E(\tilde{a}_1) \times (u(a_1) - v(a_1)) \tag{7}$$

$$H(\tilde{a}_1) = E(\tilde{a}_1) \times (u(a_1) + v(a_1)) \tag{8}$$

Definition 5 (Liu and Wang 2014). Let $\tilde{a}_1$ and $\tilde{a}_2$ are any two ILNs, then the Hamming distance is defined as follows:

$$d(\tilde{a}_1, \tilde{a}_2) = \frac{1}{2(l-1)} (|(1 + u(a_1) - v(a_1))\theta(a_1) - (1 + u(a_2) - v(a_2))\theta(a_2)|) \tag{9}$$

2.3 Optimal Combination Weights

Weight of an attribute plays an important role in decision making. Combination weighting is one of the commonly used methods which gathers the advantages of subjective and objective weighting approaches. In this paper, the optimal combination weighting method based on square sums of distance is adopted (Chen 2003; Chen and Xia 2007).

Let $W = (w_1, w_2, \ldots, w_l)$, where w_k represents the weight vector for the $k - th$ type of weighting, $k = 1, 2, \ldots, l$. $\Theta = (\theta_1, \theta_2, \ldots \theta_l)^T$, where θ_k denotes combination coefficient, $\theta_k \geq 0$ and satisfies the condition: $\sum_{k=1}^{l} \theta_k = 1$. Combination weighting coefficient vector can be defined as $W_c^* = W\Theta^*$, Θ^* is the unit eigenvector corresponding to the maximum eigenroot $\lambda_{\max}$ of the matrix $W^T B_1 W$ (Chen and Xia 2007), where,

$$\mathbf{B}_1 = \begin{bmatrix} \sum\limits_{i=1}^{m}\sum\limits_{i_1=1}^{m} d(b_{i1},b_{i_11})d(b_{i1},b_{i_11}) & \sum\limits_{i=1}^{m}\sum\limits_{i_1=1}^{m} d(b_{i1},b_{i_11})d(b_{i2},b_{i_12}) & \cdots & \sum\limits_{i=1}^{m}\sum\limits_{i_1=1}^{m} d(b_{i1},b_{i_11})d(b_{in},b_{i_1n}) \\ \sum\limits_{i=1}^{m}\sum\limits_{i_1=1}^{m} d(b_{i2},b_{i_12})d(b_{i1},b_{i_11}) & \sum\limits_{i=1}^{m}\sum\limits_{i_1=1}^{m} d(b_{i2},b_{i_12})d(b_{i2},b_{i_12}) & \cdots & \sum\limits_{i=1}^{m}\sum\limits_{i_1=1}^{m} d(b_{i2},b_{i_12})d(b_{in},b_{i_1n}) \\ \cdots & & \cdots & \cdots \\ \sum\limits_{i=1}^{m}\sum\limits_{i_1=1}^{m} d(b_{in},b_{i_1n})d(b_{i1},b_{i_11}) & \sum\limits_{i=1}^{m}\sum\limits_{i_1=1}^{m} d(b_{in},b_{i_1n})d(b_{i2},b_{i_12}) & \cdots & \sum\limits_{i=1}^{m}\sum\limits_{i_1=1}^{m} d(b_{in},b_{i_1n})d(b_{in},b_{i_1n}) \end{bmatrix}$$

$$(10)$$

2.4   Objective Weight Determination Method

In the process of group decision making, the integration of subjective and objective weights is a key step. The typical way for computing the objective weight of each attribute is to take full advantages of the distribution of judgment information. The degree of deviation between interval values of alternatives in respect of one attribute indicates the importance of that attribute to some extent. Guo and Chen (2013) offered a method for determining the objective weight of each attribute by utilizing the proportion of the sum of comprehensive deviations of alternatives on a single attribute to that on all attributes. In this paper, we extend this method to the context of ILS, the detailed steps are listed as follows.

① Under the attribute $G_j(j = 1, 2, \ldots, n)$, calculate the distance between alternative $A_i(i = 1, 2, \ldots, m)$ and other alternatives:

$$D_{ij} = \sum_{k=1}^{m} d(a_{ij}, a_{kj}) \tag{11}$$

② Under the attribute $G_j(j = 1, 2, \ldots, n)$, calculate the comprehensive distances of all alternatives $A_i(i = 1, 2, \ldots, m)$:

$$D_j = \sum_{i=1}^{m} D_{ij} = \sum_{i=1}^{m}\sum_{k=1}^{m} d(a_{ij}, a_{kj}) \tag{12}$$

③ Calculate the sum of the distances of all attributes for all alternatives:

$$D = \sum_{j=1}^{n} D_j = \sum_{j=1}^{n}\sum_{i=1}^{m}\sum_{k=1}^{m} d(a_{ij}, a_{kj}) \tag{13}$$

④ Obtain the weight of the attribute G_j

$$w_j = \frac{D_j}{D}(j = 1, 2, \ldots, n) \tag{14}$$

The greater the proportion of the sum of the distances D_j of G_j to the sum of the distances of D, the greater the weight w_j of the attributes G_j should be.

3   Model of Evaluation

Suppose the alternative of decision making is expressed as $A = \{A_i | i = 1, 2, \ldots, m\}$, with attribute collection $C = \{C_j | j = 1, 2, \ldots, n\}$ and expert team of $DM = \{DM_1, DM_2, \ldots, DM_k\}$. The value of A_i in relation to the attribute C_j assessed by the p-th decision maker is expressed as $\left\langle s_{ij}^p, \left(u_{ij}^p, v_{ij}^p\right)\right\rangle$.

Step 1: Build a decision matrix
The p-th decision making matrix is presented as follows:

$$R_p = [r_{ij}^p]_{m \times n} = \begin{bmatrix} \left\langle s_{11}^p, \left(u_{11}^p, v_{11}^p\right)\right\rangle & \left\langle s_{12}^p, \left(u_{12}^p, v_{12}^p\right)\right\rangle & \cdots & \left\langle s_{1n}^p, \left(u_{1n}^p, v_{1n}^p\right)\right\rangle \\ \left\langle s_{21}^p, \left(u_{21}^p, v_{21}^p\right)\right\rangle & \left\langle s_{22}^p, \left(u_{22}^p, v_{22}^p\right)\right\rangle & \cdots & \left\langle s_{2n}^p, \left(u_{2n}^p, v_{2n}^p\right)\right\rangle \\ \cdots & \cdots & \vdots & \cdots \\ \left\langle s_{m1}^p, \left(u_{m1}^p, v_{m1}^p\right)\right\rangle & \left\langle s_{m2}^p, \left(u_{m2}^p, v_{m2}^p\right)\right\rangle & \cdots & \left\langle s_{mn}^p, \left(u_{mn}^p, v_{mn}^p\right)\right\rangle \end{bmatrix}$$

$$(15)$$

Step 2: Compute the distance matrix of each decision maker
The p-th distance matrix is presented as follows:

$$D_p = \begin{bmatrix} d_{11}^p & d_{12}^p & \cdots & d_{1n}^p \\ d_{21}^p & d_{22}^p & \cdots & d_{2n}^p \\ \cdots & \cdots & \vdots & \cdots \\ d_{m1}^p & d_{m2}^p & \cdots & d_{mn}^p \end{bmatrix}$$

$$(16)$$

where d_{ij}^p is the sum of the distance between r_{ij}^p and other elements with respect to the j–th attribute:

$$d_{ij}^p = \sum_{i_1=1}^{m} d(r_{ij}^p, r_{i_1 j}^p)$$

$$(17)$$

Step 3: Compute the objective weight of each attribute
The objective weight $w_j (j = 1, 2, \ldots, n)$ of each attribute will be obtained by Eq. (11–14).

Step 4: Compute the integrated assessment value of each decision maker
Suppose the subject weight $\boldsymbol{W}_{sub}$ is known for attributes, and the objective weight $\boldsymbol{W}_p$ for attributes is obtained based on the results of Step 3.

$$Z_p = \beta R_p W_{sub} + (1 - \beta) R_p W_p \qquad (18)$$

Step 5: Compute the integrated assessment value of each alternative based on the decision maker's weight

$$\boldsymbol{Z} = \sum_{p=1}^{k} \lambda_p \boldsymbol{Z}_p = \sum_{p=1}^{k} (\beta \lambda_p \boldsymbol{R}_p \boldsymbol{W}_{sub} + (1 - \beta) \lambda_p \boldsymbol{R}_p \boldsymbol{W}_p)$$

$$= \beta (\lambda_1 \boldsymbol{R}_1, \lambda_2 \boldsymbol{R}_2, \ldots, \lambda_k \boldsymbol{R}_k) \begin{pmatrix} \boldsymbol{W}_{sub} \\ \boldsymbol{W}_{sub} \\ \ldots \\ \boldsymbol{W}_{sub} \end{pmatrix} + (1 - \beta)(\lambda_1 \boldsymbol{R}_1, \lambda_2 \boldsymbol{R}_2, \ldots, \lambda_k \boldsymbol{R}_k) \begin{pmatrix} \boldsymbol{W}_1 \\ \boldsymbol{W}_2 \\ \ldots \\ \boldsymbol{W}_k \end{pmatrix}$$

$$(19)$$

Suppose $\boldsymbol{W}_1^* = \begin{pmatrix} \boldsymbol{W}_{sub} \\ \boldsymbol{W}_{sub} \\ \ldots \\ \boldsymbol{W}_{sub} \end{pmatrix}$, $\boldsymbol{W}_2^* = \begin{pmatrix} \boldsymbol{W}_1 \\ \boldsymbol{W}_2 \\ \ldots \\ \boldsymbol{W}_k \end{pmatrix}$, then:

$$\boldsymbol{Z} = (\lambda_1 \boldsymbol{R}_1, \lambda_2 \boldsymbol{R}_2, \ldots, \lambda_k \boldsymbol{R}_k) \times (\beta \boldsymbol{W}_1^* + (1 - \beta) \boldsymbol{W}_2^*) \qquad (20)$$

Step 6: Compute the combination coefficient value
Step 7: Rank the alternatives.

The alternatives are sorted by the value of expectation (E), according to the rule for comparison of intuitionistic linguistic variables given in Definition 4.

4   Case Study

In this Section, five container liners are selected as samples, which are Maersk, CMA CGM (CMA CGM 2019), Yang Ming (Yang Ming 2019), Grimaldi Group, Shandong International Transportation Corporation (SITC 2020). Three industrial experts are invited to carry out intuitionistic linguistic assessment of the sustainability of the selected container liners. According to the decision-making model established in this paper, the integrated assessment values are obtained and then expectation values are calculated by the rule for comparing intuitionistic linguistic variables given in Definition 4. Linguistic scale set for expert judgement making: S = {s_0: extremely poor, s_1: poor, s_2: relatively poor, s_3: general, s_4: relatively good, s_5: good and s_6: excellent}.

4.1   Selection of Attributes and Structure of Decision Matrixes

Competitiveness Attributes. The five samples are named as Container Liner 1(CL_1), Container Liner 2(CL_2), Container Liner 3(CL_3), Container Liner 4(CL_4) and Container Liner 5(CL_5). Then an investigation sheet was prepared and sent to shipping experts for recommended key attributes to be used in sustainable competitiveness evaluation of the container liners.

Four key attributes were identified from three aspects, social, environmental, and economic. The four attributes were determined by the Cosco Sustainability Report and the linkage between SDGs adopted by the United Nations and the Maersk's Business (Cosco 2018; Maersk 2017), which including economic conditions, production status, staff state and environmental friendliness. Finally, the basic data was collected from various companies regarding the four sustainable attributes, as shown in Table 1.

Table 1. Container liners sustainable development competitiveness attribute data.

Samples		Container Liner 1(CL_1)	Container Liner 2(CL_2)	Container Liner 3(CL_3)	Container Liner 4(CL_4)	Container Liner 5(CL_5)
Attribute1	Average ship size (TEUs)[a]	5542	5366	6097	1072	1413
	No. of ships	700	476	100	123 (48 container ships)	67
	Revenue[c]	39,019; 26%	23,476; 11.1%	4674; 8.2%	3251.3; 6%	1,449; 7.5%
	Tax (USD million)	645	99.4	−25.7[b]	21.2	8.6
Attribute2	Technological upgrading	Over the past four years, invested around USD 1bn in developing and deploying energy-efficient solutions	Improve propulsion and hydrodynamic technology, invest scrubber, LNG propulsion	Fuel-saving equipment, main engine fuel consumption abnormal alert program system, fleet performance monitoring platform	Installation of scrubber systems, reblading and promas lite system, zero emissions in port project:	Introduction of new energy-efficient and environmental-friendly vessels; upgrade cylinder oil
	Safe production	7 fatalities; LTIF[d]: 1.29	1 fatality; maritime LTIF: 1.25; terminals LTIF: 13.2	1 fatality; LTIF: 1.01	5 fatalities; maritime LTIF: 10.35; shore LTIF: 30.35	1 fatality; LTIF: N/A[e]
Attribute3	Gender equality[f]	Total employee: 25%; 27%	On-shore: 13%; 45%	On-shore: N/A; 45%	Total employee: 19%; 23%	On-shore: N/A; 41%

(*continued*)

Table 1. (*continued*)

Samples		Container Liner 1(CL$_1$)	Container Liner 2(CL$_2$)	Container Liner 3(CL$_3$)	Container Liner 4(CL$_4$)	Container Liner 5(CL$_5$)
	Talent training	200 instructors involved in diversified courses for employee; TTR: N/A	TTR[g]: 212,342	On-shore employee TTR: 37,260	TTR: 24710	TTR: 29,120
Attribute4	SOx emission[h]	0.88	0.84	0.85	0.71	0.36
	NO$_x$ emission[h]	1.36	1.47	1.50	1.10	0.53
	GHG emission[h]	55.95	54.12	54.79	44.92	21.10

Note:

a Average ship size (TEUs) = Total 20-foot equivalent units/No of Ships, including owned and chartered.

b The Income Before Income Tax of this Container Liner is negative.

c Revenue: USD million; percentage increase over last year

d LTIFR (Lost time injury frequency rate) = number of work accidents/(number of hours worked) × 1,000,000

e N/A: Not available

f Gender equality: Women in leadership; female in total

g TTR: Total Training Hours

h SOx, NOx, GHG emiss

① Attribute1(A_1): Economic conditions. This attribute is measured by the average ship size, the number of ships (owned and chartered), revenue, revenue increase (decrease) and the tax.

② Attribute2(A_2): Production status. This attribute is measured by technological upgrading and safe production.

③ Attribute3(A_3): Staff state. This attribute is measured by the gender equality and the talent training.

④ Attribute4(A_4): Environmental friendliness. This attribute is measured by the SO$_x$ emissions, GHG emission and NO$_x$ emission per ship.

Experts' Weights. Expert A is a senior PSC inspector with around 18-year working experience. Expert B is a supertanker chief engineer. Expert C is a captain and professor dedicated to teaching and studying of shipping management. Before making judgement, the three experts expressed their familiarity with the attributes to be assessed. The weight vector of the three experts is $\lambda = (0.3, 0.4, 0.3)^T$.

Experts' Judgement. The data sheets of the five container liner samples are provided to the three experts who are allowed to use intuitionistic linguistic variables to evaluate these companies. Then the decision matrixes are obtained as listed in Tables 2, 3 and 4.

Table 2. Judgement matrix of expert A

	A_1	A_2	A_3	A_4
CL_1	$<s_6,(0.9, 0.1)>$	$<s_5,(0.7, 0.2)>$	$<s_5,(0.7, 0.3)>$	$<s_4,(0.7, 0.3)>$
CL_2	$<s_5,(0.8, 0.1)>$	$<s_4,(0.7, 0.2)>$	$<s_4,(0.7, 0.3)>$	$<s_4,(0.7, 0.3)>$
CL_3	$<s_2,(0.7, 0.2)>$	$<s_3,(0.6, 0.3)>$	$<s_2,(0.7, 0.2)>$	$<s_3,(0.6, 0.2)>$
CL_4	$<s_3,(0.6, 0.3)>$	$<s_2,(0.7, 0.3)>$	$<s_4,(0.6, 0.2)>$	$<s_4,(0.6, 0.2)>$
CL_5	$<s_4,(0.6, 0.3)>$	$<s_3,(0.7, 0.2)>$	$<s_3,(0.7, 0.1)>$	$<s_4,(0.5, 0.3)>$

Table 3. Judgement matrix of expert B

	A_1	A_2	A_3	A_4
CL_1	$<s_6,(0.9, 0.1)>$	$<s_5,(0.8, 0.2)>$	$<s_5,(0.8, 0.1)>$	$<s_5,(0.7, 0.2)>$
CL_2	$<s_5,(0.8, 0.2)>$	$<s_5,(0.8, 0.1)>$	$<s_5,(0.8, 0.2)>$	$<s_5,(0.8, 0.2)>$
CL_3	$<s_3,(0.7, 0.3)>$	$<s_1,(0.8, 0.2)>$	$<s_3,(0.7, 0.2)>$	$<s_3,(0.7, 0.2)>$
CL_4	$<s_4,(0.8, 0.2)>$	$<s_4,(0.7, 0.2)>$	$<s_2,(0.6, 0.4)>$	$<s_5,(0.8, 0.1)>$
CL_5	$<s_3,(0.6, 0.3)>$	$<s_3,(0.6, 0.2)>$	$<s_3,(0.6, 0.3)>$	$<s_5,(0.8, 0.1)>$

Table 4. Judgement matrix of expert C

	A_1	A_2	A_3	A_4
CL_1	$<s_6,(0.9, 0.1)>$	$<s_4,(0.8, 0.2)>$	$<s_5,(0.8, 0.1)>$	$<s_5,(0.7, 0.2)>$
CL_2	$<s_5,(0.8, 0.1)>$	$<s_4,(0.7, 0.1)>$	$<s_4,(0.7, 0.2)>$	$<s_5,(0.7, 0.3)>$
CL_3	$<s_1,(0.7, 0.2)>$	$<s_3,(0.8, 0.1)>$	$<s_1,(0.6, 0.3)>$	$<s_4,(0.8, 0.2)>$
CL_4	$<s_4,(0.8, 0.1)>$	$<s_3,(0.7, 0.2)>$	$<s_3,(0.8, 0.2)>$	$<s_5,(0.6, 0.2)>$
CL_5	$<s_3,(0.6, 0.2)>$	$<s_2,(0.7, 0.2)>$	$<s_5,(0.8, 0.1)>$	$<s_5,(0.6, 0.2)>$

4.2   Decision Making Process

① Steps 1 to 2 yield the following distance matrixes:

$$D_1 = \begin{bmatrix} 1.8833 & 1.0667 & 0.7500 & 0.1833 \\ 1.3083 & 0.6917 & 0.4000 & 0.1833 \\ 1.3667 & 0.6167 & 0.9167 & 0.4000 \\ 1.1417 & 0.8917 & 0.4000 & 0.1833 \\ 1.0333 & 0.5667 & 0.4667 & 0.2500 \end{bmatrix}$$

$$D_2 = \begin{bmatrix} 1.7250 & 1.0583 & 1.2667 & 0.4583 \\ 1.0250 & 1.1833 & 1.1417 & 0.4167 \\ 1.0750 & 1.6917 & 0.8500 & 1.2083 \\ 0.8917 & 0.8917 & 1.2750 & 0.4583 \\ 1.1500 & 1.0417 & 0.9000 & 0.4583 \end{bmatrix}$$

$$D_3 = \begin{bmatrix} 1.8500 & 0.5500 & 1.1167 & 0.2167 \\ 1.2750 & 0.5500 & 0.9083 & 0.0917 \\ 2.0250 & 0.4417 & 1.8833 & 0.2417 \\ 1.1333 & 0.4917 & 1.0083 & 0.0917 \\ 1.3500 & 0.8667 & 1.1167 & 0.0917 \end{bmatrix}$$

② Step 3 yields the following objective weights:

$$W_1 = (0.4580 \quad 0.2608 \quad 0.1995 \quad 0.0816)^T$$
$$W_2 = (0.2909 \quad 0.2909 \quad 0.2694 \quad 0.1488)^T$$
$$W_3 = (0.4412 \quad 0.1676 \quad 0.3487 \quad 0.0424)^T$$

Let subjective weight vector be:

$$W_{sub} = (0.2 \quad 0.35 \quad 0.2 \quad 0.25)^T$$

③ Steps 4 to 6 yield:

$$W^T B_1 W = \begin{pmatrix} 0.6028 & 0.4720 \\ 0.4720 & 0.3712 \end{pmatrix}$$

The maximum eigenvalue of the matrix $(W^T B_1 W)$ corresponds to $\lambda_{max} = 0.9370$, and the eigenvector $v = (0.7869 \quad 0.6171)^T$, which are further normalized to obtain $\beta = 0.5604$.
Combination weight is achieved as follows:

$W_c^* = \beta W_1^* + (1 - \beta) W_2^*$
$= (0.3446 \quad 0.3000 \quad 0.1997 \quad 0.1556 \quad 0.2509 \quad 0.3169 \quad 0.2389 \quad 0.1933 \quad 0.3352$
$\quad 0.2478 \quad 0.2833 \quad 0.1337)$

The integrated assessment value of each sample container liner (Steps 4 and 5) is given by combining the combination weight and the judgement information.

$$Z = (\lambda_1 R_1, \lambda_2 R_2, \ldots, \lambda_k R_k) \times W_c^* = \begin{pmatrix} <s_{5.1831} & 0.8160 & 0.1492 > \\ <s_{4.6439} & 0.7651 & 0.1603 > \\ <s_{2.2521} & 0.7100 & 0.2124 > \\ <s_{3.4834} & 0.7140 & 0.2043 > \\ <s_{3.4805} & 0.6614 & 0.1973 > \end{pmatrix}$$

④ Rank the container liner's sustainable competitiveness (Step 7):

According to integrated assessment value and the rules for comparing intuitionistic linguistic variables given in Definition 4, the sustainable competitiveness E for an individual container liner is obtained:

$$E(\widetilde{CL}_1, \widetilde{CL}_2, \widetilde{CL}_3, \widetilde{CL}_4, \widetilde{CL}_5) = (s_{4.3196}, s_{3.7263}, s_{1.6864}, s_{2.6294}, s_{2.5479})$$

By comparing the competitiveness E for each container liner, the ranking of sustainable competitiveness level for five sample container liners is achieved as follows:

$$CL_1 \succ CL_2 \succ CL_4 \succ CL_5 \succ CL_3$$

5 Discussion

As per A_1, CL_1 performs the best. CL_1 and CL_2 have larger average ship size and number of ships, higher revenue growth rate and income tax compared to the other three. CL_4 is similar to CL_5 in terms of the number of container ships. CL_3 is poor with lower turnover and due tax. For A_2, CL_2 and CL_4 will be more competitive in the future due to installation of scrubber and investment in ship upgrading. CL_1's investment shows its emphasis on green and safe production.

In terms of A_3, the total training hours for onshore staff of CL_3 is relatively less, training hours for seafarers is not mentioned in its CSR. It is believed that CL_3's manning is outsourced to a manning agency based in a seafarer supply country. Moreover, the investment of CL_3's shipboard equipment innovation is not satisfactory.

For A_4, CL_3's per ship emission ranks the second, and its ship fuel consumption is relatively high. It seems that CL_3 has suffered financial pressure and is unable to adopt substantial technical innovation for energy saving. Investment by CL_2 and CL_4 in equipment innovation is substantial, which needs strong financial support and complies with global emission control regulations. Hence their advantages of fuel use will make them more competitive and profitable in the future. CL_5 takes the geographical advantage, and its ships are in economical size and flexible in short sea trade. CL_5 is also low in energy consumption and operating cost.

Overall, the rank of shipping liners is generally in line with their performance (See Table 1). The ranking results would help managers in the shipping industry to identify container liners that are in the leading position in sustainable competitiveness, and to analyze their good practices in economy, production, staff and environmental friendliness. It can also provide reference for sustainable development management of international container liners.

6 Conclusion

We proposed a new hybrid method based on maximum deviation square and intu-itionistic fuzzy linguistic variables. This application study validates that an IFS is a powerful tool for assessing the competitiveness of shipping liners in the MCDM environment. We introduced a combination coefficient to fuse the objective deviation square weights and the subjective weights of each attribute. Based on subjective factors, namely expert opinions, we adopted the fuzzy weighted average operator to calculate the comprehensive evaluation value of the container liners, thereby reflecting its sustainability. We believe that the proposed method has better applicability to actions with non-quantitative or descriptive attributes (See Attribute 2 in Table 1). For the scenarios where the effect of execution needs to be reflected in the future, it is very effective for experts to use ILS to respond.

For the limitation of this study, we found that there are not many shipping companies that issue sustainability reports. The data in the reports has non-standard limitations, which makes it difficult for the authors to refine the indicators and the data. Furthermore, making experts fully understand the gist of an IFS is also a challenge to the authors, because experts unfamiliar with an IFS might compromise the validity of the evaluation given by them.

Future work will focus on the issue of competitiveness gaps between companies. To predict which companies will be more vulnerable in future uncertain shipping environments, we intend to use the latest IFS distance measures like perturbation between IFSs and divergence-based cross entropy measure. In addition, the overly complex IFS derivative model was not used in the present study in case that decision makers could hardly understand the research process. It is suggested to use IFS and its derivatives to evaluate the possibility of bankruptcy of shipping companies in the course of future researches.

References

Atanassov, K.T.: Intuitionistic fuzzy sets. Fuzzy Sets Syst. **4120**(1), 87–96 (1986)

Abbes, S.: Seaport competitiveness: a comparative empirical analysis between north and West African countries using principal component analysis. Int. J. Transp. Econ. **42**(3), 289–314 (2015)

Chen, H.: Research on optimal combination determining weights method for multiple attribute decision making. Oper. Res. Manag. Sci. **12**(02), 6–10 (2003)

Chen, W., Xia, J.: An optimal weights combination method considering both subjective and objective weight information. Math. Pract. Theory **01**, 17–22 (2007)

Cosco Shipping Lines Co., Ltd.: 2018 Sustainability report (2018). http://lines.coscoshipping.com/lines_resource/pdf/quality2018en.pdf. Accessed 29 May 2020

CMA CGM: Corporate social responsibility report 2018 (2019). https://www.cmacgm-group.com/en/csr/social. Accessed 29 May 2020

Fanam, P., Nguyen, H., Cahoon, S.: Competitiveness of the liner operators: methodological issues and implications. J. Traffic Transp. Eng. **4**(5), 231–241 (2016)

Guo, X., Chen, Y.: A deviation method for determining the weight of interval number index. Stat. Decis. **20**, 61–63 (2013)

Liu, P., Wang, Y.: Multiple attribute group decision making methods based on intuitionistic linguistic power generalized aggregation operators. Appl. Soft Comput. **17**, 90–104 (2014)

Maersk: Sustainability report (2017). https://www.maersk.com.cn/about/sustainability/reports. Accessed 29 May 2020

Magnus, U., Basedow, J., Wolfrum, R.: The Hamburg Lectures on Maritime Affairs 2009 & 2010, pp. 3–27. Springer, Berlin (2012). https://link.springer.com/book/10.1007%2F978-3-642-27419-0#toc, http://dx.doi.org/10.1007/978-3-642-27419-0. Accessed 29 May 2020

Lespier, L.P., Long, S., Shoberg, T., Corns, S.: A model for the evaluation of environmental impact indicators for a sustainable maritime transportation systems. Front. Eng. Manag. **6**(3), 368–383 (2018)

Singh, R.K., Murty, H.R., Gupta, S.K., Dikshit, A.K.: An overview of sustainability assessment methodologies. Ecol. Ind. **15**(1), 281–299 (2012)

SITC International Holdings Co., Ltd.: Social responsibility (2020). http://www.sitc.com/en/social%20responsibility/index2.asp. Accessed 8 Mar 2020

United Nations Conference on Trade and Development: Review of maritime transport 2018. United Nations, New York (2018). https://unctad.org/en/pages/PublicationWebflyer.aspx?publicationid=2245. Accessed 29 May 2020

Wang, J., Li, J.: The multi-criteria group decision making method based on multi-granularity intuitionistic two semantics. Sci. Technol. Inf. **33**, 8–9 (2009)

Wang, J., Li, H.: Multi-criteria decision-making method based on aggregation operators for intuitionistic linguistic fuzzy numbers. Control Decis. **25**(10), 1571–1574+1584 (2010)

Xu, Z.: Induced uncertain linguistic OWA operators applied to group decision making. Inf. Fusion **7**(2), 231–238 (2006)

Yang, C., Lu, C., Haider, J., Marlow, P.: The effect of green supply chain management on green performance and firm competitiveness in the context of container shipping in Taiwan. Transp. Res. Part E Logist. Transp. Rev. **55**, 55–73 (2013)

Yuen, K., Wang, X., Wong, Y., Zhou, Q.: Antecedents and outcomes of sustainable shipping practices: the integration of stakeholder and behavioural theories. Transp. Res. Part E Logist. Transp. Rev. **108**, 18–35 (2017)

Yang Ming Marine Transport Corporation: (Yang Ming) 2019 Corporate Social Responsibility Report (2019). https://www.yangming.com/files/Investor_Relations/2019CSR.pdf. Accessed 29 May 2020

2-Dimensional Interval Neutrosophic Linguistic Numbers and Their Utilization in Group Decision Making

Jiani Wu, Lidong Wang$^{(\boxtimes)}$, and Lanlan Li

College of Science, Dalian Maritime University, Dalian
116026, People's Republic of China
ldwang@hotmail.com

Abstract. Due to growing risk and complexity of modern decision-making problems, the graded information associated with linguistic terms is suitable to characterize the uncertainty during cognition process. By combining 2-dimensional linguistic variables and interval neutrosophic numbers, we put forward a new concept of 2-dimensional interval neutrosophic linguistic numbers and establish pertinent operational rules. Moreover, we develop an operator for fusing 2-dimensional interval neutrosophic linguistic numbers by means of Choquet integral, by which the relationships among the attributes can be reflected. The established aggregation operator's properties are also studied. In the sequel, a new group PROMETHEE (preference ranking organization method for enrichment evaluations) II method is created to deal with the group decision problems with 2-dimensional interval neutrosophic linguistic terms.

Keywords: 2-dimensional interval neutrosophic linguistic numbers · Group decision making · Aggregation operator · PROMETHEE II model.

1 Introduction

Decision making is one of the commonly encountered cognitive behaviors, by which a desirable solution can be identified from several potential alternatives in the presence of some attributes [21]. In decision making process, evaluation information comes mainly from the preference information provided by decision-makers. It needs to be emphasized that fuzzy set theory is beneficial to express graded information when capturing uncertainty [24]. With the aid of fuzzy set theory, humans have promoted diversity representation approaches of uncertainty information, but it is required to further explore new approaches for reflecting human cognitive mechanism because of both the inherent vagueness and the increasing complexity of external situation [7]. Thus, classical fuzzy sets have been extended in the forms of fuzzy numbers and multiple membership grades to capture the human cognitive mechanism in handling real application

© Springer Nature Singapore Pte Ltd. 2020
H. Zhang et al. (Eds.): NCAA 2020, CCIS 1265, pp. 234–246, 2020.
https://doi.org/10.1007/978-981-15-7670-6_20

problems. Neutrosophic set theory can be considered as a prominent tri-partition tactic in capturing cognitive information and solving decision problems. The concept of neutrosophic sets was originally developed by Smarandache [15,16] by considering an indeterminacy membership grade, which allows decision makers to use tri-membership grades (viz. the truth grade, falsity grade and indeterminate grade) to express uncertainty information. Wang et al. [20] explored the theory and method regarding interval form of neutrosophic sets, in which tri-membership grades were described by three intervals. The advances in these fuzzy descriptive methods of cognitive decision have resulted in the emergence of new solution strategies of practical problems.

In some practical applications, membership functions are required to be determined in advance, which brings some restrictions on membership grades-based methods in terms of keeping track of the cognitive behavior [4]. The fuzzy linguistic method based on symbols is another important strategy to manage linguistic information about the utterance, in which variable values can take natural words or sentences instead of some particular numbers. Semantically, symbolic approaches are simpler and easier to comprehend, but they may induce more loss of information than those encountered in the former methods [14]. To address this issue, linguistic variables and real numbers (or membership grades) have also been integrated to form new methods of characterizing decision-maker's preference information [10]. Along this line, different types of neutrosophic linguistic terms have been developed, which include neutrosophic trapezoid linguistic numbers [2], interval neutrosophic linguistic numbers [23], interval neutrosophic numbers in form of uncertain linguistic terms [3,6] and neutrosophic linguistic sets [11], etc. Their effectiveness and flexibility come with a number of aggregation operators for information fusion under different forms of attribute values. These results enrich the representation method of cognitive information and also offer applicable approaches for solving managerial decision-making problems.

The sources of assessment information are associated with human judgments, thus the reliability of individual's cognitive information is influenced largely by their age, experience, knowledge, and other factors [19]. As a matter of fact, it is indispensable to evaluate the credibility of individual's cognitive information. Fuzzy linguistic terms with the reliability aspect bring more effective representation methods for characterizing cognitive information. The objective and motivation of this study are to develop a new approach for handling cognitive-oriented evaluation information, in which the interval linguistic neutrosophic information is measured by a reliability interval. Moreover, we establish an extended group decision method by introducing the operational rules, comparison laws and fusion method of 2-dimensional interval neutrosophic linguistic terms.

2   2-Dimensional Interval Neutrosophic Linguistic Number

In linguistic decision making activities, the linguistic term set is usually predefined to serve as the quantitative scale for evaluating all alternatives. $S = \{s_0, s_1, s_2, \ldots, s_l\}$ is one way to define linguistic term sets, where s_i denotes a linguistic term with the following order relationship $s_i > s_j$ iff $i > j$ and negative relationship $neg(s_i) = s_{l-i}$ for $i \in \{1, 2, \ldots, l\}$ [8,9]. In general, l can take 3, 5, 7, 9 and so on. For instance, $S = \{s_0, s_1, s_2\} = \{bad, fair, good\}$. By using linguistic terms in the process of solving decision making problems, operational results are imposed to match with one element of the linguistic term set S, which brings about the loss of information. Xu extended the set $S = \{s_i | i = 0, 1, \ldots, l - 1\}$ to the continuous counterpart $\overline{S} = \{s_\alpha | \alpha \geq 0\}$ for minimizing the loss of information in terms of information fusion [22].

Inspired by the fundamental operational rules of 2-dimensional linguistic terms [12] and the interval neutrosophic numbers [25], we define the following concepts and the corresponding operational rules.

Definition 1. *A 2-dimensional interval neutrosophic linguistic number (2DINLN) is defined as follows.*

$$\hat{D}_1 = \langle (\dot{s}_{\alpha_1}, [T_1^L, T_1^U], [I_1^L, I_1^U], [F_1^L, F_1^U]), \ddot{s}_{\beta_1} \rangle, \tag{1}$$

where $[T_1^L, T_1^U]$ is interval truth-membership grade, $[I_1^L, I_1^U]$ represents interval indeterminacy membership grade, and $[F_1^L, F_1^U]$ denotes interval falsity-membership grade of $\hat{D}_1$ to the I class linguistic variable $\dot{s}_{\alpha_1} \in S_I = (\dot{s}_0, \cdots, \dot{s}_{l-1})$, while $\ddot{s}_{\beta_1} \in S_{II} = (\ddot{s}_0, \ddot{s}_1, \cdots, \ddot{s}_{g-1})$ is the II class linguistic term which describes reliability of $\langle \dot{s}_{\alpha_1}, [T_1^L, T_1^U], [I_1^L, I_1^U], [F_1^L, F_1^U] \rangle$.

For each element $\hat{D}_1$, we have $[T_1^L, T_1^U], [I_1^L, I_1^U], [F_1^L, F_1^U] \subseteq [0, 1]$ and $0 \leq T_1^U + I_1^U + F_1^U \leq 3$.

Definition 2. *Let $\hat{D}_1 = \langle (\dot{s}_{\alpha_1}, [T_1^L, T_1^U], [I_1^L, I_1^U], [F_1^L, F_1^U]), \ddot{s}_{\beta_1} \rangle$, $\hat{D}_2 = \langle (\dot{s}_{\alpha_2}, [T_2^L, T_2^U], [I_2^L, I_2^U], [F_2^L, F_2^U]), \ddot{s}_{\beta_2} \rangle$, and the parameter $\eta \geq 0$, then the following fundamental operations are introduced:*

1) $\hat{D}_1 + \hat{D}_2 = \langle (\dot{s}_{\alpha_1 + \alpha_2}, [T_1^L + T_2^L - T_1^L T_2^L, T_1^U + T_2^U - T_1^U T_2^U], [I_1^L I_2^L, I_1^U I_2^U],$
$[F_1^L F_2^L, F_1^U F_2^U]), \ddot{s}_{\min(\beta_1, \beta_2)} \rangle,$

2) $\hat{D}_1 \cdot \hat{D}_2 = \langle (\dot{s}_{\alpha_1 \cdot \alpha_2}, [T_1^L T_2^L, T_1^U T_2^U], [I_1^L + I_2^L - I_1^L I_2^L, I_1^U + I_2^U - I_1^U I_2^U],$
$[F_1^L + F_2^L - F_1^L F_2^L, F_1^U + F_2^U - F_1^U F_2^U]), \ddot{s}_{\min(\beta_1, \beta_2)} \rangle,$

3) $\eta \hat{D}_1 = \langle (\dot{s}_{\eta \alpha_1}, [1 - (1 - T_1^L)^\eta, 1 - (1 - T_1^U)^\eta], [(I_1^L)^\eta, (I_1^U)^\eta],$
$[(F_1^L)^\eta, (F_1^U)^\eta]), \ddot{s}_{\beta_1} \rangle,$

4) $\hat{D}_1^\eta = \langle (\dot{s}_{\alpha_1^\eta}, [(T_1^L)^\eta, (T_1^U)^\eta], [1 - (1 - I_1^L)^\eta, 1 - (1 - I_1^U)^\eta],$
$[1 - (1 - F_1^L)^\eta, 1 - (1 - F_1^U)^\eta]), \ddot{s}_{\beta_1} \rangle.$

According to Definition 2, we can obtain some interesting properties of 2-dimensional interval neutrosophic linguistic numbers.

Theorem 1. For $\hat{D}_1 = \langle(\dot{s}_{\alpha_1}, [T_1^L, T_1^U], [I_1^L, I_1^U], [F_1^L, F_1^U]), \ddot{s}_{\beta_1}\rangle$, $\hat{D}_2 = \langle(\dot{s}_{\alpha_2}, [T_2^L, T_2^U], [I_2^L, I_2^U], [F_2^L, F_2^U]), \ddot{s}_{\beta_2}\rangle$, and the parameters $\eta, \eta_1, \eta_2 \geq 0$, then the following properties hold:

$$1)\ \hat{D}_1 + \hat{D}_2 = \hat{D}_2 + \hat{D}_1,$$
$$2)\ \hat{D}_1 \cdot \hat{D}_2 = \hat{D}_2 \cdot \hat{D}_1,$$
$$3)\ \eta(\hat{D}_1 + \hat{D}_2) = \eta\hat{D}_1 + \eta\hat{D}_2,$$
$$4)\ \eta_1\hat{D}_1 + \eta_2\hat{D}_1 = (\eta_1 + \eta_2)\hat{D}_1,$$
$$5)\ \hat{D}_1^{\eta_1} \cdot \hat{D}_1^{\eta_2} = \hat{D}_1^{\eta_1+\eta_2},$$
$$6)\ \hat{D}_1^{\eta_1} \cdot \hat{D}_2^{\eta_1} = (\hat{D}_1 \cdot \hat{D}_2)^{\eta_1}.$$

Inspired by the work presented in [23], three characteristic functions of the 2-dimensional interval neutrosophic linguistic number are introduced.

Definition 3. The functions of score, accuracy and certainty of the 2DINL-N $\hat{D}_1 = \langle(\dot{s}_{\alpha_1}, [T_1^L, T_1^U], [I_1^L, I_1^U], [F_1^L, F_1^U]), \ddot{s}_{\beta_1}\rangle$ are defined, respectively, as follows:

$$SF(\hat{D}_1) = \frac{1}{6}(4 + T_1^L - I_1^L - F_1^L + T_1^U - I_1^U - F_1^U)(\dot{s}_{\alpha_1}, \ddot{s}_{\beta_1})$$
$$= (\dot{s}_{\frac{1}{6}\alpha_1(4+T_1^L-I_1^L-F_1^L+T_1^U-I_1^U-F_1^U)}, \ddot{s}_{\beta_1}), \tag{2}$$

$$HF(\hat{D}_1) = \frac{1}{2}(T_1^L - F_1^L + T_1^U - F_1^U)(\dot{s}_{\alpha_1}, \ddot{s}_{\beta_1})$$
$$= (\dot{s}_{\frac{1}{2}\alpha_1(T_1^L-F_1^L+T_1^U-F_1^U)}, \ddot{s}_{\beta_1}), \tag{3}$$

$$CF(\hat{D}_1) = \frac{1}{2}(T_1^L + T_1^U)(\dot{s}_{\alpha_1}, \ddot{s}_{\beta_1})$$
$$= (\dot{s}_{\frac{1}{2}\alpha_1(T_1^L+T_1^U)}, \ddot{s}_{\beta_1}). \tag{4}$$

From Definition 3, one derives that the functions of score, accuracy and certainty of $\hat{D}_1 = \langle(\dot{s}_{\alpha_1}, [T_1^L, T_1^U], [I_1^L, I_1^U], [F_1^L, F_1^U]), \ddot{s}_{\beta_1}\rangle$ are all 2-dimensional linguistic variables. In light of the related definitions in [13], a lexicographic ranking method can be determined.

Definition 4. Let $\hat{D}_1$, $\hat{D}_2$ be two 2-dimensional interval neutrosophic linguistic numbers, the ranking order is defined.
(1) If $SF(\hat{D}_1) > SF(\hat{D}_2)$, then $\hat{D}_1 > \hat{D}_2$,
(2) If $SF(\hat{D}_1) = SF(\hat{D}_2)$ and $HF(\hat{D}_1) > HF(\hat{D}_2)$, then $\hat{D}_1 > \hat{D}_2$,
(3) If $SF(\hat{D}_1) = SF(\hat{D}_2)$, $HF(\hat{D}_1) = HF(\hat{D}_2)$, and $CF(\hat{D}_1) > CF(\hat{D}_2)$, then $\hat{D}_1 > \hat{D}_2$,
(4) If $SF(\hat{D}_1) = SF(\hat{D}_2)$, $HF(\hat{D}_1) = HF(\hat{D}_2)$, and $CF(\hat{D}_1) = CF(\hat{D}_2)$, then $\hat{D}_1 = \hat{D}_2$.

Definition 5. *For* $\hat{D}_1 = \langle (\dot{s}_{\alpha_1}, [T_1^L, T_1^U], [I_1^L, I_1^U], [F_1^L, F_1^U]), \ddot{s}_{\beta_1} \rangle$ *and* $\hat{D}_2 = \langle (\dot{s}_{\alpha_2}, [T_2^L, T_2^U], [I_2^L, I_2^U], [F_2^L, F_2^U]), \ddot{s}_{\beta_2} \rangle$, *the distance measure between* $\hat{D}_1$ *and* $\hat{D}_2$ *is defined by*

$$\hat{d}(\hat{D}_1, \hat{D}_2) = \frac{1}{6l \times g} \bigg(\mid \theta_1 T_1^L - \theta_2 T_2^L \mid + \mid \theta_1 T_1^U - \theta_2 T_2^U \mid + \mid \theta_1 I_1^L - \theta_2 I_2^L \mid$$

$$+ \mid \theta_1 I_1^U - \theta_2 I_2^U \mid + \mid \theta_1 F_1^L - \theta_2 F_2^L \mid + \mid \theta_1 F_1^U - \theta_2 F_2^U \mid \bigg), \quad (5)$$

where $\theta_1 = (\alpha_1 + 1)(\beta_1 + 1)$, $\theta_2 = (\alpha_2 + 1)(\beta_2 + 1)$ *and* $\alpha_i \in [0, l]$, $\beta_i \in [0, g]$, $i = 1, 2$.

3   Choquet Integral Operator for 2-Dimensional Interval Neutrosophic Linguistic Numbers

Fuzzy Choquet integral is an important method of characterizing the interaction of attributes during information fusion, whose effectiveness and flexibility had hastened the emergence of a number of aggregation operators under different forms of attribute values. In view of the advantages of Choquet integral and actual requirements, we develop an operator for 2-dimensional interval neutrosophic linguistic variables by taking interactions among attributes into account
.

Definition 6 *[17]. The map function* $m(\cdot)$ *from* $P(X)$ *to* $[0, 1]$ *is called fuzzy measure on* X *when* m *satisfies the following requirements:*
  (1) $m(\emptyset) = 0, m(X) = 1$,
  (2) *assume that* $\Gamma_1, \Gamma_2 \in P(X)$ *and* $\Gamma_1 \subseteq \Gamma_2$, *then* $m(\Gamma_1) \leq m(\Gamma_2)$.

In Definition 6, $m(\Gamma_1)$ can be regarded as the importance of attributes set A, which can be used to reflect the interaction between attributes [18]. By judging whether $m(\Gamma_1 \cup \Gamma_2)$ equals $m(\Gamma_1) \cup m(\Gamma_2)$, the relationship of attributes can be divided into three categories, including non-interaction, positive interaction and negative interaction [18].

Definition 7 *[5]. Let* λ *($\lambda > 0$) denote a real-valued function on* $X = \{x_1, \ldots, x_q\}$. *The discrete Choquet integral of* λ *in connection with* m *is defined.*

$$C_m(\lambda) = \sum_{t=1}^{q} \lambda_{(t)} [m(\Gamma_{(t)}) - m(\Gamma_{(t+1)})], \quad (6)$$

where m represents a fuzzy measure on X, $(\cdot)$ denotes a permutation on X for arranging $\lambda_{(1)} \leq \lambda_{(2)} \leq \ldots \leq \lambda_{(q)}$, $\Gamma_{(t)} = \{x_{(t)}, \ldots, x_{(q)}\}$, $\Gamma_{(q+1)} = \emptyset$.
  In order to capture the interaction involved in decision making processes, it is necessary to extend the definition of the Choquet integral for 2-dimensional interval neutrosophic linguistic numbers.

Definition 8. *Let Ω be a collection of $\hat{D}_t$ $(t = 1, 2, \cdots, q)$, and m be a fuzzy measure on Ω. The 2-dimensional interval neutrosophic linguistic Choquet (2DINLC) integral operator is defined as follows:*

$$2DINLC(\hat{D}_1, \hat{D}_2, \cdots, \hat{D}_q) = \prod_{t=1}^{q} \hat{D}_{(t)}^{m(\Gamma_{(t)}) - m(\Gamma_{(t+1)})}, \tag{7}$$

where $\Gamma_{(t)} = \{\hat{D}_{(t)}, \ldots, \hat{D}_{(q)}\}$, $\Gamma_{(q+1)} = \emptyset$, and $\hat{D}_{(1)} \leq \hat{D}_{(2)} \leq \ldots \leq \hat{D}_{(q)}$.

To simplify the representation, we introduce the following notations in the sequel. $\coprod_{t=1}^{q} (\Upsilon_t^{\chi})^{\gamma} = 1 - \prod_{t=1}^{q} (1 - \Upsilon_t^{\chi})^{\gamma}$, $\coprod_{t=1}^{q} (\Upsilon_t^{\chi} \Upsilon_t^{\chi})^{\gamma} = 1 - \prod_{t=1}^{q} (1 - \Upsilon_t^{\chi})^{\gamma} (1 - \Upsilon_t^{\chi})^{\gamma}$, $\coprod_{t=1}^{q} (\Upsilon_t^{\chi})^{\gamma_1} (\Upsilon_t^{\chi})^{\gamma_2} = 1 - \prod_{t=1}^{q} (1 - \Upsilon_t^{\chi})^{\gamma_1} (1 - \Upsilon_t^{\chi})^{\gamma_2}$, where γ, γ_1 and γ_2 denote parameter, $\Upsilon \in \{I, F\}$, $\chi \in \{L, U\}$.

Theorem 2. *Let Ω be a collection of $\hat{D}_t = \langle (\dot{s}_{\alpha_t}, [T_j^L, T_t^U], [I_t^L, I_t^U], [F_t^L, F_t^U]), \ddot{s}_{\beta_j} \rangle$ $(t = 1, 2, \cdots, q)$. Then their aggregated value by using (7) is also a 2-dimensional interval neutrosophic linguistic number as follows:*

$$2DINLC(\hat{D}_1, \hat{D}_2, \cdots, \hat{D}_q)$$

$$= \left\langle (\dot{s}_{\prod_{t=1}^{q} \alpha_{(t)}^{m_t}}, [\prod_{t=1}^{q} (T_{(t)}^L)^{m_t}, \prod_{t=1}^{q} (T_{(t)}^U)^{m_t}], [\coprod_{t=1}^{q} (I_{(t)}^L)^{m_t}, \coprod_{t=1}^{q} (I_{(t)}^U)^{m_t}], \right.$$

$$\left. [\coprod_{t=1}^{q} (F_{(t)}^L)^{m_t}, \coprod_{t=1}^{q} (F_{(t)}^U)^{m_t}]), \ddot{s}_{\min\{\beta_{(1)}, \beta_{(2)}, \cdots, \beta_{(q)}\}} \right\rangle, \tag{8}$$

where $m_t = m(\Gamma_{(t)}) - m(\Gamma_{(t+1)})$, $\Gamma_{(t)} = \{\hat{D}_{(t)}, \ldots, \hat{D}_{(q)}\}$, $\Gamma_{(q+1)} = \emptyset$, and $\hat{D}_{(1)} \leq \hat{D}_{(2)} \leq \ldots \leq \hat{D}_{(q)}$.

Proof. We employ mathematical induction to give a proof of (8) with respect to positive integer n.

For the case of $q = 2$, according to the fundamental operators of Definition 2, we have

$$\hat{D}_{(1)}^{m_1} = \langle (\dot{s}_{\alpha_{(1)}^{m_1}}, [(T_{(1)}^L)^{m_1}, (T_{(1)}^U)^{m_1}], [\coprod (I_{(1)}^L)^{m_t}, \coprod (I_{(1)}^U)^{m_t}],$$

$$[\coprod (F_{(1)}^L)^{m_1}, \coprod (F_{(1)}^U)^{m_1}]), \ddot{s}_{\beta_{(1)}} \rangle,$$

$$\hat{D}_{(2)}^{m_2} = \langle (\dot{s}_{\alpha_{(2)}^{m_2}}, [(T_{(2)}^L)^{m_2}, (T_{(2)}^U)^{m_2}], [\coprod (I_{(2)}^L)^{m_2}, \coprod (I_{(2)}^U)^{m_2}],$$

$$[\coprod (F_{(2)}^L)^{m_2}, \coprod (F_{(2)}^U)^{m_2}]), \ddot{s}_{\beta_{(2)}} \rangle,$$

$$\hat{D}_{(1)}^{m_1} \cdot \hat{D}_{(2)}^{m_2} = \left\langle (\dot{s}_{\alpha_{(1)} \cdot \alpha_{(2)}}, [(T_{(1)}^L)^{m_1} (T_{(2)}^L)^{m_2}, (T_{(1)}^U)^{m_1} (T_{(2)}^U)^{m_2}], \right.$$

$$[\coprod (I_{(1)}^L)^{m_1} (I_{(2)}^L)^{m_2}, \coprod (I_{(1)}^U)^{m_1} (I_{(2)}^U)^{m_2}],$$

$$\left. [\coprod (F_{(1)}^L)^{m_1} (F_{(2)}^L)^{m_2}, \coprod (F_{(1)}^U)^{m_1} (F_{(2)}^U)^{m_2}], \ddot{s}_{\min(\beta_{(1)}, \beta_{(2)})} \right\rangle.$$

Therefore, (8) holds for $q = 2$.

Assume that (8) holds for the case of $q = h$, i.e.,

$$2DINLC(\hat{D}_1, \hat{D}_2, \cdots, \hat{D}_h)$$
$$= \left\langle (\dot{s}_{\prod_{t=1}^{h} \alpha_{(t)}^{m_t}}, [\prod_{t=1}^{h}(T_{(t)}^{L})^{m_t}, \prod_{t=1}^{h}(T_{(t)}^{U})^{m_t}], [\coprod_{t=1}^{h}(I_{(t)}^{L})^{m_t}, \coprod_{t=1}^{h}(I_{(t)}^{U})^{m_t}], \right.$$
$$\left. [\coprod_{t=1}^{h}(F_{(t)}^{L})^{m_t}, \coprod_{t=1}^{h}(F_{(t)}^{U})^{m_t}]), \ddot{s}_{\min\{\beta_{(1)}, \beta_{(2)}, \cdots, \beta_{(h)}\}} \right\rangle.$$

For the case of $q = h + 1$, we obtain the following equality according to Definition 2.

$$2DINLC(\hat{D}_1, \hat{D}_2, \cdots, \hat{D}_{h+1})$$
$$= \left\langle (\dot{s}_{\alpha_{(h+1)}^{m_{h+1}} \prod_{t=1}^{h} \alpha_{(t)}^{m_t}}, [\prod_{t=1}^{h}(T_{(t)}^{L})^{m_t}(T_{(h+1)}^{L})^{m_{h+1}}, \prod_{t=1}^{h}(T_{(t)}^{U})^{m_t}(T_{(h+1)}^{U})^{m_{h+1}}], \right.$$
$$[\coprod_{t=1}^{h}(I_{(t)}^{L})^{m_t}(I_{(h+1)}^{L})^{m_{h+1}}, \coprod_{t=1}^{h}(I_{(t)}^{U})^{m_t}(I_{(h+1)}^{U})^{m_{h+1}}], [\coprod_{t=1}^{h}(F_{(t)}^{L})^{m_t}(F_{(h+1)}^{L})^{m_{h+1}},$$
$$\left. \coprod_{t=1}^{h}(F_{(t)}^{U})^{m_t}(F_{(h+1)}^{U})^{m_{h+1}}]), \ddot{s}_{\min\{\beta_{(1)}, \beta_{(2)}, \cdots, \beta_{(h+1)}\}} \right\rangle$$
$$= \left\langle (\dot{s}_{\prod_{t=1}^{h+1} \alpha_{(t)}^{m_t}}, [\prod_{t=1}^{h+1}(T_{(t)}^{L})^{m_t}, \prod_{t=1}^{h+1}(T_{(t)}^{U})^{m_t}], [\coprod_{t=1}^{h+1}(I_{(t)}^{L})^{m_t}, \coprod_{t=1}^{h+1}(I_{(t)}^{U})^{m_t}], \right.$$
$$\left. [\coprod_{t=1}^{h+1}(F_{(t)}^{L})^{m_t}, \coprod_{t=1}^{h+1}(F_{(t)}^{U})^{m_t}]), \ddot{s}_{\min\{\beta_{(1)}, \beta_{(2)}, \cdots, \beta_{(h+1)}\}} \right\rangle.$$

That means that (8) holds for the case of $q = h + 1$. Therefore, (8) holds for all $q \in N$. $\qquad\square$

It is not difficult to verify that the operator defined by (7) has the following properties.

Proposition 1 *(Idempotence). Let Ω be a collection of $\hat{D}_t = \langle(\dot{s}_{\alpha_t}, [T_t^L, T_t^U], [I_t^L, I_t^U], [F_t^L, F_t^U]), \ddot{s}_{\beta_t}\rangle$. If all $\hat{a}_t$ $(t = 1, 2, \ldots, q)$ are equal, that is,*

$$\hat{D}_t = \hat{D}_1 = \langle(\dot{s}_{\alpha_1}, [T_1^L, T_1^U], [I_1^L, I_1^U], [F_1^L, F_1^U]), \ddot{s}_{\beta_1}\rangle,$$

then $2DINLC(\hat{D}_1, \ldots, \hat{D}_q) = \hat{D}_1$.

Proposition 2. *Let Ω be a collection of $\hat{D}_t = \langle(\dot{s}_{\alpha_t}, [T_t^L, T_t^U], [I_t^L, I_t^U], [F_t^L, F_t^U]), \ddot{s}_{\beta_t}\rangle$ $(t = 1, 2, \cdots, q)$. If $(\hat{D}_1', \hat{D}_2', \ldots, \hat{D}_q')$ is any permutation of $(\hat{D}_1, \hat{D}_2, \ldots, \hat{D}_q)$, then*

$$2DINLC(\hat{D}_1, \ldots, \hat{D}_q) = 2DINLC(\hat{D}_1', \hat{D}_2', \ldots, \hat{D}_q').$$

Proposition 3. *Let Ω be a collection of $\hat{D}_t = \langle(\dot{s}_{\alpha_t}, [T_t^L, T_t^U], [I_t^L, I_t^U], [F_t^L, F_t^U]), \ddot{s}_{\beta_j}\rangle$ $(t = 1, 2, \cdots, q)$. For a given $\hat{s}_r = \langle(\dot{s}_{\alpha_r}, [T_r^L, T_r^U], [I_r^L, I_r^U], [F_r^L, F_r^U]), \ddot{s}_{\beta_r}\rangle$, then*

$$2DINLC(\hat{D}_1 \cdot \hat{s}_r, \ldots, \hat{D}_q \cdot \hat{s}_r) = 2DINLC(\hat{D}_1, \ldots, \hat{D}_n) \cdot \hat{s}_r.$$

Proposition 4. *Let $\hat{D}_t = \langle(\dot{s}_{\alpha_t}, [T_t^L, T_t^U], [I_t^L, I_t^U], [F_t^L, F_t^U]), \ddot{s}_{\beta_t}\rangle$ $(t = 1, 2, \cdots, q)$. If $r > 0$, then*

$$2DINLC(\hat{D}_1^r, \ldots, \hat{D}_q^r) = (2DINLC(\hat{D}_1, \ldots, \hat{D}_q))^r.$$

4   A Group Decision Making Scheme Based on PROMETHEE II Model

Considering the aforementioned aggregation operators, we establish a group method under 2-dimensional interval neutrosophic linguistic fuzzy situation, in which PROMETHEE II [1] is applied to fuse information derived from multiple experts. The decision context involves alternatives $Al = \{Al_1, Al_2, \ldots, Al_p\}$, attributes $Z = \{z_1, z_2, \ldots, z_q\}$, experts $E = \{e_1, e_2, \ldots, e_K\}$. The details are outlined below.

Step 1: Normalize the evaluation matrices.
In general, in order to compare different alternatives for different types attributes, we need to standardize the decision making matrices by transforming the 2-dimensional interval neutrosophic linguistic fuzzy decision matrix $D_k = (a_{it}^k)_{p \times q}$ into the normalized 2-dimensional interval neutrosophic linguistic fuzzy decision matrix $R_k = (r_{it}^k)_{p \times q}$ by using (9) and (10).
(1) For benefit type attributes:

$$\hat{r}_{st}^k = \langle(\dot{s}_{\alpha_{it}^k}, [T_{st}^{kL}, T_{it}^{kU}], [I_{it}^{kL}, I_{it}^{kU}], [F_{it}^{kL}, F_{it}^{kU}]), \ddot{s}_{\beta_{it}^k}\rangle. \tag{9}$$

(2) For cost type attributes:

$$\hat{r}_{it}^k = \langle(\text{neg}(\dot{s}_{\alpha_{it}^k}), [T_{it}^{kL}, T_{it}^{kU}], [I_{it}^{kL}, I_{it}^{kU}], [F_{it}^{kL}, F_{it}^{kU}]), \ddot{s}_{\beta_{it}^k}\rangle, \tag{10}$$

where $\text{neg}(\dot{s}_{\alpha_{it}^k}) = \dot{s}_{l-1-\alpha_{it}^k}$.

Step 2: (1) Predetermine the fuzzy measures of attributes sets.
(2) For individual's decision tables, rank the $\hat{r}_{it}^k$ with respect to different attributes by virtue of calculating the functions of score, accuracy and certainty of $\hat{r}_{it}^k$ $(t = 1, 2, \ldots, q)$.
(3) Employ 2-dimensional interval neutrosophic linguistic Choquet operator to integrate each individual's information with respect to different attributes.

$$\hat{r}_i^k = 2DINLC(\hat{r}_{i1}^k, \ldots, \hat{r}_{iq}^k)$$

$$= \left\langle (\dot{s}_{\prod_{t=1}^q (\alpha_{i(t)}^k)^{m_t}}, [\prod_{t=1}^q (T_{i(t)}^{kL})^{m_t}, \prod_{t=1}^q (T_{i(t)}^{kU})^{m_t}], [\coprod_{t=1}^q (I_{i(t)}^{kL})^{m_t}, \coprod_{t=1}^q (I_{i(t)}^{kU})^{m_t}],\right.$$

$$\left. [\coprod_{t=1}^q (F_{i(t)}^{kL})^{m_t}, \coprod_{t=1}^q (F_{i(t)}^{kU})^{m_t}], \ddot{s}_{\min\{\beta_{i(1)}^k, \beta_{i(2)}^k, \cdots, \beta_{i(q)}^k\}} \right\rangle, \tag{11}$$

where $m_t = m(\Gamma_{(t)}) - m(\Gamma_{(t+1)})$, $\Gamma_{(t)} = \{z_{(t)}, \ldots, z_{(q)}\}$.

Step 3: Construct the preference function $P_k(Al_i, Al_j)$ for the k-th expert as follows:

$$P_k(Al_i, Al_j) = \begin{cases} 0, & \text{if } \hat{d}(\hat{r}_i^k, \hat{r}^{k+}) \geq \hat{d}(\hat{r}_j^k, \hat{r}^{k+}), \\ \hat{d}(\hat{r}_j^k, \hat{r}^{k+}) - \hat{d}(\hat{r}_i^k, \hat{r}^{k+}), & \text{if } \hat{d}(\hat{r}_i^k, \hat{r}^{k+}) < \hat{d}(\hat{r}_j^k, \hat{r}^{k+}), \end{cases} \quad (12)$$

where, $\hat{r}^{k+}$ denotes the positive ideal alternative,

$$\hat{r}^{k+} = \langle (\max_i(\dot{s}_i^k), [\max_i T_i^{kL}, \max_i T_i^{kU}], [\min_i I_i^{kL}, \min_i I_i^{kU}]$$
$$[\min_i F_i^{kL}, \min_i F_i^{kU}]), \max_{1 \leq i \leq p}(\ddot{s}_i^k)\rangle, \ 1 \leq i \leq p.$$

Step 4: Calculate the preference index $\tilde{H}(Al_i, Al_j)$ as follows:

$$\tilde{H}(Al_i, Al_j) = \sum_{k=1}^{K} e_k P_k(Al_i, Al_j), \quad (13)$$

where e_k denotes the weighting value of the k-th expert.

Step 5: In light of the above preference index $\tilde{H}(Al_i, Al_j)$, the positive flow $\tilde{\phi}^+(Al_i)$, negative flow $\tilde{\phi}^-(Al_i)$ and net flow $\tilde{\phi}(Al_i)$ of the i-th alternative can be formulated as follows:

$$\tilde{\phi}^+(Al_i) = \sum_{j=1}^{p} \tilde{H}(Al_i, Al_j), \quad (14)$$

$$\tilde{\phi}^-(Al_i) = \sum_{j=1}^{p} \tilde{H}(Al_j, Al_i), \quad (15)$$

$$\tilde{\phi}(Al_i) = \tilde{\phi}^+(Al_i) - \tilde{\phi}^-(Al_i). \quad (16)$$

Step 6: Compare and rank all alternatives in terms of the performance of $\tilde{\phi}(Al_i)$. The one with the largest value of net flow will be considered as the ideal alternative.

5   Numerical Study

Suppose that there are three graduation theses as alternatives $\{Al_1, Al_2, Al_3\}$, which are evaluated by three making deciders $\{e_1, e_2, e_3\}$ for ranking students' overall level of innovation ability. Three factors are taken into account including the level of innovation z_1, the practical value z_2 and the rationality and accuracy of conclusions z_3. The weighting vector of three making deciders $\{e_1, e_2, e_3\}$ involved in the decision process is set as $\mathbf{e} = (0.3, 0.25, 0.45)$. The assessment results given by the decision makers take the form of 2-dimensional interval neutrosophic linguistic numbers which are shown in Tables 1, 2, and 3, respectively. Two class linguistic sets are given to perform as a reference quantitative scale as following: $S_I = \{\dot{s}_0, \dot{s}_1, \dot{s}_2, \dot{s}_3, \dot{s}_4, \dot{s}_5, \dot{s}_6\} = \{\text{very bad, bad, below general, general, above general, good, very good}\}$ and $S_{II} = \{\ddot{s}_0, \ddot{s}_1, \ddot{s}_2, \ddot{s}_3, \ddot{s}_4\} = \{\text{very unreliable, unreliable, general, reliable, very reliable}\}$.

5.1 The Decision Making Steps

In what follows, we utilize the established approach to obtain the desirable alternative.

Step 1. Due to the attributes involved in decision context are all the same type (benefit type), the decision making matrices do not require to be normalized.

Step 2. Fuzzy measures can be determined by some methods, such as granular computing, statistical inference and subjective evaluating. In this paper, the fuzzy measures of all attributes' combinations are set to $m(z_1) = 0.25$, $m(z_2) = 0.37$, $m(z_3) = 0.3$, $m(z_1, z_2) = 0.65$, $m(z_1, z_3) = 0.5$, $m(z_2, z_3) = 0.85$, $m(z_1, z_2, z_3) = 1$, respectively.

According to Tables 1, 2, and 3, the partial evaluation $\hat{r}_{it}^k$ of the alternative A_i is rearranged such that $\hat{r}_{i(t)}^k \leq \hat{r}_{i(t+1)}^k (i = 1, 2, 3, k = 1, 2, 3)$. For every alternative Al_i $(i = 1, 2, 3)$, we utilize (11) to obtain the individual expert's overall preference values $\hat{r}_{ki}$, for example $\hat{r}_{11} = \langle (\dot{s}_{2.36}, [0.39, 0.5], [0.09, 0.2], [0.32, 0.42]), \ddot{s}_2 \rangle$.

Steps 3 and 4. By using (12), the preference indexes $\tilde{H}(Al_i, Al_j)$ are obtained. $\tilde{H}(Al_1, Al_2) = 0.23$, $\tilde{H}(Al_1, Al_3) = 0.06$, $\tilde{H}(Al_2, Al_1) = 0.29$, $\tilde{H}(Al_2, Al_3) = 0$, $\tilde{H}(Al_3, Al_1) = 0.53$, $\tilde{H}(Al_3, Al_2) = 0.27$.

Steps 5 and 6. The performances of $\tilde{\phi}^+(i)$, $\tilde{\phi}^-(i)$ and $\tilde{\phi}(i)$ are shown in Table 4, using which alternatives $Al_i(i = 1, 2, 3)$ are ranked as $Al_3 \succ Al_2 \succ Al_1$. Thus, the alternative Al_3 is identified as the best one.

Table 1. Evaluation information given by e_1

z_1	
Al_1	$\langle (\dot{s}_5, [0.4, 0.5], [0.2, 0.3], [0.3, 0.4]), \ddot{s}_2 \rangle$
Al_2	$\langle (\dot{s}_2, [0.5, 0.7], [0.1, 0.2], [0.2, 0.3]), \ddot{s}_3 \rangle$
Al_3	$\langle (\dot{s}_1, [0.3, 0.5], [0.1, 0.2], [0.3, 0.4]), \ddot{s}_3 \rangle$
z_2	
Al_1	$\langle (\dot{s}_3, [0.7, 0.8], [0, 0.1], [0.1, 0.2]), \ddot{s}_3 \rangle$
Al_2	$\langle (\dot{s}_5, [0.4, 0.6], [0.2, 0.3], [0.1, 0.2]), \ddot{s}_1 \rangle$
Al_3	$\langle (\dot{s}_2, [0.5, 0.6], [0.1, 0.3], [0.1, 0.2]), \ddot{s}_4 \rangle$
z_3	
Al_1	$\langle (\dot{s}_1, [0.2, 0.3], [0.1, 0.2], [0.5, 0.6]), \ddot{s}_4 \rangle$
Al_2	$\langle (\dot{s}_1, [0.5, 0.7], [0.2, 0.4], [0.1, 0.2]), \ddot{s}_2 \rangle$
Al_3	$\langle (\dot{s}_3, [0.3, 0.4], [0.1, 0.2], [0.1, 0.3]), \ddot{s}_4 \rangle$

Table 2. Evaluation information given by e_2

z_1	
Al_1	$\langle (\dot{s}_1, [0.3, 0.5], [0.3, 0.6], [0.4, 0.5]), \ddot{s}_2 \rangle$
Al_2	$\langle (\dot{s}_2, [0.4, 0.7], [0.1, 0.3], [0.1, 0.3]), \ddot{s}_4 \rangle$
Al_3	$\langle (\dot{s}_3, [0.2, 0.5], [0.3, 0.6], [0.3, 0.5]), \ddot{s}_3 \rangle$
z_2	
Al_1	$\langle (\dot{s}_4, [0.6, 0.7], [0.2, 0.4], [0.2, 0.3]), \ddot{s}_2 \rangle$
Al_2	$\langle (\dot{s}_2, [0.2, 0.6], [0.2, 0.3], [0.1, 0.3]), \ddot{s}_1 \rangle$
Al_3	$\langle (\dot{s}_1, [0.3, 0.5], [0.4, 0.7], [0.4, 0.6]), \ddot{s}_4 \rangle$
z_3	
Al_1	$\langle (\dot{s}_2, [0.1, 0.4], [0.3, 0.4], [0.2, 0.4]), \ddot{s}_4 \rangle$
Al_2	$\langle (\dot{s}_2, [0.6, 0.7], [0.2, 0.5], [0.5, 0.7]), \ddot{s}_2 \rangle$
Al_3	$\langle (\dot{s}_1, [0.3, 0.6], [0.4, 0.5], [0.7, 0.8]), \ddot{s}_3 \rangle$

Table 3. Assessment information given by expert e_3

z_1	
Al_1	$\langle(\dot{s}_4, [0.1, 0.2], [0.1, 0.4], [0.2, 0.5]), \ddot{s}_2\rangle$
Al_2	$\langle(\dot{s}_4, [0.3, 0.5], [0.2, 0.5], [0.1, 0.3]), \ddot{s}_4\rangle$
Al_3	$\langle(\dot{s}_5, [0.4, 0.5], [0.1, 0.2], [0.4, 0.5]), \ddot{s}_3\rangle$
z_2	
Al_1	$\langle(\dot{s}_2, [0.5, 0.8], [0.4, 0.6], [0.1, 0.3]), \ddot{s}_3\rangle$
Al_2	$\langle(\dot{s}_4, [0.3, 0.7], [0.2, 0.3], [0.2, 0.4]), \ddot{s}_2\rangle$
Al_3	$\langle(\dot{s}_2, [0.5, 0.8], [0.3, 0.4], [0.1, 0.2]), \ddot{s}_3\rangle$
z_3	
Al_1	$\langle(\dot{s}_2, [0.2, 0.3], [0.2, 0.3], [0.3, 0.6]), \ddot{s}_4\rangle$
Al_2	$\langle(\dot{s}_3, [0.4, 0.7], [0.4, 0.5], [0.3, 0.4]), \ddot{s}_2\rangle$
Al_3	$\langle(\dot{s}_3, [0.3, 0.5], [0.5, 0.6], [0.1, 0.3]), \ddot{s}_3\rangle$

Table 4. The ranking result

Alternatives	$\tilde{\phi}^+(Al_i)$	$\tilde{\phi}^-(Al_i)$	$\tilde{\phi}(Al_i)$	Ranking
1	0.1	0.26	-0.16	3
2	0.09	0.22	-0.13	2
3	0.31	0.02	0.29	1

6   Conclusions

In this study a new representation of complex cognitive information was provided, and its application was also explored. 2-Dimensional interval neutrosophic linguistic numbers are proposed by incorporating linguistic reliability into interval neutrosophic linguistic numbers, whose operations laws, distance measure and comparison methods are developed with detailed proofs. Moreover, we extended Choquet integral into the context of 2-dimensional interval neutrosophic linguistic numbers, whose properties were also examined as well. Based on these developments, we designed an extended PROMETHEE II method to solve decision management problems by taking the interactive characteristics under the contexts of 2-dimensional interval neutrosophic linguistic numbers.

Acknowledgements. This work is supported by the higher education undergraduate teaching reform project of Liaoning province in 2018, and the innovation and entrepreneurship training program for college students (No. 202010151583).

References

1. Brans, J.P., Vincke, P.: A preference ranking organization method: the PROMETHEE method for MCDM. Manage. Sci. **31**(6), 647–656 (1985)

2. Broumi, S., Smarandache, F.: Single valued neutrosophic trapezoid linguistic aggregation operators based multi-attribute decision making. Bull. Pure Appl. Sci. Math. Stat. **33**(2), 135–155 (2014)
3. Broumi, S., Ye, J., Smarandache, F.: An extended TOPSIS method for multiple attribute decision making based on interval neutrosophic uncertain linguistic variables. Neutrosophic Sets Syst. **8**, 22–31 (2015)
4. Cheng, H.D., Chen, J.R.: Automatically determine the membership function based on the maximum entropy principle. Inf. Sci. **96**(3–4), 163–182 (1997)
5. Choquet, G.: Theory of capacities. Annales del Institut Fourier **5**, 131–295 (1953)
6. Cui, W.H., Ye, J., Shi, L.L.: Linguistic Neutrosophic uncertain numbers and their multiple attribute group decision-making method. J. Intell. Fuzzy Syst. **1**, 649–660 (2019)
7. Gupta, M.M.: On fuzzy logic and cognitive computing: some perspectives. Sci. Iranica D **18**(3), 590–592 (2011)
8. Herrera, F., Herrera-Viedma, E.: A model of consensus in group decision making under linguistic assessments. Fuzzy Sets Syst. **78**(1), 73–87 (1996)
9. Herrera, F., Herrera-Viedma, E.: Linguistic decision analysis: steps for solving decision problems under linguistic information. Fuzzy Sets Syst. **115**(1), 67–82 (2000)
10. Li, X.H., Chen, X.H.: D-intuitionistic hesitant fuzzy sets and their application in multiple attribute decision making. Cogn. Comput. **10**, 496–505 (2018)
11. Li, Y.Y., Wang, J.Q., Wang, T.L.: A linguistic neutrosophic multi-criteria group decision-making approach with edas method. Arab. J. Sci. Eng. **44**, 2737–2749 (2019)
12. Liu, P., Wang, Y.: The aggregation operators based on the 2-dimension uncertain linguistic information and their application to decision making. Int. J. Mach. Learn. Cybernet. **7**(6), 1057–1074 (2015). https://doi.org/10.1007/s13042-015-0430-x
13. Liu, P.D., Teng, F.: An extended TODIM method for multiple attribute group decision-making based on 2-dimension uncertain linguistic variable. Complexity **29**(5), 20–30 (2016)
14. Martínez, L., Rodriguez, R.M., Herrera, F.: The 2-tuple Linguistic Model: Computing with Words in Decision Making. Springer, Cham (2015). https://doi.org/10.1007/978-3-319-24714-4
15. Smarandache, F.: Neutrosophy/Neutrosophic Probability, Set, and Logic. American Research Press, Rehoboth (1998)
16. Smarandache, F.: A Unifying Field in Logics: Neutrosophic Logic. American Research Press, Rehoboth (1999)
17. Sugeno, M.: Theory of fuzzy integrals and its applications. Ph.D. Dissertation, Tokyo Institute of Technology (1974)
18. Tan, C.Q.: A multi-criteria interval-valued intuitionistic fuzzy group decision making with Choquet integral-based TOPSIS. Exp. Syst. Appl. **38**(4), 3023–3033 (2011)
19. Tian, Z.P., Nie, R.X., Wang, J.Q., Luo, H.Y., Li, L.: A prospect theory-based QUALIFLEX for uncertain linguistic Z-number multi-criteria decision-making with unknown weight information. J. Intell. Fuzzy Syst. **38**(2), 1775–1787 (2020)
20. Wang, H.B., Smarandache, F., Sunderraman, R., Zhang, Y.Q.: Interval Neutrosophic Sets and Logic Theory and Applications in Computing. Hexis, Phoenix (2005)
21. Wang, Y.X., Ruhe, G.: The cognitive process of decision making. Int. J. Cogn. Inf. Nat. Intell. **1**(2), 73–85 (2007)

22. Xu, Z.S.: Uncertain linguistic aggregation operators based approach to multiple attribute group decision making under uncertain linguistic environment. Inf. Sci. **168**, 171–184 (2004)
23. Ye, J.: Some aggregation operators of interval neutrosophic linguistic numbers for multiple attribute decision making. J. Intell. Fuzzy Syst. **27**(5), 2231–2241 (2014)
24. Zadeh, L.A.: Fuzzy sets. Inf. Control **8**(3), 338–353 (1965)
25. Zhang, H., Wang, J., Chen, X.: An outranking approach for multi-criteria decision-making problems with interval-valued neutrosophic sets. Neural Comput. Appl. **27**(3), 615–627 (2015). https://doi.org/10.1007/s00521-015-1882-3

Cross-Modal N-Pair Network for Generalized Zero-Shot Learning

Biying Cui, Zhong Ji$^{(\boxtimes)}$, and Hai Wang

Tianjin University, Tianjin 300072, China
{cuibiying,jizhong,wanghai515}@tju.edu.cn

Abstract. Generalized Zero-Shot Learning (GZSL), which aims at transferring information from seen categories to unseen categories and recognizing both during the test stage, has been attracting a lot of attention in recent years. As a cross-modal task, the key to aligning visual and semantic representations is accurately measuring the distance in the projection space. Although many methods achieve GZSL by utilizing metric learning, few of them leverage intra- and inter-category information sufficiently and would suffer from domain shift problem. In this paper, we introduce a novel Cross Modal N-Pairs Network (CMNPN) to alleviate this issue. Our CMNPN firstly maps visual features and semantic prototypes into a common space with an embedding network, and then employs two N-pairs networks, VMNPN and SMNPN, to optimize an N-pair loss in both visual and semantic spaces, where we utilize a cross-modal N-pair mining strategy to mine information from all classes. Specifically, we select the hard visual representation for its corresponding semantic prototype and combines the two features with $(N-1)$ negative samples to form an N-pair. In VMNPN, the negative samples are hard visual representations of the other $(N-1)$ categories, while in SMNPN, they are the other $(N-1)$ semantic prototypes. Extensive experiments on three benchmark datasets demonstrate that our proposed CMNPN achieves the state-of-the-art results on GZSL tasks.

Keywords: Generalized Zero-Shot Learning · Deep metric learning · n-pair loss.

1 Introduction

With the development of deep learning, image classification has made a remarkable progress, but traditional supervised learning gradually shows weakness because it is expensive to collect and annotate data for all categories in reality. Recently, Zero-shot Learning (ZSL) [5,7,8,10,25], aiming at handling the classification for unseen categories which are absent during the training stage, has attracted extensive concern. Based on the semantic prototypes that are utilized as side information (e.g., class-leval attributes [11]), ZSL achieves to transfer

Supported by the National Natural Science Foundation of China under Grant 61771329.

H. Zhang et al. (Eds.): NCAA 2020, CCIS 1265, pp. 247–258, 2020.
https://doi.org/10.1007/978-981-15-7670-6_21

knowledge from seen categories to unseen categories. However, in reality, it is unclear that whether a test instance is from unseen classes or seen classes, so Generalized Zero-shot Learning (GZSL) is proposed to recognize visual instances from both the seen and unseen classes.

Although great improvement has been made by existing GZSL approaches [14,22,28], most of them focus on learning a visual-semantic embedding and leveraging the nearest neighbor search to assign label for test instances, which are easy to cause hubness problem when high-dimensional visual features are projected into a low-dimensional semantic space, and visual samples may aggregate to an ambiguous center as a result. According to DEM [27], the embedded direction is the key to alleviate hubness problem. Therefore, our proposed approach maps visual features and semantic prototypes into the same high-dimensional space to align the cross-modal features. Moreover, utilizing an appropriate metric to measure the distance between features from two modalities is critical to GZSL and metric learning could help us achieve this model. Bucher et al. [2] formulated ZSL as a distance metric problem with a Mahalanobis distance metric learning. ALE [1] proposes to utilize a triplet loss to measure the distance in the embedding space. This method employs a negative semantic prototype in a triplet, while ignores the impact of negative visual samples. Due to the slow convergence of the triplet loss, Sohn [17] proposed an N-pair-multi-class loss, which leverages $(N-1)$ negative samples. Inspired by it, we propose to utilize an N-pair loss in a cross-modal form.

To sufficiently utilize the information of both modalities and mine intra- and inter-category correction, we propose a Cross-Modal N-pairs Network (CMNPN), which consists of two parts: an embedding network and a deep metric network. The former is utilized to map visual representations and semantic prototypes into a common space to relieve the hubness problem, while the latter aims at constructing a discriminative metric space, including a Visual Modal N-pairs Network (VMNPN) and a Semantic Modal N-pairs Network (SMNPN). Furthermore, a hard sample mining strategy is leveraged in our approach. Specifically, we combine each semantic prototype with the hard visual sample that has the largest distance from it to form a positive pair. Then the positive pair and negative samples from other categories constitute N-pair samples. In VMNPN, the negative samples are hard visual representations of the other $(N-1)$ categories, while in SMNPN, the negative samples are the other $(N-1)$ semantic prototypes. Our cross-modal N-pair loss is calculated by the two sets of N-pair samples. By doing so, the information of both modalities and the correlation among categories are fully mined and utilized.

In summary, our contributions are concluded into the following three-fold.

(1) A novel cross-modal metric learning model, called Cross-Modal N-pairs Network (CMNPN), is proposed for ZSL. It consists of an embedding network and a deep metric network. The former projects visual representations and semantic prototypes into a high-dimensional common space to alleviate hubness problem, while the latter learns to construct a metric space to align the visual and semantic features.

(2) In the deep metric network, we propose to mine the hard visual sample for each category and form an N-pair together with its corresponding semantic prototype and the hard features of other categories. In addition, we propose to employ the cross-modal N-pair mining strategy in both visual and semantic modalities, which would mine intra- and inter-category correlation to leverage the information sufficiently from both spaces.

(3) Extensive experiments on three ZSL benchmark datasets demonstrate the effectiveness of our approach and it obtains the state-of-the-art performance on GZSL tasks.

2 Related Work

In this section, we present an overview of the related literature on zero-shot learning as well as metric learning. Existing approaches for GZSL could be divided into two groups, feature synthesis based approaches and mapping based approaches. Recently, Generative Adversarial Network (GAN) is applied in the feature synthesis based methods which aim to generate pseudo visual features taking as input the class semantic prototypes [12,21,23,29]. From this perspective, GZSL could be transformed into a supervised learning problem. And the mapping based approaches usually learn a mapping function between visual space and semantic space or embeds two representations into a common space. According to DEM [27], mapping semantic features into the visual space is an effective method to alleviate the hubness problem [15]. Our approach also learns an embedding network from a low-dimensionality to a high-dimensionality space, but different from them, we embed both visual features and semantic features into a common space, who has the highest dimensionality to relieve the hubness problem. What's more, on the basis of the embedding space, instead of directly calculating the loss functions, we further construct a metric space to take full advantage of intra- and inter-categories correlation from both visual and semantic information.

Metric learning aims at constructing a metric space and learning a distance function to measure similarity for different tasks. Recently, metric learning is incorporated in the Traditional ZSL (TZSL) task. Bucher et al. [2] projected the visual features into the semantic space and then learned to measure the similarity with Mahalanobis distance. ALE [1] selects a visual feature and its corresponding attribute representation to form a triplet with an attribute from a negative class. However, both of them choose negative samples from semantic representations but ignore the contribution of negative visual representations. And the underutilization of representations would cause a weak generalization in GZSL. Therefore, we utilize negative samples from multi-class and both modalities to sufficiently leverage intra- and inter-class correlation. Moreover, because of the slow convergence of the triplet loss, Sohn [17] proposed an N-pair-multi-class loss, which leverages hard negative class mining in a single modal. Inspired by it, we propose to transform the N-pair loss to our cross-modal hard sample mining and verify its efficiency on GZSL task, which would be more appropriate for transferring knowledge from seen to unseen categories.

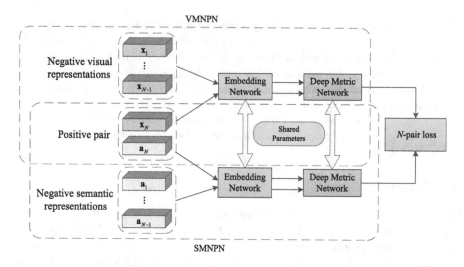

Fig. 1. The illustration of our proposed CMNPN framework.

3   Methodology

3.1   Model Architecture

We tackle the task of GZSL by CMNPN. Suppose that we have a training set $\{\mathbf{x}_i, \mathbf{y}_i, \mathbf{a}_i\}_{i=1}^N$, where $\mathbf{x}_i \in X_S \subset \mathbb{R}^D$ is the D-dimensional visual representation of visual the i-th sample, $\mathbf{a}_i \in A_S$ and $\mathbf{y}_i \in S$ are the corresponding semantic prototype and one-hot label of x_i, $A_S = \{\mathbf{a}_1, \ldots, \mathbf{a}_k\}$ and $S = \{\mathbf{y}_1, \ldots, \mathbf{y}_k\}$ denote the set of semantic representations and labels of k seen categories. For l unseen categories, $A_U = \{\mathbf{a}_{k+1}, \ldots, \mathbf{a}_{k+l}\}$ and $U = \{\mathbf{y}_{k+1}, \ldots, \mathbf{y}_{k+l}\}$ denote the set of semantic features and labels respectively, and X_U is the unseen instances. Given a test instance $\mathbf{x}_t \in \mathbb{R}^D$, the goal of Traditional Zero-Shot Learning (TZSL) is predicting its unseen class label $\mathbf{y}_t \in U$ by its semantic feature A_S, while Generalized Zero-Shot Learning (GZSL) aims at predicting its label from T and $T = S \bigcup U$. Note that the seen class set S and the unseen class set U are disjoint with each other, *i.e.* $S \bigcap U = \emptyset$.

As shown in Fig. 1, our CMNPN framework involves two modules: an embedding network and a deep metric network, and the latter could be divided into two parts: Visual Modal N-pairs Network (VMNPN) and Semantic Modal N-pairs Network(SMNPN). For the N-th seen category, the hard visual representation $\mathbf{x}_N$ and its corresponding semantic prototype $\mathbf{a}_N$ institute a positive pair. In VMNPN, $(N-1)$ hard visual representations from other $(N-1)$ categories are regarded as negative hard samples, while in SMNPN, $(N-1)$ negative semantic prototypes form an N-pair together with the positive pair. These features are input an embedding network and a deep metric network successively to optimize an N-pair loss.

3.2   Embedding Network

To alleviate the hubness problem, we propose to embed visual representation and semantic representation into a high-dimensional common space, in which the features from different modalities can be aligned with our proposed N-pairs metric network that would be introduced in Sect. 3.3. Specifically, we input the visual representation $\mathbf{x}$ and semantic prototype $\mathbf{a}$, and utilize a three-layer neural network to realize the embedding network, where each layer consists of a fully connected (FC) layer followed by a Rectified Linear Unit (ReLU). Then output $f_\theta(\mathbf{x})$ and $f_\theta(\mathbf{a})$, where f denotes the whole embedding network and θ are the parameters that need to be learned.

3.3   Deep Metric Network

After building the common space for visual and semantic modalities, we input the embedding features $f_\theta(\mathbf{x})$ and $f_\theta(\mathbf{a})$ into a three-layer metric network to align them in a metric space. And there are three FC layers, followed by a Sigmoid function, output $g_\varphi\left(f_\theta(\mathbf{x})\right)$ and $g_\varphi\left(f_\theta(\mathbf{a})\right)$. Then $s(\mathbf{x}, \mathbf{a}) = d\left(g_\varphi\left(f_\theta(\mathbf{x})\right), g_\varphi\left(f_\theta(\mathbf{a})\right)\right)$ shows the distance of $\mathbf{x}$ and $\mathbf{a}$ in the metric space, where d is the distance metric function, such as Euclidean distance and cosine distance.

VMNPN and SMNPN. Since a semantic prototype corresponds to many visual representations of the same class, there exists many positive sample pairs and negative sample pairs from the same classes and different classes, respectively. After selecting a visual representation and its corresponding semantic prototype as a positive pair, we choose $(N - 1)$ negative visual samples to take part in training process and constitute an N-pair construction together with the positive pair. We utilize multi-negative visual samples to supervise the positive pair at the same time and this network is named after Visual Modal N-pairs Network (VMNPN). Similarly, we propose a Semantic Modal N-pairs Network (SMNPN), where the N-pair construction consists of $(N - 1)$ semantic prototypes and a positive pair. It is worth noting that there are the same positive pairs in both VMNPN and SMNPN, and by doing which, the positive pairs can be supervised in both modalities and its semantic consistency could be kept better. What's more, we utilize the same structure to design VMNPN and SMNPN. Then to increase the calculating efficiency, the parameters are shared with each other. The loss function is expressed as follows:

$$
\begin{aligned}
\mathcal{L}\left(\mathbf{x}, \mathbf{a}, \{\mathbf{x}_i, \mathbf{a}_i\}_{i=1}^{N-1}\right) = {} & \log\left(1 + \sum_{i=1}^{N-1} \exp\left(s\left(\mathbf{x}_i, \mathbf{a}\right) - s(\mathbf{x}, \mathbf{a})\right)\right) \\
& + \log\left(1 + \sum_{i=1}^{N-1} \exp\left(s\left(\mathbf{x}, \mathbf{a}_i\right) - s(\mathbf{x}, \mathbf{a})\right)\right) \\
& + \lambda \cdot \left(\Omega\left(\theta\right) + \Omega\left(\varphi\right)\right),
\end{aligned}
\tag{1}
$$

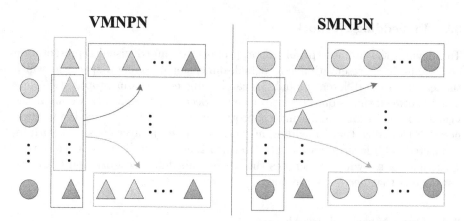

Fig. 2. The illustration of VMNPN and SMNPN. Different colors represent different categories. The circle denotes a semantic prototype, and the triangle denotes a visual feature (Color figure online).

where $\mathbf{x}$ and $\mathbf{a}$ are a visual representation and a semantic prototype from the same category, $\mathbf{a}_i$ is a negative semantic feature and $\mathbf{x}_i$ is a negative visual feature. $\Omega(\theta)$ and $\Omega(\varphi)$ are the regularizations for the embedding network and the deep metric network, where Ω denotes the Frobenius norm, θ and φ are the weights of the networks, and λ is the coefficient of the regularization term.

Hard Sample Mining Strategy. To construct an effective strategy to mine the relationship between visual and semantic modalities, we propose to employ a hard positive pair, which consists of a semantic prototype and a visual sample that is the furthest away from it. Besides, note that randomly choosing $(N-1)$ negative samples from a large number of samples for a hard positive pair would result in huge computational costs. Therefore, we employ an N-pairs hard negative sample mining strategy based on finding a hard positive pair. Specifically, as shown in Fig. 2, the needed $(N-1)$ negative samples for one class are all selected from the other $(N-1)$ classes. In VMNPN, for one category, we regard the hard positive visual representations from the other $(N-1)$ categories as the hard negative visual representations and form an N-pair together with a positive pair. As for SMNPN, the other $(N-1)$ semantic prototypes would be seen as hard negative samples and form an N-pair with a hard positive pair. Therefore, the N-pair loss function of Eq. (1) could be rewritten as:

$$\mathcal{L}\left(\{\max s\left(\mathbf{x}_i, \mathbf{a}_i\right)\}_{i=1}^{N}\right) = \frac{1}{N}\sum_{i=1}^{N}\log\left(1 + \sum_{j\neq i}\exp\left(s\left(\mathbf{x}_i, \mathbf{a}_j\right) - s\left(\mathbf{x}_i, \mathbf{a}_i\right)\right)\right)$$

$$+ \frac{1}{N}\sum_{i=1}^{N}\log\left(1 + \sum_{j\neq i}\exp\left(s\left(\mathbf{x}_j, \mathbf{a}_i\right) - s(\mathbf{x}_i, \mathbf{a}_i)\right)\right) \tag{2}$$

$$+ \lambda \cdot \left(\Omega\left(\theta\right) + \Omega\left(\varphi\right)\right),$$

where $\max s\left(\mathbf{x}_i, \mathbf{a}_i\right)$ denotes the maximum distance of the selected hard positive pair $\mathbf{x}_i$ and $\mathbf{a}_i$ and there are N pairs in total. $s\left(\mathbf{x}_i, \mathbf{a}_j\right)$ and $s\left(\mathbf{x}_j, \mathbf{a}_i\right)$ are the distances of hard negative sample pairs in which $i \neq j$.

3.4 Zero-Shot Prediction

For zero-shot prediction, we first map the visual representations and unseen semantic prototypes into the embedding space. Then the label of the test instance is predicted by nearest neighbor approach in metric space. For TZSL, a test sample is from an unseen class, while GZSL extends this range to both seen and unseen classes. Their prediction functions are expressed as:

$$\hat{\mathbf{y}}_t = \arg\min_{\mathbf{y}_n} s\left(\mathbf{x}_t, \mathbf{a}_i\right), \quad \mathbf{x}_i \in X_U, \mathbf{y}_n \in U, \mathbf{a}_i \in A_U \tag{3}$$

$$\hat{\mathbf{y}}_t = \arg\min_{\mathbf{y}_n} s\left(\mathbf{x}_t, \mathbf{a}_i\right), \quad \mathbf{x}_i \in X_T, \mathbf{y}_n \in T, \mathbf{a}_i \in A_T \tag{4}$$

where X_T and A_T are sets of visual and semantic representations of both seen and unseen categories, Eq. (3) is for TZSL and Eq. (4) is for GZSL.

4   Experiments

In this section, we conduct experiments to verify the superiority of our CMNPN on GZSL tasks on three benchmark datasets. Firstly we introduce the three datasets and experimental setups, then we select 12 state-of-the-art approaches for comparison. Finally, we conduct ablation experiments to explore the impacts of N-pairs networks of different modalities and N-pair hard sample mining strategy in CMNPN.

4.1   Datasets and Settings

Datasets. We conduct experiments on three benchmark ZSL datasets, *i.e.* Animal with Attributes1 (AwA1) [10], Caltech-UCSD Birds-200-2011 dataset (CUB)   [18], and aPascal & aYahoo (aPY) [4]. Specifically, AwA1 consists of 30,475 instances from 50 animal categories and has 85-dimensional attributes. CUB contains 11,788 images from 200 bird categories with 312-dimensional attributes. And aPY has 32 classes, 15,399 images, and 64-dimensional attributes. We use the dataset split proposed in [22]. The dataset summary is given in Table 1. For fair comparison, we follow [20] and utilize the 2048-dimensional visual representations that extracted from the last layer of the ResNet101 pooling layer [6], while the class-level attribute information is our semantic representation.

Table 1. The statistics of three benchmark datasets in terms of the dimensionality of attributes, the number of seen instances and unseen instances.

Dataset	#Attribute	#Seen	#Unseen
AwA1	85	40	10
CUB	312	150	50
aPY	64	20	12

Settings

(1) Mapping network: The embedding network maps visual representations and semantic representations into a common space with a three-layer neural network. The number of hidden units is designed as 1024, 1200 and 1600 for AwA1, CUB and aPY, respectively. And the dimensionality of common space is setted as 2400. While developing the model, we have observed that by applying the ReLU activation function for the hidden layer of the mapping network would obtain more stable and better results.

(2) Deep metric network: We set the deep metric networks of VMNPN and SMNPN as three-layer neural networks that share parameters with each other. In all datasets, the numbers of hidden units are 2400, 1024 and 2400 for three FC layers. And we share parameters of VMNPN and SMNPN for training efficiency.

(3) Hyper-parameter setting: The batch-size for all experiments is set to 32 and the model is trained for 80000 epochs. The learning rate is 1e-4 which would decay to half every 2000 epochs, and the coefficient of L2 norm is 1e-6, which is obtained by cross validation.

Evaluation Protocol. For the task of TZSL, we adopt the average per-class top-1 accuracy as metric to evaluate the approaches. For GZSL, we follow the protocol proposed in [22] to evaluate the approaches with both seen class accuracy tr and unseen class accuracy ts, as well as their harmonic mean H, which is expressed as:

$$H = 2 \times \frac{tr \times ts}{tr + ts}. \tag{5}$$

4.2 Comparing State-of-the-Art Approaches

In this section, we compare our approach with 12 competitors [1,3,5,9,11,13, 14,16,19,24,26,28]. For fair comparison, we select the performances of all the competing approaches with the same features as ours. Table 2 illustrates the performance of our CMNPN on the task of TZSL and GZSL. Firstly, for GZSL, we can draw some conclusions: 1) On the result of H, our CMNPN outperforms other methods on AwA1, CUB and aPY by 0.8%, 0.3% and 5.2% over the runners-up, which proves the effectiveness and superiority of our method. 2) Our CMNPN attains an outstanding performance on ts over all datasets, which demonstrates that it has a better generalization capability on unseen categories than the other competitors. The reason is that our proposed hard samples mining strategy would fully employ hidden information of negative samples. And the cross-modal setting could contribute to knowledge interaction between visual and semantic modalities. However, it can be observed that the performance on TZSL is inferior to that on GZSL, which may due to our N-pair construction pays more attention on the model's generalization ability and the introduced negative samples may bring much more noises than useful information from the other modality with the increase of negative samples.

Table 2. Performance comparison on TZSL and GZSL. The best results are annotated as bold font.

	AwA1				CUB				aPY			
	T	ts	tr	H	T	ts	tr	H	T	ts	tr	H
DAP [11]	44.1	0	88.7	0	40.0	1.7	67.9	3.3	33.8	4.8	78.3	9.0
IAP [16]	35.9	2.1	78.2	4.1	24.0	0.2	72.8	0.4	36.6	5.7	65.6	10.4
CONSE [13]	45.6	0.4	88.6	0.8	34.3	1.6	72.2	3.1	26.9	0	91.2	0
DEVISE [5]	54.2	13.4	68.7	22.4	52.0	23.8	53.0	32.8	39.8	4.9	76.9	9.2
SSE [28]	60.1	7.0	80.5	12.9	43.9	8.5	46.9	14.4	34.0	0.2	78.9	0.4
ESZSL [14]	58.2	6.6	75.6	12.1	53.9	12.6	63.8	21	38.3	2.4	70.1	4.6
LATEM [19]	55.1	7.3	71.7	13.3	49.3	15.2	57.3	24	35.2	0.1	73	0.2
SYNC [3]	54.0	8.9	87.3	16.2	55.6	11.5	70.9	19.8	23.9	7.4	66.3	13.3
SAE [9]	53.0	1.8	77.1	3.5	33.3	7.8	54.0	13.6	8.3	0.4	80.9	0.9
ALE [1]	59.9	16.8	76.1	27.5	54.9	23.7	62.8	34.4	39.7	4.6	73.7	8.7
DSS [24]	-	18.9	**97.6**	31.7	-	25.0	**93.2**	39.4	-	12.8	**93.6**	22.5
DVN [26]	**67.7**	34.9	73.4	48.5	**57.8**	26.2	55.1	35.5	**41.2**	13.7	72.2	23.1
Ours	64.2	**36.2**	77.3	**49.3**	49.8	**36.5**	43.5	**39.7**	30.0	**18.8**	56.9	**28.3**

4.3   Ablation Studies

In this section, we carry out additional experiments to verify the impacts of some components in CMNPN.

Evaluating Different Modalities. To explore the impacts of different modalities, we conduct experiments on models removed VMNPN and SMNPN seperately. As shown in Table 3, it is observed that most results on SMNPN are superior to those on VMNPN, which illustrates that employing semantic prototypes as negative samples would be more effective than visual features to promote knowledge transfer from seen classes to unseen classes. Moreover, our cross-modal method takes advantage of both VMNPN and SMNPN and shows the best performance, which means the setting of cross-modal could make the information communication between two modalities more sufficient.

Table 3. Ablation study on VMNPN and SMNPN. VMNPN denotes the model without SMNPN, and SMNPN denotes the model without VMNPN. The best results are annotated as bold font.

	AwA1				CUB				aPY			
	T	ts	tr	H	T	ts	tr	H	T	ts	tr	H
VMNPN	61.1	35.0	75.5	47.8	45.9	33.4	40.5	36.6	27.6	17.2	53.3	26.0
SMNPN	62.8	34.1	74.6	46.8	46.8	35.9	42.5	38.9	28.0	18.0	52.9	26.9
Ours	**64.2**	**36.2**	**77.3**	**49.3**	**49.8**	**36.5**	**43.5**	**39.7**	**30.0**	**18.8**	**56.9**	**28.3**

Evaluating Hard Sample Mining and N-pair Samples from Multi-class. To further validate the superiority of the proposed model, we compare our CMNPN against the method without hard sample mining (CMNPN-WHSM), which would randomly select positive visual representations and apply them as negative samples for other categories. Then we conduct experiments on the method without using hard negative samples from multi-class (CMNPN-WMC), which means employing hard positive pairs and randomly selected $(N-1)$ negative samples in training.

Table 4. Ablation study on CMNPN-WHSM and CMNPN-WMC. The best results are annotated as bold font.

	AwA1				CUB				aPY			
	T	ts	tr	H	T	ts	tr	H	T	ts	tr	H
CMNPN-WHSM	63.0	35.8	75.3	48.5	47.5	34.9	42.9	38.5	29.6	17.1	55.2	26.1
CMNPN-WMC	62.8	34.1	72.6	44.4	46.9	30.8	40.1	34.8	26.3	15.3	51.9	23.6
Ours	**64.2**	**36.2**	**77.3**	**49.3**	**49.8**	**36.5**	**43.5**	**39.7**	**30.0**	**18.8**	**56.9**	**28.3**

As shown in Table 4, the performances of methods without hard sample mining or hard negative sample from multi-class are inferior to our proposed CMNPN with both techniques, and the latter is more obvious to improve the model. Furthermore, directly employing the hard samples from other categories could not only reduce additional time and space cost, but also leverage the relationship among all training classes better. At the same time, utilizing hard sample mining strategy could save training time because we select the most representative positive pairs which is more effective than selecting in random.

5 Conclusion

In this paper, we have proposed a novel approach, Cross-Modal N-Pairs Network (CMNPN), for GZSL. First we embed visual and semantic features into a common high-dimensional space to alleviate the hubness problem. Then a deep metric network employs VMNPN to train an N-pair containing $(N-1)$ negative visual representations and SMNPN to train an N-pair containing $(N-1)$ negative semantic prototypes. And with the hard sample mining strategy, intra- and inter-category correction of both modalities is mined sufficiently. Extensive experiments on three benchmark datasets showed our CMNPN achieved good performance in the task of GZSL.

References

1. Akata, Z., Perronnin, F., Harchaoui, Z., Schmid, C.: Label-embedding for image classification. IEEE Trans. Pattern Anal. Mach. Intell. **38**(7), 1425–1438 (2015)
2. Bucher, M., Herbin, S., Jurie, F.: Improving semantic embedding consistency by metric learning for zero-shot classiffication. In: Leibe, B., Matas, J., Sebe, N., Welling, M. (eds.) ECCV 2016. LNCS, vol. 9909, pp. 730–746. Springer, Cham (2016). https://doi.org/10.1007/978-3-319-46454-1_44
3. Changpinyo, S., Chao, W.L., Gong, B., Sha, F.: Synthesized classifiers for zero-shot learning. In: Proceedings of the IEEE Conference on Computer Vision and Pattern Recognition, pp. 5327–5336 (2016)
4. Farhadi, A., Endres, I., Hoiem, D., Forsyth, D.: Describing objects by their attributes. In: 2009 IEEE Conference on Computer Vision and Pattern Recognition, pp. 1778–1785. IEEE (2009)
5. Frome, A., et al.: Devise: a deep visual-semantic embedding model. In: Advances in Neural Information Processing Systems, pp. 2121–2129 (2013)
6. He, K., Zhang, X., Ren, S., Sun, J.: Deep residual learning for image recognition. In: Proceedings of the IEEE Conference on Computer Vision and Pattern Recognition, pp. 770–778 (2016)
7. Ji, Z., et al.: Deep ranking for image zero-shot multi-label classification. IEEE Trans. Image Process. **29**, 6549–6560 (2020)
8. Ji, Z., Sun, Y., Yu, Y., Pang, Y., Han, J.: Attribute-guided network for cross-modal zero-shot hashing. IEEE Trans. Neural Netw. **31**(1), 321–330 (2020)
9. Kodirov, E., Xiang, T., Gong, S.: Semantic autoencoder for zero-shot learning. In: Proceedings of the IEEE Conference on Computer Vision and Pattern Recognition, pp. 3174–3183 (2017)
10. Lampert, C.H., Nickisch, H., Harmeling, S.: Learning to detect unseen object classes by between-class attribute transfer. In: 2009 IEEE Conference on Computer Vision and Pattern Recognition, pp. 951–958. IEEE (2009)
11. Lampert, C.H., Nickisch, H., Harmeling, S.: Attribute-based classification for zero-shot visual object categorization. IEEE Trans. Pattern Anal. Mach. Intell. **36**(3), 453–465 (2013)
12. Li, J., Jing, M., Lu, K., Ding, Z., Zhu, L., Huang, Z.: Leveraging the invariant side of generative zero-shot learning. In: Proceedings of the IEEE Conference on Computer Vision and Pattern Recognition, pp. 7402–7411 (2019)
13. Norouzi, M., et al.: Zero-shot learning by convex combination of semantic embeddings (2013). arXiv preprint arXiv:1312.5650

14. Romera-Paredes, B., Torr, P.: An embarrassingly simple approach to zero-shot learning. In: International Conference on Machine Learning, pp. 2152–2161 (2015)
15. Shigeto, Y., Suzuki, I., Hara, K., Shimbo, M., Matsumoto, Y.: Ridge regression, hubness, and zero-shot learning. In: Appice, A., Rodrigues, P.P., Santos Costa, V., Soares, C., Gama, J., Jorge, A. (eds.) ECML PKDD 2015. LNCS (LNAI), vol. 9284, pp. 135–151. Springer, Cham (2015). https://doi.org/10.1007/978-3-319-23528-8_9
16. Socher, R., Ganjoo, M., Manning, C.D., Ng, A.: Zero-shot learning through cross-modal transfer. In: Advances in Neural Information Processing Systems, pp. 935–943 (2013)
17. Sohn, K.: Improved deep metric learning with multi-class n-pair loss objective. In: In: Advances in Neural Information Processing Systems, pp. 1857–1865 (2016)
18. Wah, C., Branson, S., Welinder, P., Perona, P., Belongie, S.: The caltech-ucsd birds-200-2011 dataset (2011)
19. Xian, Y., Akata, Z., Sharma, G., Nguyen, Q., Hein, M., Schiele, B.: Latent embeddings for zero-shot classification. In: Proceedings of the IEEE Conference on Computer Vision and Pattern Recognition, pp. 69–77 (2016)
20. Xian, Y., Lampert, C.H., Schiele, B., Akata, Z.: Zero-shot learning–a comprehensive evaluation of the good, the bad and the ugly. IEEE Trans. Pattern Anal. Mach. Intell. 41(9), 2251–2265 (2018)
21. Xian, Y., Lorenz, T., Schiele, B., Akata, Z.: Feature generating networks for zero-shot learning. In: Proceedings of the IEEE Conference on Computer Vision and Pattern Recognition, pp. 5542–5551 (2018)
22. Xian, Y., Schiele, B., Akata, Z.: Zero-shot learning-the good, the bad and the ugly. In: Proceedings of the IEEE Conference on Computer Vision and Pattern Recognition. pp. 4582–4591 (2017)
23. Xian, Y., Sharma, S., Schiele, B., Akata, Z.: f-VAEGAN-D2: a feature generating framework for any-shot learning. In: Proceedings of the IEEE Conference on Computer Vision and Pattern Recognition, pp. 10275–10284 (2019)
24. Yang, G., Liu, J., Xu, J., Li, X.: Dissimilarity representation learning for generalized zero-shot recognition. In: Proceedings of the 26th ACM international conference on Multimedia, pp. 2032–2039 (2018)
25. Yu, Y., Ji, Z., Fu, Y., Guo, J., Pang, Y., Zhang, Z.M., et al.: Stacked semantics-guided attention model for fine-grained zero-shot learning. In: Advances in Neural Information Processing Systems, pp. 5995–6004 (2018)
26. Zhang, H., Long, Y., Yang, W., Shao, L.: Dual-verification network for zero-shot learning. Inf. Sci. 470, 43–57 (2019)
27. Zhang, L., Xiang, T., Gong, S.: Learning a deep embedding model for zero-shot learning. In: Proceedings of the IEEE Conference on Computer Vision and Pattern Recognition, pp. 2021–2030 (2017)
28. Zhang, Z., Saligrama, V.: Zero-shot learning via semantic similarity embedding. In: Proceedings of the IEEE International Conference on Computer Vision, pp. 4166–4174 (2015)
29. Zhu, Y., Elhoseiny, M., Liu, B., Peng, X., Elgammal, A.: A generative adversarial approach for zero-shot learning from noisy texts. In: Proceedings of the IEEE Conference on Computer Vision and Pattern Recognition, pp. 1004–1013 (2018)

A Bi-directional Relation Aware Network for Link Prediction in Knowledge Graph

Kairong Hu, Hai Liu, Choujun Zhan, Yong Tang, and Tianyong Hao[✉]

School of Computer Science, South China Normal University, Guangzhou, China
{2018022615,ytang,haoty}@m.scnu.edu.cn, namelh@gmail.com,
zchoujun2@gmail.com

Abstract. Knowledge graph embedding technique aims to represent elements in knowledge graph, such as entities and relations, with numerical embedding vectors in semantic spaces. In general, an existing knowledge graph has relatively stable number of entities and directional relations before being updated. Though existing research has utilized relations of entities for link predication in knowledge graph, the relational directivity feature has not been fully exploited. Therefore, this paper proposes a bi-directional relation aware network (BDRAN) for representation learning, mining information based on directivity of relations in existing knowledge graphs. BDRAN leverages an encoder to capture features of entities in different patterns with diverse directional relations in entity representation level and semantic representation level. Besides, decoder is used to simulate interactions between entities and relations for precise representation learning. Experiments are conducted with widely used standard datasets including WN18RR, FB15k-237, NELL-995 and Kinship. The results present the improvement of BDRAN on the datasets, demonstrating the effectiveness of our model for link prediction.

Keywords: BDRAN · Knowledge graph · Link prediction

1 Introduction

Derived from knowledge base, knowledge graph is a graph form data essentially and it can also be viewed as textual data, consisting of triples (h, r, t) in which h, r and t present head entity, relation and tail entity, respectively. Knowledge graph emerges in various of hot research topics, such as information extraction, question answering and recommender system [19]. Describing interconnected relations among concepts, valuable information in knowledge graph is hidden in graph form data, which benefits downstream tasks. In order to mine information from a knowledge graph, knowledge graph embedding techniques are explored to map text form triples (e.g. "entity-relation-entity" pairs) into a continuous vector space [19] so as to acquire numerical representations of entities and relations.

Different knowledge graph embedding models exploit diverse methods to embed entities and relations. Typical models like TransE [1], TransH [20] and

© Springer Nature Singapore Pte Ltd. 2020
H. Zhang et al. (Eds.): NCAA 2020, CCIS 1265, pp. 259–271, 2020.
https://doi.org/10.1007/978-981-15-7670-6_22

TransD [6] simulated entities and relations with translational distance constraints. Based on tensor factorization, RESCAL [13], ComplEx [16] and SimplE [7] captured interactions between entities and relations by designing different semantic score functions. In recent years, convolutional operation has been studied to capture semantic information between embeddings of entities and relations [4,11,12], which achieves promising efficiency.

However, node connectivity information in knowledge graph was not considered in these models. To improve the models, R-GCN [14] (an extended version of GCN [8]) encoded neighborhood entity information for link prediction task. Utilizing graph attention mechanism, GAT [18] and KBAT [10] captured neighbor entity representations by allocating attention weights so as to bring forth new entity embeddings. Nevertheless, these models attempted to leverage neighborhood entity information for aggregation without considering directivity of relations among entities. From our observation, an entity may be connected by relations in different directions (in or out), expressing diverse semantic meanings that can be adopted for integration of entity representations. In this paper, a bi-directional relation aware network (BDRAN) is proposed to aggregate neighborhood entity information by perceiving relational directions for representation learning in knowledge graph. As an encoder-decoder architecture, BDRAN applies an encoder to encode semantics of entities in different relation directions while decoder is boosted by encoded representations of entities and relations. In order to verify our model, four standard datasets are used. Compared with the state-of-the-art models, the results show an improvement of 1.8% of MRR, 3.5% of Hits@1 and 1.3% of Hits@3 on average of our model in link prediction, which demonstrates relational directivity features may have crucial impacts on representations in knowledge graph.

Our contribution mainly lies in following aspects:

- We propose that the semantic meanings of an entity can be represented from different directions of relations and propose a strategy to integrate the relational directivity information into entity representations in knowledge graph.
- A bi-directional relation aware network (BDRAN) is proposed for representation learning in knowledge graph on basis of encoder-decoder architecture. The proposed model consists of an encoder that encodes neighborhood features in entity representation level and semantic meanings in semantic representation level. A convolution-based model plays a role as decoder for effective representation learning in knowledge graph.
- Experiments on link prediction task with standard datasets demonstrate the effectiveness of our model in representing entities and relations in knowledge graph.

2   Related Work

Knowledge graph embedding techniques aim to project entities and relations into a low dimension continuous vector space so as to obtain numerical embedding vectors for efficient computation. As a typical translational distance model,

TransE [1] imposed translational semantic constraint $h + r \approx t$ on embedding vectors of entities and relations given a valid triple (h, r, t) in knowledge graph. Despite of its simplicity and effectiveness, modified models, such as TransH [20] and TransD [6], designed strategies for entity projection, focusing on solving 1-to-N, N-to-1 and N-to-N problems in TransE. Capturing latent semantics between head entity and tail entity, semantic matching model RESCAL [13] simulated latent interactions between head entity and tail entity with relation associated matrices instead of translational distance constraints in translational distance models. On basis of tensor factorization, ComplEx [16] explored to extend representations of entities and relations from a real space to a complex space. As an expressive model for knowledge graph completion, SimplE [7] simplified ComplEx by reducing parameters and computation. In order to generate high-quality negative samples, there are some models [2,5,9] that leverage generative adversarial network for performance improvement. Different from tensor factorization based model, ConvE [4], ConvKB [11] and CapsE [12] took advantage of convolutional operation with diverse convolution methods to capture hidden semantic features between entities and relations through applying convolution on different shapes of embedding vectors and selecting features by different strategies.

Despite the promising effectiveness of convolutional models, node's local structural features (connectivity information) haven't been considered for node representation by aforementioned knowledge graph embedding models. R-GCN [14], an extended model of GCN [8], was proposed to encode structural features by message passing scheme for relational data in entity classification and link prediction task. Besides, GAT [18] and KBAT [10] exploited graph attention mechanism to allocate attention weights to neighborhood entities for feature aggregation so as to generate representation of a given central entity. Diving into relational features in representation learning, M-GNN [21] was proposed to encode hierarchical features from various of coarsened graphs generated by graph coarsening scheme. Complex relation patterns between entities and relations were addressed in [24], in which TransMS was proposed to leverage a non-linear function for multidirectional semantics transmission. However, the semantic features of bi-directional relations are not leveraged in aforementioned works. Hence, a bi-directional relation aware network (BDRAN) is proposed to mine bi-directional relation features of entities for precise representations in this paper.

3 Model

3.1 Notation

Knowledge graph $\mathcal{G}$ is a specific graph composed of elements containing nodes (entities) and edges (relations), in which relations are r_k and entities are denoted as e_i (or h_i and t_j for head and tail entities). The subscripts i, j and k of this elements are the indexes used to indicate a particular relation or entity in relation set $\mathcal{R}$ or entity set $\mathcal{E}$. In addition, the numerical representations of relations and entities are denoted as embedding vectors r_k and e_i (or h_i and t_j for head

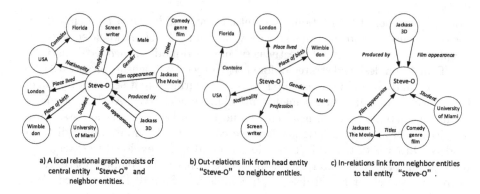

a) A local relational graph consists of central entity "Steve-O" and neighbor entities.

b) Out-relations link from head entity "Steve-O" to neighbor entities.

c) In-relations link from neighbor entities to tail entity "Steve-O".

Fig. 1. Figure 1(b) and (c) are derived from Fig. 1(a), a local graph of a central entity "Steve-O", which is a part of FB15k-237 dataset. Relations in different directions of a central entity express different meanings that can be utilized to represent the central entity from different perspectives in knowledge graph representation learning. 1-hop relations (in-relations and out-relations like "Student" and "Nationality") and 2-hop relations (stepwise relations like out-out relations (e.g. "Nationality"-"Contains") and in-in relations (e.g. "Titles"-"Film appearance")) are considered in this paper.

and tail entity, respectively), given an "entity-relation-entity" pair in triple form (e_i, r_k, e_j) (or (h_i, r_k, t_j)) from a knowledge graph. An entity is regarded as a central entity in a sub-graph derived from a knowledge graph if the entity is surrounded by other entities (neighbor entities) with in-relations (the relations link from head entities to the central entity) and out-relations (the relations connect from the central entity to tail entities). In-relations and out-relations of a central entity e_i are defined as $\mathcal{R}_{\text{in}}^i = \{r_k | (h_j, r_k, e_i) \in \mathcal{T}, e_i, h_j \in \mathcal{E}, r_k \in \mathcal{R}\}$ and $\mathcal{R}_{\text{out}}^i = \{r_k | (e_i, r_k, t_j) \in \mathcal{T}, e_i, t_j \in \mathcal{E}, r_k \in \mathcal{R}\}$, respectively. The neighbor entity set of a central entity e_i is $N(e_i) = \{t_j | (e_i, r_k, t_j) \in \mathcal{T}, e_i, t_j \in \mathcal{E}, r_k \in \mathcal{R}_{\text{out}}^i\} \cup \{h_j | (h_j, r_k, e_i) \in \mathcal{T}, e_i, h_j \in \mathcal{E}, r_k \in \mathcal{R}_{\text{in}}^i\}$, where $\mathcal{T}$ is a set of triples.

3.2 Encoder

As is known to all, a knowledge graph consists of entities and relations, demonstrating relationship between an entity and others. It is worth noting that there exist relations with different directions that describe facts of an entity. As illustrated in Fig. 1(a), a sub-graph from widely used standard dataset FB15k-237, comedy actor "Steve-O" acts as the central entity surrounded by neighbor entities like "USA" and "University of Miami", and connected by out-relations (e.g. "Nationality") and in-relations (e.g. "Student"). Figure 1(b) and Fig. 1(c) are derived from Fig. 1(a), which are solely composed of out-relations or in-relations between the central entity and its neighbors. Relations with different directions may describe diverse attributes of a central entity. For instance, neighbor entities "Wimbledon", "USA" and "Male" with out-relations "Place lived", "Nationality" and "Gender" imply that central entity "Steve-O" is likely to be a man with multiple nationality. The in-relations connect from head entities to the central entity, as Fig. 1(c), indicate that central entity "Steve-O" who is or used to be a student of "University of Miami", has the attribute of being an actor of "Jackass" movie series. As depicted in Fig. 1, some n-hop (e.g. 2-hop) entities may have auxiliary information to represent an unique entity [10]. Besides, the relations and neighbor entities in such sub-graph patterns may express various of semantic meanings, which present different contributions to represent the central entity in knowledge graph.

Therefore, inspired by GAT [18] and KBAT [10], a graph attention mechanism is leveraged in this paper to capture semantic meanings of entities in different relation directions. Generally, the encoder in our model BDRAN have two representation levels, entity representation level and semantic representation level. The former means that attention weights are allocated to different hidden triple representations under certain directional pattern (e.g. out-relation pattern or in-relation pattern) so as to bring forth aggregated representation that integrates neighborhood features expressing a central entity. In latter level, encoder allocates representation weights to different semantics according to directional patterns, which may have different expressiveness to represent an entity.

Firstly, in entity representation level, entity representations are brought forth via graph attention mechanism that allocates attention weights to different hidden triple representations for neighbor entities importance measurement. Triple representations are first generated by simple concatenation with embedding vectors of entities and relations. Given a central entity e_i, Eq. 1 and Eq. 2 are representations of triple (e_i, r_k, e_j) or (e_j, r_k, e_i) in out-relations and in-relations sub-graph, respectively.

$$v_{ijk}^{\text{out}} = [e_i \| e_j \| r_k] \tag{1}$$

$$v_{jik}^{\text{in}} = [e_j \| e_i \| r_k] \tag{2}$$

Note that the embedding vectors of triples are slightly different in this two sub-graph of central entity, since the roles of central entity and its neighbor entities (e.g. head entity and tail entity) are exchanged.

Following KBAT [10], hidden features of triple (e_i, r_k, e_j) in out-relation subgraph is denoted as u_{ijk}^{out}, which is formulated as Eq. 3.

$$u_{ijk}^{\text{out}} = W_1 v_{ijk}^{\text{out}} \tag{3}$$

The attention importance of neighbor entities are measured by graph attention mechanism given out-relations which are the same as in-relations, shown as Eq. 4.

$$a_{ijk}^{\text{out}} = \text{LeakyReLU}\left(W_2 u_{ijk}^{\text{out}}\right) \tag{4}$$

a_{ijk}^{out} presents importance of triple (e_i, r_k, e_j) when e_i acts as central entity (also as head entity h_i), while W_1, W_2 and LeakyReLU are linear transformation matrices and non-linear activation function.

Intuitively, attention weight $\alpha_{ijk}^{\text{out}}$ presents interaction impact of central entity e_i (or h_i) imposed by neighbor entity e_j (or t_j) under the pattern of out-relations, as Eq. 5.

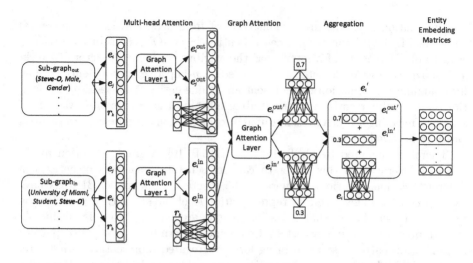

Fig. 2. The overall network of encoder in BDRAN. As an encoder-decoder architecture, BDRAN is composed of an encoder and a decoder, where the encoder is utilized to capture bi-directional relation features of entities through graph attention mechanism and decoder is a convolution-based knowledge graph embedding model that learns precise representations of entities and relations in knowledge graph.

$$\alpha_{ijk}^{\text{out}} = \frac{\exp\left(a_{ijk}^{\text{out}}\right)}{\sum_{l \in \mathcal{E}_N} \sum_{r \in \mathcal{R}_{il}} \exp\left(a_{ilr}^{\text{out}}\right)} \tag{5}$$

In which the neighbor entity set of central entity e_i is denoted by $\mathcal{E}_N$ while $\mathcal{R}_{il}$ presents out-relation set composed of out-relations that link from head entity e_i to its neighbors e_l.

In entity representation level, the encoder not solely generates attention weights under out-relation pattern but also those under in-relation pattern before generating representations of central entities according to different perspectives of both in-relations and out-relations in semantic representation level. Following Transformer [17] and GAT [18], multi-head attention is utilized to stabilize the training process and capture features from neighbors and specific direction of relations given a central entity. Specifically, aggregated embedding vector of central entity e_i with multi-head attention mechanism is defined as Eq. 6.

$$e_i^{\text{out}} = \overset{M}{\underset{m=1}{\|}} \sigma\left(\sum_{j \in \mathcal{E}_N, k \in \mathcal{R}_{ij}} \alpha_{ijk,m}^{\text{out}} \boldsymbol{u}_{ijk,m}^{\text{out}}\right) \tag{6}$$

$m = 1, 2, \ldots M$ presents heads of multi-head attention and the output embedding e_i^{out} is concatenation of aggregated embedding vectors from different semantic spaces of attention. Intuitively, the new embedding vector consists of weighted features of triples with out-relations from attention layers, where the most representative triple may have large proportion that influences output embeddings of central entity intensely. Linear transformation of relation embedding vectors are obtained to generate input vectors of final attention layer as shown in Fig. 2. The output of final attention is denoted as $e_i^{\text{out}'}$ in Eq. 7, where $u_{ijk}^{\text{out}'}$ presents triple hidden representation of final attention layer.

$$e_i^{\text{out}'} = \sigma \left(\sum_{j \in \mathcal{E}_N, k \in \mathcal{R}_{ij}} \alpha_{ijk}^{\text{out}'} u_{ijk}^{\text{out}'} \right) \tag{7}$$

However, the output embedding vectors $e_i^{\text{out}'}$ and $e_i^{\text{in}'}$ capture features from neighbors of the central entity only under the patterns of out-relations and in-relations in entity representation level. Thus, aggregation steps in semantic representation level are proposed to bring forth final embeddings of entities combining distinguishing semantic meanings derived from different patterns of relations. In order to quantify importance of semantic meanings given a central entity, multi-layer perceptrons (MLPs) and a softmax function are utilized to allocate weights for representations of different meanings in out-relation and in-relations patterns, as shown in Eq. 8. Aggregating initial semantic information of entities [10] and semantics of directional relation patterns, the final embeddings of a central entity is a weighted sum of embedding vectors using Eq. 9.

$$\beta_i^{\text{out}} = \frac{\exp \left(MLP \left(e_i^{\text{out}'} \right) \right)}{\exp \left(MLP \left(e_i^{\text{out}'} \right) \right) + \exp \left(MLP \left(e_i^{\text{in}'} \right) \right)} \tag{8}$$

$$e_i' = \beta_i^{\text{out}} e_i^{\text{out}'} + \left(1 - \beta_i^{\text{out}} \right) e_i^{\text{in}'} + W_e e_i \tag{9}$$

Linear transformation is performed on original entity embeddings which is denoted as $W_e e_i$.

Finally, the output embedding vectors of entities and relations are constrained by translational distance calculated by Eq. 10 and marginal loss function is regarded as objective function, shown as Eq. 11.

$$f(e_i, r_k, e_j) = \| e_i + r_k - e_j \| \tag{10}$$

$$L = \sum_{(e_i, r_k, e_j) \in \mathcal{T}} \sum_{(e_i', r_k, e_j') \in \mathcal{T}'} \max \left(0, f(e_i, r_k, e_j) - f(e_i', r_k, e_j') + \gamma \right) \tag{11}$$

In which $\mathcal{T}$ and $\mathcal{T}'$ present positive and negative triple sets, respectively.

3.3 Decoder

As an encoder-decoder model, ConvKB [11] is regarded as a decoder which is used to learn representations of entities and relations in knowledge graph on basis of entity features encoded by the encoder. The score function is expressed as Eq. 12 and the objective function with L_2 regularization presents as Eq. 13.

$$f(v) = \left(\overset{N}{\underset{n=1}{\|}} \text{ReLU} \left(v * \omega^n \right) \right) W \tag{12}$$

Table 1. The statistics of the datasets used in the experiments.

Datasets	#Entity	#Relation	#Training set	#Validation set	#Testing set
WN18RR	40,943	11	86,835	3,034	3,134
NELL-995	75,492	200	149,678	543	3992
FB15k-237	14,541	237	272,115	17,535	20,466
Kinship	104	25	8544	1068	1074

Table 2. The statistics of degree of entities in training sets.

Datasets	Mean in-degree	Mean out-degree
WN18RR	2.72	2.19
NELL-995	4.21	2.77
FB15k-237	20.34	19.75
Kinship	82.15	82.15

$$L = \sum_{(e_i,r_k,e_j)\in T \cup T'} \log\left\{1 + \exp\left[-l \cdot f(v)\right]\right\} + \frac{\lambda}{2}\|w\|_2^2 \qquad (13)$$

where v is an embedding vector of triple v consisted of concatenation of entity and relation vectors while $*$ is convolution operator with filter ω. Linear transformation matrix is denoted as W and l is label for positive and negative triple, while λ and w are regularization hyper-parameter and embedding weights of the model.

4 Experiments

4.1 Datasets

Four widely applied standard datasets WN18RR, FB15k-237, NELL-995 and Kinship are used in the experiments as a link prediction task. The detailed statistics of the datasets are shown in Table 1 and Table 2 with brief introduction as follows:

- **FB15k-237** is derived from FB15k (a subset of Freebase knowledge base). This dataset, the same as FB15k, consists of triples that describe facts about actor, movies, locations, etc. FB15k-237 is released by [15] after removing the redundant triples, to enhance the difficulty for models to learn representations of entities and relations.
- **WN18RR** is a subset of dataset WN18 (derived from WordNet knowledge base), released by [4]. WN18RR has 11 types of relations, such as "hyponym" and "hypernym" that describe lexical facts between words.
- **NELL-995** is a dataset extracted from NELL [3] by [22], containing general facts about agricultural products, books, animals, etc.

– **Kinship** is a small-scale dataset that describes personal relationship in Alyawarra tribe from Central Australia.

4.2   Baselines

Our experiments are conducted with aforementioned standard datasets on the link prediction task, comparing with following baseline methods:

– **TransE** is a typical translational distance model proposed by [1], which imposes translational distance constraint on triple (h, r, t).
– **DistMult** [23] is a simplified version of RESCAL, reducing parameters of relation matrices by restraining matrices into diagonal matrices.
– **ComplEx** [16], as a semantic matching model, projects entities and relations into a complex space instead of a real space.
– **ConvE** [4] is convolution-based knowledge graph embedding model, utilizing 2D convolution to capture features from embedding vectors of entities and relations.
– **ConvKB** [11] is also a convolutional model in which different score function and loss function are developed, comparing with ConvE.
– **R-GCN** [14] takes advantage of graph convolutional network to encode features from neighborhood entities.
– **KBAT** [10] is a state-of-the-art method, in which graph attention mechanism is applied to capture features from multi-hop neighborhood entities.

4.3   Evaluation Metrics

The evaluation metrics for the link prediction task are the frequently used mean reciprocal rank (MRR) and Hits@N, as Eq. 14 and Eq. 15. Following the filtered settings in [1], the rank of a tail or head entity in each testing triple (h_i, r_k, t_j) are calculated within filtered entity set (without valid tail or head entities t and h that may generate valid triples (h_i, r_k, t) or (h, r_k, t_j) in training set) when testing our model. Large MRR indicates knowledge graph embedding models are capable to represent head and tail entities precisely, while Hits@N presents a rate for head and tail entities that ranks within N (N can be 1, 3 or 10, etc.).

$$MRR = \frac{1}{2*|\mathcal{T}_t|} \sum_{(h,r,t)\in\mathcal{T}_t} \frac{1}{rank_h} + \frac{1}{rank_t} \quad (14)$$

$$Hits@N = \frac{1}{2*|\mathcal{T}_t|} \sum_{(h,r,t)\in\mathcal{T}_t} I\left(rank_h \leq N\right) + I\left(rank_t \leq N\right) \quad (15)$$

where $|\mathcal{T}_t|$ is size of testing triple set $\mathcal{T}_t$ and $I(\cdot)$ is indicator function while $rank_h$ and $rank_t$ present values of rank given head entity h and tail entity t during prediction with model.

Table 3. The performances of BDRAN on WN18RR and NELL-995 datasets, comparing with other baselines.

	WN18RR				NELL-995			
	MRR	H@1	H@3	H@10	MRR	H@1	H@3	H@10
TransE	0.243	4.27	44.1	53.2	0.401	34.4	47.2	50.1
DistMult	0.444	41.2	47.0	50.4	0.485	40.1	52.4	61.0
ComplEx	0.449	40.9	46.9	53.0	0.482	39.9	52.8	60.6
ConvE	**0.456**	**41.9**	47.0	53.1	0.491	40.3	53.1	61.3
ConvKB	0.265	5.82	44.5	55.8	0.430	37.0	47.0	54.5
R-GCN	0.123	8.0	13.7	20.7	0.12	8.2	12.6	18.8
KBAT	0.440	36.1	48.3	58.1	**0.530**	44.7	56.4	**69.5**
BDRAN	0.446	36.8	**49.0**	**58.2**	**0.530**	**45.0**	**57.4**	68.0

Table 4. The result comparisons of BDRAN on FB15k-237 and Kinship datasets.

	FB15k-237				Kinship			
	MRR	H@1	H@3	H@10	MRR	H@1	H@3	H@10
TransE	0.279	19.8	37.6	44.1	0.309	0.9	64.3	84.1
DistMult	0.281	19.9	30.1	44.6	0.516	36.7	58.1	86.7
ComplEx	0.278	19.4	29.7	45.0	0.823	73.3	89.9	97.1
ConvE	0.312	22.5	34.1	49.7	0.833	73.8	91.7	**98.1**
ConvKB	0.289	19.8	32.4	47.1	0.614	43.6	75.5	95.3
R-GCN	0.164	10.0	18.1	30.0	0.109	3.0	8.8	23.9
KBAT	0.518	46.0	54.0	**62.6**	0.904	85.9	**94.1**	98.0
BDRAN	**0.539**	**49.4**	**55.5**	**62.6**	**0.919**	**89.2**	93.2	97.8

4.4   Results

The experiment results on standard datasets for link prediction task are shown in Table 3 and Table 4. The overall performance of BDRAN on all datasets have 1.8% relative improvement of MRR, 3.5% of Hits@1 and 1.3% of Hits@3 on average except for Hits@10. The performance of BDRAN on WN18RR and NELL-995 datasets outperforms KBAT, especially using MRR and Hits@1, as shown in Table 3. However, the results of BDRAN using MRR and Hits@1 on WN18RR dataset are slightly worse than ConvE, indicating that the encoded neighborhood features in encoder may be inaccurate enough for representation in WN18RR dataset. Nevertheless, the performance is still better than that of KBAT. Note that, in Table 2, the mean in-degree and out-degree in WN18RR dataset are small and the difference between them is also a relatively small value (e.g. 0.53), indicating that the existing neighborhood entities do not have enough information for central entity representation and it is hard for model to distinguish semantic meanings from neighbors in diverse directional patterns. This insufficient information may result in inadequate features for representing semantic patterns and cause the encoder to allocate similar weights to represent different semantic meanings from relational patterns. Different from WN18RR, mean in-degree value of entities in training set is larger than out-degree relatively on NELL-995, showing larger MRR and Hits@N than WN18RR. The weights for representation from in-relation patterns perhaps achieve large values compared with weights for out-relation patterns such that rich features in in-relation patterns can be exploited for precise representation. In addition, both FB15k-237 and Kinship datasets have relatively large mean in-degree and out-degree than WN18RR and NELL-995, providing diverse semantic meanings. Therefore, leveraging graph attention mechanism, the encoder in BDRAN captures features from both in-relation and out-relation patterns and aggregates entity representation according to expression of meanings in different patterns on the central entity. As a result, BDRAN outperforms KBAT on MRR (about 4% in FB15k-237 and 1.6% in Kinship relatively) and Hits@1 (about 7.4% in FB15k-237 and 3.8% in Kinship). Therefore, semantic meanings expressed by different directional relations can be captured through graph attention mechanism and integrated by encoder while decoder learns precise representations of entities and relations in knowledge graph on basis of encoded entity and relation embeddings.

5   Conclusion

This paper proposes a bi-directional relation aware network (BDRAN) for link prediction in knowledge graph. Leveraging encoder-decoder architecture, the encoder in BDRAN exploits graph attention mechanism to capture neighborhood features in entity representation level and aggregate different semantic meanings from in-relation and out-relation patterns. The decoder takes advantage of output embeddings from encoder that incorporates semantic information to learn precise representation. Further more, experiments on standard datasets, such as WN18RR, FB15k-237, NELL-995 and Kinship, are conducted to verify our model on link prediction task. The results present improvement that demonstrates effectiveness of representation learning to aggregate semantic meanings from different directional patterns of relations in knowledge graph.

Acknowledgement. This work is supported by National Natural Science Foundation of China (No. 61772146, No. 61772211, No. U1811263) and Natural Science Foundation of Guangdong (No. c20140500000225).

References

1. Bordes, A., Usunier, N., Garcia-Duran, A., Weston, J., Yakhnenko, O.: Translating embeddings for modeling multi-relational data. In: Advances in Neural Information Processing Systems, pp. 2787–2795 (2013)
2. Cai, L., Wang, W.Y.: KBGAN: adversarial learning for knowledge graph embeddings (2017). arXiv preprint arXiv:1711.04071

3. Carlson, A., Betteridge, J., Kisiel, B., Settles, B., Hruschka, E.R., Mitchell, T.M.: Toward an architecture for never-ending language learning. In: Twenty-Fourth AAAI Conference on Artificial Intelligence (2010)

4. Dettmers, T., Minervini, P., Stenetorp, P., Riedel, S.: Convolutional 2d knowledge graph embeddings. In: Thirty-Second AAAI Conference on Artificial Intelligence (2018)

5. Hu, K., Liu, H., Hao, T.: A knowledge selective adversarial network for link prediction in knowledge graph. In: Tang, J., Kan, M.-Y., Zhao, D., Li, S., Zan, H. (eds.) NLPCC 2019. LNCS (LNAI), vol. 11838, pp. 171–183. Springer, Cham (2019). https://doi.org/10.1007/978-3-030-32233-5_14

6. Ji, G., He, S., Xu, L., Liu, K., Zhao, J.: Knowledge graph embedding via dynamic mapping matrix. In: Proceedings of the 53rd Annual Meeting of the Association for Computational Linguistics and the 7th International Joint Conference on Natural Language Processing (Volume 1: Long Papers), pp. 687–696 (2015)

7. Kazemi, S.M., Poole, D.: Simple embedding for link prediction in knowledge graphs. In: Advances in Neural Information Processing Systems, pp. 4284–4295 (2018)

8. Kipf, T.N., Welling, M.: Semi-supervised classification with graph convolutional networks (2016). arXiv preprint arXiv:1609.02907

9. Liu, H., Hu, K., Wang, F.L., Hao, T.: Aggregating neighborhood information for negative sampling for knowledge graph embedding. Neural Comput. Appl. (2020)

10. Nathani, D., Chauhan, J., Sharma, C., Kaul, M.: Learning attention-based embeddings for relation prediction in knowledge graphs (2019). arXiv preprint arXiv:1906.01195

11. Nguyen, D.Q., Nguyen, T.D., Nguyen, D.Q., Phung, D.: A novel embedding model for knowledge base completion based on convolutional neural network (2017). arXiv preprint arXiv:1712.02121

12. Nguyen, D.Q., Vu, T., Nguyen, T.D., Nguyen, D.Q., Phung, D.: A capsule network-based embedding model for knowledge graph completion and search personalization (2018). arXiv preprint arXiv:1808.04122

13. Nickel, M., Tresp, V., Kriegel, H.P.: A three-way model for collective learning on multi-relational data. ICML 11, 809–816 (2011)

14. Schlichtkrull, M., Kipf, T.N., Bloem, P., van den Berg, R., Titov, I., Welling, M.: Modeling relational data with graph convolutional networks. In: Gangemi, A., et al. (eds.) ESWC 2018. LNCS, vol. 10843, pp. 593–607. Springer, Cham (2018). https://doi.org/10.1007/978-3-319-93417-4_38

15. Toutanova, K., Chen, D., Pantel, P., Poon, H., Choudhury, P., Gamon, M.: Representing text for joint embedding of text and knowledge bases. In: Proceedings of the 2015 Conference on Empirical Methods in Natural Language Processing, pp. 1499–1509 (2015)

16. Trouillon, T., Welbl, J., Riedel, S., Gaussier, É., Bouchard, G.: Complex embeddings for simple link prediction. In: International Conference on Machine Learning, pp. 2071–2080 (2016)

17. Vaswani, A., et al.: Attention is all you need. In: Advances in Neural Information Processing Systems, pp. 5998–6008 (2017)

18. Veličković, P., Cucurull, G., Casanova, A., Romero, A., Lio, P., Bengio, Y.: Graph attention networks (2017). arXiv preprint arXiv:1710.10903

19. Wang, Q., Mao, Z., Wang, B., Guo, L.: Knowledge graph embedding: a survey of approaches and applications. IEEE Trans. Knowl. Data Eng. 29(12), 2724–2743 (2017)

20. Wang, Z., Zhang, J., Feng, J., Chen, Z.: Knowledge graph embedding by translating on hyperplanes. In: Twenty-Eighth AAAI Conference on Artificial Intelligence (2014)
21. Wang, Z., Ren, Z., He, C., Zhang, P., Hu, Y.: Robust embedding with multi-level structures for link prediction. In: Proceedings of the 28th International Joint Conference on Artificial Intelligence, pp. 5240–5246. AAAI Press (2019)
22. Xiong, W., Hoang, T., Wang, W.Y.: DeepPath: a reinforcement learning method for knowledge graph reasoning (2017). arXiv preprint arXiv:1707.06690
23. Yang, B., Yih, W., He, X., Gao, J., Deng, L.: Embedding entities and relations for learning and inference in knowledge bases (2014). arXiv preprint arXiv:1412.6575
24. Yang, S., Tian, J., Zhang, H., Yan, J., He, H., Jin, Y.: TransMS: knowledge graph embedding for complex relations by multidirectional semantics. In: Proceedings of the 28th International Joint Conference on Artificial Intelligence, pp. 1935–1942. AAAI Press (2019)

Deep K-Means: A Simple and Effective Method for Data Clustering

Shudong Huang[1(✉)], Zhao Kang[2], and Zenglin Xu[2,3,4]

[1] College of Computer Science, Sichuan University, Chengdu 610065, China
`huangsd@scu.edu.cn`
[2] School of Computer Science and Engineering,
University of Electronic Science and Technology of China,
Chengdu 611731, China
`zkang@uestc.edu.cn`
[3] School of Computer Science and Technology,
Harbin Institute of Technology, Shenzhen 518055, China
`xuzenglin@hit.edu.cn`
[4] Center for Artificial Intelligence, Peng Cheng Lab, Shenzhen 518055, China

Abstract. Clustering is one of the most fundamental techniques in statistic and machine learning. Due to the simplicity and efficiency, the most frequently used clustering method is the k-means algorithm. In the past decades, k-means and its various extensions have been proposed and successfully applied in data mining practical problems. However, previous clustering methods are typically designed in a single layer formulation. Thus the mapping between the low-dimensional representation obtained by these methods and the original data may contain rather complex hierarchical information. In this paper, a novel deep k-means model is proposed to learn such hidden representations with respect to different implicit lower-level characteristics. By utilizing the deep structure to conduct k-means hierarchically, the hierarchical semantics of data is learned in a layerwise way. The data points from same class are gathered closer layer by layer, which is beneficial for the subsequent learning task. Experiments on benchmark data sets are performed to illustrate the effectiveness of our method.

Keywords: Clustering · Matrix factorization · Deep learning

1 Introduction

The goal of clustering is to divide a given dataset into different groups such that similar instances are allocated into one group [22,33,34]. Clustering is one of the most classical techniques that has been found to perform surprisingly well [15,19,21]. Clustering has been successfully utilized in various application areas, text mining [16,36], voice recognition [1], image segmentation [38], to name a few. Up to now, myriads of clustering methods have been designed under the framework of different methodologies and statistical theories [30,31],

© Springer Nature Singapore Pte Ltd. 2020
H. Zhang et al. (Eds.): NCAA 2020, CCIS 1265, pp. 272–283, 2020.
https://doi.org/10.1007/978-981-15-7670-6_23

like k-means clustering [26], spectral clustering [29], information theoretic clustering [9], energy clustering [37], discriminative embedded clustering [11], multi-view clustering [12,20,23,32], *etc.* Among them, k-means, as one of the most popular clustering algorithms, has received considerable attention due to its efficiency and effectiveness since it was introduced in 1967 [26]. Furthermore, it has been categorized as one of the top ten data mining algorithms in term of usage and clustering performance [39]. There is no doubt that k-means is the most popularly used clustering method in various practical problems [13].

Recently, the Nonnegative Matrix Factorization (NMF) draws much attention in data clustering and achieves promising performance [2,14,18]. Previous work indicated that NMF is identical to k-means clustering with a relaxed condition [8]. Until now several variants of k-means have been presented to improve the clustering accuracy. Inspired by the principal component analysis (PCA), [6] shown that principal components actually provide continuous solutions, which can be treated as the discrete class indicators for k-means clustering. Moreover, the subspace separated by the cluster centroids can be obtained by spectral expansion. [5] designed a spherical k-means for text clustering with good performance in terms of both solution quality and computational efficiency by employing cosine dissimilarities to perform prototype-based partitioning of weight representations. [10] extended a kernel k-means clustering to handle multi-view datasets with the help of multiple kernel learning. To explore the sample-specific attributes of data, the authors focused on combining kernels in a localized way. [27] proposed a fast accelerated exact k-means, which can be considered as a general improvement of existing k-means algorithms with better estimates of the distance bounds. [28] assumed that incorporates distance bounds into the mini-batch algorithm, data should be preferentially reused. That is, data in a mini-batch at current iteration is reused at next iteration automatically by utilizing the nested mini-batches.

Although the aforementioned k-means methods have shown their effectiveness in many applications, they are typically designed in a single layer formulation. As a result, the mapping between the low-dimensional representation obtained by these methods and the original data may still contain complex hierarchical information. Motivated by the development of deep learning that employs multiple processing layers to explore the hierarchical information hidden in data [3], in this paper, we propose a novel deep k-means model to learn the hidden information with respect to multiple level characteristics. By utilizing the deep structure to conduct k-means hierarchically, the data hierarchical semantics is learned in a layerwise way. Through the deep k-means structure, instances from same class are pushed closer layer by layer, which is beneficial for the subsequent learning task. Furthermore, we introduce an alternative updating algorithm to address the corresponding optimization problem. Experiments are conducted on benchmark data sets and show promising results of our model compared to several state-of-the-art algorithms.

2   Preliminaries

As mentioned before, NMF is essentially identical to relaxed k-means algorithm [8]. Before introducing our deep k-means, first we briefly review NMF [25]. Denote a nonnegative data matrix as $\mathbf{X} = [x_1, x_2, \cdots, x_n] \in \mathbb{R}^{m \times n}$, where n is the number of instances and m is the feature dimension. NMF tries to search two nonnegative matrices $\mathbf{U} \in \mathbb{R}^{m \times c}$ and $\mathbf{V} \in \mathbb{R}^{n \times c}$ such that

$$J_{NMF} = \sum_{i=1}^{n} \sum_{j=1}^{m} \left(\mathbf{X}_{ij} - (\mathbf{UV}^T)_{ij} \right)^2 = \|\mathbf{X} - \mathbf{UV}^T\|_F^2 \tag{1}$$

$$\text{s.t. } \mathbf{U} \geq 0, \mathbf{V} \geq 0,$$

where $\|\cdot\|_F$ indicates a Frobenius norm and $\mathbf{X}_{ij}$ is the (i, j)-th element of $\mathbf{X}$. [25] proved that Eq. (1) is not jointly convex in $\mathbf{U}$ and $\mathbf{V}$ (i.e., convex in $\mathbf{U}$ or $\mathbf{V}$ only), and proposed the following alternative updating rules to search the local minimum:

$$\mathbf{U}_{ij} \quad \leftarrow \quad \mathbf{U}_{ij} \frac{(\mathbf{XV})_{ij}}{(\mathbf{UV}^T\mathbf{V})_{ij}},$$

$$\mathbf{V}_{ij} \quad \leftarrow \quad \mathbf{V}_{ij} \frac{(\mathbf{X}^T\mathbf{U})_{ij}}{(\mathbf{VU}^T\mathbf{U})_{ij}},$$

where $\mathbf{V}$ denotes the class indicator matrix in unsupervised setting [17], $\mathbf{U}$ denotes the centroid matrix, and c is cluster number. Since $c \ll n$ and $c \ll m$, NMF actually tries to obtain a low-dimensional representation $\mathbf{V}$ of the original input $\mathbf{X}$.

Real-world data sets are rather complex that contain multiple hierarchical modalities (i.e., factors). For instance, face data set typically consists several common modalities like pose, scene, expression, etc. Traditional NMF with single layer formulation is obviously unable to fully uncover the hidden structures of the corresponding factors. Therefore, [35] proposed a multi-layer deep model based on semi-NMF to exploit hierarchical information with respect to different modalities. And models a multi-layer decomposition of an input data $\mathbf{X}$ as

$$\mathbf{X} \approx \mathbf{U}_1 \mathbf{V}_1^T,$$

$$\mathbf{X} \approx \mathbf{U}_1 \mathbf{U}_2 \mathbf{V}_2^T,$$

$$\vdots \tag{2}$$

$$\mathbf{X} \approx \mathbf{U}_1 \mathbf{U}_2 \cdots \mathbf{U}_r \mathbf{V}_r^T,$$

where r means the number of layers, $\mathbf{U}_i$ and $\mathbf{V}_i$ respectively denote the i-th layer basis matrix and representation matrix. It can be seen that deep semi-NMF model also focuses on searching a low-dimensional embedding representation that targets to a similar interpretation at the last layer, i.e., $\mathbf{V}_r$. The deep model in Eq. (2) is able to automatically search the latent hierarchy by further factorizing $\mathbf{V}_i(i < r)$. Furthermore, this model is able to discover representations suitable for clustering with respect to different modalities (e.g., for face data set, $\mathbf{U}_3$ corresponds to the attributes of expressions, $\mathbf{U}_2\mathbf{U}_3$ corresponds to the attributes of poses, and $\mathbf{U} = \mathbf{U}_1\mathbf{U}_2\mathbf{U}_3$ finally corresponds to identities mapping of face images). Compared to traditional single layer NMF, deep semi-NMF provides a better ability to exploit the hidden hierarchical information, as different modalities can be fully identified by the obtained representations of each layer.

3 The Proposed Method

In this section, we introduce a novel deep k-means model for data clustering, followed with its optimization algorithm.

3.1 Deep K-Means

Traditional k-means methods are typically designed in a single layer formulation. Thus the mapping between the obtained low-dimensional representation and the original data may contain complex hierarchical information corresponding to the implicit modalities. To exploit such hidden representations with respect to different modalities, we propose a novel deep k-means model by utilizing the deep structure to conduct k-means hierarchically. The hierarchical semantics of the original data in our model is comprehensively learned in a layerwise way. To improve the robustness of our model, the sparsity-inducing norm, $l_{2,1}$-norm, is used in the objective. Since $l_{2,1}$-norm based residue calculation adopts the l_2-norm within a data point and the l_1-norm among data points, the influence of outliers is reduced by the l_1-norm [24]. Moreover, the non-negative constraint on $\mathbf{U}_i$ is removed such that the input data can consist of mixed signs, thus the applicable range of the proposed model is obviously enlarged. Since the non-negativity constraints on V_i make them more difficult to be optimized, we introduce new variables V_i^+ to which the non-negativity constraints are applied, with the constraints $V_i = V_i^+$. In this paper, we utilize the alternating direction method of multipliers (ADMM) [4] to handle the constraint with an elegant way, while maintain the separability of the objective. As a result, the non-negativity constraints are effectively incorporated to our deep k-means model. Our deep k-means model (DKM) is stated as

$$J_{DKM} = \|\mathbf{X} - \mathbf{Y}\|_{2,1}$$

$$\text{s.t.} \, \mathbf{Y} = \mathbf{U}_1\mathbf{U}_2\cdots\mathbf{U}_r\mathbf{V}_r^T, (\mathbf{V}_r)_{.c} = \{0,1\}, \sum_{c=1}^{C}(\mathbf{V}_r)_{.c} = 1, \tag{3}$$

$$\mathbf{V}_i = \mathbf{V}_i^+, \mathbf{V}_i^+ \geq 0, i \in [1,\ldots,r-1].$$

While $\|\mathbf{X} - \mathbf{Y}\|_{2,1}$ is simple to minimize with respect to $\mathbf{Y}$, $\|\mathbf{X} - \mathbf{U}_1\mathbf{U}_2\cdots\mathbf{U}_r\mathbf{V}_r^T\|_{2,1}$ is not simple to minimize with respect to $\mathbf{U}_i$ or $\mathbf{V}_i$. Multiplicative updates implicitly address the problem such that $\mathbf{U}_i$ and $\mathbf{V}_i$ decouple.

In ADMM context, a natural formulation would be to minimize $\|\mathbf{X} - \mathbf{Y}\|_{2,1}$ with the constraint $\mathbf{Y} = \mathbf{U}_1\mathbf{U}_2 \cdots \mathbf{U}_r\mathbf{V}_r^T$. This is the core reason why we choose to solve such a problem like Eq. (3). In addition, each row of $\mathbf{V}_r$ in Eq. (3) is enforced to satisfy the 1-of-C coding scheme. Its primary goal is to ensure the uniqueness of the final solution $\mathbf{V}_r$.

The augmented Lagrangian of Eq. (3) is

$$\mathcal{L}(\mathbf{Y}, \mathbf{U}_i, \mathbf{V}_i, \mathbf{V}_i^+, \boldsymbol{\mu}, \boldsymbol{\lambda}_i) = \|\mathbf{X} - \mathbf{Y}\|_{2,1} + <\boldsymbol{\mu}, \mathbf{Y} - \mathbf{U}_1\mathbf{U}_2 \cdots \mathbf{U}_r\mathbf{V}_r^T>$$

$$+\frac{\rho}{2}\|\mathbf{Y} - \mathbf{U}_1\mathbf{U}_2 \cdots \mathbf{U}_r\mathbf{V}_r^T\|_F^2 + \sum_{i=1}^{r-1} <\boldsymbol{\lambda}_i, \mathbf{V}_i - \mathbf{V}_i^+> + \frac{\rho}{2}\sum_{i=1}^{r-1}\|\mathbf{V}_i - \mathbf{V}_i^+\|_F^2, \quad (4)$$

where $\boldsymbol{\mu}$ and $\boldsymbol{\lambda}_i$ are Lagrangian multipliers, ρ is the penalty parameter, and $<\cdot, \cdot>$ represents the inner product operation.

The alternating direction method for Eq. (4) is derived by minimizing $\mathcal{L}$ with respect to $\mathbf{Y}, \mathbf{U}_i, \mathbf{V}_i, \mathbf{V}_i^+$, one at a time while fixing others, which will be discussed below.

3.2 Optimization

In the following, we propose an alternative updating algorithm to solve the optimization problem of the proposed objective. We update the objective with respect to one variable while fixing the other variables. This procedure repeats until convergence.

Before the minimization, first we perform a pre-training by decomposing the data matrix $\mathbf{X} \approx \mathbf{U}_1\mathbf{V}_1^T$, where $\mathbf{V}_1 \in \mathbb{R}^{n \times k_1}$ and $\mathbf{U}_1 \in \mathbb{R}^{m \times k_1}$. The obtained representation matrix $\mathbf{V}_1$ is then further decomposed as $\mathbf{V}_1 \approx \mathbf{U}_2\mathbf{V}_2^T$, where $\mathbf{V}_2 \in \mathbb{R}^{n \times k_2}$ and $\mathbf{U}_2 \in \mathbb{R}^{k_1 \times k_2}$. We respectively denote k_1 and k_2 as the dimensionalities of the first layer and the second layer[1]. Continue to do this, finally all layers are pre-trained, which would greatly improve the training time as well as the effectiveness of our model. This trick has been applied favourably in deep autoencoder networks [4].

Optimizing Eq. (4) is identical to minimizing the formulation as follows

$$\mathcal{L}(\mathbf{Y}, \mathbf{U}_i, \mathbf{V}_i, \mathbf{V}_i^+, \boldsymbol{\mu}, \boldsymbol{\lambda}_i) = \mathrm{Tr}\left((\mathbf{X} - \mathbf{Y})\mathbf{D}(\mathbf{X} - \mathbf{Y})^T\right)$$

$$+\frac{\rho}{2}\mathrm{Tr}\left((\mathbf{Y} - \mathbf{U}_1\mathbf{U}_2 \cdots \mathbf{U}_r\mathbf{V}_r^T)(\mathbf{Y} - \mathbf{U}_1\mathbf{U}_2 \cdots \mathbf{U}_r\mathbf{V}_r^T)^T\right)$$

$$+ <\boldsymbol{\mu}, \mathbf{Y} - \mathbf{U}_1\mathbf{U}_2 \cdots \mathbf{U}_r\mathbf{V}_r^T> + \sum_{i=1}^{r} <\boldsymbol{\lambda}_i, \mathbf{V}_i - \mathbf{V}_i^+> \quad (5)$$

$$+ \frac{\rho}{2}\sum_{i=1}^{r}\mathrm{Tr}\left((\mathbf{V}_i - \mathbf{V}_i^+)(\mathbf{V}_i - \mathbf{V}_i^+)^T\right),$$

---

[1] For simplicity, the layer size (dimensionalities) of layer 1 to layer r is denoted as $[k_1 \cdots k_r]$ in the experiments.

where $\mathbf{D}$ is a diagonal matrix and its j-th diagonal element is

$$d_j = \frac{1}{2\|\mathbf{e}_j\|_2}. \tag{6}$$

and $\mathbf{e}_j$ is the j-th column of the following matrix

$$\mathbf{E} = \mathbf{X} - \mathbf{Y}. \tag{7}$$

Updating $\mathbf{U}_i$ Minimizing Eq. (4) w.r.t. $\mathbf{U}_i$ is identical to solving

$$\begin{aligned}\mathcal{L}_U &= <\boldsymbol{\mu}, \mathbf{Y} - \mathbf{U}_1 \mathbf{U}_2 \cdots \mathbf{U}_r \mathbf{V}_r^T> \\ &+ \frac{\rho}{2} \mathrm{Tr}\left(\left(\mathbf{Y} - \mathbf{U}_1 \mathbf{U}_2 \cdots \mathbf{U}_r \mathbf{V}_r^T\right) \left(\mathbf{Y} - \mathbf{U}_1 \mathbf{U}_2 \cdots \mathbf{U}_r \mathbf{V}_r^T\right)^T \right).\end{aligned} \tag{8}$$

Calculating the derivative of $\mathcal{L}_U$ w.r.t. $\mathbf{U}_i$ and setting it to zero, then we have

$$\mathbf{U}_i = \left(\boldsymbol{\Phi}^T \boldsymbol{\Phi}\right)^{-1} \left(\boldsymbol{\Phi}^T \mathbf{Y} \widetilde{\mathbf{V}}_i + \frac{\boldsymbol{\Phi}^T \boldsymbol{\mu} \widetilde{\mathbf{V}}_i}{\rho}\right) \left(\widetilde{\mathbf{V}}_i^T \widetilde{\mathbf{V}}_i\right)^{-1}, \tag{9}$$

where $\boldsymbol{\Phi} = \mathbf{U}_1 \mathbf{U}_2 \cdots \mathbf{U}_{i-1}$ and $\widetilde{\mathbf{V}}_i$ denotes the reconstruction of the i-th layer's centroid matrix.

Updating $\mathbf{V}_i$ $(i < r)$ Minimizing Eq. (4) w.r.t. $\mathbf{V}$ is identical to solving

$$\begin{aligned}\mathcal{L}_V &= < \boldsymbol{\mu}, \mathbf{Y} - \mathbf{U}_1 \mathbf{U}_2 \cdots \mathbf{U}_r \mathbf{V}_r^T > \\ &+ \frac{\rho}{2} \mathrm{Tr}\left(\left(\mathbf{Y} - \mathbf{U}_1 \mathbf{U}_2 \cdots \mathbf{U}_r \mathbf{V}_r^T\right) \left(\mathbf{Y} - \mathbf{U}_1 \mathbf{U}_2 \cdots \mathbf{U}_r \mathbf{V}_r^T\right)^T \right) \\ &+ \sum_{i=1}^{r} <\boldsymbol{\lambda}_i, \mathbf{V}_i - \mathbf{V}_i^+> + \frac{\rho}{2} \sum_{i=1}^{r} \mathrm{Tr}\left(\left(\mathbf{V}_i - \mathbf{V}_i^+\right) \left(\mathbf{V}_i - \mathbf{V}_i^+\right)^T \right).\end{aligned} \tag{10}$$

Similarly, calculating the derivative of $\mathcal{L}_V$ w.r.t. $\mathbf{V}_i$, and setting it to zero, we obtain

$$\mathbf{V}_i = \left(\mathbf{Y}^T \boldsymbol{\Phi} \mathbf{U}_i + \mathbf{V}_i^+ + \frac{\boldsymbol{\mu}^T \boldsymbol{\Phi} \mathbf{U}_i}{\rho} - \frac{\boldsymbol{\lambda}_i}{\rho}\right) \left(\mathbf{I} + \mathbf{U}_i^T \boldsymbol{\Phi}^T \boldsymbol{\Phi} \mathbf{U}_i\right)^{-1}, \tag{11}$$

where $\mathbf{I}$ represents an identity matrix.

Updating $\mathbf{V}_r$ $\left(\text{i.e., } \mathbf{V}_i, (i = r)\right)$ We update $\mathbf{V}_r$ by solving

$$J_{V_r} = \min_{\mathbf{V}_r} \|\mathbf{Y} - \mathbf{U}_1 \mathbf{U}_2 \cdots \mathbf{U}_r \mathbf{V}_r^T\|_{2,1} = \min_{\mathbf{v}} \sum_{j=1}^{n} d_j \|\mathbf{x}_j - \mathbf{U}_1 \mathbf{U}_2 \cdots \mathbf{U}_r \mathbf{v}_j\|_2^2 \tag{12}$$

$$\text{s.t. } (\mathbf{V}_r)_{.c} = \{0, 1\}, \sum_{c=1}^{C} (\mathbf{V}_r)_{.c} = 1,$$

where $\mathbf{x}_j$ denotes the j-th data sample of $\mathbf{X}$, and $\mathbf{v}_j$ denotes the j-th column of $\mathbf{V}_r^T$. Taking a closer look at Eq. (12), we can see that it is independent between different j. Thus we can independently solve it one by one:

$$\min_{\mathbf{v}} \left(d \| \mathbf{x} - \mathbf{U}_1 \mathbf{U}_2 \cdots \mathbf{U}_r \mathbf{v} \|_2^2 \right)$$

$$\text{s.t.} v_c = \{0, 1\}, \sum_{c=1}^{C} v_c = 1. \tag{13}$$

Since $\mathbf{v}$ is coded by 1-of-C scheme, there exists C candidates that could be the solution of Eq. (13). And each individual solution is exactly the c-th column of identity matrix $\mathbf{I}_C = [\mathbf{f}_1, \mathbf{f}_2, \cdots, \mathbf{f}_C]$. Thus the optimal solution can be obtained by performing an exhaustive search, i.e.,

$$\mathbf{v}^* = \mathbf{f}_c, \tag{14}$$

where c is given by

$$c = \arg\min_{\bar{c}} \left(d \| \mathbf{x} - \mathbf{U}_1 \mathbf{U}_2 \cdots \mathbf{U}_r \mathbf{f}_{\bar{c}} \|_2^2 \right). \tag{15}$$

Updating Y Minimizing Eq. (4) w.r.t. $\mathbf{Y}$ is identical to solving

$$\mathcal{L}_Y = \text{Tr}\left((\mathbf{X} - \mathbf{Y}) \mathbf{D} (\mathbf{X} - \mathbf{Y})^T \right)$$

$$+ \frac{\rho}{2} \text{Tr}\left((\mathbf{Y} - \mathbf{U}_1 \mathbf{U}_2 \cdots \mathbf{U}_r \mathbf{V}_r^T)(\mathbf{Y} - \mathbf{U}_1 \mathbf{U}_2 \cdots \mathbf{U}_r \mathbf{V}_r^T)^T \right) \tag{16}$$

$$+ <\boldsymbol{\mu}, \mathbf{Y} - \mathbf{U}_1 \mathbf{U}_2 \cdots \mathbf{U}_r \mathbf{V}_r^T>.$$

Calculating the derivative of $\mathcal{L}_Y$ w.r.t. $\mathbf{Y}$ and setting it to 0, we have

$$\mathbf{Y} = \left(2\mathbf{X}\mathbf{D} + \rho \mathbf{U}_1 \mathbf{U}_2 \cdots \mathbf{U}_r \mathbf{V}_r^T - \boldsymbol{\mu} \right) (2\mathbf{D} + \rho \mathbf{I})^{-1}. \tag{17}$$

Updating $\mathbf{V}_i^+$ Minimizing Eq. (4) w.r.t. $\mathbf{V}_i^+$ is identical to solving

$$\mathcal{L}_{V+} = \sum_{i=1}^{r} <\boldsymbol{\lambda}_i, \mathbf{V}_i - \mathbf{V}_i^+> + \frac{\rho}{2} \sum_{i=1}^{r} \text{Tr}\left((\mathbf{V}_i - \mathbf{V}_i^+)(\mathbf{V}_i - \mathbf{V}_i^+)^T \right). \tag{18}$$

Calculating the derivative of $\mathcal{L}_{V+}$ w.r.t. $\mathbf{V}_i^+$ and setting it to 0, we get

$$\mathbf{V}_i^+ = \mathbf{V}_i + \frac{\boldsymbol{\lambda}_i}{\rho}. \tag{19}$$

In summary, we optimize the proposed model by orderly performing the above steps. This procedure repeats until convergence.

Fig. 1. COIL dataset.

4   Experiments

To validate the effectiveness of our method. We compare it with the classical *k*-means [26], NMF [25], Orthogonal NMF (ONMF) [8], Semi-NMF (SNMF) [7], $l_{2,1}$-NMF [24] and the deep Semi-NMF (DeepSNMF) [35].

4.1   Data Sets

We empirically evaluate the proposed method on six benchmark data sets[2]. As a demonstration, Fig. 1 shows the dataset COIL. Table 1 shows the specific characteristics of all datasets. The number of instances is ranged from 400 to 4663, and feature number is ranged from 256 to 7511.

Table 1. Characteristics of experimental data sets.

Data sets	# samples	# features	# classes
ORL32	400	1024	40
COIL	1440	1024	20
Semeion	1593	256	10
Text	1946	7511	2
Cranmed	2431	462	2
Cacmcisi	4663	348	2

---

[2] https://archive.ics.uci.edu/ml/datasets.html.

4.2   Parameter Setting

For k-means algorithm, we perform k-means on each data set until it convergence. And the corresponding results are treated as the final result of k-means. We also use this result as the initialization of all other compared methods for fairness. For each compared method, the optimal value is solved based on the parameter setting range recommended by the relevant literature, and then the result under this condition is regarded as the final result output. For the proposed deep method, the layer sizes (as described in Sect. 3.2) are set as $[100\ c]$, $[50\ c]$ and $[100\ 50\ c]$ for simplicity. As for parameter ρ, we search it from $\{1e-5,$ $1e-4, 1e-3, 0.01, 0.1, 1, 10, 100\}$.

Under each parameter setting, we repeat the experiments 10 times for all methods, and the average results are reported for fair comparison.

4.3   Results and Analysis

The clustering performance measured by clustering accuracy (ACC) and normalized mutual information (NMI) of all methods are given in Tables 2-3. It is obvious that our method has better results than other algorithms. The superiority of DKM verifies that it could better explore cluster structure by uncovering the hierarchical semantics of data. That is, by utilizing the deep framework to conduct k-means hierarchically, the hidden structure of data is learned in a layerwise way, and finally, a better high-level, final-layer representation can be obtained for clustering task. By leveraging the deep framework and k-means model, the proposed DKM can enhance the performance of data clustering in general cases.

Table 2. Clustering results of ACC on all data sets.

Datasets	Kmeans	NMF	ONMF	L21NMF	SNMF	DeepSNMF	DKM
ORL32	50.30 ± 2.2	51.97 ± 2.8	49.90 ± 3.1	53.40 ± 4.1	51.78 ± 3.5	49.86 ± 2.0	$\mathbf{54.50 \pm 1.2}$
COIL	59.43 ± 6.8	62.24 ± 3.1	58.35 ± 6.0	63.49 ± 4.4	63.78 ± 5.9	66.36 ± 6.2	$\mathbf{68.03 \pm 3.8}$
Semeion	51.93 ± 2.8	42.46 ± 3.5	48.50 ± 2.8	42.47 ± 2.0	43.95 ± 2.8	44.77 ± 1.2	$\mathbf{53.97 \pm 3.3}$
Text	91.84 ± 2.1	93.85 ± 3.9	92.47 ± 2.9	90.21 ± 4.0	90.99 ± 2.0	90.67 ± 4.5	$\mathbf{93.88 \pm 5.7}$
Cranmed	74.58 ± 0.1	80.13 ± 8.8	77.31 ± 1.2	77.39 ± 2.5	76.49 ± 4.9	80.23 ± 3.1	$\mathbf{82.31 \pm 3.9}$
Cacmcisi	91.99 ± 0.2	89.75 ± 5.4	94.96 ± 0.6	95.37 ± 0.8	92.22 ± 0.3	92.80 ± 3.5	$\mathbf{97.12 \pm 7.7}$

Table 3. Clustering results of NMI of on all data sets.

Datasets	Kmeans	NMF	ONMF	L21NMF	SNMF	DeepSNMF	DKM
ORL32	71.06 ± 1.3	72.10 ± 1.3	70.11 ± 1.7	72.70 ± 1.8	71.76 ± 1.9	68.83 ± 1.3	$\mathbf{72.94 \pm 1.5}$
COIL	74.53 ± 2.8	73.12 ± 1.7	72.84 ± 2.6	74.04 ± 2.3	74.91 ± 3.0	77.52 ± 7.4	$\mathbf{78.99 \pm 1.6}$
Semeion	57.34 ± 4.7	49.64 ± 5.6	53.87 ± 5.8	47.88 ± 2.7	49.72 ± 5.4	56.07 ± 2.0	$\mathbf{62.07 \pm 5.9}$
Text	61.31 ± 6.5	61.21 ± 1.5	60.88 ± 1.9	60.81 ± 3.0	57.85 ± 5.2	60.01 ± 4.2	$\mathbf{61.55 \pm 5.4}$
Cranmed	18.79 ± 0.3	31.67 ± 5.6	20.74 ± 0.3	20.84 ± 1.2	24.66 ± 5.4	25.05 ± 2.4	$\mathbf{32.89 \pm 2.3}$
Cacmcisi	58.47 ± 0.1	60.01 ± 2.5	70.42 ± 0.2	72.05 ± 0.4	70.52 ± 0.2	70.07 ± 3.3	$\mathbf{73.06 \pm 2.4}$

Based on the theoretical analysis and empirical results presented in this paper, it would be interesting to combine deep structure learning and classical machine learning models into a unified framework. By taking advantages of the both learning paradigms, more promising results in various learning tasks can be expected.

5 Conclusion

In this paper, we propose a novel deep k-means model to learn the hidden information with respect to multiple level characteristics. By utilizing the deep structure to conduct k-means hierarchically, the data hierarchical semantics is learned in a layerwise way. Through the deep k-means structure, instances from same class are pushed closer layer by layer, which benefits the subsequent clustering task. We also introduce an alternative updating algorithm to address the corresponding optimization problem. Experiments are conducted on six benchmark data sets and show promising results of our model against several state-of-the-art algorithms.

Acknowledgments. This work was partially supported by the National Key Research and Development Program of China under Contract 2017YFB1002201, the National Natural Science Fund for Distinguished Young Scholar under Grant 61625204, the State Key Program of the National Science Foundation of China under Grant 61836006, and the Fundamental Research Funds for the Central Universities under Grant 1082204112364.

References

1. Ault, S.V., Perez, R.J., Kimble, C.A., Wang, J.: On speech recognition algorithms. Int. J. Mach. Learn. Comput. **8**(6) (2018)
2. Badea, L.: Clustering and metaclustering with nonnegative matrix decompositions. In: Gama, J., Camacho, R., Brazdil, P.B., Jorge, A.M., Torgo, L. (eds.) ECML 2005. LNCS (LNAI), vol. 3720, pp. 10–22. Springer, Heidelberg (2005). https://doi.org/10.1007/11564096_7
3. Bengio, Y.: Learning deep architectures for AI. Found. Trends® in Mach. Learn. **2**(1), 1–127 (2009)
4. Boyd, S., Parikh, N., Chu, E., Peleato, B., Eckstein, J., et al.: Distributed optimization and statistical learning via the alternating direction method of multipliers. Found. Trends® Mach. Learn. **3**(1), 1–122 (2011)
5. Buchta, C., Kober, M., Feinerer, I., Hornik, K.: Spherical k-means clustering. J. Stat. Softw. **50**(10), 1–22 (2012)
6. Ding, C., He, X.: K-means clustering via principal component analysis. In: Proceedings of the Twenty-First International Conference on Machine Learning, pp. 29–37 (2004)
7. Ding, C., Li, T., Jordan, M.I.: Convex and semi-nonnegative matrix factorizations. IEEE Trans. Pattern Anal. Mach. Intell. **32**(1), 45–55 (2010)
8. Ding, C., Li, T., Peng, W., Park, H.: Orthogonal nonnegative matrix t-factorizations for clustering. In: Proceedings of the 12th ACM SIGKDD International Conference on Knowledge Discovery and Data Mining, pp. 126–135 (2006)

9. Gokcay, E., Principe, J.C.: Information theoretic clustering. IEEE Trans. Pattern Anal. Mach. Intell. **24**(2), 158–171 (2002)
10. Gönen, M., Margolin, A.A.: Localized data fusion for kernel k-means clustering with application to cancer biology. In: Advances in Neural Information Processing Systems,. pp. 1305–1313 (2014)
11. Hou, C., Nie, F., Yi, D., Tao, D.: Discriminative embedded clustering: a framework for grouping high-dimensional data. IEEE Trans. Neural Netw. Learn. Syst. **26**(6), 1287–1299 (2015)
12. Huang, S., Kang, Z., Xu, Z.: Self-weighted multi-view clustering with soft capped norm. Knowl. Based Syst. **158**, 1–8 (2018)
13. Huang, S., Ren, Y., Xu, Z.: Robust multi-view data clustering with multi-view capped-norm k-means. Neurocomputing **311**, 197–208 (2018)
14. Huang, S., Wang, H., Li, T., Li, T., Xu, Z.: Robust graph regularized nonnegative matrix factorization for clustering. Data Min. Knowl. Disc. **32**(2), 483–503 (2018)
15. Huang, S., Xu, Z., Kang, Z., Ren, Y.: Regularized nonnegative matrix factorization with adaptive local structure learning. Neurocomputing **382**, 196–209 (2020)
16. Huang, S., Xu, Z., Lv, J.: Adaptive local structure learning for document co-clustering. Knowl.-Based Syst. **148**, 74–84 (2018)
17. Huang, S., Xu, Z., Wang, F.: Nonnegative matrix factorization with adaptive neighbors. In: International Joint Conference on Neural Networks, pp. 486–493 (2017)
18. Huang, S., Zhao, P., Ren, Y., Li, T., Xu, Z.: Self-paced and soft-weighted nonnegative matrix factorization for data representation. Knowl.-Based Syst. **164**, 29–37 (2018)
19. Kang, Z., Peng, C., Cheng, Q.: Kernel-driven similarity learning. Neurocomputing **267**, 210–219 (2017)
20. Kang, Z., et al.: Multi-graph fusion for multi-view spectral clustering. Knowl. Based Syst. **189**, 105102 (2020)
21. Kang, Z., Wen, L., Chen, W., Xu, Z.: Low-rank kernel learning for graph-based clustering. Knowl.-Based Syst. **163**, 510–517 (2019)
22. Kang, Z., Xu, H., Wang, B., Zhu, H., Xu, Z.: Clustering with similarity preserving. Neurocomputing **365**, 211–218 (2019)
23. Kang, Z., et al.: Partition level multiview subspace clustering. Neural Netw. **122**, 279–288 (2020)
24. Kong, D., Ding, C., Huang, H.: Robust nonnegative matrix factorization using l21-norm. In: Proceedings of the 20th ACM International Conference on Information and Knowledge Management, pp. 673–682. ACM (2011)
25. Lee, D.D., Seung, H.S.: Algorithms for non-negative matrix factorization. In: Advances in Neural Information Processing Systems, vol. 13, pp. 556–562 (2001)
26. Macqueen, J.: Some methods for classification and analysis of multivariate observations. In: Proceedings of Berkeley Symposium on Mathematical Statistics and Probability, pp. 281–297 (1967)
27. Newling, J., Fleuret, F.: Fast k-means with accurate bounds. In: International Conference on Machine Learning, pp. 936–944 (2016)
28. Newling, J., Fleuret, F.: Nested mini-batch k-means. In: Advances in Neural Information Processing Systems, pp. 1352–1360 (2016)
29. Ng, A.Y., Jordan, M.I., Weiss, Y.: On spectral clustering: analysis and an algorithm. In: Advances in Neural Information Processing Systems, pp. 849–856 (2002)
30. Ren, Y., Domeniconi, C., Zhang, G., Yu, G.: Weighted-object ensemble clustering: methods and analysis. Knowl. Inf. Syst. **51**(2), 661–689 (2017)
31. Ren, Y., Hu, K., Dai, X., Pan, L., Hoi, S.C., Xu, Z.: Semi-supervised deep embedded clustering. Neurocomputing **325**, 121–130 (2019)

32. Ren, Y., Huang, S., Zhao, P., Han, M., Xu, Z.: Self-paced and auto-weighted multi-view clustering. Neurocomputing **383**, 248–256 (2020)
33. Ren, Y., Kamath, U., Domeniconi, C., Xu, Z.: Parallel boosted clustering. Neurocomputing **351**, 87–100 (2019)
34. Ren, Y., Que, X., Yao, D., Xu, Z.: Self-paced multi-task clustering. Neurocomputing **350**, 212–220 (2019)
35. Trigeorgis, G., Bousmalis, K., Zafeiriou, S., Schuller, B.W.: A deep matrix factorization method for learning attribute representations. IEEE Trans. Pattern Anal. Mach. Intell. **39**(3), 417–429 (2017)
36. Tunali, V., Bilgin, T., Camurcu, A.: An improved clustering algorithm for text mining: multi-cluster spherical k-means. Int. Arab J. Inf. Technol. **13**(1), 12–19 (2016)
37. Wang, J., et al.: Enhancing multiphoton upconversion through energy clustering at sublattice level. Nat. Mater. **13**(2), 157 (2014)
38. Wang, L., Pan, C.: Robust level set image segmentation via a local correntropy-based k-means clustering. Pattern Recogn. **47**(5), 1917–1925 (2014)
39. Wu, X., et al.: Top 10 algorithms in data mining. Knowl. Inf. Syst. **14**(1), 1–37 (2008)

Brain Storm Optimization Algorithms: A Brief Review

Sicheng Hou[1]([✉]), Mingming Zhang[2,3], Shi Cheng[2], and Zhile Yang[3]

[1] Graduate School of Information, Production and Systems, Waseda University,
Kitakyushu, Japan
sicheng_hou@fuji.waseda.jp
[2] School of Computer Science, Shaanxi Normal University,
Xi'an 710119, China
cheng@snnu.edu.cn
[3] Shenzhen Institutes of Advanced Technology,
Chinese Academy of Sciences Chinese, Shenzhen 518055, China
zl.yang@siat.ac.cn

Abstract. Nowadays, many real-world optimization problems, which are separated and non-differentiable, could not be solved by traditional optimization algorithms efficiently. To deal with those complex problems, Swarm Intelligence (SI) algorithms have been widely applied due to their powerful search ability based on population. Brain Storm Optimization (BSO) algorithm is a young and promising SI algorithm inspired by human beings' behavior of the brain-storming process. Through the continuous use of convergence and divergence process, individuals in BSO move towards optima over iterations, exploration and exploitation ability of algorithm could also achieve optimal balance. In this paper, the historical development, state-of-the-art, the different variants, and the real-world applications of BSO algorithms are reviewed. Besides, three typical optimization problems and the key points to solve them are also discussed. In the future, BSO algorithms could be used to solve more complex optimization problems, the strength and limitation of various BSO algorithms will be revealed furthermore.

Keywords: Brain storm optimization · Exploration/exploitation · Swarm Intelligence · Multi-objective/multimodal/dynamic optimization

1 Introduction

Optimization problems require algorithms to find the "best available" solutions satisfying some particular constrains in acceptable computing time. Many real-

This work is partially supported by National Natural Science Foundation of China (Grant No. 61806119), Natural Science Basic Research Plan In Shaanxi Province of China (Grant No. 2019JM-320), Natural Science Foundation of Guangdong Province under grants 2018A030310671 and Outstanding Young Researcher Innovation Fund of SIAT, CAS (201822).

© Springer Nature Singapore Pte Ltd. 2020
H. Zhang et al. (Eds.): NCAA 2020, CCIS 1265, pp. 284–295, 2020.
https://doi.org/10.1007/978-981-15-7670-6_24

world optimization problems are generally complex and have mutually conflicting objectives to be optimized simultaneously. Although many classic mathematical optimization methods such as climbing-method and gradient descent methods, which are based on some strict mathematical operations like differentiation, have been widely and successfully applied to solve some optimization problems, however, they cannot get satisfactory solutions for some separated and non-differentiable problems. To solve those complex optimization problems efficiently, Swarm Intelligence (SI) algorithms, which possess powerful search ability based on population, have attracted much attention in the optimization area and deserve more consideration in future research.

The brain storm optimization (BSO) algorithm was proposed in 2011 [14], which is a young and promising algorithm of SI. BSO algorithm is inspired by the brainstorming process, which is an efficient process created by a group of human beings to solve difficult problems. Two major processes: convergence and divergence are included in the BSO algorithm. Specifically, in the BSO algorithm, the convergence process refers to solutions that are clustered into several groups from an initial random distribution, the divergence process indicates that new individuals are generated by the mutation of existing groups. Moreover, two important issues are affecting the performance of BSO algorithms: exploration ability and exploitation ability. On the one hand, exploration refers to the ability of an algorithm to explore or distribute along with different promising areas that may contain potential better solutions, the diversity of population could be improved when the exploration ability is enhanced. On the other hand, exploitation indicates the ability of an algorithm to exploit or focus on one promising searching region, which means candidate promising solutions could be refined when the exploitation ability is enhanced. In order to converge population into true optima of the problem finally, the exploitation ability and exploration ability are required to achieve the optimal balance, however, since the complex searching environment of different optimization problems, it is hard to balance those two abilities well, therefore, some efficient strategies are necessary to modify BSO algorithm. According to different categories of optimization problems, those modified strategies need to be investigated carefully and the mechanism of different BSO variants should be clarified to improve their performance furthermore. Based on the above motivation, this paper aims to give a brief review of the BSO algorithm in terms of the basic procedure, various modified strategies, improved operations, typical optimization problems with mechanism analysis, and related applications. With the help of the detailed survey and mechanism analysis, the BSO algorithm could be utilized to solve a more complex problem, the strength and limitation of the BSO algorithm will also be revealed furthermore.

The remaining paper is organized as follows. In Sect. 2, the original BSO algorithm is first reviewed. Followed by the modified strategies and theoretical mechanism analysis in Sect. 3. In Sect. 4, three typical optimization problems are introduced, then the state of the art BSO variant algorithms designed to solve them are investigated. Finally, conclusions and future work of the BSO algorithm are given in Sect. 5.

2   Brain Storm Optimization Algorithms

2.1   The Original Brain Storm Optimization Algorithm

In 2011, the BSO algorithm was proposed by Shi under the inspiration of the brain storming process [14]. The flowchart proposed BSO algorithm is given in Fig. 1. BSO implement four main operations: Initialization, Clustering, Generation, and Selection. After initializing the population randomly and clustering operation, new individuals are produced by adding mutation value based on current clusters, then the selection operation is used to pick the better individuals into the next iteration.

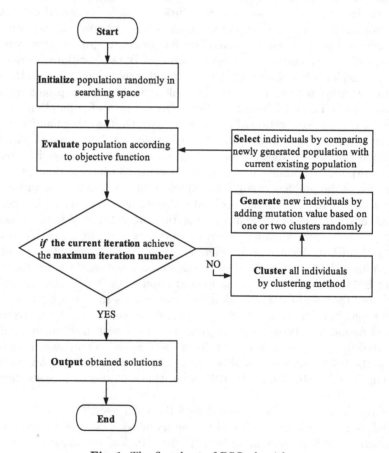

Fig. 1. The flowchart of BSO algorithm

2.2   Initialization Operation

The random initialization operation of BSO algorithms is similar to initialization operation in other SI methods. A population of solutions is initialized randomly and uniformly in the range of search space. Normally, the population size is set to be a fixed number before the algorithm running.

2.3   Clustering Operation

Convergence and divergence are two common processes in SI algorithms. The convergence process means that the whole population is gathered into several promising areas of the searching space. In the BSO algorithms, clustering operation is utilized to converge the individuals of the population into several smaller regions which may have the potential to generate better solutions. The original BSO chooses the k-means method to cluster all solutions in a swarm. Note that the number of clusters: k needs to be determined according to the scale of the problem before algorithm running. After a population has been divided into several clusters, the individual, whose fitness value is the best among all individuals from the same cluster, is picked up as the center of the corresponding cluster. Compared with other individuals, the cluster center is given a higher probability to generate new solutions, thus, the population can be guided towards better and better areas over iterations.

2.4   Generation Operation

As mentioned above, in an SI algorithm, the divergence process refers to all individuals diverge to different areas of searching space. As for the BSO algorithm, the divergence process mainly occurs when new individuals are generated by mutation operation based on different clusters. The flowchart of generation operation is given in Fig. 2 in which the newly generated individuals could be produced by adding mutation value based on one or two existing clusters, a probability value $p_{generation}$, which has been determined before algorithm running, is used to decide whether the new individual is generated from one or two current clusters. Then, two probability values, termed as $p_{oneCluster}$ and $p_{twoCluster}$, are utilized to decide whether the individual selected to generate new one is from the centers of selected clusters or other normal solutions in selected clusters, respectively. When the generation operation is based on one cluster case, the exploitation ability of the BSO algorithm could be enhanced. Meanwhile, in the case of generation operation based on two clusters, when a new individual is produced based on the combination of normal individuals from different clusters, the population diversity and exploration ability of the BSO algorithm, could be improved. The new individuals are generated based on Eq. (1) and (2).

$$x_{new}^d = x_{selected}^d + \xi(t) \times N(\mu, \sigma^2) \tag{1}$$

$$\xi(t) = \log \text{sig}(\frac{0.5 \times T - t}{k}) \times \text{rand}() \tag{2}$$

$$x_{selected}^d = w_1 \times x_{selected1}^d + w_2 \times x_{selected2}^d \tag{3}$$

$$w_1 + w_2 = 1 \tag{4}$$

where x_{new}^d and $x_{selected}^d$ are the dth dimension of the new individual and that of the selected individual from one cluster, respectively. Note that when the generation operation is based on two clusters case, $x_{selected}^d$ is the combination of two individuals according to the Eq. (3) and (4). Parameter T is the pre-determined maximum iteration number and t is the current iteration number, the value of logarithmic sigmoid transfer function $\log \text{sig}()$ is in the range of $(0, 1)$, k is a coefficient to adjust the convergent and divergent speed of BSO algorithm.

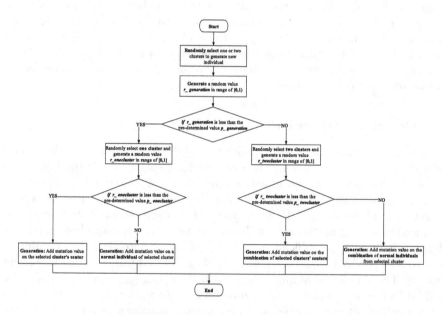

Fig. 2. The flowchart generation operation

2.5   Selection Operation

The selection operation aims to pick good individuals with higher fitness value into the next iteration after generation operation. From the viewpoint of population, selection operation works together with generation operation to update the population. With efficient updating strategies, the diversity and robustness of the population could be improved [15].

3   Brain Storm Optimization: State-of-the-art

BSO is a young and promising algorithm in the SI algorithms and has attracted many research attentions for recent years, in conclusion, the development of BSO algorithm concentrates on the following areas:

- To improve BSO algorithms' performance during iteration, i.e., higher accuracy of optima, faster convergent speed, and less computation burden.
- To apply BSO algorithms in typical optimization problems like multi-objective optimization problem (MOP), multimodal optimization problems (MMOP), dynamic optimization problem (DOP), and large-scale optimization problems (LSOP), etc.
- To apply BSO algorithms optimizing real-world problems like manufacturing scheduling problems, energy dispatch scheduling problems, vehicle routing problems, and data mining problems, just to name a few.

3.1   Mechanism Analyses

BSO algorithm has obtained many achievements since its introduction, to improve the performance furthermore, some theoretical mechanism analyses are necessary to clarify the merits of BSO and more efficient strategies could be designed based on related analyses.

As mentioned above, two abilities of BSO algorithms: the exploration and exploitation should achieve optimal balance during the searching process. Generally speaking, to avoid algorithm been trapped into prematurely convergent to local optima, the exploration ability of an algorithm should be given a higher priority at the early phase of the searching process, therefore the population could be distributed along with the promising search space as much as possible. However, with the development of iterations, exploitation ability should be enhanced, thus, several promising areas searched by the former process could be refined and the population could be converged into global optima finally. In BSO algorithms, the exploitation ability is improved when the new individuals are close to the optima of an optimization problem, meanwhile, when the newly generated individuals are produced randomly or from the two-clusters case, exploration ability is enhanced. Moreover, with efficient searching strategies derived from population searching environment, exploration and exploitation could achieve optimal balance, BSO algorithm could be more flexible to deal with different problems.

To validate the influence of different parameters set on the BSO performance, Zhan et al. investigated the impacts of parameters on BSO in [22]. Cheng et al. [6] analyzed the solution clustering strategy and concluded that the exploration and exploitation ability of BSO could achieve optimal balance by utilizing an efficient updating strategy. Zhou [25] built a discrete Markov chain model which could approximate the searching process of a BSO algorithm in continuous space. Cheng has defined the population diversity of BSO algorithms in [7]. Population diversity, which reflects the changing condition of exploitation and exploration ability of BSO algorithms, could be used to guide the convergent and divergent activity of algorithms during the searching process. In [18], Wang analyzed the interaction performance of BSO algorithm by population interaction network to build the connection of different solutions in the population.

3.2 Modification of Clustering Operation

In BSO algorithms, clustering operation is used to divide the population into several groups/areas which have a higher potential to search for better solutions. It could be used to reveal the landscapes of optimization problems and guide the population moving toward better and better searching areas [14].

From the viewpoint of computational complexity analysis, the computational burden of k-means is heavy since it is executed in each generation iteration, which results in a huge computational time cost. To reduce the computation cost efficiently, one simpler clustering method called a simple grouping method is utilized in BSO algorithm by Zhan et al. [23] and a random grouping strategy is utilized by Cao [1].

From the viewpoint of analyzing the clustering strategy itself, another important topic is how to determine the suitable number of clusters. Chen et al. [3] proposed an enhanced BSO algorithm using the affinity propagation clustering method, compared with the k-means method, the clusters' numbers of AP method are decided adaptively according to the numerical characteristics of every individual in each generation. Xie et al. [19] proposed the modified multi-objective BSO algorithm utilizing another clustering method called DBSCAN, which adjusts the number of clusters according to the distributed density of clusters in decision space at each iteration. Both the above-mentioned clustering strategies improve the exploitation of BSO algorithm well since information among the population during the searching process could be made full use.

3.3 Modification of Generation Operation

After clustering the population into promising searching areas, generation operation is used to produce new individuals from the above-mentioned clusters.

To improve the exploration ability of BSO algorithm, a new individual generation strategy using modified search step size is proposed by Zhou et al. in [24], the modified algorithm generates individuals in a batch-mode way, which means there are $3n$ new individuals, rather than n individuals (n is the population size), are generated at each iteration and n individuals are selected into the next generation. Chen et al. [4] proposed a modified BSO algorithm utilizing the improved generating operator, which combines the information from different clusters to enhance the local searching ability of BSO algorithm. Yang et al. [21] have introduced the discussion mechanism into BSO algorithm called DBSO, which uses separate inter- and intra-cluster discussions at each iteration of a individual's generation, could improve the accuracy of obtained solutions efficiently and balance the exploration and exploitation ability well.

3.4 Hybrid Algorithm

To improve the performance of BSO algorithm when solving a more complex problem, combining other classic intelligent algorithms with the BSO algorithm is a natural way. The hybrid algorithm tries to take advantages of two or more algorithms and eliminate their disadvantages, thus a better hybrid algorithm could be formed. Cao et al. [2] proposed an improved BSO algorithm called BSODE, which combines DE strategy to improve both exploitation ability and population diversity simultaneously. BSO combining with the teaching-learning-based algorithm [11] is proposed to solve the multi-objective optimal power flow problem.

4 Different Optimization Problems and Application

4.1 Different Optimization Problems

Many real-world optimization problems are generally complex and have mutually conflicting optimized objectives simultaneously. Generally speaking, optimization problems can be categorized into the single-objective problem and multi-objective problem (MOP), unimodal optimization problem and multimodal optimization problem, static optimization problem and dynamic/uncertain optimization problem, small scale problem and large-scale problem, etc. Moreover, different optimization problem could be combined into more complex problems like a multimodal and multi-objective problem (MMO), and dynamic multimodal optimization (DMMO) problem. To solve different kinds of optimization problems efficiently, it is necessary to modify the original BSO algorithm to satisfy the requirements of different problems. There are many variants of BSO algorithms to handle different optimization problems, such as MOP in [10,12,17,19,20], MMOP in [13], just to name a few.

4.2 Multi-objective Optimization Problems

The multi-objective optimization problem (MOP) refers to those problems which have more than two objectives that need to be optimized simultaneously. The main difficulty to solve MOP lies in that the improvement of performance in one objective/dimension always results in the deterioration of other objectives/ dimensions performance, which indicates that in MOP, it is difficult to find one best solution which could dominate other all solutions for each objective. Thus, instead of finding the best one solution, one natural way to solve MOP is trying to find a set of trade-off points in decision space. Necessary modification based on the Pareto dominance concept needs to be implemented into the original BSO algorithm when dealing with MOP. In 2013, Shi [17] proposed multi-objective brain storming optimization (MBSO), which makes use of information of solutions in objective space to guide the population convergence process and the fitness value of solutions obtained by the algorithm is evaluated by Pareto dominance concept.

Another BSO variant algorithm is the BSO in the objective space (BSO-OS) algorithm proposed in [16]. BSO-OS divided a swarm into two different individual sets termed as "elitists" and "normal" according to the fitness value in the objective space, the individuals in "elitists" set, which is given a higher probability to generate new individual, have better fitness value than individuals in "normal" set. From the comparison experiment result, it could be found that the computation time of BSO-OS decreases efficiently since the convergent time of BSO-OS is just determined by population size while the convergent time of the original BSO is determined by the population size and problem dimension simultaneously. It is beneficial for BSO-OS to solve large scale MOP. To improve the diversity and accuracy of non-dominated solutions, Guo et al. [10] proposed a modified self-adaptive multi-objective BSO algorithm. To solve the sparse MOP, a novel multi-objective BSO algorithm in objective space, termed as MOBSO-OS, is proposed by Liang et al. [12]. To avoid BSO been trapped into premature convergence when solving MOP, MBSO/D, which combines the idea of decomposition with BSO algorithm, is proposed by Dai et al. [8].

4.3 Multi-objective and Multimodal Optimization Problems

MMOP, multi-objective and multimodal optimization problems, are proposed based on MOP and refer to those problems of which there are at least two Pareto sets (PS) in decision space reflecting the same one Pareto front (PF) in objective space. Without loss of generality, one MOP will be an MMOP if the former satisfies one of the following two conditions:

1. There exist at least one local optima in the decision space.
2. There exist at least two or more global optima in the decision space.

MMOP aims to find multiple optima in each iteration and try to maintain these found optima during all iterations [13]. To solve MMOP efficiently, SBSO, which utilizes the max-fitness grouping cluster method to improve the exploration ability of algorithm, is proposed by Guo et al. [9]. Moreover, niching technology, which could avoid the algorithm been trapped into premature convergence, is introduced to balance the exploration and exploitation ability when solving MMOP.

4.4 Dynamic Multimodal Optimization Problems

Dynamic optimization indicates those problems involving a series of events that need to be optimized in dynamic environments, i.e., some changing characteristics are contained in the searching process [5]. In 2018, Cheng et al. [5] introduced a new combination problem of dynamic and multimodal optimization termed as dynamic multimodal optimization (DMMO) problem, involving multiple changing global optima during the searching process. When solving DMMO problems, it is necessary to search for multiple changing global optima at each iteration and try to maintain these optima during all iterations. To validate the effectiveness of BSO algorithms, BSO-OS proposed in [16] has been used to solve

DMMO problems defined in [5], the result showed that from the overall trends, the searching error of optima obtained by BSO-OS keeps decreasing over iteration. Some SI algorithms such as PSO utilize the explicit memory mechanism to guide the searching process by prior searching experience. However, since the global optima of DMMO problems are changing dynamically, this memory mechanism may increase the searching error over iterations. BSO algorithm, which uses the distribution of solutions, has a natural advantage to solve DMMO. In the future, more researches and modifications could be implemented on BSO algorithms to improve performance when solving DMMO problems.

5   Conclusion

This paper has comprehensively reviewed the brain storm optimization algorithm in terms of the basic BSO procedure, modifications of different operations, and various BSO algorithm with related mechanism analysis. Moreover, three typical optimization problems have also been discussed. It can be concluded that BSO is not only a young promising algorithm but also a framework of the SI method using a convergent and divergent process to obtain the optima of an optimization problem. It is clear that at the early phase of searching process, the exploration ability of BSO algorithm should be given higher priority to diverge population into promising areas which have higher potential to contain global optima, meanwhile, with the development of iterations, the exploitation ability should be improved, therefore BSO algorithms could converge to the global optima finally.

BSO algorithms have obtained many achievements, but there are still several issues that need to be further researched in the future, like the optimal combination between meta-heuristic based exploration ability and knowledge-based local exploitation ability, the design of self-learning BSO algorithm driven by knowledge, the connection among algorithm, mechanism analysis, and data parsing, etc. Besides, BSO algorithm could also be used to solve more complex real-world applications, the strength and limitation of BSO algorithms will be found furthermore.

References

1. Cao, Z., Shi, Y., Rong, X., Liu, B., Du, Z., Yang, B.: Random grouping brain storm optimization algorithm with a new dynamically changing step size. In: Tan, Y., Shi, Y., Buarque, F., Gelbukh, A., Das, S., Engelbrecht, A. (eds.) ICSI 2015. LNCS, vol. 9140, pp. 357–364. Springer, Cham (2015). https://doi.org/10.1007/978-3-319-20466-6_38
2. Cao, Z., Wang, L., Hei, X., Shi, Y., Rong, X.: An improved brain storm optimization with differential evolution strategy for applications of ANNs. Math. Problems Eng. **2015**, 1–18 (2015)
3. Chen, J., Cheng, S., Chen, Y., Xie, Y., Shi, Y.: Enhanced brain storm optimization algorithm for wireless sensor networks deployment. In: Tan, Y., Shi, Y., Buarque, F., Gelbukh, A., Das, S., Engelbrecht, A. (eds.) ICSI 2015. LNCS, vol. 9140, pp. 373–381. Springer, Cham (2015). https://doi.org/10.1007/978-3-319-20466-6_40

4. Chen, J., Xie, Y., Ni, J.: Brain storm optimization model based on uncertainty information. In: Proceedings of the 2014 Tenth International Conference on Computational Intelligence and Security, Kunming, China, pp. 99–103, November 2014
5. Cheng, S., Lu, H., Song, W., Chen, J., Shi, Y.: Dynamic multimodal optimization using brain storm optimization algorithms. In: Qiao, J., et al. (eds.) BIC-TA 2018. CCIS, vol. 951, pp. 236–245. Springer, Singapore (2018). https://doi.org/10.1007/978-981-13-2826-8_21
6. Cheng, S., Shi, Y., Qin, Q., Gao, S.: Solution clustering analysis in brain storm optimization algorithm. In: Proceedings of The 2013 IEEE Symposium on Swarm Intelligence (SIS 2013), pp. 111–118. IEEE, Singapore (2013)
7. Cheng, S., Shi, Y., Qin, Q., Ting, T.O., Bai, R.: Maintaining population diversity in brain storm optimization algorithm. In: Proceedings of 2014 IEEE Congress on Evolutionary Computation (CEC 2014), pp. 3230–3237. IEEE, Beijing (2014)
8. Dai, C., Lei, X.: A multiobjective brain storm optimization algorithm based on decomposition. Complexity **2019**, 1–11 (2019)
9. Guo, X., Wu, Y., Xie, L.: Modified brain storm optimization algorithm for multimodal optimization. In: Tan, Y., Shi, Y., Coello, C.A.C. (eds.) ICSI 2014. LNCS, vol. 8795, pp. 340–351. Springer, Cham (2014). https://doi.org/10.1007/978-3-319-11897-0_40
10. Guo, X., Wu, Y., Xie, L., Cheng, S., Xin, J.: An adaptive brain storm optimization algorithm for multiobjective optimization problems. In: Tan, Y., Shi, Y., Buarque, F., Gelbukh, A., Das, S., Engelbrecht, A. (eds.) ICSI 2015. LNCS, vol. 9140, pp. 365–372. Springer, Cham (2015). https://doi.org/10.1007/978-3-319-20466-6_39
11. Krishnanand, K.R., Hasani, S.M.F., Panigrahi, B.K., Panda, S.K.: Optimal power flow solution using self–evolving brain–storming inclusive teaching–learning–based algorithm. In: Tan, Y., Shi, Y., Mo, H. (eds.) ICSI 2013. LNCS, vol. 7928, pp. 338–345. Springer, Heidelberg (2013). https://doi.org/10.1007/978-3-642-38703-6_40
12. Liang, J.J., Wang, P., Yue, C.T., Yu, K.J., Li, Z.H., Qu, B.Y.: Multi-objective brainstorm optimization algorithm for sparse optimization. In: Proceedings of 2018 IEEE Congress on Evolutionary Computation (CEC 2018), pp. 1–8, July 2018
13. Liang, J.J., Yue, C.T., Qu, B.Y.: Multimodal multi-objective optimization: a preliminary study. In: 2016 IEEE Congress on Evolutionary Computation (CEC), pp. 2454–2461, July 2016
14. Shi, Y.: Brain storm optimization algorithm. In: Tan, Y., Shi, Y., Chai, Y., Wang, G. (eds.) ICSI 2011. LNCS, vol. 6728, pp. 303–309. Springer, Heidelberg (2011). https://doi.org/10.1007/978-3-642-21515-5_36
15. Shi, Y.: An optimization algorithm based on brainstorming process. Int. J. Swarm Intell. Res. (IJSIR) **2**(4), 35–62 (2011)
16. Shi, Y.: Brain storm optimization algorithm in objective space. In: Proceedings of 2015 IEEE Congress on Evolutionary Computation (CEC 2015), pp. 1227–1234. IEEE, Sendai (2015)
17. Shi, Y., Xue, J., Wu, Y.: Multi-objective optimization based on brain storm optimization algorithm. Int. J. Swarm Intell. Res. (IJSIR) **4**(3), 1–21 (2013)
18. Wang, Y., Gao, S., Yu, Y., Xu, Z.: The discovery of population interaction with a power law distribution in brain storm optimization. Memet. Comput. **11**(1), 65–87 (2017). https://doi.org/10.1007/s12293-017-0248-z
19. Xie, L., Wu, Y.: A modified multi-objective optimization based on brain storm optimization algorithm. In: Tan, Y., Shi, Y., Coello, C.A.C. (eds.) ICSI 2014. LNCS, vol. 8795, pp. 328–339. Springer, Cham (2014). https://doi.org/10.1007/978-3-319-11897-0_39

20. Xue, J., Wu, Y., Shi, Y., Cheng, S.: Brain Storm optimization algorithm for multi-objective optimization problems. In: Tan, Y., Shi, Y., Ji, Z. (eds.) ICSI 2012. LNCS, vol. 7331, pp. 513–519. Springer, Heidelberg (2012). https://doi.org/10.1007/978-3-642-30976-2_62
21. Yang, Y., Shi, Y., Xia, S.: Advanced discussion mechanism-based brain storm optimization algorithm. Soft. Comput. **19**(10), 2997–3007 (2014). https://doi.org/10.1007/s00500-014-1463-x
22. Zhan, Z., Chen, W.N., Lin, Y., Gong, Y.J., Li, Y.l., Zhang, J.: Parameter investigation in brain storm optimization. In: 2013 IEEE Symposium on Swarm Intelligence (SIS), pp. 103–110, April 2013
23. Zhan, Z., Zhang, J., Shi, Y., Liu, H.l.: A modified brain storm optimization. In: Proceedings of the 2012 IEEE Congress on Evolutionary Computation (CEC 2012), Brisbane, QLD, pp. 1–8 (2012)
24. Zhou, D., Shi, Y., Cheng, S.: Brain storm optimization algorithm with modified step-size and individual generation. In: Tan, Y., Shi, Y., Ji, Z. (eds.) ICSI 2012. LNCS, vol. 7331, pp. 243–252. Springer, Heidelberg (2012). https://doi.org/10.1007/978-3-642-30976-2_29
25. Zhou, Z., Duan, H., Shi, Y.: Convergence analysis of brain storm optimization algorithm. In: Proceedings of 2016 IEEE Congress on Evolutionary Computation (CEC 2016), pp. 3747–3752. IEEE, Vancouver (2016)

A Binary Superior Tracking Artificial Bee Colony for Feature Selection

Xianghua Chu[1,2], Shuxiang Li[1], Wenjia Mao[1], Wei Zhao[1], and Linya Huang[1(✉)]

[1] College of Management, Shenzhen University, Shenzhen, China
17718150336@163.com
[2] Institute of Big Data Intelligent Management and Decision,
Shenzhen University, Shenzhen, China

Abstract. Feature selection is a NP-hard combinatorial problem of selecting the effective features from a given set of original features to reduce the dimension of dataset. This paper aims to propose an improved variant of learning algorithm for feature selection, termed as Binary Superior Tracking Artificial Bee Colony (BST-ABC) algorithm. In BST-ABC, a binary learning strategy is proposed to enable each bee to learn from the superior individuals in each dimension for exploitation capacity enhancement. Ten datasets from UCI repository are adopted as test problems, and the results of BST-ABC are compared with particle swarm optimization (PSO) and original ABC. Experimental results demonstrate that BST-ABC could obtain the optimal classification accuracy and the minimum number of features.

Keywords: Feature selection · Artificial bee colony · Superior tracking strategy

1 Introduction

Feature engineering is the cornerstone of machine learning, and feature selection is an important aspect of feature engineering [1]. Feature selection refers to selecting the best combination of features to best describe the target concept. Through feature selection, the useful features are retained while the redundant features and irrelevant features are removed. However, the number of possible combinations increases exponentially with the number of candidate features. In addition, appropriate features can achieve satisfactory learning performance, reduce computational complexity and improve knowledge discovery. Therefore, efficient methods for feature selection are always in need.

Swarm intelligence algorithm is suitable for solving complex optimization problem based on its advantages of parallel computation and structure [2]. Artificial bee colony (ABC), as an efficient swarm intelligence meta-heuristic [3], was applied to address feature selection problems. Kaya Keles et al. [4] employed ABC to conduct feature selection on SCADI dataset with 70 samples and 206 attributes. Seven features were finally selected from 206 features to classify the dataset with various classification methods, which significantly improve the classification accuracy. Bamit et al. [5]

© Springer Nature Singapore Pte Ltd. 2020
H. Zhang et al. (Eds.): NCAA 2020, CCIS 1265, pp. 296–305, 2020.
https://doi.org/10.1007/978-981-15-7670-6_25

proposed ABC to select features on z-alizadeh Sani dataset with 303 samples and 56 attributes, and the classification accuracy is enhanced on the original data.

Although ABC has better optimization ability, its convergence speed and search quality is at the early stage [6]. P. Shunmugapriya et al. [7] improved the ABC algorithm by recording the global optimal solutions and the previously abandoned solutions, the feature selection performance was improved, and the computational cost did not increase significantly. Zeynep Banu ÖZGER [8] implemented 8 variants of binary ABC algorithm to solve the feature selection problem. Experimental results show that the proposed algorithm using bit-by-bit operator has a better global search capability. Wang, H. et al. [9] proposed an improved algorithm for the initial food source. The initial source generated by the algorithm is almost short period or fixed point, which makes the initial source evenly distributed. Simulation results show that the method can get high classification accuracy and smaller feature subset. Liu, M. [10] proposed an improved multi-objective ABC algorithm based on Knee Points to improve the convergence speed. The experimental results show that the algorithm is effective in feature selection.

In order to improve convergence speed and exploitation capability of ABC for feature selection, this paper proposes the binary superior tracking artificial bee colony (BST-ABC). Compared with original ABC, BST-ABC is improved in two aspects: (1) bees learn from other bees in each dimension in each iteration, instead of only updating one dimension in each iteration; (2) the learning samples are not selected randomly, and the bees select individuals with better fitness to learn. The superior tracking strategy improves the efficiency of information change between bees. We use 10 datasets with different types of UCI to test BST-ABC on feature selection problems. ABC and particle swarm optimization (PSO) are added for comparison. Experimental results show that BST-ABC achieves the best results on classification accuracy and the minimum number of features, while the average computational cost is comparable to ABC and PSO. BST-ABC provides a new idea for complex combinatorial optimization problems.

This paper is organized as follows: Sect. 2 introduces the basic ABC algorithm and the propose BST-ABC algorithm. Section 3 shows the process and results of the experiment. Section 4 make conclusions.

2 Binary Superior Tracking Artificial Bee Colony

2.1 Basic Artificial Bee Colony

The ABC framework is divided into employed bee stage, onlooker bee stage, scout bee stage [11–15]. Firstly, initialize the food source. The initialization method is as follows (1).

$$x_{ij} = lb_j + rand(0, 1)(ub_j - lb_j) \tag{1}$$

$i = 1, 2, \cdots, SN$, where SN refers to the number of food source, $j = 1, 2, \cdots, D$, D refers to problem dimension, ubj, lbj refer to the maximum and minimum values of the j-dimension of source respectively.

1) Employed bee stage, the amount of employed bee is half of the initial food source, and it is attached to half of the better source. A new food source is mined near the attached food source. The mining method is shown in Eq. (2).

$$v_{ij} = x_{ij} + R_{ij}(x_{ij} - x_{kj}) \tag{2}$$

v_{ij} refers to newly mined food source, R_{ij} is a random number between $[-1, 1]$, x_{ij}, x_{kj} refer to the j dimension of food source i, k respectively.

Calculating and updating the objective function value of the new food source, after the employed bee found a better food source, calculating the new fitness according to (3).

$$fit(x_i) = \begin{cases} \dfrac{1}{1+f(x_i)}, f(x_i) \geq 0 \\ 1 + |f(x_i)|, f(x_i) < 0 \end{cases} \tag{3}$$

$f(x_i)$ refers to objective function value of food source x_i.

2) Onlooker bee stage, calculating the probability by (4) and then selecting food source, the better food sources are retained by roulette method. Further exploring food source by (5)

$$P_i = \frac{fit(x_i)}{\displaystyle\sum_{i=1}^{D} fit(x_i)} \tag{4}$$

$$v_{ij} = x_{kj} + R_{ij}(x_{kj} - x_{gj}) \tag{5}$$

x_{kj} refers to the j-dimension of food source selected by roulette, x_{gj} is the food source different from k.

3) Scout bee stage, if the food source is not replaced by a better one, then starting scout bee and generating new food source randomly according to (1).

2.2   Binary Superior Tracking Artificial Bee Colony

In each iteration of ABC, each food source only updates one dimension, and the neighbor tracking strategy simply guides the bees to learn another randomly selected bee, resulting in the slow convergence speed and insufficient exploitation capability of ABC in the process of optimization. We improved ABC based on superior tracking strategy and proposed Binary Superior Tracking Artificial Bee Colony (BST-ABC) [16, 17].

Compared with ABC, BST-ABC has two differences: (1) In each iteration, bees learn from other bees in each dimension, instead of only learning one dimension in each iteration. (2) The learning samples were not randomly selected, and the bees chose the ones with better fitness to learn. In BST-ABC, the food sources are updated as following equation:

$$v_i = x_i + r_i(x_i - SN_i) \tag{6}$$

SN_i denotes the superior neighbor for guidance, it is a D-dimensional vector of which elements are constructed by itself (position of the food source) for $i - th$ food source and other superior food sources with two probabilistic selection methods (roulette selection and tournament selection) (Table 1).

Table 1. Pseudo code for generating SN_i

Input parameters
For $d = 1 : D$
If $rand < Pr$
$A_1 = FS(randi_1(ps)$
$A_2 = FS(randi_2(ps)$
If $FV(A_1) < FV(A_2)$
$SN_{i,d} = A_{1,d}$
Else $SN_{i,d} = A_{2,d}$
End
Else $SN_{i,d} = A_{i,d}$
End
End
Return $SN_{i,d}$

Pr is the initialized probability threshold, which determines whether the current individual learns from his superior neighbor or itself. *randi* is a function used to produce an integer from a uniform discrete distribution, FS refers to the position of food source, FV is function value.

After updating the food source according to (6), we convert the continuous food source to the binary one. The position of food source is transformed into probability value according to (7) firstly, and then the position food source is updated as (8).

$$P(v_i) = \frac{1}{1 + \exp(-v_i + 1)} \tag{7}$$

$$V_i = \begin{cases} 1, & rand < P(v_i) \\ 0, & rand > P(v_i) \end{cases} \tag{8}$$

where v_i refers to continuous food source and V_i refers to binary food source. *rand* is a random number uniformly distributed between 0 and 1.

In BST-ABC, each dimension of food source is updated in each iteration and learns from the better food source, which ensures the timely information exchange between food sources. The flowchart of BST-ABC is given in Fig. 1.

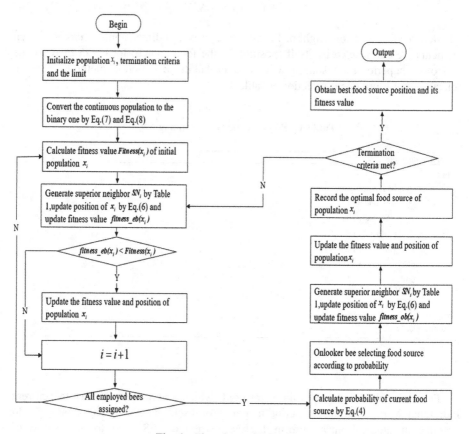

Fig. 1. Flowchart of BST-ABC

3   Experimental Comparison

Feature selection is a NP hard combinatorial problem; the number of possible solutions increases exponentially with the number of features. Therefore, an exhaustive search that requires an average high computational complexity is unrealistic. We applied BST-ABC to the feature selection problem, and evaluated the performance of BST-ABC based on the optimal classification accuracy, average classification accuracy and the number of selected features, and ABC and PSO are used in performance comparison. KNN ($k = 5$) is selected as the classification algorithm. As for feature selection, the solution is represented in binary form, if the feature is taken as 1, it means the feature is selected; if the feature is taken as 0, it means the feature is not selected.

3.1   Dataset Parameter Setting

Considering the number of instances and dimensions of different datasets, ten datasets were collected from the UCI. The number of instances and dimensions of the datasets are shown in Table 2.

Table 2. The ten utilized benchmark datasets

UCI dataset	Number of instances	Number of features	Number of classes
Breastcancerwisconsin	699	9	2
Diabetic retinopathy	1151	19	2
Glass identification	214	10	6
Ionosphere	351	34	2
Movementlibras	360	90	15
Musk_1	476	166	2
BreastCancerCombiadataR2	116	9	2
Lungcancer	32	55	3
Pdspeechfeatures	756	753	2
Seeds	210	7	3

In order to make the results convincing, the number of population and the number of iteration of each algorithm are set to 10 and 100, and each algorithm runs 10 times independently. All the analysis was done in MATLABR2017b software by using a computer with Intel Core i7 2.6 GHz and 16.0 GB RAM.

3.2   Experimental Results

KNN algorithm is a common classification method in machine learning [18]. Without feature selection, the classification accuracy of KNN (K = 5) algorithm is shown in Table 3.

Table 3. Classification accuracy without feature selection

UCI dataset	Classification accuracy without feature selection
Breastcancerwisconsin	97.36
Diabetic retinopathy	64.73
Glass identification	67.29
Ionosphere	84.62
Movementlibras	76.11
Musk_1	87.61
BreastCancerCombiadataR2	51.72
Lungcancer	56.25
Pdspeechfeatures	72.22
Seeds	88.10

We use BST-ABC do feature selection of the ten datasets. The experimental results are shown in Table 4. The best value of the three methods is shown in bold.

Table 4. Experimental results of three methods

Dataset		Best accuracy	Worst accuracy	Mean accuracy	STD	Feature size
Breastcancerwisconsin	PSO	97.66	97.36	97.45	0.0013	8.20
	ABC	97.66	97.36	97.54	0.0012	6.80
	BST-ABC	97.66	**97.66**	**97.66**	**0.0000**	**6.60**
Diabetic retinopathy	PSO	72.81	66.38	68.76	0.0241	12.40
	ABC	70.55	66.12	68.65	0.0212	11.00
	BST-ABC	**72.89**	**70.98**	**71.78**	**0.0065**	**7.00**
Glass identification	PSO	71.03	70.56	70.84	0.0026	7.20
	ABC	71.03	71.03	71.03	0.0000	7.20
	BST-ABC	71.03	71.03	71.03	**0.0000**	**6.00**
Ionosphere	PSO	86.89	86.32	86.72	**0.0025**	25.40
	ABC	89.46	87.18	88.15	0.0108	15.60
	BST-ABC	**90.03**	**88.03**	**88.89**	0.0073	**13.20**
Movementlibras	PSO	79.17	77.78	78.72	**0.0058**	80.20
	ABC	80.56	78.61	79.39	0.0077	49.60
	BST-ABC	**80.83**	**78.89**	**79.94**	0.0077	**44.00**
Musk_1	PSO	90.34	89.29	89.92	0.0039	155.80
	ABC	89.92	88.87	89.45	0.0040	91.00
	BST-ABC	**90.97**	**90.13**	**90.42**	**0.0035**	**86.60**
BreastCancerCombiadataR2	PSO	81.90	77.59	78.45	0.0193	7.80
	ABC	81.90	77.59	81.03	0.0193	6.60
	BST-ABC	**81.90**	**81.90**	**81.90**	**0.0000**	**5.40**
Lungcancer	PSO	71.88	62.50	69.37	0.0407	51.00
	ABC	71.88	65.63	69.37	**0.0261**	34.80
	BST-ABC	**81.25**	**71.88**	**74.38**	0.0407	**26.40**
Pdspeechfeatures	PSO	74.21	72.75	73.60	**0.0058**	598.60
	ABC	74.34	73.28	73.70	0.0044	383.60
	BST-ABC	**77.12**	**74.34**	**75.93**	0.0140	**369.60**
Seeds	PSO	95.71	94.76	95.14	0.0040	4.00
	ABC	95.71	93.81	95.14	0.0078	4.80
	BST-ABC	95.71	**95.71**	**95.71**	**0.0000**	**2.00**

It can be concluded that the optimal classification accuracy obtained by BST-ABC is greater than that obtained by PSO and ABC in 7 datasets, the average classification accuracy of BST-ABC is greater than that obtained by PSO and ABC, BST-ABC gets the lowest STD value in 6 datasets. After using BST-ABC for feature selection in all 10 datasets, the classification accuracy increased by 15.31% on average, and BST-ABC removed the most redundant features, which accounted for 50.43% of the original data set. The increased classification accuracy of the ten datasets by the three algorithms is shown in Fig. 2.

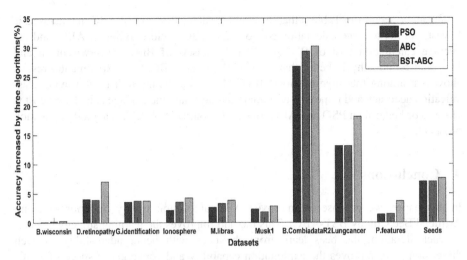

Fig. 2. Accuracy increased by three algorithms

The classification accuracy is the average classification accuracy obtained from 10 runs. For most of the ten datasets, the classification accuracy increased by the three algorithm within 10%. Because the classification accuracy without feature selection of BreastCancerCombiadataR2 and Lungcancer is low, the increased classification accuracy obtained by the three algorithms is highest. The increased classification accuracy obtained by BST-ABC is higher than obtained by ABC and PSO. The results indicate BST-ABC enhances ABC's exploitation capability.

The experimental results show that the performance of BST-ABC is better than the other two algorithms, and it can keep high consistency. Compared with the original dataset, BST-ABC can ensure a higher classification accuracy after feature selection, and greatly reduce the number of features and computational complexity. Table 5 shows the average computational cost of the three methods.

Table 5. The average computational cost of five feature selection methods.

Dataset	PSO	ABC	BST-ABC
Breastcancerwisconsin	15.797	15.860	15.585
Diabetic retinopathy	46.522	46.798	45.952
Glass identification	1.707	1.775	1.834
Ionosphere	3.607	3.602	3.704
Movementlibras	3.782	3.747	3.798
Musk_1	8.200	8.087	8.6046
BreastCancerCombiadataR2	0.756	0.781	0.784
Lungcancer	0.168	0.160	0.161
Pdspeechfeatures	26.016	23.766	23.867
Seeds	1.658	1.955	1.782

It can be seen from Table 5 that the average calculation cost of ABC in 6 datasets is lowest, but the average calculation cost of BST-ABC is close to that of ABC, and the average calculation cost of BST-ABC in 2 datasets of Breastcancerwisconsin and Diagonal retinopathy is better than that of PSO and ABC. The experimental results show that among the three methods, BST-ABC achieves the best in improving classification accuracy and removing redundant features, and the average calculation cost is close to or better than PSO and ABC. In conclusion, BST-ABC has a good application prospect.

4 Conclusions

In this paper, we propose an improved ABC algorithm: Binary Superior Tracking Artificial Bee Colony BST-ABC). This method improves the learning strategy of bees. In each iteration, the bees learn from the bees with better adaptability in each dimension, and improves the exploitation capability and convergence speed of ABC. Ten datasets were extracted from UCI for feature selection test, and the results of BST-ABC were compared with PSO and ABC. KNN (k = 5) is selected as the classification algorithm. The experimental results showed that BST-ABC could obtain the optimal classification accuracy and the minimum number of features among the three methods. Thus, it is concluded BST-ABC can be useful in complex combinatorial optimization problem.

Regarding the future work, we will examine the adaptability of BST-ABC to more complex problems and apply BST-ABC to practical problems.

Acknowledgement. This work was partially supported by the National Natural Science Foundation of China (Grant No. 71971142 and 71701079).

References

1. Cui, H., Xu, S., Zhang, L., Welsh, R.E., Horn, B.K.P.: The key techniques and future vision of feature selection in machine learning. J. Beijing Univ. Posts Telecommun. **41**, 1–12 (2018)
2. Wang, Y.: Research on feature selection methods based on swarm intelligent algorithm. Jilin University (2019)
3. Chu, X., Wu, T., Weir, J.D., Shi, Y., Niu, B., Li, L.: Learning–interaction–diversification framework for swarm intelligence optimizers: a unified perspective. Neural Comput. Appl. **32**(6), 1789–1809 (2020). https://doi.org/10.1007/s00521-018-3657-0
4. Keleş, M.K., Kılıç, Ü.: Artificial bee colony algorithm for feature selection on SCADI dataset. In: 2018 3rd International Conference on Computer Science and Engineering, UBMK, pp. 463–466. IEEE (2018)
5. Kiliç, Ü., Keleş, M.K.: Feature selection with artificial bee colony algorithm on Z-Alizadeh Sani dataset. In: 2018 Innovations in Intelligent Systems and Applications Conference, ASYU, pp. 1–3. IEEE (2018)

6. Chu, X., Xu, S.X., Cai, F., Chen, J., Qin, Q.: An efficient auction mechanism for regional logistics synchronization. J. Intell. Manuf. **30**(7), 2715–2731 (2019). https://doi.org/10.1007/s10845-018-1410-2

7. Shunmugapriya, P., Kanmani, S., Supraja, R., Saranya, K.: Feature selection optimization through enhanced artificial bee colony algorithm. In: 2013 International Conference on Recent Trends in Information Technology, ICRTIT, pp. 56–61. IEEE (2013)

8. Özger, Z.B., Bolat, B., Dırı, B.: A comparative study on binary artificial bee colony optimization methods for feature selection. In: 2016 International Symposium on INnovations in Intelligent SysTems and Applications, INISTA, pp. 1–4. IEEE (2016)

9. Wang, H., Yu, H., Zhang, Q., Cang, S., Liao, W., Zhu, F.: Parameters optimization of classifier and feature selection based on improved artificial bee colony algorithm. In: 2016 International Conference on Advanced Mechatronic Systems, ICAMechS, pp. 242–247. IEEE (2016)

10. Liu, M.: Improved multi-objective artificial bee colony algorithm and its application in feature selection. Anhui University (2018)

11. Chao, X., Li, W.: Feature selection method optimized by artificial bee colony algorithm. J. Front. Comput. Sci. Technol. **13**(02), 300–309 (2019)

12. Too, J., Abdullah, A.R., Mohd Saad, N.: A new co-evolution binary particle swarm optimization with multiple inertia weight strategy for feature selection. Informatics **6**(2), 21 (2019)

13. Karaboga, D., Basturk, B.: A powerful and efficient algorithm for numerical function optimization: artificial bee colony (ABC) algorithm. J. Global Optim. **39**(3), 459–471 (2007)

14. Li, Y.: Research and application of multi-objective artificial bee colony algorithm. Northeastern University (2012)

15. Qin, Q., Cheng, S., Li, L., Shi, Y.: A review of artificial bee colony algorithm. CAAI Trans. Intell. Syst. **9**(02), 127–135 (2014)

16. Chu, X., et al.: Adaptive differential search algorithm with multi-strategies for global optimization problems. Neural Comput. Appl. **31**(12), 8423–8440 (2019). https://doi.org/10.1007/s00521-019-04538-6

17. Chu, X., Hu, G., Niu, B., Li, L., Chu, Z.: An superior tracking artificial bee colony for global optimization problems. In: 2016 IEEE Congress on Evolutionary Computation, CEC, pp. 2712–2717. IEEE (2016)

18. Li, S., Yang, L., Chen, P., Yang, C.: Natural neighbor algorithm for breast cancer diagnosis. J. Phys. Conf. Ser. **1395**(1), 13–19 (2019)

A Hybrid Neural Network RBERT-C Based on Pre-trained RoBERTa and CNN for User Intent Classification

Yuanxia Liu[1], Hai Liu[1], Leung-Pun Wong[2], Lap-Kei Lee[2], Haijun Zhang[3], and Tianyong Hao[1(✉)]

[1] School of Computer Science, South China Normal University, Guangzhou, China
{liuyuanxia, haoty}@m.scnu.edu.cn, liuhai@scnu.edu.cn
[2] School of Science and Technology,
The Open University of Hong Kong, Ho Man Tin, Hong Kong SAR, China
{s1243151, lklee}@ouhk.edu.hk
[3] School of Computer Science and Technology, Harbin Institute of Technology, Shenzhen, China
hjzhang@hit.edu.cn

Abstract. User intent classification plays a critical role in identifying the interests of users in question-answering and spoken dialog systems. The question texts of these systems are usually short and their conveyed semantic information are frequently insufficient. Therefore, the accuracy of user intent classification related to user satisfaction may be affected. To address the problem, this paper proposes a hybrid neural network named RBERT-C for text classification to capture user intent. The network uses the Chinese pre-trained RoBERTa to initialize representation layer parameters. Then, it obtains question representations through a bidirectional transformer structure and extracts essential features using a Convolutional Neural Network after question representation modeling. The evaluation is based on the publicly available dataset ECDT containing 3736 labeled sentences. Experimental result indicates that our model RBERT-C achieves a F1 score of 0.96 and an accuracy of 0.96, outperforming a number of baseline methods.

Keywords: User intent classification · RoBERTa · CNN · Hybrid network

1 Introduction

User intent classification is essential in the identification and analysis of users' intents from conversion text data accumulated in question-answering systems and spoken dialog systems. Spoken dialogue systems [1] allow human to communicate with computers more conveniently and commonly contain an initial step Spoken Language Understanding (SLU), which further contain user intent classification as a sub-task [2]. Given a question sentence from a user, user intent classification predicts the intent label of the question to understand what the user truly wants [3]. For example, the predicted intent label of a sentence 'play signe anderson chant music that is newest' is 'play

© Springer Nature Singapore Pte Ltd. 2020
H. Zhang et al. (Eds.): NCAA 2020, CCIS 1265, pp. 306–319, 2020.
https://doi.org/10.1007/978-981-15-7670-6_26

music' in spoken dialogue systems, while the predicted label of a question 'who was the founder of Apple?' is 'corporate information' in question-answering systems.

Recently, vast neural networks methods [4–6] are proposed to classify question text for analyzing user intent. These models generally use distributed word vectors to represent words in sentences. This representation method maps words to a high-dimensional semantic space. The similarities of words are obtained according to the spatial distances of the vectors corresponding to the words [7]. However, distributed word vectors may not be able solve the problem of polysemy. A context-dependent sentence representation is provided by the pre-trained BERT [8] model and RoBERTa [9] is built on BERT language masking strategy and modifies key hyperparameters in BERT. Currently, pre-trained language models are usually based on large-scale English unlabeled corpus. In terms of the evaluation of user intent classification models, two English datasets, including ATIS and Snips, are commonly used while Chinese datasets are extremely lacking. To promote Chinese natural language processing, the Joint Laboratory of HIT and iFLYTEK Research publishes Chinese pre-trained RoBERTa with Whole Word Masking [10] to model Chinese sentences according to transfer learning and provides a better pre-trained model for Chinese natural language processing tasks. User intent classification is a problem working on the sentence-level to assign intent labels to sentences. However, user intent classification is mostly based on English corpus and there is less research on user intent classification in Chinese.

Therefore, this paper proposes a model RBERT-C for user intent classification of Chinese text. The model explores the combination of pre-trained model and Convolutional Neural Networks (CNN) for user intents classification. The Chinese pre-trained RoBERTa is used to obtain context-dependent sentence embeddings. The sentence embeddings are based on token-level, which map Chinese characters to a high-dimensional semantic space. Words composed of multiple characters often have different meanings in Chinese and more useful for Chinese sentence understanding than characters. Thus, the CNN model is used to capture key word-level features of Chinese question sentences through multiple convolution kernels of different sizes. A publicly available dataset from Social Media Processing (SMP) conference is applied to assess the performance of the model. The result shows that our model improves user intent classification by a F1 score of 1.5% on the dataset.

Major contributions of this work lie on: 1) a hybrid model RBERT-C is proposed for user intent classification of Chinese texts, 2) a convolution kernel is utilized to extract word-level key features based on Chinese pre-trained RoBERTa of character-level sentence representation, 3) compared with baseline methods in F1 score and accuracy on a publicly available dataset, our model obtains the best performance and demonstrates its effectiveness.

In the rest of this paper, Sect. 2 introduces related work on user intent classification. The details of our model are described in Sect. 3. Section 4 demonstrates the experimental dataset, parameters and results, and Sect. 5 draws conclusions of this work.

2 Related Work

User intent classification is a sub-task in question answering and dialogue systems. The early text classification is based on traditional machine learning methods by utilizing manually extracted features [11, 12]. For large amount datasets or the complexity of semantic features, the performance of these methods is unsatisfactory [13]. In recent years, due to the capability in extracting latent semantic features automatically, deep learning methods [14–16] has been extensively applied to address this problem and have made outstanding achievements in text classification.

For improving the accuracy of user intent classification, deep neural networks models were proposed in the past few years. The models include CNN [4], LSTM [16], attention based on CNN [17], hierarchical attention networks [18], and others. Recently, the joint learning of intention classification and semantic slot filling had become a popular method to have a better effect compared to relative independent methods for user intent classification. A joint model [5] was proposed using CNN-based triangular CRF. Guo et al. [19] explored the hierarchical representations of input text for the joint task. Liu et al. [20] described the joint model with RNN. Based on BERT, Chen et al. [21] proposed a joint model to obtain the better accuracy of user intent classification.

The ATIS [22] dataset and Snips [23] dataset were widely used in the SLU task. The ATIS dataset included audio recording for flight reservations and the Snips dataset was collected from Snips personal voice assistant. He et al. [24] proposed the method by combining CNN with BERT on the ATIS dataset for intent determination. Compared with the user intent classification on English datasets, researches focused on other language datasets were relatively few. Khalil et al. [25] explored multilingual transfer capability based on English and French for intent classification. A Chinese dataset ECDT [26] included 31 sub-categories involved two aspects namely chat and task-driven dialogue, containing 3736 labeled sentences. In order to classify the user intent in single utterance accurately, Xie et al. [27] proposed a method named ST-UIC to use the multiple semantic features. An architecture [28] integrating Bidirectional-GRU and Capsule networks [29] was proposed for Chinese question target classification.

However, these methods with distributed word vectors only obtained context-independent embedding and lacked context information in sentence representations. Pre-trained models applied transfer learning to obtain contextually dependent language representations based on large-scale unsupervised text corpora. The effectiveness of the pre-trained model was often seen in English natural language understanding tasks, but rarely in the Chinese task. We try to explore the applicability of pre-trained models on Chinese question sentences in this paper. Our proposed model leverages a Chinese pre-trained model to obtain context-dependent sentence representations and combines CNN for improving the performance of Chinese user intent classification.

3 The RBERT-C Model

We propose a hybrid model named RBERT-C that uses pre-trained model RoBERTa and CNN for user intention classification. The RBERT-C model consists of 4 layers: embedding layer, representation layer, convolution layer, and output layer. Figure 1 demonstrates the architecture of our model. The embedding layer maps each Chinese character to a multidimensional space vector combining position information. The position information encodes character positions in sentence sequences. The representation layer as encoder based on Chinese pre-trained RoBERTa with Whole Word Masking and through transfer learning computes a context-dependent sentence representation with rich semantic information. The convolution layer extracts sentence key features applying multiple different sizes convolution kernels and captures the most important feature by maximum pooling operation. The output layer maps the sentence semantic feature to the sample label space and uses a softmax function to obtain the probability distribution of category labels.

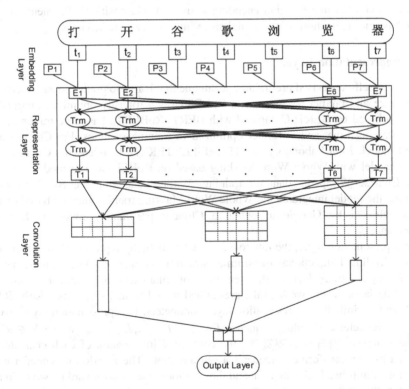

Fig. 1. The architecture of our proposed RBERT-C model.

3.1 Embedding Layer

For BERT, the input embedding representation is a combination of WordPiece embedding [30], positional embedding, and segment embedding. The input sequence uses special characters [CLS] to indicate the beginning of a sentence and [SEP] to indicate the end of a sentence. In our model, the segmentation is the same as BERT-Chinese and the token vocabulary contains almost all Chinese characters. For instance, the segmentation sequences are '打', '开', '谷', '歌', '浏', '览', '器' of the sentence '打开谷歌浏览器' (*Open Google browser*). In the sentence, the i-th token is denoted as character embedding $t_i \in R^d$ which is the d-dimensional vector. We denote position embedding as $P_i \in R^d$ and segment embedding as $S_i \in R^d$. We obtain input embedding sequences $E = (E_1, E_2, \ldots, E_{m-1}, E_m)$ where $E_i \in R^d$ is the d-dimensional vector, which is a summation of character embedding, positional embedding, and the segment embedding and m is the max-length of the sentence. The position embedding is calculated by *sine* and *cosine* functions to encode where every character appears in the sequence. In the position encoding process, *sine* function encoding is used at even positions, while *cosine* function encoding is used at odd positions. Segment embeddings have no distinction in our model since input is a single sentence.

3.2 Representation Layer

BERT, is a self-supervised pre-training technique that learns to predict masked sections of sentence and predict next sentence. RoBERTa is a robustly optimized pretraining approach based on BERT. Compared with BERT, RoBERTa has more training data, longer training time, and is further optimized in training methods. For Chinese pre-trained model, Joint Laboratory of HIT and iFLYTEK (HFL) released a Chinese pre-trained model with Whole Word Masking based on RoBERTa. The model used the latest Chinese Wikipedia data for training. Without changing the BERT network structure, the model integrated the Whole Word Masking training method based on the BERT released by Google to obtain a Chinese pre-training model with better performance.

In representation layer, the raw token sequence is firstly represented as embedding vectors, feeding bidirectional transformer structure as encoder. The input sequence $E = (E_1, E_2, \ldots, E_{m-1}, E_m)$ is the sequence of characters of length m from the embedding layer. Then, we load the pre-trained model from HEL, named RoBERTa-WWM-ext, to initialize representation layer parameters. In representation layer, given an input character embedding sequence $E = (E_1, E_2, \ldots, E_{m-1}, E_m)$ where $E_i \in R^{m \times d}$ for the sentence '打开谷歌浏览器', the contextual information of each character is captured by the transformer encoder via self-attention. The transformer encoder consists of the multi-head self-attention and the position-wise feed-forward network. In the first part, the multi-head means that the input is divided into h parts after passing through a linear mapping and every part is calculated through a scaled dot product attention function. The self-attention input consists of query $Q \in R^{m \times d}$, key $K \in R^{m \times d}$, as well as the value $V \in R^{m \times d}$. The calculation of the scaled dot product attention function is as follows:

$$Attention(Q, K, V) = softmax\left(\frac{QK^T}{\sqrt{d}}\right) \cdot V \tag{1}$$

The attention weights $A_{i:j}$ enable to better represent about the importance of the i-th key-value pair with the j-th query in generating the sentence representation. The output of multi-head self-attention part is concatenated to acquire final values.

$$Multi\_Head(Q, K, V) = Concat(head_1, \ldots, head_h)W^o \tag{2}$$

$head_i$ equals to $Attention(QW_i^Q, KW_i^K, VW_i^V)$. W_i^Q, W_i^K, and $W_i^V \in R^{d \times dhead}$ ($dhead = d \div h$) are parameter matrices. In the second part, two linear transformations and a ReLU activation constitutes the feed-forward neural network, where M is the output of multi-head self-attention part. f_1 and f_2 are linear transformation functions while W_1 and W_2 are the weights of both linear transformation functions, respectively.

$$Y_i = f_2(RELU(f_1(M, W_1))W_2) \tag{3}$$

The transformer encoder is composed of L encoder layers. The lower encoder layers capture phrase-level information based on Chinese characters embedding, the intermediate transformer encoder layers capture syntactic features and at the top transformer encoder layers capture sentence semantic features [31]. The last layer produces sequences of contextual embedding $T = (T_1, T_2, \ldots, T_{m-1}, T_m)$. In $T_i \in R^h$, h is the dimension of the hidden feature space to represent the sentence.

3.3   Convolution Layer

The representation layer output $T = (T_1, T_2, \ldots, T_{m-1}, T_m)$ is used as the input of the convolution layer. Different from the traditional context-independent sentence representation based on word2vec, T is a high-quality deep context-dependent sentence embedding, encoding syntactic and semantic information through transfer learning. In the sentence sequence '打开谷歌浏览器', $T_{i:j}$ represents the splice of the embeddings of the i-th token to the j-th token. A kernel $w_i \in R^{l \times d}$ is a window of l words to generate a new feature by the convolution operation. Multiple different sizes of convolution kernels are used to extract word-level key features since the output of representation layer is a character-based embedding sentence. The calculation for convolution layer is as Eq. (4), where f is a non-linear function and b is the bias term.

$$c_i = f(wT_{i:i+l-1} + b) \tag{4}$$

This kernel represents every possible word window in a sentence and generates a feature map $C = [c_1, c_2, \ldots, c_{m-l+1}]$. We obtain three different C values from convolution kernels, which include the heights 2, 3, and 4, respectively. Then, a maximum pooling operation calculates the maximum value $c^{max} = max\{C_i\}$ as the feature representation of the i-th kernel. The idea is to capture the most important feature for each feature map. Finally, the maximum pooling operation obtains three different c^{max} as different features.

3.4 Output Layer

Three results of maximum pooling operation are spliced together as feature embedding $X \in R^d$. The input of the output layer X maps the key feature representation from the higher-dimensional hidden space to the lower-dimensional label space through the fully connected operation in Eq. (5).

$$O = f(WX + b) \tag{5}$$

We obtain the output $O \in R^n$ mapping to the n redefined label categories space. In multi-category classification problem, a softmax function assigns each category a probability expressed as a decimal. These probabilities, expressed as decimals, add up to 1.0. We take the probability of the intent label corresponding to the Chinese sentence by softmax function. The probability calculation is as the Eq. (6).

$$p_i = Softmax(O^n) \tag{6}$$

4   Experiments and Results

4.1   Dataset

A publicly available dataset ECDT [26] from SMP conference is used in the experiment. This dataset includes 31 sub-categories involved 'chat' and 'task-driven' dialogue. The original training and developing dataset contain 3069 sentences with predefined labels and the test dataset contains 667 labeled sentences. Table 1 shows that the statistical information of the dataset.

Table 1. The statistics of the SMP ECDT dataset.

Data	Count
Training dataset	2299
Developing dataset	770
Testing dataset	667
Total of the dataset	3736
Average sentence length	8.38
Number of categories	31

4.2 Evaluation Metrics

We apply four widely used evaluation metrics in classification tasks to test our model including Precision, Recall, F1score (F1), and Accuracy. The calculations of the metrics are as Eq. (7)–(10).

$$Precision = \frac{TP}{TP + FP} \tag{7}$$

$$Recall = \frac{TP}{TP + FN} \tag{8}$$

$$F1 = \frac{2 \times Precision \times Recall}{Precision + Recall} \tag{9}$$

$$Accuracy = \frac{TP + TN}{TP + FP + TN + FN} \tag{10}$$

Precision is a measure of proportion of relevant predictions among all predictions, while Recall is a measure of proportion of relevant predictions among all relevant cases. F1 score can be considered as a harmonic average of model precision and recall. Accuracy is the percentage of a model that predicts correct outcomes.

4.3 Baseline Methods

Our model was compared with a set of baseline methods on the same ECDT dataset to explore the effectiveness for user intent classification based on pre-trained model. These baseline methods were chosen from the best performed models in the shared task and were described as follows:

1) CNN + domain template: The two-level system with domain template and Convolutional Neural Network combined expert knowledge and external knowledge to assist the neural network model to make multi-domain classifications.
2) Lib-SVM +1-gram + 2-gram + 3-gram + 4-gram: This method designed a multi-feature user intent classification system based on the Lib-SVM classifier while feature selection adopted 1-gram + 2-gram + 3-gram + 4-gram.
3) CNN + rules: Two kinds of user intent classification methods were proposed, in which one intent classification method was based on semantic rules and the other method was based on deep Convolutional Neural Network.
4) CNN + ensemble learning: The method consisted of multiple self-suppression residual convolution modules in series alternating with the maximum pooling layer and ensemble learning was used to train a group of classification models with the same structure.

5) LSTM + domain dictionary: The classification method adopted Long-Short Term Memory network (LSTM), while external domain knowledge table was constructed according to the limited number of sample training sets.
6) LR + character + POS tags + target words + semantic tags: The method used a traditional Logistic Regression with characters, non-key-noun part-of-speech tags, target words, semantic tags as feature expansion.

4.4   Parameter Settings

In the experiment, we applied a Chinese pre-trained RoBERTa with Whole Word Masking model and this pre-trained model was trained with the wiki + ext dataset containing 5.4 billion words. The parameter settings in the training were shown in Table 2.

Table 2. The parameter settings of this model.

	Parameter	Value
Pre-trained model	Hidden layers	12
	Hidden states	768
	Attention heads	12
	Vocab size	21128
Our model	Sentence length	128
	Batch size	32
	Optimization	Adam
	Learning rate	2e−5
	Dropout rate	0.1

4.5   The Results

To evaluate the hybrid neural network RBERT-C for user intent classification, an experiment on ECDT was conducted. Table 3 listed the comparison of performance for multiple different combinations to seek a better model for Chinese user intent classification. The training data contained 3069 sentences from original training data and developing data. The testing data contained 667 sentences with the predefined label for evaluating the performance of user intent classification.

Table 3. The performance of different methods on ECDT.

Methods	Precision	Recall	F1	Accuracy
Bert-Chinese	0.960	0.955	0.956	0.954
RoBERTa-Chinese-WWM	0.962	0.952	0.956	0.955
Bert-Chinese+CNN	0.957	0.956	0.955	0.952
RoBERTa-Chinese-WWM+RNN	0.964	0.952	0.957	0.955
RoBERTa-Chinese-WWM+CNN	0.968	0.956	0.960	0.960

Table 3 showed that the classification results using the pre-trained model achieved a F1 score of 0.96 and an accuracy of 0.96, among which the performance obtained by BERT+Roberta-Chinese-WWM + CNN was the best. The performance of the optimal model improved by 0.6% on accuracy than the model using Bert-Chinese only as initial parameters for transformer encoder. It illustrated that using Roberta-Chinese-WWM as pre-trained model could obtain better sentence representation and combining CNN as feature extraction layer could improve the performance of user intent classification.

The training data included 31 categories and there were some differences in distribution of categories. To evaluate the classification performance on each category with our proposed model, we computed the recall value of each category obtained by the optimized model, which achieved a recall of 1.000 on 17 sub-categories such as 'health', 'datetime', 'flight' etc., but lower recall on other categories, as shown in Table 4. For example, the category label of a query '你的QQ号是多少' (*What is your QQ number*) was 'chat' but it was wrongly classified into 'app'. The possible reasons of the results lie in: these categories were unevenly distributed; some categories were difficult to distinguish; these sentences have less contextual information.

Table 4. The performance in terms of recall on different categories.

Category	Recall	Category	Recall
App	0.723	Novel	0.941
Chat	0.875	Poetry	0.875
Contacts	0.989	Riddle	0.900
Cookbook	0.833	Schedule	0.958
Email	0.972	Translation	0.958
Match	0.952	Channel	0.902
Message	0.864	Weather	0.889

We set up several sub-datasets to study the relationship between the number of samples in the training dataset and model performance. The training datasets contained 1000, 1500, 2000, 2500 and 3000 sentences respectively. Figure 2 showed the results in terms of precision, recall, F1 score and accuracy on these training datasets.

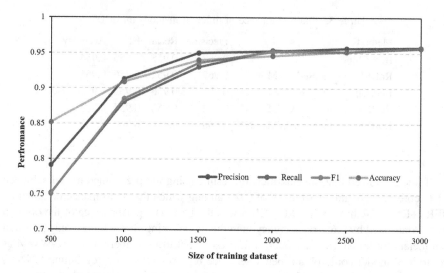

Fig. 2. The performance of our model with different sizes of training datasets.

Figure 2 showed how these evaluation metrics change. With the size of the train data changing, the performance of our model was improved. When the size of training data was less than 1500, the performance had improved dramatically. The performance was very close when the size of training data was between 2000 and 3000, indicating that our model was relatively stable.

Table 5. The F1 scores of different methods.

Methods	F1 score
CNN + domain template	0.899
Lib-SVM + 1-gram + 2-gram + 3-gram + 4-gram	0.912
CNN + rules	0.926
CNN + ensemble learning	0.929
LSTM + domain dictionary	0.941
LR + character + POS tags + target words + semantic tags	0.945
RBERT-C	0.960

The F1 score was shown in Table 5, in which there are six different methods to consist of the compared baselines applied for Chinese user intent classification. The comparison was conducted on the SMP2017 ECDT dataset containing 3736 sentences. These methods were all tested on the original SMP2017 ECDT test data. From the results, the neural networks obtained an optimal F1 score of 0.941 and the machine learning method LR with four kinds of features obtained a F1 score of 0.945. Our proposed model performed better than the baseline methods and achieved a F1 score of 0.960, a relative 1.5% improvement. The use of pre-trained language model could obtain context-dependency sentence representation and the CNN captured richer key features from representation layer. The result proved the effectiveness of our proposed model for Chinese user intent classification.

5 Conclusions

This paper proposed a model to represent sentence using Chinese-RoBERTa-WWM pre-trained model, and CNN with three different convolution kernels to extract more features. Experimental results have shown that our model outperforms these baseline methods and demonstrates the effectiveness of our proposed architecture for Chinese user intent classification. In future, in addition to incorporating additional information for sentence representation, we will explore the category imbalance problem for Chinese user intent classification.

Acknowledgements. This work was supported in part by the National Key R&D Program of China (2018YFB1003800, 2018YFB1003805), China University Production Innovation Research Fund Project (2018A01007), Philosophy and Social science planning Project of Guangdong Province (GD18CJY05), and National Natural Science Foundation of China (61772146, 61832004).

References

1. Jokinen, K., Mctear, M.F.: Spoken dialogue systems. Synth. Lect. Hum. Lang. Technol. **2** (1), 151 (2009)
2. Liu, J., Pasupat, P., Wang, Y., Cyphers, S., Glass, J.: Query understanding enhanced by hierarchical parsing structures. In: IEEE Workshop on Automatic Speech Recognition and Understanding, pp. 72–77 (2013)
3. Hao, T., Xie, W.X., Wu, Q.Y., Weng, H., Qu, Y.Y.: Leveraging question target word features through semantic relation expansion for answer type classification. Knowl.-Based Syst. **133**, 43–52 (2017)
4. Kim, Y.: Convolutional neural networks for sentence classification. In: Proceedings of the 2014 Conference on Empirical, pp. 1746–1751 (2014)
5. Xu, P., Sarikaya, R.: Convolutional neural network based triangular CRF for joint intent detection and slot filling. In: IEEE Workshop on Automatic Speech Recognition and Understanding, pp. 78–83 (2013)
6. Liu, B., Lane, I.: Attention-based recurrent neural network models for joint intent detection and slot filling. In: INTERSPEECH, pp. 685–689 (2016)
7. Hinton, G.E.: Learning distributed representations of concepts. In: Proceedings of the Eighth Annual Conference of the Cognitive Science Society, vol. 1, p. 12 (1989)
8. Devlin, J., Chang, M.W., Lee, K., et al.: BERT: pre-training of deep bidirectional transformers for language understanding. arXiv preprint arXiv:1810.04805 (2018)
9. Liu, Y., Ott, M., Goyal, N., et al.: RoBERTa: a robustly optimized BERT pretraining approach. arXiv preprint arXiv:1907.11692 (2019)
10. Cui, Y., Che, W., Liu, T., et al.: Pre-training with whole word masking for Chinese BERT. arXiv preprint arXiv:1906.08101 (2019)

11. Hao, T., Xie, W., Xu, F.: A WordNet expansion-based approach for question targets identification and classification. In: Sun, M., Liu, Z., Zhang, M., Liu, Y. (eds.) CCL/NLP-NABD -2015. LNCS (LNAI), vol. 9427, pp. 333–344. Springer, Cham (2015). https://doi.org/10.1007/978-3-319-25816-4_27

12. Xie, W., Gao, D., Hao, T.: A feature extraction and expansion-based approach for question target identification and classification. In: Wen, J., Nie, J., Ruan, T., Liu, Y., Qian, T. (eds.) CCIR 2017. LNCS, vol. 10390, pp. 249–260. Springer, Cham (2017). https://doi.org/10.1007/978-3-319-68699-8_20

13. Liu, J., Yang, Y., Lv, S., et al.: Attention based BiGRU CNN for Chinese question classifcation. J Ambient Intell. Hum. Comput. https://doi.org/10.1007/s12652-019-01344-9 (2019)

14. Johnson, R., Zhang, T.: Deep pyramid convolutional neural networks for text categorization. In: Proceedings of the 55th Annual Meeting of the Association for Computational Linguistics, pp. 562–570 (2017)

15. Liu, P., Qiu, X., Huang, X.: Recurrent neural network for text classification with multi-task learning. arXiv preprint arXiv:1605.05101 (2016)

16. Ravuri, S.V., Stolcke, A.: Recurrent neural network and LSTM models for lexical utterance classification. In: INTERSPEECH-2015, pp. 135–139 (2015)

17. Chen, Z., Tang, Y., Zhang, Z., et al.: Sentiment-aware short text classification based on convolutional neural network and attention. In: IEEE 31st International Conference on Tools with Artificial Intelligence (ICTAI), pp. 1172–1179 (2019)

18. Pappas, N., Popescu-Belis, A.: Multilingual hierarchical attention networks for document classification. arXiv preprint arXiv:1707.00896 (2017)

19. Guo, D., Tur, G., Yih, W., Zweig, G.: Joint semantic utterance classification and slot filling with recursive neural networks. In: IEEE Spoken Language Technology Workshop, pp. 554–559 (2014)

20. Liu, B., Lane, I.: Joint online spoken language understanding and language modeling with recurrent neural networks. arXiv preprint arXiv:1609.01462 (2016)

21. Chen, Q., Zhuo, Z., Wang, W.: BERT for joint intent classification and slot filling. arXiv preprint arXiv:1902.10909 (2019)

22. Tur, G., Hakkani-Tur, D., Heck, L.: What is left to be understood in ATIS? In: Spoken Language Technology Workshop, pp. 19–24 (2010)

23. Coucke, A., Saade, A., Ball, A., et al.: Snips voice platform: an embedded spoken language understanding system for private-by-design voice interfaces. arXiv preprint arXiv:1805.10190 (2018)

24. He, C., Chen, S., Huang, S., et al.: Using convolutional neural network with BERT for intent determination. In: International Conference on Asian Language Processing (IALP), pp. 65–70 (2019)

25. Khalil, T., Kielczewski, K., Chouliaras, G.C., et al.: Cross-lingual intent classification in a low resource industrial setting. In: International Joint Conference on Natural Language Processing, pp. 6418–6423 (2019)

26. Zhang, W.-N., Chen, Z., Che, W., Hu, G., Liu, T.: The first evaluation of Chinese human computer dialogue technology. arXiv preprint arXiv:1709.10217 (2017)

27. Xie, W., Gao, D., Ding, R., Hao, T.: A feature-enriched method for user intent classification by leveraging semantic tag expansion. In: Zhang, M., Ng, V., Zhao, D., Li, S., Zan, H. (eds.) NLPCC 2018. LNCS (LNAI), vol. 11109, pp. 224–234. Springer, Cham (2018). https://doi.org/10.1007/978-3-319-99501-4_19

28. Chen, S., Zheng, B., Hao, T.: Capsule-based bidirectional gated recurrent unit networks for question target classification. In: Zhang, S., Liu, T.-Y., Li, X., Guo, J., Li, C. (eds.) CCIR 2018. LNCS, vol. 11168, pp. 67–77. Springer, Cham (2018). https://doi.org/10.1007/978-3-030-01012-6_6
29. Zhao, W., Ye, J., Yang, M., et al.: Investigating capsule networks with dynamic routing for text classification. arXiv preprint arXiv:1504.00538 (2018)
30. Wu, Y., Schuster, M., Chen, Z., et al.: Google's neural machine translation system: bridging the gap between human and machine translation. arXiv preprint arXiv:1609.08144 (2016)
31. Jawahar, G., Sagot, B., Seddah, D., et al.: What does BERT learn about the structure of language? In: Annual Meeting of the Association for Computational Linguistics (2019)

Image Registration Algorithm Based on Manifold Regularization with Thin-Plate Spline Model

Anna Dai, Huabing Zhou[✉], Yulu Tian, Yanduo Zhang, and Tao Lu

School of Computer Science and Engineering, Wuhan Institute of Technology,
Wuhan 430205, China
zhouhuabing@gmail.com

Abstract. In this paper, we propose a new method based on manifold regularization technology with a thin-plate spline for image registration, which is used to remove the outlier by approximating the transformation function. Under a Bayesian framework, we use a latent variable to indicate whether a correspondence is an inlier that should satisfy a mapping function; and then, we formulate the problem as optimizing a posterior problem related to the mapping function. The initial correspondences discard some mismatching points because of the similarity constraint, which may contain important information; however, the manifold regularization (MR) term utilizes all of the feature points and preserve this information as a constraint. In addition, we use the thin-plate spline (TPS) model, which is composed of a global affine transformation and local bending function, to construct the transformation function. Finally, we obtain the solution using the expectation-maximization algorithm. Extensive experiments show that our method outperforms with other comparable state-of-art methods.

Keywords: Image registration · Manifold regularization · Thin-plate spline

1 Introduction

Image registration is an important preprocessing and has a long history. It can be used in a range of fields, such as computer vision, medicine, and remote sensing [1–4]. For instance, the real-world tasks involved in image registration in remote sensing are environmental monitoring, change detection, image mosaic, image fusion, and so on [5–9].

This work was supported in part by the National Natural Science Foundation of China under Grants 61771353 and 41501505, in part by Hubei Technology Innovation Project 2019AAA045, and in part by the Guangdong Provincial Department of Education 2017 "Innovation and Strong School Project" Scientific Research Project: Natural Science Characteristic Innovation Project 2017GKTSCX014.

© Springer Nature Singapore Pte Ltd. 2020
H. Zhang et al. (Eds.): NCAA 2020, CCIS 1265, pp. 320–331, 2020.
https://doi.org/10.1007/978-981-15-7670-6_27

The problem of image registration is nontrivial due to the wide variability of the imaging process [10]. For instance, the noise will make the feature descriptors inaccurate, and this error invariably gives poor results. Thus, researchers turn their goals to distinguish inliers from outliers in putative correspondences, where inliers should satisfy the transformation function but outliers not. RANSAC [11] iterates the steps of obtaining the set of inliers and approximating transformation function, until it convergence. Based on the Bayes framework, VFC [12–14] applies the vector field to remove outliers. The L_2E is adopted to estimate the image transformation in a robust way [15,16], and the spatial clustering can seek the inlier correspondences with multiple motion patterns [17], whereas GS [18] exploits the common visual pattern.

However, the problem of image registration is still a challenge because the putative correspondences, on which these methods are based, are obtained under the similarity constraint and do not contain all feature points. Sometimes, these mismatched points may contain very important information that could improve the performance of methods [19]. In our method, we introduce the manifold regularization technology (MR) which is composed of all feature points [20]. On the one hand, it can provide more information about the image that helps us to distinguish the inliers from outliers. On the other hand, the manifold regularization term can control the complexity of the transformation function. Besides, we apply the thin-plate spline model (TPS) into our method. This is a nonrigid transformation model that contains a global affine transformation and local bending function. More special, it set in a reproducing kernel hilbert space (RKHS) [21,22]. In our method, we use a latent variable to indicate whether this point is inlier and formulate the problem as maximum a posteriori problem solved by the Expectation-Maximization algorithm.

Our contribution in this paper includes the following three aspects: 1) For solving the problem of merely using a part of feature points to approximate the transformation function, we introduce the manifold regularization technology which is composed by the whole of feature points; 2) Combining the TPS model with manifold regularization, we propose a novel method for approximating transformation function, which can learn transformation from putative correspondences and remove outliers; 3) We name our method as MR-TPS and compare it with other methods on SYM [23] and Oxford [24] datasets, and the results show that our method is better than others.

In the rest of this paper, we get details of the proposed method in Sect. 2. Experimental results and conclusion will be presented in Sect. 3 and Sect. 4 respectively.

2   Method

In this paper, we use the $M \times 2$ matrix $X_M = (x_1, \ldots, x_m)$ and $N \times 2$ matrix $Y = (y_1, \ldots, y_n)$ to denote the position of feature points that come from the reference and sensed image, where $M > N$. By using the similarity constraint, we could obtain the correspondences $S = \{(x_n, y_n)\}_{n=1}^N$ and a $N \times 2$ matrix

$X = (x_1, \ldots, x_n)$. The goal is to find out the transformation function that can map point x_n to the corresponding point y_n, i.e. $y_n = f(x_n)$.

2.1 Problem Formulation

Considering the complex variables of image registration, we use a mixture model to fit the distribution of correspondences for solving this problem. Assuming that we have obtained the correspondences $S = \{(x_n, y_n)\}_{n=1}^N$, and they contain the inliers and outliers due to the error. We introduce a latent parameter $z_n \in \{0, 1\}$ to denote the nature of the putative correspondence, i.e.,

$$z_n = \begin{cases} 1, & \text{if } (x_n, y_n) \text{ is inlier} \\ 0, & \text{if } (x_n, y_n) \text{ is outlier} \end{cases} \tag{1}$$

and apply the Gaussian Mixture Model (GMM) to build the probability distribution. Where the inliers should follow the isotropic Gaussian distribution with zero mean and covariance $\sigma^2 I$, and I is the identity matrix; in contrast, the outliers are disordered and follow the uniform distribution. Thus, the probability model can be expressed as

$$\begin{aligned} p(y_n|x_n, \theta) &= \sum_{z_n} p(y_n, z_n|x_n, \theta) \\ &= p(z = 1)p(y_n|x_n, \theta, z_n = 1) \\ &\quad + p(z = 0)p(y_n|x_n, \theta, z_n = 0) \\ &= \frac{\gamma}{2\pi\sigma^2} e^{\frac{\|y_n - f(x_n)\|^2}{2\sigma^2}} + \frac{1 - \gamma}{a} \end{aligned} \tag{2}$$

where γ is the mixing coefficients specifying the marginal distribution over z, i.e., $p(z_n = 1) = \gamma$; a is the parameter of the uniform distribution which is dependent on the range of y_n; and $\theta = \{f, \sigma^2, \gamma\}$ contains all the unknown parameters needed to be solved. Under the assumption of independent and identically (i.i.d.) data distribution, our aim is to solve this likelihood function $p(Y|X, \theta) = \prod_{n=1}^N p(y_n|x_n, \theta)$, and obtain the set of parameters $\theta^* = \text{argmax}_\theta p(Y|X, \theta)$. To simplify the calculation, we add the log manipulation on the function, and obtain the following energy function:

$$E(\theta) = -\ln p(Y|X, \theta) = -\sum_{n=1}^N \ln p(y_n|x_n, \theta). \tag{3}$$

The EM algorithm is a general way to solve the problem with latent variables, and alternates with Expectation-step and Maximization step. In the expectation-step, we will fix the parameter set θ to estimate the range of the inlier; and in the maximization-step, we will update θ according to the current estimation. Following the standard notation and omitting the terms independent of θ. We can simplify the function and obtain the complete-data log-likelihood function like this:

$$Q(\theta, \theta^{old}) = -\frac{1}{2\sigma^2} \sum_{n=1}^{N} p_n \parallel y_n - f(x_n) \parallel^2 - \ln \sigma^2 \sum_{n=1}^{N} p_n$$
$$+ \ln \gamma \sum_{n=1}^{N} p_n + \ln(1-\gamma) \sum_{n=1}^{N} (1 - p_n) \tag{4}$$

where $p_n = P(z_n = 1 | x_n, y_n, \theta^{old})$ is a posterior probability indicating the intensity of (x_n, y_n) to be an inlier.

E-Step: We design a diagonal matrix P which is described as $P = \mathrm{diag}(p_1, \ldots, p_N)$; and base on the Bayes rule, we fix the current parameter set θ^{old} to calculate p_n for every correspondences. Thereby, p_n can be written as:

$$p_n = \frac{\gamma e^{-\frac{\parallel y_n - f(x_n)^2 \parallel^2}{2\sigma^2}}}{\gamma e^{-\frac{\parallel y_n - f(x_n)^2 \parallel^2}{2\sigma^2}} + \frac{2\pi\sigma^2(1-\gamma)}{a}} \tag{5}$$

M-step: In this step, we according the $\theta^{new} = \mathrm{argmax}_\theta Q(\theta, \theta^{old})$ to update the set of parameters. Taking derivatives of Q with respect to σ^2 and γ, and setting to zero. Then we obtain the function:

$$\sigma^2 = \frac{\mathrm{tr}((Y-F)^T P(Y-F))}{2 \cdot \mathrm{tr}(P)}$$
$$\gamma = \frac{\mathrm{tr}P}{N} \tag{6}$$

where $F = (f(x_1), \ldots, f(x_n))^T$. The form of transformation function f will be discussed in the later.

Until the EM iteration converges, we can distinguish the inlier and outlier through a predefined threshold t, where the set of inliers IS satisfy:

$$IS = \{(x_n, y_n) : p_n > \tau, n \in \mathbb{Z}\} \tag{7}$$

2.2   Manifold Regularization

The transformation function obtained before is based on the putative correspondences; but due to the similarity constraint, the putative correspondences have discarded mismatched points. There is no doubt that it will cause the loss of information, sometimes the mismatched points contain greatly important information. Thus, for overcoming this problem, we introduce the manifold regularization technique. The manifold regularization term is comprised of all feature points extracted from the reference image and it can exploit the intrinsic structure of the image as the posterior distribution [25–27]. Graph Laplacian is a discrete analog of the manifold Laplacian. Under the assumption that the feature points are drawn i.i.d. from the manifold, we use the weighted neighbor-hood graph for the data. $\{x_i\}_{i=1}^m$ is the feature points in reference image with edges (x_i, x_j) if and only if $\| x_i - x_j \|^2 < \varepsilon$, and assigning to edge (x_i, x_j) the weight

$$G_{ij} = e^{-\frac{1}{\varepsilon}\|x_i - x_j\|^2} \tag{8}$$

The graph Laplacian is the matrix L which is given by

$$L_{ij} = \Lambda_{ij} - G_{ij} \tag{9}$$

where $\Lambda = \text{diag}(\sum_{j=1}^n G_{ij})_{i=1}^m$ is the diagonal matrix whose i-th entry is the sum of the weights of edges leaving x_i. Let $F = (f(x_1), \ldots, f(x_m))^T$, the the manifold regularization term can be defined as

$$\Psi(f) = \sum_{i=1}^n \sum_{j=1}^n G_{ij}(F_i - F_j)^2 = \text{tr}(T^T L T) \tag{10}$$

where $\text{tr}(\cdot)$ denotes the trace. Under this constraint, we combine it with (4), the function in the M-step then rewrite as:

$$\hat{Q}(\theta, \theta^{\text{old}}) = Q(\theta, \theta^{\text{old}}) - \lambda \text{tr}(F^T L F) \tag{11}$$

where $\lambda > 0$ controls the degree of the influence that the manifold regularization have on the final result. There is no doubt that if λ is too large will cause a bad result, and if λ is too small, it cannot have obvious improvement in the algorithm. To obtain the solution of f, we can get the following energy function:

$$E(f) = \| P^{1/2}(Y - f(X)) \|_F^2 + \lambda \text{tr}(f(X_M)^T L f(X_M)) \tag{12}$$

where the first term is the putative correspondences loss, and the second is the regularization term. The manifold regularization term controls the complexity of transformation function.

2.3 TPS Model

Thin plate spline is a spline-based technique for smoothing and has closed-form solution; thus, we introduce the TPS model to approximate mapping function well. It is a non-rigid model which contains a global affine transformation and a local bending function and has the following form:

$$f(x) = A\bar{x} + g_x(x)$$
$$g_x(x) = \sum_{i=1}^m K(x, x_i)\bar{c}_i \tag{13}$$

where A is an 3×3 affine matrix, $\bar{x}$ is a homogeneous coordinates matricx defined as $(x^T, 1)$. The thin plate spline has a natural representation in terms of radial basis functions, thus we define K (a radial basis function) as a TPS kernel matrix.

$$\phi(r) = r^2 \log r$$
$$K(x, x_i) = \| x - x_i \|^2 \log \| x - x_i \| \tag{14}$$

We design $C = (\bar{c}_1, \ldots, \bar{c}_m)^T$ is a $M \times 3$ as bending coefficient matrix. As shown in [21], we lie the TPS model in the Reproducing Kernel Hilbert Space (RKHS) - $\mathcal{H}$; thus, the regularization term ϕ has the following form:

$$\phi(f) = \| f \|_{\mathcal{H}} \tag{15}$$

where $\| \cdot \|_{\mathcal{H}}$ denote the norm of space $\mathcal{H}$. And we can get the regularization term:

$$\| f_x \|_H^2 = \sum_{i=1}^{M} \sum_{j=1}^{M} < K(x_i, x_j)\bar{c}_i, \bar{c}_j > = \text{tr}(C^T K C) \tag{16}$$

Based on the solution of [21], the affine and bending calculation can be divided into two-part. After combining (13), (16) and the manifold regularization term, the error function can be written as:

$$\begin{aligned} E(A, C) = \| P^{1/2}(\bar{Y} - \bar{X}A^T - JKC) \|^2 &+ \lambda_1 \text{tr}(C^T K C) \\ &+ \lambda_2 \text{tr}((\bar{X}_M A^T)^T L(\bar{X}_M A^T)) \\ &+ \lambda_3 \text{tr}((KC)^T L(KC)) \end{aligned} \tag{17}$$

where $\bar{Y} = (\bar{y}_1, \ldots, \bar{y}_n)^T$, $\bar{X} = (\bar{x}_1, \ldots, \bar{x}_n)^T$, $\bar{X}_M = (\bar{x}_1, \ldots, \bar{x}_m)^T$ and $J = [I_{N \times N}, 0_{N \times (M-N)}]$. For get the solutions of A and C, we set

$$\tilde{Y} = P^{1/2}\bar{Y}$$
$$\tilde{X} = P^{1/2}\bar{X} \tag{18}$$

and we use QR decomposition on matrix $\tilde{X}$:

$$\tilde{X} = [Q_1 | Q_2] \begin{bmatrix} R \\ 0 \end{bmatrix} \tag{19}$$

where Q_1 and Q_2 are the orthogonal matrixes of $N \times N$ and $N \times (N - 3)$, respectively. R is a 3×3 upper triangular matrix. Assuming $C = Q_2\tilde{C}$, $\tilde{C}$ is a $(N - 3) \times 3$ matrix, then we obtain the following equation:

$$\begin{aligned} E(A, \tilde{C}) = \| Q_1^T \tilde{Y} - RA^T - Q_1^T P^{\frac{1}{2}} JKC \|^2 \\ + \| Q_2^T \tilde{Y} - Q_2^T P^{\frac{1}{2}} JKC \|^2 \\ + \lambda_1 \text{tr}(\tilde{C}^T Q_2^T K Q_2 \tilde{C}) + \lambda_2 \text{tr}((\bar{X}_M A)^T L(\bar{X}_M A)) \\ + \lambda_3 \text{tr}((KC)^T L(KC)) \end{aligned} \tag{20}$$

Minimizing the function (21), the result of $\tilde{C}$ and A are:

$$C = Q_2\tilde{C} = (S^T S + \lambda_1 K + \lambda_3 KLK)^{-1} S^T \tilde{Y}$$
$$A = (\tilde{Y} - SC)^T Q_1 R(R^T R + \lambda_2 \bar{X_M}^T L \bar{X_M})^{-T} \tag{21}$$

where $S = P^{\frac{1}{2}} JKC$. Until now, we summarize the parameters in M-step of the MR algorithm for the TPS model in Algorithm 1.

Algorithm 1: The MR algorithm for TPS model

Input: Correspondences set $S = \{(x_n, y_n : n \in N)\}$, whole feature points set X_M, parameters λ, ε, τ
Output: Inlier set IS
1 Initialize γ, $C = 0$, $A = I_{3 \times 3}$, $P = I_{N \times N}$;
2 Set a to the volume of the output space;
3 Initialize σ^2 by Eq. (6);
4 Construct the Kernel matrix K using the definition of K;
5 Compute the L of the reference image;
6 **repeat**
7 | E-step:
8 | Update P by Eq. (5);
9 | M-step:
10 | Update A, C by Eqs. (32);
11 | Update σ^2 and γ by Eq. (6);
12 **until** Q *converges;*;
13 The consensus set IS is determined by Eq. (7).

---

3   Experiment

In this part, we test the performance of our algorithm on the SYM [23] and Oxford datasets [24], and the experimental results are evaluated by precision and recall. The precision is equal to the ratio of inlier number and correspondence number, while the recall is defined as the ratio of inlier number and inlier contained in the putative correspondence set. We also compare our algorithm with RANSAC [11], VFC [12] and GS [18]. The experiments have been conducted on a laptop with 2.5-GHz Intel Core CPU, 8-GB memory, and MATLAB code. The details of datasets are described as following:

1) SYM: This dataset contains 46 pairs of the image, of which most are architectural scenes, that is seen as the challenging model for feature matching. The dataset contains a dramatic appearance change, like the photo to painting, or the obvious illumination change.

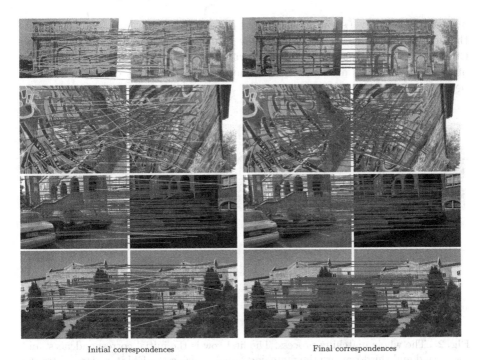

Initial correspondences   Final correspondences

Fig. 1. Qualitative matching results on typical image pairs. The first row is that a picture matches the painting, the second row is multi-view, and the third row is illumination change, and the last row is the quality of the image decreased. The left is the initial correspondences and the right is the final correspondences. The blue line denotes the right correspondence and the red denotes the wrong one. (Color figure online)

2) Oxford dataset: This dataset contains 8 groups of scenes. We obtain 210 pairs of images from this dataset. It contains the illumination change and view change. Besides, some images exist mosaic which degrades the quality of the image and increases the difficulty of matching.

As shown in Fig. 1, the blue lines denote the correct correspondences but red means wrong. The first column is the results of putative correspondences and the second column is our method's results. The first row is from the SYM dataset, we can find that the challenge is matching the photo to painting. The photo does not keep good quality, and the differences between imaging ways make registration difficult. The second, third, and fourth rows are the data of the Oxford dataset. The second row suffers from a big view change and that makes the shape of lines changed. The photos on third-row are taken at different times causing the intensity of light decreased and some details disappeared. And in the last row, the mosaic appears on left one image but the right not. For visibility, we merely show 50 pairs of correspondences, and our method gets good performances on these image pairs that it finds more right correspondences with removing the outliers. In Fig. 2, we show the matching process of our method. The first row is the original matching and the last is the result of our method. The second and third-row show the matching process during the EM algorithm and the intensity of the blue line with an arrow means the probability of this correspondence to be an inlier.

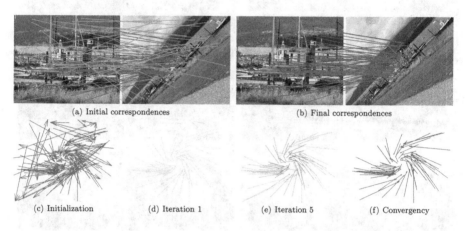

(a) Initial correspondences   (b) Final correspondences

(c) Initialization   (d) Iteration 1   (e) Iteration 5   (f) Convergency

Fig. 2. The whole matching process. The first row is the initial correspondences, and the second and third-row are the matching process during the EM iteration, where the blue line with an arrow denotes the correspondence and the intensity of blue means the possibility of this correspondence to be inlier. The last row is the final correspondence obtained by MR-TPS. (Color figure online)

The results of the four methods are summarized in Fig. 3, two columns are the cumulative distribution with precision and recall, respectively. According to the challenge of matching in the image pairs, we divide data into four groups corresponding to these four rows. From top to bottom, they are view change, the quality degrades, illumination change, and different imaging processes. We can observe that all methods get the best performance in the second row, i.e. quality degrade will not highly decrease the efficiency of methods. This is because there is no big change in view, and the transformation function is simple. The worst result on the fourth row, the challenge is matching the photo to painting. For example, there is a tower's photo match to the sketch of it, the sketch does not have some details that the photo has, like the texture of the building. GS always gets the worst results compared with other methods, and the low precision and high recall mean that it finds many inaccurate inliers. RANSAC has a better precision-recall tradeoff than GS, but it still can not get satisfying results. VFC is better than the other two methods and it can get more best results in the single image pairs, but our methods get higher mean value than it. The average precision-recall pairs are summarized in Table 1. As we can see, our method outperforms other methods in three groups except for the group of quality degrade.

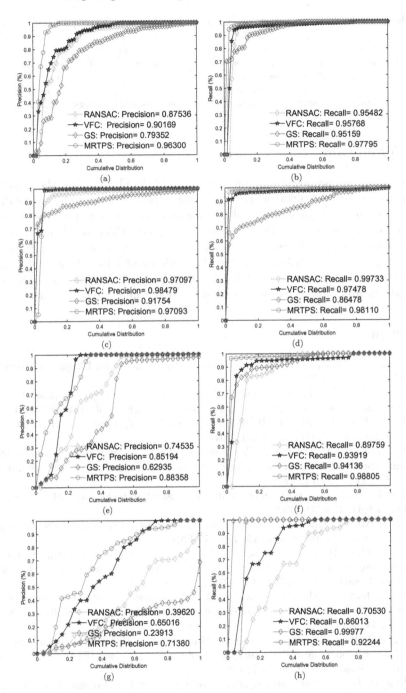

Fig. 3. Quantitative comparisons results of MRTPS, RANSAC [11], GS [18], and VFC [12] on datasets. We divide the 165 image pairs into four groups (top to bottom): view change, quality degrade, illumination change and different imaging process. And the two columns is the cumulative distribution function of precision and recall, respectively.

Table 1. Comparison of average precision-recall pairs on four groups of image pairs

	View change	Quality degrade	Illumination change	Different imaging processes
RANSAC	(87.54%, 95.48%)	(97.10%, **99.73%**)	(74.54%, 89.76%)	(39.62%, 70.53%)
VFC	(90.17%, 95.77%)	(**98.48%**, 97.48%)	(85.19%, 93.92%)	(65.02%, 86.01%)
GS	(79.35%, 95.16%)	(91.75%, 86.48%)	(62.93%, 94.14%)	(23.91%, 99.98%)
MR-TPS	(**96.30%, 97.80%**)	(97.09%, 98.11%)	(**88.36%, 98.81%**)	(**71.28%, 92.24%**)

4   Conclusion

In this paper, we proposed a new method called MR-TPS. We use the manifold regularization term to construct the constraint term, which can sufficiently utilize the information to achieve the goal of improving the performance. Considering the error of the feature detection method, the methods which merely use the local feature cannot satisfy the robustness of the matching. Moreover, the TPS model can be described as a global affine transformation and a local bending function can better fit the change in the global and local parts. The result shows that our method can find more correspondences while maintaining high precision.

References

1. Zitova, B., Flusser, J.: Image registration methods: a survey. Image Vis. Comput. **21**(11), 977–1000 (2003)
2. Ma, J., Jiang, X., Jiang, J., Zhao, J., Guo, X.: LMR: learning a two-class classifier for mismatch removal. IEEE Trans. Image Process. **28**(8), 4045–4059 (2019)
3. Ma, J., Zhao, J., Jiang, J., Zhou, H., Guo, X.: Locality preserving matching. Int. J. Comput. Vis. **127**(5), 512–531 (2019)
4. Ma, J., Jiang, J., Liu, C., Li, Y.: Feature guided Gaussian mixture model with semi-supervised EM and local geometric constraint for retinal image registration. Inf. Sci. **417**, 128–142 (2017)
5. Lisa Gottesfeld Brown: A survey of image registration techniques. ACM Comput. Surv. **24**(4), 325–376 (1992)
6. Ma, J., Ma, Y., Chang, L.: Infrared and visible image fusion methods and applications: a survey. Inf. Fusion **45**, 153–178 (2019)
7. Ma, J., Zhou, H., Zhao, J., Gao, Y., Jiang, J., Tian, J.: Robust feature matching for remote sensing image registration via locally linear transforming. IEEE Trans. Geosci. Remote Sens. **53**(12), 6469–6481 (2015)
8. Ma, J., Jiang, J., Zhou, H., Zhao, J., Guo, X.: Guided locality preserving feature matching for remote sensing image registration. IEEE Trans. Geosci. Remote Sens. **56**(8), 4435–4447 (2018)
9. Jiang, X., Jiang, J., Fan, A., Wang, Z., Ma, J.: Multiscale locality and rank preservation for robust feature matching of remote sensing images. IEEE Trans. Geosci. Remote. Sens. **57**(9), 6462–6472 (2019)
10. Moravec, H.P.: Rover visual obstacle avoidance. In: Proceedings of the International Joint Conference on Artificial Intelligence, pp. 785–790 (1981)
11. Fischler, M.A., Bolles, R.C.: Random sample consensus: a paradigm for model fitting with applications to image analysis and automated cartography. Commun. ACM **24**(6), 381–395 (1981)

12. Ma, J., Zhao, J., Tian, J., Yuille, A., Zhuowen, T.: Robust point matching via vector field consensus. IEEE Trans. Image Process. **23**(4), 1706–1721 (2014)
13. Ma, J., Zhao, J., Tian, J., Bai, X., Zhuowen, T.: Regularized vector field learning with sparse approximation for mismatch removal. Pattern Recognit. **46**(12), 3519–3532 (2013)
14. Zhao, J., Ma, J., Tian, J., Ma, J., Zhang, D.: A robust method for vector field learning with application to mismatch removing. In: Proceedings of IEEE Conference on Computer Vision and Pattern Recognition, pp. 2977–2984 (2011)
15. Ma, J., Zhao, J., Tian, J., Tu, Z., Yuille, A.L.: Robust estimation of nonrigid transformation for point set registration. In: Proceedings of the IEEE Conference on Computer Vision and Pattern Recognition, pp. 2147–2154 (2013)
16. Ma, J., Qiu, W., Zhao, J., Ma, Y., Yuille, A.L., Tu, Z.: Robust L_2E estimation of transformation for non-rigid registration. IEEE Trans. Signal Process. **63**(5), 1115–1129 (2015)
17. Jiang, X., Ma, J., Jiang, J., Guo, X.: Robust feature matching using spatial clustering with heavy outliers. IEEE Trans. Image Process. **29**, 736–746 (2020)
18. Liu, H., Yan, S.: Common visual pattern discovery via spatially coherent correspondences. In: Proceedings of the IEEE Conference on Computer Vision and Pattern Recognition, pp. 1609–1616 (2010)
19. Ma, J., Zhao, J., Jiang, J., Zhou, H.: Non-rigid point set registration with robust transformation estimation under manifold regularization. In: Proceedings of AAAI Conference on Artificial Intelligence, pp. 4218–4224 (2017)
20. Ma, J., Wu, J., Zhao, J., Jiang, J., Zhou, H., Sheng, Q.Z.: Nonrigid point set registration with robust transformation learning under manifold regularization. IEEE Trans. Neural Netw. Learn. Syst. **30**(12), 3584–3597 (2019)
21. Zhou, H., et al.: Image deformation with vector-field interpolation based on MRLS-TPS. IEEE Access **6**, 75886–75898 (2018)
22. Chen, J., Ma, J., Yang, C., Ma, L., Zheng, S.: Non-rigid point set registration via coherent spatial mapping. Signal Process. **106**, 62–72 (2015)
23. Hauagge, D.C., Snavely, N.: Image matching using local symmetry features. In: Proceedings of the IEEE Conference on Computer Vision and Pattern Recognition, pp. 206–213 (2012)
24. Mikolajczyk, K., et al.: A comparison of affine region detectors. Int. J. Comput. Vis. **65**(1), 43–72 (2005)
25. Zhou, H., Ma, J., Zhang, Y., Yu, Z., Ren, S., Chen, D.: Feature guided non-rigid image/surface deformation via moving least squares with manifold regularization. In: IEEE International Conference on Multimedia and Expo, pp. 1063–1068 (2017)
26. Tikhonov, A.N.: On the solution of ill-posed problems and the method of regularization. In: Proceedings of the Russian Academy of Sciences, pp. 501–504 (1963)
27. Belkin, M., Niyogi, P., Sindhwani, V.: Manifold regularization: a geometric framework for learning from labeled and unlabeled examples. J. Mach. Learn. Res. **7**(11), 2399–2434 (2006)

Optimal Control of Nonlinear Time-Delay Systems with Input Constraints Using Reinforcement Learning

Jing Zhu[1,2]([✉]) [iD], Peng Zhang[1], and Yijing Hou[1]

[1] College of Automation Engineering,
Nanjing University of Aeronautics and Astronautics, Nanjing, China
drzhujing@nuaa.edu.cn
[2] Key Laboratory of Navigation, Control and Health-Management Technologies
of Advanced Aerocraft, Ministry of Industry and Information Technology,
Nanjing University of Aeronautics and Astronautics, Nanjing, China

Abstract. In this paper, input-constrained optimal control policy for nonlinear time delay system is proposed in virtue of Lyapunov theories and adaptive dynamic programming method. The stability on delayed nonlinear systems is investigated based on linear matrix inequalities, upon which a sufficient stability condition is proposed. To implement the feedback control synthesis, a single neural network is constructed to work as critic and actor network simultaneously, which consequently reduces the computation complexity and storage occupation in programs. The weights of NN are online tuned and the weight estimate errors are proved to be convergent. Finally, simulation results are demonstrated to illustrate our results.

Keywords: Time delay system · Adaptive control · Reinforcement learning · Input saturation

1 Introduction

In large engineering applications, there are various input constraints such as saturation, dead-band, to name a few, in control systems which may cause performance degradation, response inaccuracy and even system instability [1–3]. The existence of these input limitations, however, in most cases are inevitable and sometimes are indeed essential. Hence it is of extraordinary importance to investigate sorts of control systems with input constraints.

As is known to all, input constraints pose tremendous technical difficulties to the performance optimization on feedback control systems due to the nonlinearity that saturation brought. To cope with this problem, a variety of optimal control strategies by means of *maximum principle* [1, 4], *dynamic programming* [4, 5],

This research was supported in part by the National Natural Science Foundation of China under Grant 61603179, and in part by the China Postdoctoral Science Foundation under Grant 2016M601805, 2019T120427.

© Springer Nature Singapore Pte Ltd. 2020
H. Zhang et al. (Eds.): NCAA 2020, CCIS 1265, pp. 332–344, 2020.
https://doi.org/10.1007/978-981-15-7670-6_28

convex optimization [6], etc., has been developed for different control systems with actuator saturation in recent work. On the one hand, the optimal control of linear systems with input constraints has been well studied by computing *Riccati equations* and amounts of classical results are readily applicable (see [7–10], and the references therein). On the other hand, the results on nonlinear optimal control strategy are relatively inadequate, not to mention ready results on nonlinear systems with input constraints. From a perspective of optimal control theories, the difficulty mainly comes from the Hamilton-Jacobi-Bellman (HJB) equation, that is nonlinear inherently, giving rise to no handy analytical solution. The fact is, analytical results for nonlinear HJB equations are derived for merely simple nonlinear systems with known system dynamics by adopting a variety of approximation linearization methods and addressing as its linear counterpart with the use of *state dependent Riccati equation* [11], *alternative frozen Riccati equation* [12], to name a few; while several useful adaptive control schemes are proposed for uncertain nonlinear systems with input constraints [2,13], etc.

The last two decades have witnessed significant achievements on nonlinear optimal control in virtue of the rapid development on machine learning technology of reinforcement learning (RL). RL concerns with how software agents ought to determine a policy such that the notion of cumulative reward designed according to the actual demand is maximized. The algorithm starts with an initial state $s^{(0)}$, the agent typically receive an reward $r^{(t)}$ at each time t, which in turn help to establish an optimal policy $\pi^{(t)}$ for the current state $s^{(t)}$ transferring to the future state $s^{(t+1)}$ toward the direction of maximizing the total reward. The idea of long-term versus short-term reward trade-off is also well-suited in the field of control. In this case, the reward and the optimal policy are replaced by the performance index and the iterative control law. an admissible control $u^{(0)}$, we could calculate the performance index $V^{(1)}$ by solving the HJB function. Then, the iterative control policy $u^{(1)}$ is updated based on the current scheme. Through multiple iterations, the accumulated value of performance index can be maximized. The RL based optimal control policies are of closed-loop state and consequently more efficient than those open-loop control strategies derived through modern control theories, which have to recompute system states all the time. The result establishes a foundation for studying the dynamic characteristics of vehicle and provides a reference for researching of the rule of shifting and constituting a control strategy.

Furthermore, the RL based optimal control policies are applicable to various nonlinear systems, no matter the parameters therein are known or not. Motivated by above two advantages, we address the optimal control problem on nonlinear time delay systems with input constraints with the use of RL techniques. Our work, however, is not trivial and faces three challenges. First, the input signal suffers saturation limit, consequently, it becomes considerably difficult to find the close-loop control policy [14]. Second, the nominal plant is with time delays in the input signal, which brings additional dynamics to the feedback system and renders the transfer function infinite-dimensional compared with delay-free systems [15]. Third, the programming or implementation of closed-loop controllers using RL techniques remains harsh given technical and practical restrictions [16].

In this paper, we investigate the optimal control problem on nonlinear time-delay systems with saturation actuator by using RL techniques. Our contribution is three-fold. First, we construct a new performance index function that bridges the gap between unstrained control strategies to constrained strategies. A novel input-constrained nonlinear optimal control policy is developed. In the next, casting the time delay problem into the robust control problem, we analyze the stability of the delayed nonlinear system with our obtained control scheme with the use of model transformation. By means of Lyapunov stability theories, sufficient delay-dependent stability condition by virtue of linear matrix inequalities (LMIs) is derived, along with an explicit, easy to compute delay margin. An approximate neural network (NN) is utilized, where NN weights are tuned such that the error between optimal and estimated performance function is sufficiently small, and convergent to zero as soon as possible. Unlike the majority of ADP controller synthesis that requires two NNs, our synthesis scheme uses a single NN instead. As such, the computation complexity and storage requirement are reduced. We also prove mathematically the convergence of tuning weights, indicating the approximated RL-based control policy actually approximate infinitely to the real control strategy. Numerical example shows the effectiveness of our RL-based control strategy.

The rest of this paper is as followed. In Sect. 2, input-constrained nonlinear optimal control policy and sufficient stability condition are obtained, respectively. Section 3 provides the specific optimal control design strategy using RL techniques, and the NN weights are proved to be convergent. Simulation results are presented in Sect. 4 to show the efficiency of our nonlinear optimal control approach. Finally, in Sect. 5 the conclusion is presented.

2   Input-Constrained Nonlinear Optimal Control

Let $(\cdot)^T$ denotes the transpose of a real matrix or a real function, while $(\cdot)^{-1}$ refers to the inverse of any real invertible matrix and function. For any real matrix A, $A \geq 0$, and $A > 0$ implies that A is non-negative definite and positive definite, respectively. $A \leq 0$, and $A < 0$ implies that A is non-positive definite and negative definite, respectively. For any real function $x(t)$, $\|x(t)\|$ denotes its L_2 norm, which is calculated by

$$\| x(t) \| = \left(x^T(t)x(t) \right)^{-1}.$$

For any multi-variable function $J(x, y)$,

$$\nabla J(x) = \frac{\partial J(x, y)}{\partial x}.$$

2.1   Nonlinear Optimal Control Policy

The following delayed nonlinear system is investigated in this paper

$$\dot{x}(t) = \Lambda x(t - \tau_1) + \sigma(t, x(t), x(t - \tau_2))\mathrm{sat}(u_t), \tag{1}$$

where Λ is a real matrix with approriate dimention, $\sigma(\cdot)$ is a nonlinear *Lipschitz continuous* function with the initial state $\sigma(t,0,0) = 0$. In other words, there is a positive real constant K such that for any $\xi_1, \xi_2 \in \mathrm{dom}\{f\}$, the following holds

$$|\sigma(\xi_1) - \sigma(\xi_2)| \leq K \|\xi_1 - \xi_2\|. \tag{2}$$

The original system has two constant but unknown delay parameters. τ_1 represents the dead time in the linear part of control systems, also termed as leakage delay in NN community. τ_2 refers to the dead time in the nonlinear part, which is known as transmission or processing delay as well. The system matrix A is real constant with appropriate dimension. The saturation function $\mathrm{sat}(u_t)$ is equal to

$$\mathrm{sat}(u_t) = \begin{cases} \mathrm{sign}(u(t))\Xi, & |u(t)| \geq \Xi \\ u(t), & |u(t)| < \Xi \end{cases} \tag{3}$$

Ξ herein is some positive number. Our optimal control goal is to seek for the best $u^*(t)$ satisfying the saturation constraint (3) to drive $x(t)$ to zero quickly, meanwhile minimize the function of $J(\cdot)$,

$$J\left(x(t), \mathrm{sat}(u_t)\right) = \int_t^\infty \mathscr{L}(x(s), \mathrm{sat}(u_s))ds. \tag{4}$$

where

$$\mathscr{L}(x(s), \mathrm{sat}(u_s)) = x(s)^T Q x(s) + U(u_t). \tag{5}$$

Q is symmetric and positive definite with $n \times n$ dimension, and $U(u_t)$ is some function which is quadratic and positive definite in terms of unconstrained input u_t,

$$U(u_t) = 2 \int_0^{u_t} \beta^{-1}(\hat{U}^{-1}v)\hat{U}Pdv. \tag{6}$$

Herein, v and P are positive real number and matrix, respectively, and $\hat{U}$ is a constant matrix satisfying

$$0 < \hat{U} < \bar{U}(x),$$

for all $x \in \mathrm{dom}\{U\}$.

$$\beta^{-1}(u_t) = [\nu^{-1}(u_{1t}), \nu^{-1}(u_{2t}), ...\nu^{-1}(u_{mt})]^{-T}.$$

$\nu(\cdot)$ is some function that is bounded, monotonic and odd, satisfying

$$|\varphi| < 1,$$
$$|\dot{\varphi}| < \delta, \ \delta \in R^+.$$

Consequently, we are led to the following constrained nonlinear control strategy.

Theorem 1. *Consider the delayed nonlinear system (1) with input constraint (3). The optimal control is achieved by using the following feedback policy*

$$u^*(t) = -\hat{U} \tanh\left(\frac{1}{2}(\hat{U}P)^{-1}\sigma^T \frac{\partial J^*}{\partial x}\right). \tag{7}$$

Moreover, the corresponding least cost function is

$$J^*(x(t), sat(u_t)) = \min \left(\int_t^\infty x^T Q x + 2 \int_0^{u_t} \beta^{-T}(\hat{U}^{-1}v)\hat{U}Pdv \right), \qquad (8)$$

where $\hat{U}, P, Q, \beta$ are as defined previously.

Proof. Let φ be the tangent function and P be a scaler diagonal matrix with

$$P = diag(\sigma_1, \sigma_2, ...\sigma_m),$$

with $\sigma_i \in R^+$, $i = 1, 2, ...m$. The Hamilton function is expressed as

$$\mathcal{H}(x, u, t) = \mathcal{L}(x(t), u(t)) + \nabla J^T(x)(\Lambda x + \sigma u). \qquad (9)$$

Recall the Bellman's principle [17]. $J(x(t), sat(u_t))$ is derived according to Eq. (4–6).

$$J(x(t), sat(u_t)) = \int_t^\infty x^T Q x + 2 \int_0^{u_t} \beta^{-T}(\hat{U}^{-1}v)\hat{U}Pdv, \qquad (10)$$

Thus the least cost function (8) is obtained. Taking $J^*(x(t), u(t))$ with respect to $u(t)$, the following is derived.

$$\frac{\partial J^*(x(t), sat(u_t))}{\partial u(t)} = 2 \tanh^{-1}(\hat{U}u)\hat{U}P + \sigma^T \frac{\partial J}{\partial x}$$
$$= 0$$

Consequently, we are led to Theorem 1. ∎

Remark 1. Theorem 1 tackles the first challenge on input-constrained nonlinear control policy. By constructing the function in (6), we build the bridge between unconstrained nonlinear control strategies to constrained ones.

Assumption 1. The initial constraint control $u^{(0)}$ is assumed as an admissible control [18] on Ω with respect to (6), which is denoted by $u \in \Phi(\Omega)$.

Theorem 2. *If the initial input $u^{(0)}$ is an admissible control, the iterative control law $u^{(i)}$ and the iterative performance index $V^{(i)}$ will converge to the optimal uniformly on Ω as $i \to \infty$.*

Proof. It has been proved in [19], we will not expand the redundant details due to the space limitation.

2.2   Delay-Dependent Stability Analysis

In this section, robust stability analysis on the input-constrained nonlinear time delay system is demonstrated and sufficient delay-dependent stability condition is developed relying on Lyapunov functionals.

In terms of thel control policy in Theorem 1. The delayed system (1) is expressed as

$$\dot{x}(t) = \Lambda x(t - \tau_1) - \sigma(t, x(t), x(t - \tau_2))\hat{U} \tanh \left(\frac{1}{2}(\hat{U}P)^{-1}\sigma^T \frac{\partial J^*}{\partial x} \right). \quad (11)$$

From a robust control perspective, we rewrite the system (1) by employing model transformation [20]

$$\frac{d}{dt}\left(x(t) + \Lambda \int_{t-\tau_1}^t x(u)du \right) = \Lambda x(t) - \sigma(t, x(t), x(t - \tau_2))\hat{U} \tanh \left(\frac{1}{2}(\hat{U}P)^{-1}\sigma^T \frac{\partial J^*}{\partial x} \right). \quad (12)$$

To simplify above function, we define a nonlinear function of $\Pi(\cdot)$ as

$$\Pi(t, x(t), x(t - \tau_2)) = -\sigma(t, x(t), x(t - \tau_2))\hat{U} \tanh \left(0.5(\hat{U}P)^{-1}\sigma^T \frac{\partial J^*}{\partial x} \right). \quad (13)$$

Equation (12) is thus expressed as

$$\frac{d}{dt}\left(x(t) + \Lambda \int_{t-\tau}^t x(u)du \right) = \Lambda x(t) + \Pi(t, x(t), x(t - \tau_2)). \quad (14)$$

Assumption 2. Let $t, \eta_1, \eta_2, \zeta_1, \zeta_2 \in R^n$ and $\alpha, \beta \in R^+$. The nonlinear function $\Pi(t, \eta, \zeta)$ satisfies

$$\| \Pi(t, \eta_1, \zeta_1) - \Pi(t, \eta_2, \zeta_2) \| \le \alpha \| \eta_1 - \eta_2 \|^2 + \beta \| \zeta_1 - \zeta_2 \|^2. \quad (15)$$

Upon Assumption 2, the system (1) has unique equilibrium under the condition that the system matrix Λ satisfies

$$\|\Lambda^{-1}\| \le (\alpha + \beta)^{-1/2}. \quad (16)$$

Based on Lyapunov stability theories, the following theorem is derived.

Theorem 3. *Assume matrix P is symmetric positive definite, I is the identity matrix, α, β, γ_1 and γ_2 are some real scalars. The delayed nonlinear system (1) with input constraint (3) is asymptotically stable if the following LMIs hold*

$$\Omega(\tau_1) < 0 \quad (17)$$

where

$$\Omega(\tau_1) = K + \tau_1 Q + G,$$
$$K = P\Lambda + \Lambda^T P + \gamma_1^{-1}P^2 + \alpha(\gamma_1 + \tau_1\gamma_2)I + \tau\Lambda^T P\Lambda,$$
$$Q = \Lambda^T P\Lambda + \gamma_2^{-1}\Lambda^T P^2\Lambda,$$
$$G = \beta(\gamma_1 + \tau_1\gamma_2)I.$$

Proof. In terms of Lyapunov stabilities, we firstly construct $V(t)$ as

$$V(t) = \Gamma_1(t) + \Gamma_2(t) + \Gamma_3(t),$$

where

$$\Gamma_1(t) = \left(x(t) + \Lambda \int_{t-\tau_1}^{t} x(v)dv\right)^T P\left(x(t) + \Lambda \int_{t-\tau_1}^{t} x(v)dv\right), \qquad (18)$$

$$\Gamma_2(t) = \int_{t-\tau_1}^{t} \int_{z}^{t} x^T(v)Qx(v)dvdz, \qquad (19)$$

and

$$\Gamma_3(t) = \int_{t-\tau_2}^{t} x^T(v)Gx(v)dv. \qquad (20)$$

Function $V(t)$ is positive definite for that matrices P, Q, and G are some positive matrix. In light of model transformation, the derivative of (18) yields to

$$\dot{\Gamma}_1(t) = 2\left(x(t) + \Lambda \int_{t-\tau_1}^{t} x(v)dv\right)^T P(\Lambda x(t) + \Pi(t, x(t), x(t-\tau_2))), \qquad (21)$$

where $\Pi(\cdot)$ is defined in (13). $\dot{\Gamma}_1(t)$ is further bounded by using the integral inequalities in [21],

$$\dot{\Gamma}_1(t) \leq x^T(t)Kx(t) + \int_{t-\tau_1}^{t} x^T(u)Qx(v)dv + x^T(t-\tau_2)Gx(t-\tau_2).$$

The detailed derivation of above inequality is similar to that in [22], thus is omit. In the similar manner, $\dot{\Gamma}_2$ and $\dot{\Gamma}_3$ are computed as

$$\dot{\Gamma}_2(t) \leq x^T(t)(\tau_1 Q + G)x(t) - \int_{t-\tau_1}^{t} x^T(u)Qx(v)dv,$$

$$\dot{\Gamma}_3(t) \leq -x^T(t-\tau_2)Gx(t-\tau_2).$$

Hence, the following inequality is obtained in terms of $\Omega(\tau_1)$

$$\dot{V}(t) \leq x^T(t)\Omega(\tau_1)x(t),$$

where $\Omega(\tau_1)$ is negative definite. As a result, $\dot{V}(t)$ is negative definite as well. This completes the proof. ∎

Remark 2. An explicit and computable delay margin of the nonlinear system (1) is $\tau_1 \leq \|\Lambda\|^{-1}$, which is obtained similarly in [22]. Note that this delay margin only depends on the delay parameter of the linear plant, which implies the nonlinear time delay can be arbitrary large under this circumstance. The delay margin is easy to compute bu conservative to some extent, which will be showed by our simulation in Sect. 4. By taking enhance integral inequalities and matrix scaling method, however, the conservatism can be reduced gradually. Since that it is not the main result in this paper, we shall left it to interested readers.

3   RL Based Optimal Controller Synthesis

In this section, we complete the RL based optimal controller synthesis by virtue of online ADP. The RL based controller synthesis is to learn the ideal optimal control policy in Theorem 1 iteratively by interacting with the environment. More specifically, we derive a novel NN-based control synthesis policy utilizing a single NN. Unlike those results using two NNs, the single NN performs as the acttor NN and critic NN at the mean time. As a result, there is large freedom in the NN structure.

Let $W_c, \hat{W}_c \in R^{l \times p}$ refers to the ideal and the real weight matrix of NN, respectively. $\varepsilon_a(x)$ denotes the approximation error, which is bounded by some positive real number ε_{aM}, i.e., $\|\varepsilon_a(x)\| \leq \varepsilon_{aM}$. $\varepsilon_r(x)$ denotes the residual error, which is assumed to be bounded by some positive real number ε_{rM}, i.e., $\|\varepsilon_r(x)\| \leq \varepsilon_{rM}$. $\Phi(x)$ denotes activation function, which is upper and lower bounded, i.e., $\Phi_m \leq \|\Phi\| \leq \Phi_M$. According to ADP theory, $J(x)$ is nearly approximated by

$$J(x) = W_c^T \Phi(x) + \varepsilon_a(x). \tag{22}$$

Consequently, the HJB equation is calculated by

$$\mathscr{H}(x(t), u(t), W_c) = W_c^T \bigtriangledown \Phi(x)\dot{x} + x^T Q x + 2 \int_0^u \tanh\left(\frac{1}{2}\left(\hat{U}v\right)^{-1} \hat{U} P\right) dv + \bigtriangledown \varepsilon_a(x)\dot{x}. \tag{23}$$

Thus, the nonlinear Hamilton function is equivalent to the following

$$W_c^T \bigtriangledown \Phi(x)\dot{x} + x^T Q x + 2 \int_0^u \tanh\left(\frac{1}{2}(\hat{U}v)^{-1}\hat{U}P\right) dv = -\bigtriangledown \varepsilon_a(x)\dot{x}$$
$$= \varepsilon_r. \tag{24}$$

Let $\hat{J}(x)$ denote the estimation of $J(\cdot)$

$$\hat{J}(x) = \hat{W}_c^T \Phi(x). \tag{25}$$

The estimation of Hamilton function is

$$\mathscr{H}(x(t), u(t), \hat{W}_c) = \hat{W}_c^T \bigtriangledown \Phi(x)\dot{x} + x^T Q x + 2 \int_0^u \tanh\left(\frac{1}{2}(\hat{U}v)^{-1}\hat{U}P\right) dv. \tag{26}$$

Let ε_E denote the estimation error of HJB equation

$$\varepsilon_E = \mathscr{H}(x(t), u(t), W_c) - \mathscr{H}(x(t), u(t), \hat{W}_c).$$

It is also equivalent to

$$\varepsilon_E = \mathscr{H}(x(t), u(t), \widetilde{W}_c),$$

where $\widetilde{W}_c = W_c - \hat{W}_c$ refers to the estimation error of NN weights.

To realize approximate optimal control strategy, the estimation error is required to be minimized. In other words, our objective is to minimize the function $\min\{E_o(t)\}$, where

$$E_o(t) = \frac{1}{2}\varepsilon_E^2. \tag{27}$$

To implement the RL based optimal control policy, a simple back propagation (BP) NN is employed [23,24]. The update law of NN weight is

$$
\begin{aligned}
\dot{\hat{W}}_c &= -\xi\frac{\partial\varepsilon_E}{\partial\hat{W}_c} \\
&= -\xi\frac{\theta\varsigma^2}{\theta^T\theta+\varsigma^2}\left(\theta^T\hat{W}_c + x^TQx + \hat{U}\right),
\end{aligned} \tag{28}
$$

where $\xi \in R^+$ is the learning rate, U is as defined in (6), $\theta = \nabla\Phi(\dot{x})$, and $\varsigma = \theta^T\theta + 1$. The approximate optimal control law is

$$\hat{u}^*(t) = -\hat{U}\tanh\left(\frac{1}{2}\left(\hat{U}P\right)^{-1}f^T(\nabla\Phi)^T\hat{W}_c\right). \tag{29}$$

As a result, we are led to the following theorem.

Theorem 4. *Assume $\hat{U}, P, f, \Phi, \hat{W}_c$ are as defined previously. The approximate optimal control of delayed nonlinear system (1) with input constraint (3) is achieved by feedback policy (29), and the update law of the NN weight is as described in (28).*

Remark 3. Noting that we merely utilize a single NN to approximate and evaluate the ideal control policy, other than using two NNs as in other work. By doing so, the computing complexity and the storage occupied of the program is largely reduced such that the synthesis can be done quickly. Therefore, the result is applicable to various engineering and social control systems. We will show the efficiency of the proposed policy by simulation results.

4 Simulation Results

Example 1. Consider the following second-order delayed nonlinear system

$$
\begin{aligned}
\dot{x}(t) &= \begin{bmatrix} -4 & 1.8 \\ 1.9 & -3 \end{bmatrix}x(t-\tau_1) \\
&+ \begin{bmatrix} 12\sin x_2(t-\tau_2) + 0.5e^{-x_2(t-\sigma)} \\ \sin x_1(t-\tau_2) + \cos x_1(t-\tau_2) + e^{-x_1(t-\sigma)}x_1(t-\tau_2) \end{bmatrix}u(t),
\end{aligned} \tag{30}
$$

with input constraint

$$|u(t)| \leq \varXi = 0.2. \tag{31}$$

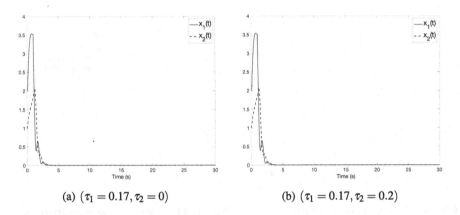

(a) $(\tau_1 = 0.17, \tau_2 = 0)$ (b) $(\tau_1 = 0.17, \tau_2 = 0.2)$

Fig. 1. Stable state response of the system (30) with input constraint (31).

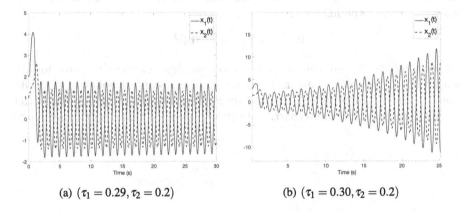

(a) $(\tau_1 = 0.29, \tau_2 = 0.2)$ (b) $(\tau_1 = 0.30, \tau_2 = 0.2)$

Fig. 2. unstable state response of the system (30) with input constraint (31).

Since that $\|A\| = 5.6$, an explicit delay margin can be computed by Remark 1, $\tau_1 \leq 0.18$. It implies that the nominal system (30) is stable under the condition that τ_1 is smaller than 0.18, no matter what is the value of τ_2. Based on Fig. 1(a) and Fig. 1(b), the state responses of system (30) converge to 0 at $\tau_1 = 0.17$, that is to say, the system is indeed stable. On the other side, the system becomes unstable when $\tau = 0.3$ as showed in Fig. 2. Noting that the system is still stable when $\tau = 0.29$ as showed in Fig. 2(a), we conclude that the delay margin in this paper is somehow conservative. It can be further improved by employing enhanced integral inequalities [25], matrix scaling methods [26], etc, which is not the key development herein, thus is left for interested readers.

In the next, we examine our online ADP-based optimal control algorithm. We consider the system (30) under input constraint (31) with the delay parameter being ($\tau_1 = 0.19$, $\tau_2 = 0.2$). Let $Q = 1$, $R = 1$.

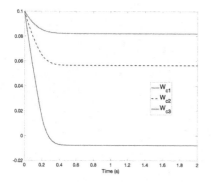

Fig. 3. NN weights trajectories.

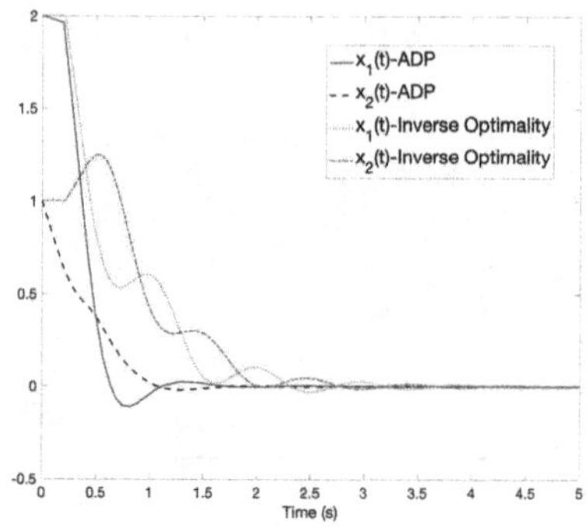

Fig. 4. The state responses comparision.

Let the learning rate ξ be 0.1. Based on Algorithm 1, we first initialize the NN weights. Assume the initial weights are

$$\hat{W}_c^{(1)} = [0.1 \quad 0.1 \quad 0.1]^T.$$

Figure 3 indicates that the NN estimation weights converge to zero quickly, within about 0.4 s. We compare the inverse optimal control policy in [27] with our RL based control policy. Figure 4 shows that our control policy has better performance than its counterpart. Both the states of $x_1(t)$ and $x_2(t)$ converge to

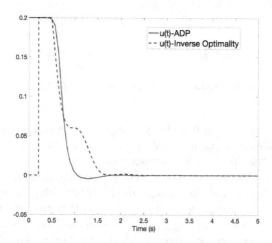

Fig. 5. The control trajectory of the nominal plant under different optimal control schemes.

zero within 2 s under our control policy, while the state convergence of the inverse optimal policy takes more than twice the time. Moreover, the state vibration of our result is better than that in inverse control policy as well. We also compare the control trajectory of system (30) with input constraint (31). As shown in Fig. 5, we observe that both the constrained control policies work, whereas, our RL based control synthesis costs half the time to achieve the convergence.

5 Conclusions

A novel optimal control policy for delayed nonlinear systems subject to input constraint is proposed in this paper using RL techniques. The distinct advantage is that our proposed nonlinear control strategy is input-constrained and can be realized easily by a single NN. Numerical simulations show the effectiveness of our results.

References

1. Cao, Y.-Y., Lin, Z.: Stability analysis of discrete-time systems with actuator saturation by a saturation-dependent Lyapunov function. Automatica **39**(7), 1235–1241 (2003)
2. Chen, M., Ge, S.S., Ren, B.: Adaptive tracking control of uncertain MIMO nonlinear systems with input constraints. Automatica **47**(3), 452–465 (2011)
3. Saberi, A., Lin, Z., Teel, A.R.: Control of linear systems with saturating actuators. IEEE Trans. Autom. Control **41**(3), 368–378 (1996)
4. Bensoussan, A.: Maximum principle and dynamic programming approaches of the optimal control of partially observed diffusions. Stoch.: Int. J. Probab. Stoch. Process. **9**(3), 169–222 (1983)
5. Himmelberg, C.J., Parthasarathy, T., VanVleck, F.S.: Optimal plans for dynamic programming problems. Math. Oper. Res. **1**(4), 390–394 (1976)
6. Angelov, V.G.: A converse to a contraction mapping theorem in uniform spaces. Nonlinear Anal.: Theory Methods Appl. **12**(10), 989–996 (1988)
7. Branicky, M.S., Borkar, V.S., Mitter, S.K.: A unified framework for hybrid control: model and optimal control theory. IEEE Trans. Autom. Control **43**(1), 31–45 (1998)
8. Ross, I.M., Karpenko, M.: A review of pseudospectral optimal control: from theory to flight. Annu. Rev. Control **36**(2), 182–197 (2012)
9. Shen, J., Lam, J.: On the algebraic Riccati inequality arising in cone-preserving time-delay systems. Automatica **113**, 108820 (2020)
10. Wu, Z., Li, Q., Wu, W., Zhao, M.: Crowdsourcing model for energy efficiency retrofit and mixed-integer equilibrium analysis. IEEE Trans. Ind. Inform. **16**(7), 4512–4524 (2019)
11. Manousiouthakis, V., Chmielewski, D.J.: On constrained infinite-time nonlinear optimal control. Chem. Eng. Sci. **57**(1), 105–114 (2002)
12. Huang, Y., Lu, W.-M.: Nonlinear optimal control: alternatives to Hamilton-Jacobi equation. In: Proceedings of 35th IEEE Conference on Decision and Control, vol. 4, pp. 3942–3947. IEEE (1996)

13. Li, R., Chen, M., Qingxian, W.: Adaptive neural tracking control for uncertain nonlinear systems with input and output constraints using disturbance observer. Neurocomputing **235**, 27–37 (2017)
14. Kurtz, M.J., Henson, M.A.: Feedback linearizing control of discrete-time nonlinear systems with input constraints. Int. J. Control **70**(4), 603–616 (1998)
15. Gu, K., Chen, J., Kharitonov, V.L.: Stability of Time-Delay Systems. Springer, Cham. https://doi.org/10.1007/978-1-4612-0039-0
16. Kamalapurkar, R., Rosenfeld, J.A., Dixon, W.E.: Efficient model-based reinforcement learning for approximate online optimal control. Automatica **74**, 247–258 (2016)
17. Dierks, T., Thumati, B.T., Jagannathan, S.: Optimal control of unknown affine nonlinear discrete-time systems using offline-trained neural networks with proof of convergence. Neural Netw. **22**(5–6), 851–860 (2009)
18. Liu, D., Wei, Q., Wang, D., Yang, X., Li, H.: Adaptive Dynamic Programming with Applications in Optimal Control—Value Iteration ADP for Discrete-Time Nonlinear Systems. AIC. Springer, Cham (2017). https://doi.org/10.1007/978-3-319-50815-3
19. Abu-Khalaf, M., Lewis, F.L.: Nearly optimal control laws for nonlinear systems with saturating actuators using a neural network HJB approach. Automatica **41**(5), 779–791 (2005)
20. Zhu, J., Chen, J.: Stability of systems with time-varying delays: an $\mathscr{L}_1$ small-gain perspective. Automatica **52**, 260–265 (2015)
21. Seuret, A., Gouaisbaut, F., Fridman, E.: Stability of systems with fast-varying delay using improved Wirtinger's inequality. In: 2013 IEEE 52nd Annual Conference on Decision and Control (CDC) (2013)
22. Zhu, J., Hou, Y., Li, T.: Optimal control of nonlinear systems with time delays: an online ADP perspective. IEEE Access **7**, 145574–145581 (2019)
23. Li, Y.U.: Optimal guaranteed cost control of linear uncertain system: an LMI approach. Control Theory Appl. **3** (2000)
24. Zhang, M.-Y., Lu, Z.-D.: Lyapunov-based analyse of weights' convergence on back-propagation neural networks algorithm. Mini-Micro Syst. **1**, 93–95 (2004)
25. Reddy, K.N.: Integral inequalities and applications. Bull. Aust. Math. Soc. **21**(1), 13–20 (1980)
26. Cohen, M.B., Madry, A., Tsipras, D., Vladu, A.: Matrix scaling and balancing via box constrained newton's method and interior point methods. In: 2017 IEEE 58th Annual Symposium on Foundations of Computer Science (FOCS) (2017)
27. Rodriguez-Guerrero, L., Santos-Sanchez, O., Mondie, S.: A constructive approach for an optimal control applied to a class of nonlinear time delay systems. J. Process Control **40**, 35–49 (2016)

Disaggregated Power System Signal Recognition Using Capsule Network

Liston Matindife[1] , Yanxia Sun[1] , and Zenghui Wang[2]([📧])

[1] Department of Electrical and Electronic Engineering Science,
University of Johannesburg, Auckland Park 2006, South Africa
[2] Department of Electrical and Mining Engineering,
University of South Africa, Florida 1710, South Africa
wangzengh@gmail.com

Abstract. Non-intrusive-load-monitoring (NILM) is normally based on power series analysis. In the load classification stage we use an image based deep convolutional neural network (DCNN) which is modelled on the biological visual cortex thereby achieving extremely high levels of object recognition and classification. However, the downsize to the DCNN is the requirement of a large image training dataset, translational invariance and loss during max pooling of information captured in small signal perturbations. In this paper to reduce the training dataset, provide appliance signature equivariance recognition and replace max pooling with routing by agreement for improved NILM recognition and classification we use Hinton's capsule network (CapsNet). Disaggregated appliance current, real power and power factor signals are converted to two-dimensional (2D) images and then complementary fused together for increased recognition accuracy before final input into the CapsNet. We implement the Discrete Wavelet Transform (DWT) since it is able to transform within a large frequency band portfolio and fuses well low pixel images. By using image fusion technique we show that for only fifteen images per appliance and with no data augmentation we are able to achieve average prediction accuracies of up to 93.75% and hence consolidate the validity of the CapsNet in the NILM recognition and classification scheme for limited data memory and improved recognition.

Keywords: Non-intrusive-load-monitoring · Disaggregated power systems signal · Deep learning · Signal to 2D image transformations · Image fusion · Capsule network

1 Introduction

Non-intrusive-load-monitoring (NILM) [1] has typically been based on appliance power (time) series signal disaggregation and classification. We note the large data volume required by deep learning neural networks (DLNNs) which is normally in the form of very long power series signal lengths. In this paper we compact the features in the power series signal into image form. We then use deep

© Springer Nature Singapore Pte Ltd. 2020
H. Zhang et al. (Eds.): NCAA 2020, CCIS 1265, pp. 345–356, 2020.
https://doi.org/10.1007/978-981-15-7670-6_29

learning convolutional neural networks (DCNNs or ConvNets) that have achieved extremely high levels of object recognition as exemplified in computer two dimensional (2D) vision [2,3], to focus on improving only the NILM classification strategy. In literature, signal to 2D image transformation has been achieved through a number of techniques that include ThemeRiver, Horizon Graphs, Short-Time Fourier Transform (STFT), Moving Average Mapping (MAM), Recurrence Plots (RP), Gramian Angular Fields (GAFs) and Markov Transition Fields (MTF) [4–8]. GAFs have a special property of bijection where there is a one-to-one relationship between each time series value with each generated 2 D image [5] element within the normalized range [0–1]. A function is bijective if it is defined as in Definition 1 [9]. Also, the temporal dependency of gramian angular felds (GAFs) gives a direct correlation between the time series, polar, and image representations of the GAFs [5]. The time series can inversely be reconstructed with much accuracy from the GAF images.

Definition 1. *Given* $f: A \rightarrow B$

1. *f is one-to-one (Short hand 1-1) or is injective if preimages are unique. In this case* $(a \neq b) \rightarrow (f(a) \neq f(b))$.
2. *f is onto or surjective if every* $(y \in B)$ *has a preimage. In this case, the range of f is equal to the codomain.*
3. *f is bijective if it is surjective and injective (one-to-one and onto).*

In this paper three different time series electrical signals namely the current (I_rms), power (Watt) and power factor (PF) have been disaggregated for each individual appliance from the aggregate signal. We then transform using GAFs the I_rms, Watt and PF time series signals into three separate images. We perform a number of disaggregation's and transformations to have a number of images per parameter per appliance. For each appliance three parallel 2 D I_rms, Watt and PF images can be input into three independent DLNNs that are subsequently concatenated together.

However, by alternatively employing complementary fusion [10,11] method we considerably increase the accuracy of our model and at the same time reduce the number of DLNNs to one. From the high number of image fusion techniques available in literature that include the PCA and the Pyramid, the Discrete Wavelet Transform (DWT) is of particular interest in this paper [11]. The DWT is able to transform within a large frequency band portfolio and fuses well low pixel images [11]. Straight DWT can result in a decrease in the peak-signal-to-noise ratio (PSNR), hence modified DWTs are developed to improve on this PSNR [12]. The sole PCA technique is either modified or combined with other methods to reduce the spatial distortion [13].

CNN based DLNNs suffer loss of detail through max pooling and require a very large training dataset [6,14–16]. The capsule network (CapsNet) that was proposed by Hinton et al. in 2017 [17] addressed the short comings of the CNN based network to achieve improved recognition. The CapsNet uses a reduced dataset for training, retains all input data through the network (dynamic routing against max pooling), and sufficiently reproduces the pose relationships among the various elements within the image [18] from input to output providing what is known as "equivariance".

The remaining sections are structured as follows. Section 2 describes the data set creation from the measured data. Section 3 shows the generation of the 2D images from the profiles, whilst Sect. 4 gives a review of the capsnet. Section 5 details the proposed NILM CapsNet based architecture. Section 6 gives the results and Sect. 7 gives the conclusion.

2   Dataset Creation

The data was acquired at a sampling rate of 1 Hz by using a 600 V 20 A Tektronix PA1000 Power Analyzer [19] in three experiments in-house numbered two to four. Data handling, time series image generation and neural network simulation was done in an environment of the python 3.6.3 64 bit (AMD64) on win32 software with keras 2.2.4, tensorflow 1.5.0 backend, numpy 1.17.0, pandas 0.20.3, pyts 0.8.0, scipy 1.3.1, and scikit-learn 0.20.1 packages, on an Intel(R)Celeron(R) CPU @ 1.60 GHz 4.00 GB Ram 64bit HP 250 G4 Notebook PC laptop with Windows 10. Experiment 2 was made up of a Hisense refrigerator, an SMW20E Salton microwave oven, a Philips 5 W (60 W) LED lamp, a Radiant 12 W (100 mA) CFL lamp and a Radiant 14 W (110 mA) CFL lamp. Experiment 3 was composed of a Dell Optiplex 7010 desktop computer, an Estia quick heating two solid plate stove, a logic kettle, and two Philips 5 W (60 W) LED lamps. Finally Experiment 4 was made up of three Philips 5 W (60 W) LED lamps, and a Radiant 5.5 W B22 LED lamp. Figures 1, 2 and 3 show the complete aggregate I_rms appliance test operation cycles for Experiment 2, 3 and 4 respectively. An example of the signal to be recognized from Fig. 2 Experiment 3 composite profile is shown in Fig. 4. Our study is however, only limited to the classification of the disaggregated (process not shown) Hisense refrigerator I_rms, Watt and PF load signals in Experiment 2 from the aggregate load profile shown in Fig. 2.

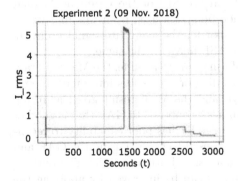
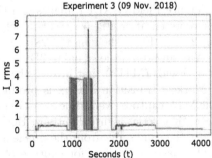

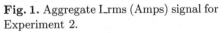

Fig. 1. Aggregate I_rms (Amps) signal for Experiment 2.

Fig. 2. Aggregate I_rms (Amps) signal for Experiment 3.

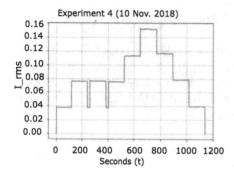

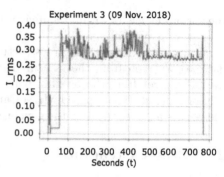

Fig. 3. Aggregate I_rms (Amps) signal for Experiment 4.

Fig. 4. Disaggregated I_rms (Amps) signal of the desktop Experiment 3.

3   2D Images and Fusion

3.1   Gramian Angular Fields

The precursor to the generation of GAFs is the Gram Matrix (Gramian or Metric). The resultant of the operations on a combination of any number of vectors for the entire vector space is contained in this matrix [20]. We first encode the appliance signals to GAF as done by Wang and Oates [5,21], rescaling the signals $X = x_1, x_2, \ldots\ldots\ldots, x_n$ to lie in the ranges of -1 to 1or 0 to 1. After rescaling, the time series is converted to polar coordinate, where the value is the angular cosine and the time stamp (t_i) is the radius r. ϕ is the polar coordinates angle and N is the regularization constant factor for the span of the polar coordinate system [5,8]. On the polar plot advancing time scale concentric circles are accompanied by time scale values that warp through the various angular points. The angular limit for the scale $[0, 1]$ is $[0, \pi]$, and for $[-1, 1]$ is $[0, \pi/2]$ [5]. A Gramian Matrix is realized from the polar coordinate vectors. Either, Gramian Angular Summation Field (GASF) or the Gramian Angular Difference Field (GADF) images can then be synthesized from this matrix [8]. For representation Fig. 5 gives the polar plot for the aggregate signal profile in Fig. 2, with Fig. 6 showing the 28 × 28 GASF image of the same.

We can extend the signal to 2D transformations to include, the Markov Transition Fields (MTFs) that involve the encoding of time series into quartile bins. A Markov Transition Matrix is produced and the MTF result given as in [5]. The MTF captures well, time series dynamics as opposed to GAF that is good at static time series transformations. However, MTF has poor capability to reconstruct the time series from the image as opposed to GAF. Fusing the GAF and MTF images will allow for capture of more signal attributes [5]. The outlook including contrast of poorly outlined images can be improved by applying the logarithmic transform to the image values [4].

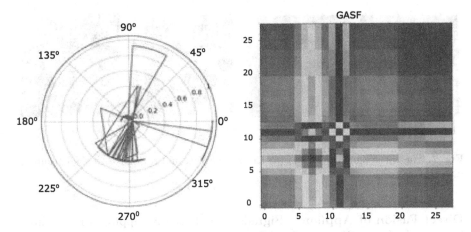

Fig. 5. Experiment 3 I_rms (Amps) aggregate polar plot.

Fig. 6. Experiment 3 I_rms (Amps) aggregate GASF image.

3.2   Image Fusion Technique

Wavelets are able to analyze signal changes and capture image characteristics more accurately than Fourier Transforms. Wavelets are localized in time and frequency (translation and scale) whereas Fourier Transforms are composed of sinusoidal parts that are only frequency dependent. Equation (1) gives the continous wavelet transform (CWT) of a function f over a scale s and time t by an analyzing function $\overline{g}(s(t - \sigma))$. The CWT is defined over the global time-scale plane $\Re \times (\Re \backslash \{0\})$. Naturally the discrete wavelet transform (DWT) is obtained when sampling is done on the CWT [22].

$$(W_g f)(t, s) = |s|^{1/2} \int_R f(\sigma) \overline{g}(s(t - \sigma)) d(\sigma) \tag{1}$$

DWT transforms premised on high pass filtering decompose the images into low and high frequency components. Using what are called decimated or undecimated algorithms further low level sub images are generated containing frequency components representing their unique characteristics. These low level sub images are then fused together to form a new image with combined features [23].

Experiment Fused Images. The three refrigerator images of I_rms, Watt and PF shown in Fig. 7 are fused together through the use of the wavemenu (Wavelet Toolbox) in Matlab R2015a or the PyWavelets software package in python. As can be seen the resultant (fusion) image in Fig. 7 contains all the marked features of each preceding image. The fusion image becomes the sole input into the Capsnet.

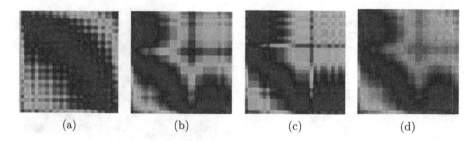

Fig. 7. The three signal images of (a) I_rms, (b) Watt and (c) PF, with their resultant fused image (d) Fusion.

Direct Fusion of Appliance Signals. An alternative approach is the direct fusion of the I_rms, Watt and PF signals together (Information Fusion) [24]. This achieves the best raw signal representation of each appliance characteristics and then perform a time series to image conversion of the resultant fused signal.

4   CapsNet Structure

The basic capsnet as shown in Fig. 8 has a convolutional layer (ReLU Conv1) with 256 kernels size 9×9 and stride 1. A second convolution layer follows with 256 kernels size 9×9 and stride 2. In the third layer are the Primary Capsules (PrimaryCaps) where the $6 \times 6 \times 256$ 1D scalars produced are reshaped to $6 \times 6 \times 32$ 8D vectors. A total of 1152 8D vectors (capsules) are output from the PrimaryCaps. In the PrimaryCaps all the constituent parts of an image are clearly articulated. The next layer known as the Digit Capsules (DigitCaps) is comprised of 10 capsules each capable of outputting a maximum of 16 vectors [15,17].

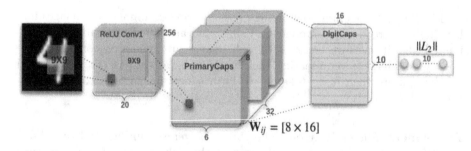

Fig. 8. The basic capsnet structure [15,17].

Each DigitCaps contains relative magnitudes of all the image information per class. Finally fully connected layers will classify the DigitCaps outputs. Here, the network output is the longest vector. Dynamic routing by agreement allows lower level neurons to freely choose which higher level neurons to work with so as to achieve the most positive result. In this case this is represented by routing of information from the PrimaryCaps to the DigitCaps vector capsules. The output from the PrimaryCaps is rescaled to between 0 and 1 through a squash function [15,17]. Within the DigitCaps, an improvement in the pose encoding, can be achieved by incorporating a reconstruction loss during network training. This is implemented through a reconstruction regularizer of the digit capsules and fully connected layers. The capsule network loss function is due to a component associated with classifying a number of output digits in an image (Margin Loss), and a component due to the reconstruction loss [15,17,25]. The Margin loss per digit capsule is given as:

$$L_c = T_c max(0, m^+ - ||V_c||^2) + \lambda(1 - T_c)max(0, ||V_c|| - m^-)^2 \qquad (2)$$

where $T_c = 1$ if digit c exists in the image and $m^+ = 0.9$ and $m^- = 0.1$, v_c is capsule representing digit c, and $\lambda = 0.5$.

The reconstruction loss on the other hand is given as [15,17,25]:

$$L_r = \sum_{i=0}^{n}(y_i - h(x_i))^2 \qquad (3)$$

where y_i is the input image and x_i is the reconstruction.

$$TotalLoss = \sum_{c} L_c + 0.0005 L_r \qquad (4)$$

The effect of L_r is minimized by the factor 0.0005.

5   Proposed NILM CapsNet Structure

The proposed structure is shown in Fig. 9. Each of Experiment 2 and Experiment 3 has a total of five (5) appliances to be recognized. However, in Experiment 4 the total number of LED lamps to be recognized is four (4). Consequently the total number of possible digit capsule outputs in any developed model is five (5), and the total number of trained independent networks is three (3) for the three experiments. However, our focus here is one network for Experiment 2. The input image is the fused image of Fig. 7.

5.1   Methodology

In the proposed structure the maximum resized dimensions of the input images was 81 × 81. This was as an attempt to increase further the depth of the initial feature extraction part of the model. However, the model was also trained with an image dimension of 28 × 8 this time without convolutional networks ConvA and ConvB. In the second case the network simplifies to the basic Capsnet of Hinton [17]. The primary capsules and the digitcaps have been preserved as

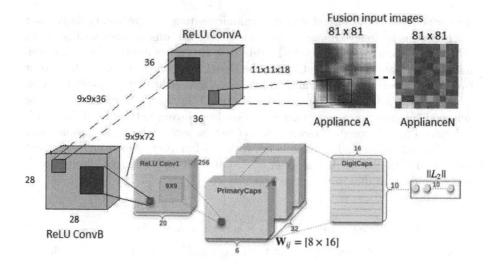

Fig. 9. Proposed NILM CapsNet structure.

original [17] so as to evaluate their classification and reconstruction (prediction) capabilities for the appliance dissagregated signals.

During model training the Adam's optimizer learning rate was varied from 0.0001 to 0.1 with sample batch size of 2 and the performance of the model observed. Filter numbers used in the models varied from 16, 18, 32, 36, 64, 72, 128 and 256. The size of the filters were selected to include 11 × 11, 3 × 3, 5 × 5 and 9 × 9 sizes. The depth of the initial feature extraction part and the filter numbers have been limited due to the slow CPU platform used to train the model. Train to Test dataset ratio is 3:1.

6   Results

6.1   Capsule Model with Reconstruction Capabilities

The reconstruction rate varies from 87.5% to 100% for the various filter sizes and numbers, a good value considering that the reconstructed images should exactly be the same as the label images as shown in Fig. 10. However, in some cases that we achieved 100% prediction the reconstructed images where not exactly the same as the label images. In this case there is need to fine tune the models. The positive aspect is that the model achieves very good classification for the extremely limited datasets that have been used for a total sample size of just fifteen fused images per appliance. In essence we have a total of forty five images per appliance, thanks to fusion. Our coding approach is borrowed from the AILearner CapsNet coding example in [26]. Model performance is evaluated and the predicted image reconstructed and assigned the correct label value.

Figure 11 shows that the capsnet model can achieve stable operation for an input image of 81 × 81, an adams learning rate of 0.001 and a network

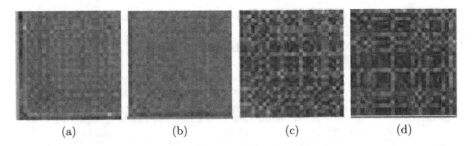

(a)   (b)   (c)   (d)

Fig. 10. (a) True label: X, (b) reconstruction: X, (c) true label: Y, and (d) reconstruction: Y fused gasf images.

specification exactly as the one in Fig. 9. When the input is a 28 × 28 image with a learning rate of 0.001 and a network structure that excludes ConvA (Model 4) in Fig. 9 we obtain good mean absolute error and root mean square error as shown in Fig. 12. During training the validation accuracy trend of the 28 × 28 input model was also fairly stable, however, with short lived instability around epochs 150 to 200. The 28 × 28 input model validation accuracy (diagram not included) peaks at an average value of 93.75%. However, Model 4 which is the best performing model configuration in our case achieves 100% prediction accuracy on test data.

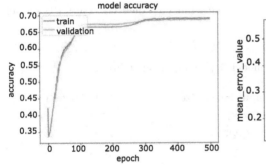

Fig. 11. Validation accuracy for the 81 × 81 input image Capsnet.

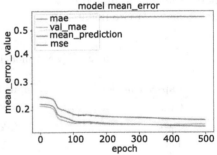

Fig. 12. Prediction errors for the 28 × 28 Model 4 input image Capsnet.

Summary of Different CapsNet Configurations. Six model configurations where evaluated and the four (One, Two, Three and Four) best performing configurations are given in Table 1. The number of filters is denoted by NF and FS denotes the kernel dimensions in each case. The code used to estimate the running time (computation cost) of each algorithm is given in the blog Reference

[27]. In Model 1 we have Dropout (0.1) between ConvA and ConvB, Dropout (0.2) between ConvB and Conv1ReLU and between Conv1ReLU and Conv2 Dropout (0.3) for CNN network regularization.

Table 1. CapsNet model setup.

Model	ConvA NF FS	ConvB NF FS	Conv1 ReLU NS FS	Conv2 NF SF	Image size	Epochs	Computation cost hh:mm:ss.ms
One	18 11 × 11	36 9 × 9	72 9 × 9	256 9 x 9	81 × 81	500	4:55:40.246678
Two	–	–	265 9 × 9	256 9 x 9	28 × 28	500	2:49:41.935765
Three	–	–	16 9 × 9	256 9 x 9	28 × 28	500	0:49:08.825828
Four	–	12 5 × 5	24 5 × 5	256 9 x 9	28 × 28	500	1:39:55.135174

Comparison With State-of-the Art. The performance of our CapsNet models to accurately recognise appliance signals in the NILM strategy is compared with the state-of-the art in Table 2. The % prediction capability gives the model's ability to accurately predict and reconstruct the appliance signal from the measured composite signal. For the given dataset our models perform at par or better compared with state-of-the art Capsnet image recognition schemes for the same image format. In Table 2 Ref [17] shows the performance of our Model Two on a comparatively large 60K simpler in form MNIST digit image training dataset. Naturally the performance here will be a bit higher than our proposed model that has a limited dataset on a some what curtailed HP 250 CPU model development environment.

Table 2. CapsNet model % prediction capabilities.

CapsNet model	Input image size	% Prediction
One	81 × 81	87.5
Two	28 × 28	87.5
Three	28 × 28	100.00
Four	28 × 28	100.00
Our average (One to Four)	–	93.75
Ref [6]	14 × 14	77.00
Ref [15]-Average	64 × 64	90
Ref [17]-MNIST	28 × 28	99.23
Ref [28]	30 × 30 × 30 (volume grid)	88.3–91.30

7   Concusion

In this paper we evaluate the capsule network's ability to accurately recognize the disaggregated power system signals for appliances in the non-intrusive-load-monitoring scheme. By using image fusion techniques we present the capsule network with a rich set of features for each load signal where the data collected is very limited. For only fifteen images per appliance and with no data augmentation (no change in viewpoint) we are able to achieve prediction accuracies of up to 93.75%. CNN requires a large dataset with a large viewpoint count to achieve the same accuracies. None the less, the CPU platform used in our experiments limited the depth and training speed of our models.

Acknowledgment. This research is supported partially by South African National Research Foundation Grants (Nos. 112108 and 112142), and South African National Research Foundation Incentive Grant (No. 95687 and 114911), Eskom Tertiary Education Support Programme Grants, Research grant from URC of University of Johannesburg.

References

1. Esa, N.F., Abdullah, M.P., Hassan, M.Y.: A review disaggregation method in non-intrusive appliance load monitoring. Renew. Sustain. Energy Rev. **66**, 163–173 (2016)
2. De Baets, L., Ruyssinck, J., Develder, C., Dhaene, T., Deschrijver, D.: Appliance classification using VI trajectories and convolutional neural networks. Energy Build. **158**, 32–36 (2018)
3. Yamashita, R., Nishio, M., Do, R.K.G., Togashi, K.: Convolutional networks: an overview and application in radiology. Insights Imaging **9**(4), 611–629 (2018)
4. Damaševicius, R., Maskeliunas, R., Wozniak, M., Polap, D.: Visualization of physiologic signals based on Hjorth parameters and Gramian angular fields. In: IEE 16th World Symposium of Applied Machine Intelligence and Informatics, pp. 000091–000096 (2018)
5. Wang, Z., Oates, T.: Imaging time-series to improve classification and imputation. In: International Joint Conference on Artificial Intelligence, pp. 3939–3945 (2015)
6. Ha, K.W., Jeong, J.W.F.: Decoding two-class motor imagery EEG with capsule networks. In: IEEE International Conference on Big Data and Smart Computing (Big Comp) (2019). https://doi.org/10.1109/BIGCOMP.2019.8678917
7. Hatami, N., Gavet, Y., Debayle, J.: Classification of time-series images using deep convolutional neural networks. In: Tenth International Conference on Machine Vision (ICMV), Vienna, Austria (2017). https://doi.org/10.1117/12.2309486
8. Yang, C.L., Yang, C.Y., Chen, Z.X., Lo, N.W.: Multivariate time series data transformation for convolutional neural network. In: Proceeding of the IEEE/SICE International Symposium on System Integration, pp. 188–192 (2019)
9. Properties of Functions. http://www.math.fsu.edu/~pkirby/mad2104/SlideShow/s4-2.pdf. Accessed 4 Aug 2019
10. Alofi, A., et al.: A review of data fusion techniques. Int. J. Comput. Appl. **167**(7), 0975–8887 (2017)

11. Shi-tong, Y.: An effective image fusion rule based on consistency checking. In: International Conference on Computer Science and Network Technology, pp. 2443–2446 (2011)
12. Chacko, B., Agrwal, S.L., et al.: Performance of image fusion technique using 4x4 block wavelet cosine transformation contribution title. In: 7th International Conference on Cloud Computing, Data Science and Engineering - Confluence, pp. 618–622 (2017)
13. Hamid, I., Nawaz, Q., Zia, M.A., Mumtaz, I.: Pixel-level multi-focus image fusion algorithm based on 2DPCA. In: IEEE 11th International Conference on Communication Software and Networks, pp. 362–365 (2019)
14. Katarya, R., Arora, Y.: Study on text classification using capsule networks. In: 5th International Conference on Advanced computing and Communication Systems (ICACCS), pp. 501–505 (2019)
15. Gumusbas, D., Yildrim, T.: Offline signature identification and verification using capsule network. In: IEEE International Symposium on Innovations in Intelligent Systems and Applications (INISTA) (2019). https://doi.org/10.1109/INISTA.2019.8778228
16. Dombetzki, L.A.: An overview over capsule networks. In: Seminar Innovative Internet Technologies and Mobile Communications, September 2018, pp. 89–95 (2018). https://doi.org/10.2313/NET-2018-11-1_12
17. Sabour, S., Frost, N., Hinton, G.E.: Dynamic routing between capsules. In: Advances in Neural Information Processing, pp. 3856–3866 (2017)
18. Sreelaksmi, K., Akarsh, S., Vinayakumar, R., Soman, K.P.: Capsule neural networks and visualization for segregation of plastic and non-plastic wastes. In: 5th International Conference on Advanced Computing & Communication Systems (ICACCS), pp. 631–636 (2019)
19. Tektronix. https://download.tek.com/manual/PA1000-Power-Analyzer-07709120 1.pdf. Accessed 25 Feb 2020
20. Rosenfeld, M., Graham, R.L., et al. (eds.): In Praise of the Gram Matrix. The Mathematics of Paul Erdos I, pp. 551–557 (2013). https://doi.org/10.1007/978-3-642-60406-5
21. Marttinez-Arellano, G., Terrazas, G., Ratchev, S.: Tool wear classification using time series imaging and deep learning. Int. J. Adv. Manuf. Technol. **104**, 3647–3662 (2019)
22. ScienceDirect Wavelet Analysis. https://www.sciencedirect.com/topics/earth-and-planetary-sciences/wavelet-analysis/. Accessed 26 Feb 2020
23. Pajares, G., Manuel, J.: A wavelet-based image fusion tutorial. Pattern Recogn. **37**(9), 1855–1872 (2004)
24. Wang, H., Zhang, C., Song, Y., Pang, B.: Information-fusion methods based simultaneous localization and mapping for robot adapting to search and rescue postdisaster environments. J. Robot. **2018**, 1–13 (2018)
25. Youtube. https://www.youtube.com/watch?v=0YYdo3vJnzU/. Accessed 19 Sep 2019
26. The AIlearner. https://theailearner.com/tag/capsule-network/. Accessed 1 Dec 2019
27. Stackoverflow. https://stackoverflow.com/questions/7370801/measure-time-elapse d-in-python/. Accessed 5 Mar 2020
28. Wang, Z., Oates, T.: 3D capsule networks for object classification from 3D model data. In: IEEE 52nd Asilomar Conference on Signals, Systems and Computers, ACSSC, pp. 2225–2229 (2018)

Design of Echo State Network with Coordinate Descent Method and l_1 Regularization

Cuili Yang, Zhanhong Wu$^{(\boxtimes)}$, and Junfei Qiao

Faculty of Information Technology, Beijing Key Laboratory of Computational
Intelligence and Intelligence System, Beijing University of Technology,
Beijing 100124, People's Republic of China
clyang5@bjut.edu.cn, 1036685809@qq.com

Abstract. The echo state networks (ESNs) have been applied in many
applications. For an ESN, its training error and network size are closely
related with output weight matrix. In this paper, the coordinate descent
method based ESN (CD-ESN for short) is designed to deal with the
relationship between training error and network size. In CD-ESN, the
l_1 regularization is used to penalty the non-important values of output
weight into 0. Moreover, the coordinate descent method is used to update
the output weights of ESN. Experimental results imply that the proposed
CD-ESN has better prediction accuracy and more sparse network topology than original ESN.

Keywords: Echo state network · Coordinate descent method ·
Predictive accuracy · l_1 regularization

1 Introduction

The time series prediction have existed in many applications, such as the prediction of disease in the medical field [1], oil production prediction [2], temperature
prediction [3] and so on. Recently, to predict time series, some artificial neural networks have been proposed, among which the recurrent neural networks
(RNNs) are widely used [4]. However, the RNNs have to face gradient explosion
and gradient disappearance problems. Then the echo state networks (ESNs) was
proposed [5] to work out this problem. The ESN is constructed by an input layer,
reservoir and output layer. In ESN, only the output weights between reservoir
and output layer should be calculated in training procedure, thus the training
speed of ESN is fast the computational complexity is low.

The performance of ESN is mainly determined by reservoir size. A larger size
of reservoir may generate smaller training error. However, if the reservoir size is
too large, it may result in the overfitting problem and decrease network generalization performance. Therefore, it is important to balance the trade-off between
reservoir size and network performance. In [6], a growing echo-state network

© Springer Nature Singapore Pte Ltd. 2020
H. Zhang et al. (Eds.): NCAA 2020, CCIS 1265, pp. 357–367, 2020.
https://doi.org/10.1007/978-981-15-7670-6_30

(GESN) has been proposed, in which the hidden units are added to the existing reservoir group by group. In [7], a pruning method based on variance sensitivity analysis is proposed to obtain a compact ESN. In [8], a pruning algorithm is proposed to delete the unimportant connection weights of reservoir, such that the resulted network structure is sparse.

Recently, regularization techniques are widely used to deal with the relationship between training error and reservoir size. In [9], l_2 regularization is used to solve the problem of ill-posed solution. However, it is a biased estimation method and could not generate sparse network topology. In [10], the l_1 regularization is used to control the model size of extreme learning machine. In [11], the l_0 regularization is used to obtain the sparse solutions of linear equations. Since the l_0-norm is NP-hard, which is difficult to solve out. Thus, the l_1 regularization becomes our focus in this paper.

Actually, the output weights training problem of an ESN is equal to a linear regression problem, which can be easily solved by gradient descent algorithms [12]. In the field of gradient descent algorithms, the batch gradient descent algorithm (BGD) and stochastic gradient descent algorithm (SGD) are widely used. However, the training speed of BGD is very slow since all training data are used in each iteration [13]. In SGD, only one sample is used in each iteration, while it has to face the problem of search blindness [14]. On the other hand, the coordinate descent algorithm have been used in applications for many years, it has the advantage of fast convergence speed and less calculation burden [15].

In this paper, the coordinate descent method based ESN (CD-ESN) is designed to make a balance of training error and reservoir size. In CD-ESN, the l_1 regularization of output weighs is added into the object function to decrease network size. Moreover, the coordinate descent method is used to generate output weights, which can penalty the non-important elements as 0, such that the reservoir nodes are decreased. Finally, simulation results illustrate that the CD-ESN has more sparse reservoir size and smaller testing error than the original ESN.

The remaining content is designed as below. In Sect. 2, the introduction of ESN is described. The proposed CD-ESN is given in Sect. 3. The design procedures of CD-ESN and simulation results are discussed in Sect. 4 and 5, respectively. Finally, some conclusions are written in Sect. 6.

2   Original ESN

The core of ESN is its reservoir which contains hundreds of sparely connected nodes [5], which is plotted in Fig. 1. The typical ESN has a n-neurons-input layer, a N-nodes-hidden layer, and an output layer with 1 neurons. In Fig. 1, $\mathbf{W}^{in} \in \mathbb{R}^{n \times N}$ and $\mathbf{W} \in \mathbb{R}^{N \times N}$ represent the input matrix and internal weight matrix, respectively. The output weight matrix is defined as $\mathbf{W}^{out} = [w_1, w_2, ..., w_{N+n}] \in \mathbb{R}^{(N+n) \times 1}$, which is the only parameter to be updated in the ESN training process. With L training samples $\{\mathbf{u}(k), t(k)\}_{k=1}^{L}$,

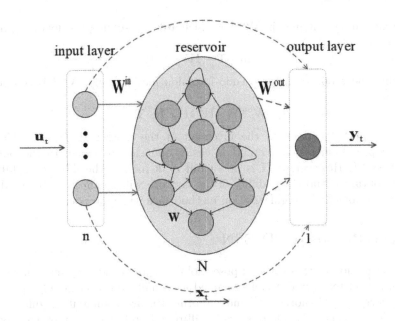

Fig. 1. The structure of an ESN.

where $\mathbf{u}(k) = [u_1(k), u_2(k), ..., u_n(k)]^T \in \mathbb{R}^{n \times 1}$ are inputs and $t(k)$ denote outputs, the reservoir states are denoted as $\mathbf{x}(k) = [x_1(k), x_2(k), ..., x_N(k)] \in \mathbb{R}^{1 \times N}$, which are calculated as below:

$$\mathbf{x}(k) = \mathbf{f}(\mathbf{u}(k)\mathbf{W}^{in} + \mathbf{x}(k-1)\mathbf{W}) \tag{1}$$

where $\mathbf{f}(\cdot) = [f_1(\cdot), ..., f_N(\cdot)]^T$ stand for the activation functions. If one denotes $\mathbf{X}(k) = [\mathbf{x}(k), \mathbf{u}(k)] \in \mathbb{R}^{1 \times (N+n)}$ as concatenation of reservoir states and input vectors, the outputs $\mathbf{y}(k)$ of ESN in the current time step k are computed as:

$$\mathbf{y}(k) = \mathbf{X}(k)\mathbf{W}^{out} \tag{2}$$

Define the state matrix as $\mathbf{H} = [\mathbf{X}(1), \mathbf{X}(2), ..., \mathbf{X}(L)]^T$, which is represented as below,

$$\mathbf{H} = \begin{bmatrix} \mathbf{X}^T(1) \\ \mathbf{X}^T(2) \\ \vdots \\ \mathbf{X}^T(L) \end{bmatrix} = \begin{bmatrix} X_{11} & X_{12} & \cdots & X_{1N} \\ X_{21} & X_{22} & \cdots & X_{2N} \\ \vdots & \vdots & \ddots & \vdots \\ X_{L1} & X_{L2} & \cdots & X_{LN} \end{bmatrix} \tag{3}$$

with

$$X_{ki} = \begin{cases} x_{ki}, & \text{if } 1 \leq i \leq N \\ u_{ki}, & \text{if } N+1 \leq i \leq N+n \end{cases} \tag{4}$$

and the target output matrix is represented as $\mathbf{T} = [\mathbf{t}(1), \mathbf{t}(2), ..., \mathbf{t}(L)]^T$, one has

$$\mathbf{H}\mathbf{W}^{out} = \mathbf{T} \tag{5}$$

If the output weight matrix $\mathbf{W}^{out}$ is calculated by solving the following minimization problem,

$$\min ||\mathbf{T} - \mathbf{H}\mathbf{W}^{out}||_2^2 \qquad (6)$$

By using the Moore-Penrose pseudo-inversion method [16], $\mathbf{W}^{out}$ is computed as below:

$$\mathbf{W}^{out} = (\mathbf{H}^T\mathbf{H})^{-1}\mathbf{H}^T\mathbf{T} \qquad (7)$$

where $(\mathbf{H}^T\mathbf{H})^{-1}$ represents the inversion of the matrix $\mathbf{H}^T\mathbf{H}$. The Moore-Penrose pseudo-inversion method is easily operated to calculate the output weight $\mathbf{W}^{out}$. However, if $\mathbf{H}$ is not a full rank metric, the ill-posed solutions may be obtained and the stability of the network will be affected. To avoid the ill-posed problem, the regularization methods are widely applied.

3   The Proposed CD-ESN

In regularization methods, the ill-posed solutions are avoided by introducing the regularization term, which is also named as penalty term. Generally speaking, the l_2-norm [17], l_1-norm [18] and l_0-norm [19] are commonly regularization term. The l_2 penalty is able to solve the ill-posed problem and improve network generalization ability simultaneously. However, it cannot generate the sparse network. Both l_1 and l_0 regularization are able to make the output weights of irrelevant nodes approach zero, such that a compact ESN is resulted. However, the l_0 penalty is NP-hard, which is difficult to calculate. Thus, in this paper, the l_1-norm of output weights is introduced, which is described as below:

$$E = \frac{1}{2}||\mathbf{T} - \mathbf{H}\mathbf{W}^{out}||_2^2 + \beta||\mathbf{W}^{out}||_1 \qquad (8)$$

where $||\mathbf{W}^{out}||_1 = \sum_{i=1}^{N+n} |w_i|$ represent the l_1-norm of $\mathbf{W}^{out}$. β is a regularization parameter that balances the trade-off between training error term and sparsity term. If β is small, the role of training error is more important than the sparsity term. Otherwise, the sparsity term is more important.

In Eq. (8), the training error term and sparsity term should be minimized simultaneously. It is observed that $||\mathbf{T} - \mathbf{H}\mathbf{W}^{out}||$ and $||\mathbf{W}^{out}||_1$ are differentiable, thus the gradient descent method can be used. The coordinate descent algorithms have been widely studied in recently years [20], which decompose a N-dimensional optimization problem into N one-dimensional subproblems. Each parameter is updated in the optimal direction, which saves the calculation time. Thus, the coordinate descent algorithm is introduced to solve the Eq. (8).

Suppose the ith variable in $\mathbf{W}^{out}$, i.e. w_i, is updated with $i = 1, 2, ..., N + n$, while other coordinates of $\mathbf{W}^{out}$ are fixed at their last update values. Then, the Eq. (8) can be rewritten as:

$$E(w_i) = \frac{1}{2}\sum_{k=1}^{L} [t(k) - X_{ki}w_i - \sum_{j\neq i}^{N+n} X_{kj}w_j]^2 + \sum_{j=1}^{N+n} \beta|w_j| \qquad (9)$$

Deriving from the above formula, one obtains

$$E(w_i) = \frac{1}{2} \sum_{k=1}^{L} (X_{ki}w_i)^2 - \sum_{k=1}^{L} (t(k) - \sum_{j \neq i}^{N+n} X_{kj}w_j)(X_{ki}w_i)$$
$$+ \frac{1}{2} \sum_{k=1}^{L} [(t(k) - \sum_{j \neq i}^{N+n} X_{kj}w_j]^2 + \sum_{j=1}^{N+n} \beta |w_j| \tag{10}$$

Since $\frac{1}{2} \sum_{k=1}^{L} [(t(k) - \sum_{j \neq i}^{N+n} X_{kj}w_j]^2$ is irrelevant to w_i, minimizing $E(w_i)$ in Eq. (10) is equal to minimizing $J(w_i)$ in the following equation,

$$J(w_i) = \frac{1}{2} \sum_{k=1}^{L} (X_{ki}w_i)^2 - \sum_{k=1}^{L} (t(k) - \sum_{j \neq i}^{N+n} X_{kj}w_j)(X_{ki}w_i) + \sum_{j=1}^{N+n} \beta |w_j| \tag{11}$$

To find the minimum value of $J(w_i)$ in Eq. (11), the derivative of $J(w_i)$ respect to w_i should be calculated. Firstly, the sub-gradient of l_1-norm is given as below,

$$\partial(||w_i||_1) = \begin{cases} 1, & \text{if } w_i > 0 \\ -1, & \text{if } w_i > 0 \\ \alpha \in [-1, 1], & \text{if } w_i = 0 \end{cases} \tag{12}$$

Based on Eq. (12), the sub-gradient of $J(w_i)$ is calculated,

$$\partial J(w_i) = \sum_{k=1}^{L} (X_{ki}w_i) - \sum_{k=1}^{L} (t(k) - \sum_{j \neq i}^{N+n} X_{kj}w_j)X_{ki} + \partial(||w_i||_1) \tag{13}$$

For the sake of simplicity of computation, two parameters D and C are introduced as below,

$$D = (\sum_{k=1}^{L} X_{ki})^T (\sum_{k=1}^{L} X_{ki}) \tag{14}$$

$$C = (\sum_{k=1}^{L} X_{ki})^T (\sum_{k=1}^{L} (t(k) - \sum_{j \neq i}^{N+n} X_{kj}w_j)) \tag{15}$$

So, the sub-gradient of $J(w_i)$ given as below,

$$\partial J(w_i) = \begin{cases} Dw_i - C + \beta, & \text{if } w_i > 0 \\ Dw_i - C - \beta, & \text{if } w_i < 0 \\ [-C - \beta, -C + \beta], & \text{if } w_i = 0 \end{cases} \tag{16}$$

By setting $\partial J(w_i) = 0$, the optimal value w_i can be obtain,

$$
w_i = \begin{cases}
\frac{C-\beta}{D}, & \text{if } C > \beta \\[2mm]
\frac{C+\beta}{D}, & \text{if } C < -\beta \\[2mm]
0, & \text{if } -\beta \le C \le \beta
\end{cases} \tag{17}
$$

Algorithm 1. Training of the CD-ESN Model

Input: Training set $\{u(k), t(k)\}_{k=1}^{k=L}$, β and $gen_{\max}$

Output: $\mathbf{W}^{out}$

Step 1: Generate the input weight $\mathbf{W}^{in}$ and internal weight matrix $\mathbf{W}_0$

Step 2: Calculate $\mathbf{X}(k)$ by Eq. (1)

Step 3: Set $gen = 1$ and initialize $W^{out} = 0$

Step 4: For each element of W^{out}, update the value of w_i $\{i = 1, 2, ..., N+n\}$ by Eq. (17)

Step 5: Increase $gen = gen + 1$. Go back to Step 3 until gen reaches to $gen_{\max}$

Step 6: Test the obtained network

---

4   The Design Procedures of CD-ESN

The design process of the proposed CD-ESN are described as below:

- Step 1. Input the training samples $\{(u(k), t(k))|u(k) \in R^n, k = 1, 2, ..., L\}$, the activation function $\mathbf{f}(\cdot)$ and the maximum updating iteration $gen_{\max}$, the regularization parameter β. Initial the input weight matrix $\mathbf{W}^{in}$ and internal weight matrix $\mathbf{W}_0$.
- Step 2. Calculate the echo states $\mathbf{X}(k)$ as Eq. (1).
- Step 3. Initialize the output weight matrix $W^{out} = 0$. Set the first algorithm iteration $gen = 1$.
- Step 4. For each element of W^{out}, i.e. w_i with $i = 1, 2, ..., N+n$, update its value according to Eq. (17) with the predefined regularization coefficient β.
- Step 5. Increase $gen = gen + 1$. If gen is less than $gen_{\max}$, go to Step 4, otherwise, turn to Step 6.
- Step 6. Testing the performance of the resulted ESN.

5   Simulation and Discussion

To verify the performance of ESNs, two benchmark problems are used, including the Lorenz chaotic time series prediction [21] and Mackey-Glass chaotic time series prediction [22]. They are the most commonly used test functions. For algorithm comparison, the performance of CD-ESN is compared with the original ESN (OESN).

The input weighs $\mathbf{W}^{in}$ of all ESNs are randomly generated from the range $[-0.1, 0.1]$. In CD-ESN, the regularization parameter β varies from 0 to 1 by step of 0.2, the optimal value of β is found by the grid search algorithm.

The model prediction performance is evaluated by using NRMSE. Its formula is given as below,

$$NRMSE = \sqrt{\sum_{k=1}^{L} \frac{[\mathbf{y}(k) - \mathbf{t}(k)]^2}{L\sigma^2}} \tag{18}$$

where $y(k)$ is the corresponding predicted output and $t(k)$ is the target output, σ^2 is the variance of the target outputs, and L is the total number of $t(k)$. The smaller NRMSE, the better training or prediction result.

5.1   Lorenz Chaotic Time Series Prediction

Firstly, the Lorenz time series prediction used to test network performance. Its ordinary differential equation is expressed as Eq. (19):

$$\frac{dx}{dt} = \sigma(y_t - x_t)$$

$$\frac{dy}{dt} = -x_t z_t + r(x_t - y_t) \tag{19}$$

$$\frac{dz}{dt} = x_t y_t - b_t z_t$$

where $\sigma = 10, r = 28, b = 8/3$. The Lorenz chaotic time series in y dimension is generated by Runge-Kutta method. 2400 training samples are used, of which the first 1200 were set as training dataset and the remaining 1200 were used as testing dataset. In the simulation, the reservoir size is 100 and the maximum algorithm iteration is $gen_{\max} = 1000$.

In CD-ESN, with different regularization parameter β, the training and testing NRMSE values and network size are shown on Table 1. It can be easily found that when β is small (take $\beta = 0$ for example), the network size is largest with the best training result. When the regularization parameter β increases, the network size decreases and the training error increases. Therefore, it is important to determine a perfect value of β.

For CD-ESN, the evolving process of training NRMSE versus algorithm iterations is presented in Fig. 2 (a), in which β is set as 0.4. When the iteration increases, the training error decrease gradually until it reaches a constant. On the other hand, the number of non-zero values in output weight W^{out} versus algorithm iteration is shown in Fig. 2 (b). Obviously, the number of reservoir nodes is gradually decreased until it reaches to a stable value. It is obvious that the reservoir size is smaller than 100, thus l_1 regularizer generates the spare ESN.

The testing results and errors of OESN and CD-ESN are shown in Fig. 3 (a) and (b), respectively. Obviously, the output of CD-ESN is fit to the target very well. Also, CD-ESN with l_1 regularization parameter has less training error than the OESN. Thus, the testing performance of CD-ESN is improved by using l_1 regularization technique.

The performance comparison between CD-ESN and OESN are given in Table 2, including the training and testing NRMSE values, the training time (s) as well as the non-zero number of output weights. It is obvious that the proposed CD-ESN has smaller prediction error and more compact reservoir.

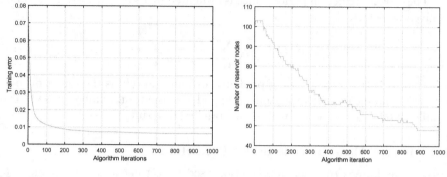

(a) The   training   NRMSE   values   evolving process.

(b) The network size.

Fig. 2. The training NRMSE values and network size versus algorithm iterations.

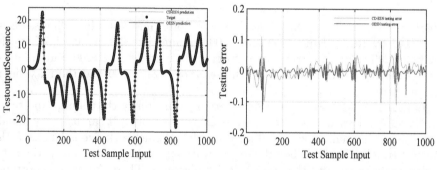

(a) Prediction result of OESN and CD-ESN.

(b) Testing errors of OESN and CD-ESN.

Fig. 3. Prediction of OESN and CD-ESN.

Table 1. Different β comparisons for Lorenz chaotic time series prediction

Approach	β	Final reservoir neuron	Training NRMSE	Testing NRMSE
CD-ESN	0	103	0.0029	0.0038
	0.2	64	0.0038	0.0047
	0.4	42	0.0052	0.0056
	0.6	28	0.0058	0.0059
	0.8	19	0.0068	0.0071
	1	10	0.0092	0.0096

Table 2. Lorenz chaotic time series prediction with different algorithms

Algorithm	Non-zero number of $\mathbf{W}^{out}$	Training time(s)	Training NRMSE	Testing NRMSE
OESN	100	4.32	0.0084	0.0087
CD-ESN ($\beta = 0.4$)	42	231.06	0.0052	0.0056

5.2 Mackey-Glass Chaotic Time Series Prediction

The Mackey-Glass chaotic time series prediction is used to test the performance of different algorithms. Its form is as below (20),

$$\frac{\partial s(t)}{\partial t} = \frac{as(t - \tau)}{1 + s^{10}(t - \tau)} + bs(t) \tag{20}$$

where $b = 0, a = 0.2, \tau = 17$. 2400 training samples are used, of which the first 1200 group and the remaining 1200 are used in training and testing stages, respectively. The reservoir size is 200, and the maximum iteration is set as perform 3000.

The prediction curve and prediction error of OESN and CD-ESN are presented in Figs. 4 (a) and (b). Obviously, the CD-ESN can better fit the Mackey-Glass time series than OESN. Simultaneously, the obtained testing error of CD-ESN is much smaller than that of OESN.

The performance of CD-ESN is compared with OESN in Table 3. As compared with original ESN, the CD-ESN has better training accuracy and smaller reservoir size than OESN.

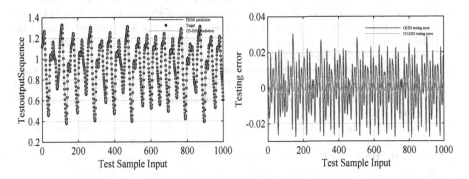

(a) Prediction result of OESN and CD-ESN.

(b) Testing errors of OESN and CD-ESN.

Fig. 4. Prediction of OESN and CD-ESN.

Table 3. Mackey-Glass time series prediction with different algorithms

Algorithm	Final reservoir neurons	Training time(s)	Training NRMSE	Testing NRMSE
OESN	100	4.32	0.0590	0.0595
CD-ESN ($\beta = 0.8$)	90	534.908	0.0102	0.0110

6   Conclusions

How to deal with the relationship between training error and reservoir size is important for an ESN. To solve this problem, the CD-ESN is designed in this paper. In CD-ESN, the l_1 regularization is added into the object function, which can decrease the reservoir nodes. Besides, the coordinate descent algorithm is introduced to update the output weight matrix. The experiment results illustrate that the proposed CD-ESN achieves more compact reservoir and better prediction performance than OESN.

Acknowledgements. This work was supported by the National Natural Science Foundation of China under Grants 61973010, 61533002 and 61890930, the Major Science and Technology Program for Water Pollution Control and Treatment of China (2018ZX07111005), and the National Key Research and Development Project under Grants 2018YFC1900800-5.

References

1. Sarkar, M., Leong, T.Y.: Characterization of medical time series using fuzzy similarity-based fractal dimensions. Artif. Intell. Med. **27**(2), 201–222 (2003)
2. Zheng, S.Q., Zhang, H.F., Bao, J.W.: Application of chaotic theory to oil production rate time series prediction. Paper Presented at the Proceeding of the 2009 IEEE International Conference on Intelligent Computing and Intelligent Systems, Shanghai, 20–22 November (2009)
3. Wei, K.X., Du, M.X.: A temperature prediction method of IGBT based on time series analysis. Paper Presented at the Proceeding of 2010 the 2nd International Conference on Computer and Automation Engineering (ICCAE), Singapore, 26–28 February (2010)
4. Connor, J.T., Martin, R.D., Atlas, L.E.: Recurrent neural networks and robust time series prediction. IEEE Trans. Neural Netw. **5**(2), 240–254 (1994)
5. Jaeger, H., Haas, H.: Harnessing nonlinearity: predicting chaotic systems and saving energy in wireless communication. Science **304**(5667), 78–80 (2004)
6. Qiao, J.F., Li, F.J., Han, H.G., Li, W.J.: Growing echo-state network with multiple subreservoirs. IEEE Trans. Neural Netw. Learn. Syst. **28**(2), 391–404 (2017)
7. Thomas, P., Suhner, M.C.: A new multilayer perceptron pruning algorithm for classification and regression applications. Neural Process. Lett. **42**(2), 437–458 (2015)
8. Li, D.Y., Liu, F., Qiao, J.F., Li, R.: Structure optimization for echo state network based on contribution. Tsinghua Sci. Technol. **24**(1), 97–105 (2019)
9. Dutoit, X., Schrauwen, B., Campenhout, J.V., Stroobandt, D., Brussel, H.V., Nuttin, M.: Pruning and regularization in reservoir computing. Neurocomputing **72**(7–9), 1534–1546 (2009)

10. Han, M.-S., Ren, W.J., Xu, M.L.: An improved echo state network via l1-norm regularization. Acta Autom. Sinica **40**(11), 2428–2435 (2014)

11. Mohimani, G.H., Babaie-Zadeh, M., Jutten, C.: Fast sparse representation based on smoothed L0 norm. In: Davies, M.E., James, C.J., Abdallah, S.A., Plumbley, M.D. (eds.) ICA 2007. LNCS, vol. 4666, pp. 389–396. Springer, Heidelberg (2007). https://doi.org/10.1007/978-3-540-74494-8_49

12. Jaeger, H., Lukosevicius, M., Popovici, D., Siewert, U.: Optimization and applications of echo state networks with leaky integrator neurons. Neural Netw. **20**(3), 335–352 (2007)

13. Wilson, D.R., Martinez, T.R.: The general inefficiency of batch training for gradient descent learning. Neural Netw. **16**(10), 1429–1451 (2004)

14. Ketkar, N.: Stochastic gradient descent. Deep Learning with Python (2017)

15. Friedman, J.H., Hastie, T., Hfling, H., Tibshirani, R.: Pathwise coordinate optimization. Ann. Appl. Stat. **1**(2), 302–332 (2007)

16. Yang, C.L., Qiao, J.F., Han, H.G., Wang, L.: Design of polynomial echo state networks for time series prediction. Neurocomputing **290**(17), 148–160 (2018)

17. Luo, X., Chang, X.H., Ban, X.J.: Regression and classification using extreme learning machine based on L1-norm and L2-norm. Neurocomputing **174**(22), 179–186 (2016)

18. Tan, K., Li, W.C., Huang, Y.S., Yang, J.M.: A regularization imaging method for forward-looking scanning radar via joint L1–L2 norm constraint. Paper Presented at the Proceeding of the 2017 IEEE International Geoscience and Remote Sensing Symposium (IGARSS), 1 July (2017)

19. Mancera, L., Portilla, J.: L0-norm-based sparse representation through alternate projections. Paper Presented at the Proceeding of the IEEE International Conference on Image Processing, Atlanta, GA, 8–11 October (2006)

20. Tseng, P.: Convergence of a block coordinate descent method for nondifferentiable minimization. J. Optim. Theory Appl. **109**(3), 475–494 (2001)

21. Lorenz, E.N.: Deterministic nonperiodic flow. J. Atmos. Sci. **20**(2), 130–141 (1963)

22. Thissen, U., Brakel, R.V., Weijer, A.P.D., Melssen, W.J., Buydens, L.M.C.: Using support vector machines for time series prediction. Chemom. Intell. Lab. Syst. **69**(1–2), 35–49 (2003)

An Advanced Actor-Critic Algorithm for Training Video Game AI

ZhongYi Zha[1], XueSong Tang[2], and Bo Wang[1(✉)]

[1] Key Laboratory of Ministry of Education for Image Processing and Intelligent
Control, Artificial Intelligence and Automation School,
Huazhong University of Science and Technology, Wuhan 430074, China
wb8517@hust.edu.cn

[2] College of Information and Science, Donghua University, Shanghai, China

Abstract. This paper presents an improved deep reinforcement learning
(DRL) algorithm, namely Advanced Actor Critic (AAC), which is based
on Actor Critic (AC) algorithm, to the video game Artificial intelligence
(AI) training. The advantage distribution estimator, Normal Constraint
(NC) function and exploration based on confidence are introduced to
improve the AC algorithm, in order to achieve accurate value estimate,
select continuous spatial action correctly, and explore effectively. The
aim is to improve the performance of conventional AC algorithm in a
complex environment. A case study of video game StarCraft II mini-game
AI training is employed. The results verify that the improved algorithm
effectively improves the performance in terms of the convergence rate,
maximum reward, average reward in every 100 episode, and time to reach
a specific reward, etc. The analysis on how these modifications improve
the performance is also given through interpretation of the feature layers
in the mini-games.

Keywords: Deep reinforcement learning · Video game AI · Advanced
Actor Critic · Advantage distribution estimator · Normal constraint

1 Introduction

In 2016, Alpha Go [1], the computer program developed by Google's DeepMind
team, defeated the 18-time world champion Lee Sedol in a five-game Go match
and manifested the power of the artificial intelligence (AI). In 2017, Alpha Go
Zero [2], which adopted the deep reinforcement learning algorithm (DRL) and
defeated the previous generation of Go in a short time, dominated the Go match
and revealed the power of the DRL. Nowadays, DRL has been widely applied
to industries in different fields, e.g., manufacture, natural language processing,
medical image processing, intelligent drive and video game AI design, etc. Among
these areas, AI design of video games is of particular interest to researchers, as it
provides a convenient test platform for investigating machine learning algorithm
performance in complex environments.

© Springer Nature Singapore Pte Ltd. 2020
H. Zhang et al. (Eds.): NCAA 2020, CCIS 1265, pp. 368–380, 2020.
https://doi.org/10.1007/978-981-15-7670-6_31

With the development of DRL, the end-to-end feature extraction method breaks the conventional manual feature extraction method, so that AI can complete the game task without intervention of humans. In 2013, Mnih et al. employed DRL to play video games. The DRL based AI even surpassed top human players in some Atari games [3]. However, these Atari games are generally too simple compared with practical applications in real life. Before long, Deep Mind team opened the StarCraft II learning environment (SC2LE) [4] as a new test platform for DRL studies. The SC2LE involves dynamic perception and estimation, incomplete information game and multi-agent cooperation problems, resulting in a more practical test environment, where decision making becomes much closer to the real-life situations.

Although the latest StarCraft II AI has been able to beat top human players in certain circumstances [5], the AI training required a vast number of episodes to achieve satisfying performance, due to the limits of the employed conventional Actor Critic (AC) algorithm ([6–8]). For example, the advantage estimator returns single value that results in limited accuracy, and the random action selection lacks effectiveness, etc. In this paper, an improved design of Actor Critic (AC) algorithm, namely Advanced Actor Critic (AAC), is proposed for better convergence of AI training. A distributional advantage estimator is employed to estimate the distribution of advantage in a certain state. In addition, a normal constraint, which means the loss function of spatial selection based on the normal distribution, as well as an exploration method based on confidence are introduced. As to the neural networks framework, the fully-convolution with long short-term memory networks (CNN LSTM) [9] is selected according to the Deep Mind approach [4]. Additionally, the sparse reward [10] is addressed by reward shaping. Finally, a combined model based on Rainbow [11] (a combination of Double Q Learning [12], dueling DQN [13], Priori-tized Experience Replay [14], distributional reward [15], noisy net [16]) is adopted, where the proposed algorithm is embedded. Such an approach is applied to the SC2LE mini-games to verify that AI can learn faster, play better and have stronger robustness with the ACC based training program. The main contribution of this paper is to improve the conventional AC algorithm, which leads to better performance in complex environments, and to propose a regional updating action mode instead of independent point updating for intelligent control in high-dimensional space.

2   Background

The SC2LE is integrated by Deep Mind [4]. In DRL, state, action and reward are the most important elements. In StarCraft II mini-games, these elements are represented by map features and mini-map features.

2.1   Environment

Reward. The reward and penalty in each mini-game are shown in detail in Table 1.

Table 1. Reward and Penalty in mini-games.

Mini-map name	Reward	Penalty
Move to beacon	+1 when the marine reaches a beacon	None
Collect mineral shards	+1 when the marine collects the mineral shards	None
Find and defeat Zerglings	+5 when the marine kills the Zergling	−1 when the marine is killed
Defeat Zerglings and Banelings	+5 when the marine kills the Zergling or Baneling	−1 when the marine is killed

State. In StarCraft II games, the environment has been processed by the interface, outputting several feature layers about the current state (see Fig. 1).

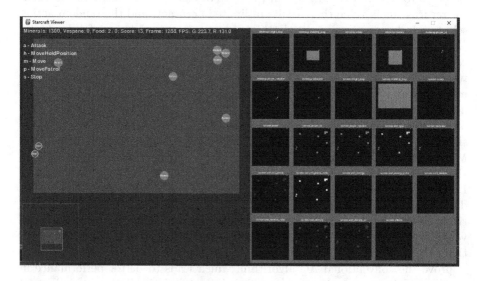

Fig. 1. Feature layers of the current state, which includes all simplified information of the mini-games.

During training, the size is the resolution of these feature images. Generally, the height and width are set to N, i.e., the spatial pixel observed and operated is $N \times N$. Deep Mind pointed out that $N \geq 64$ [4] is required when performing any micro operation in the game. However, for mini-games, $N = 32$ is proved sufficient.

Among these $N \times N$ pixels, each feature layer represents specific feature, such as player ID, unit ID, war fog, unit attribute, etc. Some belong to scale, while others belong to catalog.

In addition, the difference between the mini-map and the screen is that the screen is only a part of the mini-map camera, but the analysis of the unit is clearer. After processing, the mini-map is transformed into $F_{minimap} \times N \times N$, and the screen is transformed into $F_{screen} \times N \times N$. $F_{minimap}$ and F_{screen} represent the quantity of features that mini-map or screen has.

Action. The instruction set of StarCraft II is very large. Through the interface, each action is divided into two parts (Spatial Action and Non-Spatial Action) and output to the environment. The whole action is: $A = (A_{non-spatial}, [A_{add}, A_{spatial}])$. Among them, Spatial Action represents that AI clicks on a position in the screen, and its value range belongs to the pixel size of the environment, that is, $A_{spatial} = (x, y) \in N \times N$. Non-Spatial Action represents operation instructions, such as select, move, attack, build, view, etc. Its value range is the game instruction set of the whole StarCraft II pairs, and at the same time, it is subject to the actions available in the current environment, that is, $A_{non-spatial} \in A_{available} \cap A_{all}$. $A_{available}$ involves actions that can be performed in a certain state and A_{all} contains all action instructions in Star-Craft II. Especially, non-spatial actions are necessary while spatial actions are optional. For example, if an unit is selected by AI, only the non-spatial action is output, which means the spatial action becomes needless. A_{add} is the type of $A_{non-spatial}$ operation. When $A_{non-spatial}$ and $A_{spatial}$ are determined, A_{add} is generated automatically.

2.2   Actor Critic and Fully-Convolution with LSTM

Actor Critic. The policy adopted by the agent is the distribution of all state and action trajectories interacting with the environment. How this distribution takes the trajectory depends on the parameter θ. One of the trajectories is $\tau = (s_1, a_1, s_2, a_2, \ldots\ldots, a_t, s_t)$. The probability of getting the trajectory is $p_\theta(\tau) = \prod_{t=1}^{T} p_\theta(a_t \mid s_t)$. The relationship between the generation of environment and the parameter θ of trajectory can be ignored.

In order to get the best performance of an agent, the optimal goal is: $\theta = \arg\max_\theta E_{\tau \sim p_\theta}(R_\theta(\tau))$. $R_\theta(\tau)$ is the cumulative reward of τ, and the gradient is:

$$\nabla R_\theta(\tau) = \frac{1}{N} \sum_{n=1}^{N} \sum_{t=1}^{T_n} (\sum_{s=t}^{T_n} A_t \nabla log(p_\theta(a_t^n \mid s_t^n))) \tag{1}$$

In this paper, spatial policy net and non-spatial policy net are independent (π_θ represents the non-spatial policy net while π_θ' represents spatial policy net, and ρ indicates whether spatial action is valid or not), therefore the loss function in each episode is:

$$Loss_{\pi_\theta} = \sum_{t=1}^{n} A_t[log(\pi_\theta(a_t \mid s_t)) + \rho log(\pi_\theta'(a_t \mid s_t))] \tag{2}$$

Combine Actor Critic with CNN LSTM. The following figure shows the framework of interaction between the intelligent algorithm and StarCraft II Environment(see Fig. 2).

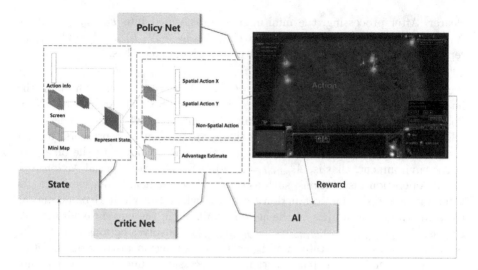

Fig. 2. Interaction between DRL based AI and StarCraft II environment. The inputs of the network are (1) available action (2) screen features (3) mini-map features. Outputs are (1) X and Y of the spatial action (2) non-spatial action (3) estimate of advantage. These outputs are all distributions.

3 Reprocessing of Environment Feature

Several feature layers have been extracted from the environment of the StarCraft II games. However, some of these layers are scale, while some are catalog, which means the distribution of values is very uneven. In addition, some feature layers are almost useless in mini-games. In order to train neural network better and faster, it is necessary and significant to reprocess these feature layers.

These layers can be simply divided into catalog value layers and scalar value layers. The catalog value layers include player ID, friend, visibility, creep, etc. Generally, these values are small. The scalar value layers include unit type, unit health, terrain level, and armor value. The prior knowledge of a StarCraft II player can be exploited to choose more helpful map feature layers and discard useless ones. Among these layers, the feature layers of mini-map are chosen as: visibility, camera, player ID, player relative, selected and the screen feature layers are: visibility, player ID, player relative, selected, HP, energy.

After the layer selection, the layers are numerically processed. In the catalog value layers, the values are standardized between [0, 1] to facilitate the training efficiency of neural networks. In the scalar value layers, the threshold is set to half of its maximum value. A sigmoid function is adopted to change the value classification into a state value. For example, if a unit has more than 50% of its maximum hit point (HP), it is considered to be healthy, otherwise be weak. After this reprocessing, the ranges of all layer values are compressed to be [0, 1].

4   Advanced Actor Critic (AAC)

The AC algorithm is improved from several perspectives: the updated data can break the correlation of time; the strategy network is more robust; the evaluation network is more accurate. The improvements are explained as follows.

4.1   Distributional Advantage

In AC algorithm, critic network is responsible for outputting the advantage estimate based on a certain policy. The accuracy of the estimate determines the update efficiency of policy net. As a result, a distributional network is designed to get more information. In this network, the maximum and minimum values are determined by the background conditions, and the probability of the advantage is handled by the last layer of convolution network plus softmax function. When updating, the sample $\widehat{R}$ obtained from Monte Carlo difference (MD) or temporal difference (TD) is transformed into a pulse function $D(\widehat{R})$ (avoiding None by using Clamp function) and then KL divergence is used to evaluate the loss of advantage estimates (see Fig. 3).

In this case, the advantage A_t and the value loss function $Loss_{V_\theta}$ can be formulated as follows:

$$A_t = E_{p(V_t)\sim V_\theta}(V_\theta(s_t)) \tag{3}$$

$$Loss_{V_\theta} = D_{KL}[D(\widehat{R}) \parallel V_\theta(s_t)] \tag{4}$$

4.2   Exploration Based on Confidence

One of the important problems in DRL is exploration. In mini-games, the exploration ability of agent in the early stage significantly influences its performance. Without exploration, the agent keeps taking the same action if the reward is always positive. Thus, it might be trapped by the local optimal solution and ignore other actions that may result in huge potential reward. Therefore, a good agent needs to have proper exploration ability. In contrast to the conventional AC algorithm exploration method, the maximum probability of the non-spatial policy, namely the confidence, is adopted to explore the environment. ω represents the lowest level of confidence in the conduct of random exploration and AI still explores based on $\epsilon - greedy$ to prevent falling into the local optimal solution. The less confidence the AI has, the more exploration is encouraged, and vice versa.

$$A_{non-spatial} = \begin{cases} \arg \max_{\theta} \pi_\theta(a_t \mid s_t), else \\ random\,choice, max(\pi_\theta(a_t \mid s_t)) < \omega \; or \; \epsilon < 0.1 \end{cases} \tag{5}$$

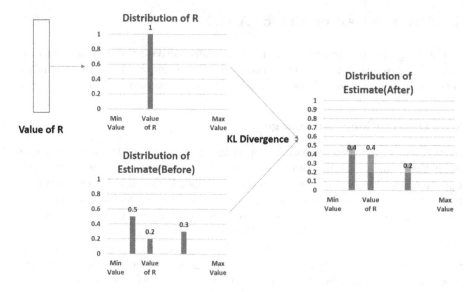

Fig. 3. The process of converting sample $\widehat{R}$ into pulse function and computing KL divergence with advantage estimate distribution.

4.3   Normal Constraint

In the environment of StarCraft II games, the most difficult problem is the huge continuous action space. In AAC algorithm, the policy net of AAC consists of three parts: spatial policy net X, spatial policy net Y and non-spatial policy net. The update mode of non-spatial policy net is introduced in Sect. 2.2. Given the particularity of action space in StarCraft II games, the update of spatial policy net X and spatial policy net Y introduce an additional function for the sake of convergence speed, namely the Normal Constraint. To enhance the exploration ability, loss entropy is usually adopted as the loss function in AC network. In the case of large continuous action space and sparse reward, if the loss entropy weight is not well designed, it is easy to enter policy net with the same output probability. In order to solve this problem, a Normal Constraint is hereby designed. In StarCraft II games, the actions are selected in a continuous space, and the selection probability around a specific point should be considered similar. Therefore, when updating an action, the nearby actions should be updated accordingly. In this case, AI can understand the high-dimensional action space regionally. Finally, when converging, the output of spatial policy is a stable distribution that approximately satisfies the Normal Distribution (see Fig. 4, Fig. 5).

The probability of the selected spatial action, i.e., the coordinate in the environment, is used as the weight of the NC loss. The greater probability implies more centralized policy, and the probability of its surrounding will be further enhanced or reduced when updated. Combined with formula (3), the NC loss function is:

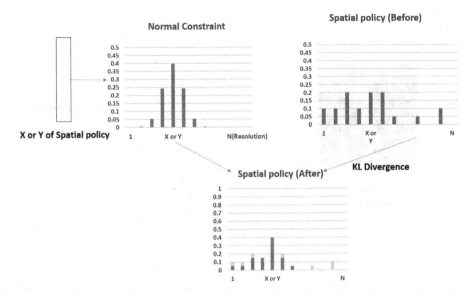

Fig. 4. How NC works.

$$Loss_{NC} = \sum_{t=1}^{n} A_t max(\pi'_\theta) D_{KL}[NC(a_t) \parallel \pi'_\theta(s_t)] \tag{6}$$

Finally, the whole network is updated according to formula (2) (4) (6).

4.4   Reward Shaping

The reward of StarCraft II is very sparse, which requires the agent to have a stronger exploration ability. During the implemented trainings, a negative reward is added to every moment to motivate AI to take actions. Besides, the reward values are standardized to be within the range [0, 1]. Such a reward shaping is very effective in a sparse environment.

5   Result

The agent based on AAC network or A3C network[1] is applied to 4 mini-games, and then the average score in every 100 episode as well as maximum score in 20K steps are compared. In the initial stage of training, i.e., observation episodes, the update of policy net is stopped. Since the training of policy net depends on critic net, critic net must be trained first in this observation.

A part of hyper parameters are shown in Table 2, and the Adam optimizer is adopted [17].

---

[1] The adopted A3C network refers to the project at https://github.com/xhujoy/pysc2-agents.

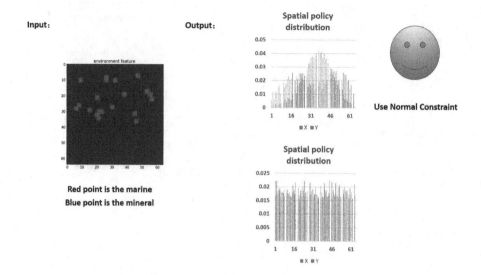

Fig. 5. Comparison of policy net with and without Normal Constraint (the figure above uses NC, while the figure below does not).

Table 2. Hyper parameters.

Hyper-parameter	Value	Notes
Observation episodes	200	Stop updating policy net
Learning rate	10^{-5}	
γ	0.99	Discount of reward
N	32	Resolution of mini-map and screen
Step	60	Steps in each iteration
Distributional advantage	$[-1, 24]$	Max value and Min value (different in each mini-game)
Distributional atoms	51	Numbers in advantage distribution
Adam epsilon	10^{-8}	
σ	2	Sigma of normal constraint
Addition reward	$\frac{-0.0001}{step}$	Reward shaping
ω	0.1	Base of confidence

6   Discussion

Mini-games in StarCraft II are divided into two categories based on their characteristics to test the performance of AAC and A3C. One category mainly focuses on spatial action and another is small scale combat game(see Fig. 6, Fig. 7). After 20K episodes, the average and max score achieved by agent based on AAC are both higher than agent based on A3C across the four mini games (see Table 3).

The key to get a better score depends on the performance of critic in Actor Critic algorithm, namely the accuracy of the advantage estimator. In a certain state (see Fig. 8), the Critic net in AAC estimates the reward by the expected

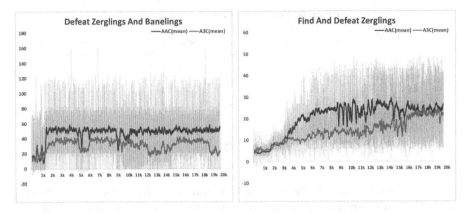

Fig. 6. Performance across 2 mini-games which mainly focus on spatial action for AAC and A3C.

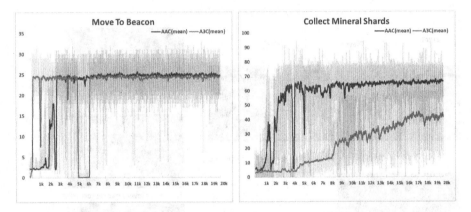

Fig. 7. Performance across 2 small scale combat games for AAC and A3C.

value 16.4 and the true value is 25. On the contrary, the Critic net in A3C estimates the reward by 6.1 and the true value is 24. Thus, the distributional estimator in critic net collects and computes the recent expected reward of a state, which contains more information and is closer to true reward than the estimate of single-valued estimator in A3C. During the experiment, it is proved that the more atoms distribution has, the more accurate the estimate is, because the distribution is more elaborate. However, more atoms need more neurons in the critic net and more time in training. In this paper, 256 feature neurons with 51 atoms are chosen [15].

The convergence speed of AAC and A3C is shown as the episode that the agent achieves scores around the best. It reveals like that the convergence speed of AAC is faster than that of A3C except in the Move to Beacon game at first

Table 3. Best score in each mini-game.

Agent	Metric	Move to beacon	Collect mineral shards	Find and defeat Zerglings	Defeat Zerglings and Banelings
AAC	BEST MEAN	26	68	29	59
	MAX	33	94	49	159
	Std of last 1k episodes	1.8	7.3	6.9	21.5
A3C	BEST MEAN	24	45	25	43
	MAX	31	76	45	118
	Std of last 1k episodes	2.4	13.6	8.6	24.5

glance. However, it should be noticed that the agent based on A3C achieved extremely bad score throughout the middle of the experiment (from episode 5k to 6k), and the reason is that the relatively small probability of spatial action leads to the hesitation but not confidence of agent (see Fig. 8).

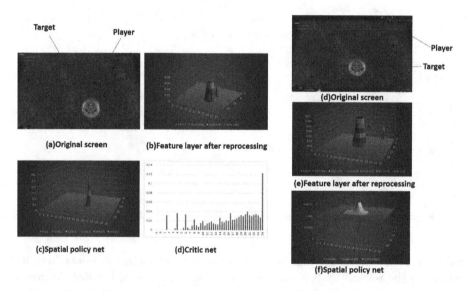

Fig. 8. Performance of AAC and A3C on Move to Beacon in episode 5k, (a) (b) (c) (d) is generated by AAC while (d) (e) (f) is generated by A3C.

In conclusion, around episode 5k, the agent based on A3C indecisively chose the right target due to the max but not dominant probability, and it indicates that the policy net has not converged. As a result, the convergence speed depends not only on when the agent performs a stable and good score, but also on the distribution of policy net which represents the confidence of the agent. In AAC, Normal Constraint encourages and guides the agent in the right direction. Additionally, this positive feedback allows the agent to explore in a certain area of huge action space with more confidence, so as to speed up the convergence. As a result, agent based on AAC converges faster than A3C across all games. Besides, probability distribution of policy net has not converged in two small scale combat games because training episodes are not enough.

In future work, more episodes and distribution atoms in advantage estimator will be applied in more difficult mini-games, full games and other fields. In addition, whether AAC can perform even better than other AC algorithm in more complex environments will also be tested.

Funding Statement. This research was supported by National Key Research and Development Program of China (Project No. 2018YFF0300300) and National Natural Science Foundation of China (Project No. 61803162).

References

1. Anthony, T., Tian, Z., Barber, D.: Thinking fast and slow with deep learning and tree search. In: Advances in Neural Information Processing Systems, pp. 5360–5370 (2017)
2. Silver, D., et al.: Mastering chess and shogi by self-play with a general reinforcement learning algorithm. arXiv preprint arXiv:1712.01815 (2017)
3. Mnih, V., et al.: Playing Atari with deep reinforcement learning. arXiv preprint arXiv:1312.5602 (2013)
4. Vinyals, O., et al.: StarCraft II: a new challenge for reinforcement learning. arXiv preprint arXiv:1708.04782 (2017)
5. Vinyals, O., et al.: Grandmaster level in StarCraft II using multi-agent reinforcement learning. Nature **575**(7782), 350–354 (2019)
6. Schulman, J., Wolski, F., Dhariwal, P., Radford, A., Klimov, O.: Proximal policy optimization algorithms. arXiv preprint arXiv:1707.06347 (2017)
7. Mnih, V., et al.: Asynchronous methods for deep reinforcement learning. In: International Conference on Machine Learning, pp. 1928–1937 (2016)
8. Wu, Y., Mansimov, E., Grosse, R.B., Liao, S., Ba, J.: Scalable trust-region method for deep reinforcement learning using kronecker-factored approximation. In: Advances in Neural Information Processing Systems, pp. 5279–5288 (2017)
9. Xingjian, S., Chen, Z., Wang, H., Yeung, D.Y., Wong, W.K., Woo, W.C.: Convolutional LSTM network: a machine learning approach for precipitation nowcasting. In: Advances in Neural Information Processing Systems, pp. 802–810 (2015)
10. Grześ, M., Kudenko, D.: Online learning of shaping rewards in reinforcement learning. Neural Netw. **23**(4), 541–550 (2010)
11. Hessel, M., et al.: Rainbow: combining improvements in deep reinforcement learning. In: Thirty-Second AAAI Conference on Artificial Intelligence (2018)
12. Van Hasselt, H., Guez, A., Silver, D.: Deep reinforcement learning with double q-learning. In: Thirtieth AAAI Conference on Artificial Intelligence (2016)
13. Wang, Z., Schaul, T., Hessel, M., Van Hasselt, H., Lanctot, M., De Freitas, N.: Dueling network architectures for deep reinforcement learning. arXiv preprint arXiv:1511.06581 (2015)
14. Hou, Y., Liu, L., Wei, Q., Xu, X., Chen, C.: A novel DDPG method with prioritized experience replay. In: 2017 IEEE International Conference on Systems, Man, and Cybernetics (SMC), pp. 316–321. IEEE (2017)

15. Bellemare, M.G., Dabney, W., Munos, R.: A distributional perspective on reinforcement learning. In: Proceedings of the 34th International Conference on Machine Learning, vol. 70, pp. 449–458. JMLR.org (2017)
16. Fortunato, M., et al.: Noisy networks for exploration. arXiv preprint arXiv:1706.10295 (2017)
17. Kingma, D.P., Ba, J.: Adam: a method for stochastic optimization. arXiv preprint arXiv:1412.6980 (2014)

Neural Network-Based Adaptive Control for EMS Type Maglev Vehicle Systems with Time-Varying Mass

Yougang Sun[1(✉)], Junqi Xu[1], Wenyue Zhang[2], Guobin Lin[1], and Ning Sun[3]

[1] National Maglev Transportation Engineering R&D Center, Tongji University, Shanghai 201804, China
1989yoga@tongji.edu.cn
[2] CRRC Zhuzhou Locomotive Co., Ltd., Zhuzhou 412001, Hunan, China
[3] College of Artificial Intelligence, Nankai University, Tianjin 300350, China

Abstract. The Electromagnetic Suspension (EMS) type maglev vehicles suffer from various control complexities such as strong nonlinearity, open-loop instability, parameter perturbations, which make the control of magnetic levitation system (MLS) very challenging. In actual operation, ever-changing passengers will cause the mass of vehicle body to deviate from the nominal mass of the levitation controller design, which will deteriorate the controller performance significantly. Therefore, the passenger carrying capacity of the maglev vehicle is strictly restricted, which restricts the further promotion of the maglev traffic. In this paper, an airgap control strategy based on the neural network (NN) is proposed for maglev vehicles with time-varying mass. Specifically, firstly, a nonlinear controller is proposed to guarantee the asymptotic stability of the closed-loop system. Next, to tackle the unknown or time-varying mass of vehicle body, an adaptive radial basis function (RBF) NN is utilized together with the proposed nonlinear controller. The stability analysis of the overall system is provided by Lyapunov techniques without any linearization to the original nonlinear model. Finally, a series of simulation results for the maglev vehicle are included to show the feasibility and superiority of the proposed control approach.

Keywords: Maglev vehicles · Magnetic Levitation System (MLS) · Adaptive control · RBF neural network

1 Introduction

The EMS Type maglev vehicles, as shown in Fig. 1, are new type of urban transportation [1–3]. The maglev vehicle has the comprehensive advantages of low vibration noise, strong climbing ability, low cost and little pollution. At present, with the increasing demand for speed and quality of transportation, maglev train will have a broad space for future railway development and application. The magnetic levitation

© Springer Nature Singapore Pte Ltd. 2020
H. Zhang et al. (Eds.): NCAA 2020, CCIS 1265, pp. 381–394, 2020.
https://doi.org/10.1007/978-981-15-7670-6_32

control technology is one of the core and key technologies for maglev vehicle. The stability of MLS of maglev vehicle is the basis and prerequisite for realizing normal operation of maglev train [4].

(a)   (b)   (c)   (d)

Fig. 1. Maglev train line: (a) Incheon line, (b) Changsha line, (c) Shanghai line, (d) Emsland line

The MLS of maglev vehicle has numerous characteristics such as strong nonlinearity, open loop instability, time-varying parameters and external disturbances, which bring great challenges to control design. The transport capacity problem of maglev trains has been plagued maglev technology experts and engineers. Most of the currently commercial levitation control algorithms are based on linear control theory [4–6], however, the MLS is a typical non-linear system. When the control system based on linear simplified design is subject to large mass variation and moves away from the equilibrium point, the system may lose stability. Therefore, the maglev trains have a strict limit on the number of passengers, and its transport capacity is lower than traditional transportation methods such as subways and light rails. How to deal with the time-varying mass becomes the key to improve the passenger transport capacity.

In recent years, researchers around the world have actively explored the levitation control system and tried to apply different control algorithms to the MLS in maglev vehicles. MacLeod [5] *et al.* presented a LQR-based levitation optimal controller based on frequency domain weighting method to improve the ability to suppress disturbance during the operation. Sinha [6, 7] *et al.* adopted an adaptive control algorithm based on model reference to control maglev trains to reduce the impact of nonlinearity. Sun [8] *et al.* proposed a novel saturated learning-based control law with adaptive notch filter to deal with external disturbance. Wai [9] *et al.* developed an observer-based adaptive fuzzy-neural-network control method for hybrid maglev transportation system with only the position state feedback. Sun [10] *et al.* presented a new nonlinear tracking controller to achieve simultaneous tracking control and mass identification for magnetic suspension systems. Morales [11] *et al.* combined precise linearization with output feedback control to propose a GPI controller for the laboratory magnetic levitation machine. In addition, many artificial intelligence algorithms, such as fuzzy logic [12], neural network [13], genetic algorithm [14], support vector machine [15] and so on, have also emerged in the field of magnetic levitation control. However, these control strategies cannot strictly guarantee the stability of the closed-loop system theoretically. On the other side, when the model parameters change greatly, these methods must be re-adjusted or re-learned, which causes a lot of inconvenience to practical application. Furthermore, the ever-changing passengers can degrade the control performance.

Therefore, the levitation control technology has gradually developed from only focusing on stability initially to focus on stability while paying more attention to the improvement of comprehensive performance.

To counter the problems above, in this paper, the dynamic model of minimum levitation unit of the maglev vehicle is deduced firstly. Then, a nonlinear controller is designed directly based on nonlinear equations without any linearization. Next, tackle the unknown or time-varying mass of vehicle body, an adaptive RBF neural network is utilized to identify the time-varying mass on-line. The stability of the closed-loop system is analyzed based on Lyapunov techniques. A series of simulations of MLS of maglev vehicle with time-varying mass are performed.

2 Model of MLS of Maglev Vehicle

The structure of an EMS type maglev vehicle can be illustrated in Fig. 2. Electromagnets are arranged along the left and right sides of the train. However, due to the structural decoupling function of the suspension chassis [16], each levitation controller can be considered independent of each other within a certain range. Hence, it is typical to study a single-point levitation system. Therefore, we can consider the single point levitation system as the minimum levitation unit for controller design.

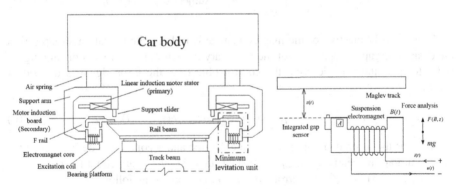

Fig. 2. Maglev vehicles structure and minimum levitation unit.

The electromagnetic force can be expressed as follows:

$$F_m = \frac{dW_m}{dz} = \frac{d}{dz}\int w_m dv = \frac{d}{dz}\left(\frac{1}{2}BHV\right) = \frac{d}{dz}\left(\frac{B^2 Az}{\mu_0}\right) = \frac{B^2 A}{\mu_0} \qquad (1)$$

where, F_m denotes electromagnetic attraction force; w_m is the magnetic field energy density; W_m is the magnetic field energy of volume V; A is the magnetic pole area of the electromagnet; μ_0 is air permeability.

Because the magnetic flux density can be expressed as $B = \mu_0 Ni(t)/2z(t)$, if inserted into (1) the following is achieved:

$$F_m = \frac{\mu_0 A N^2}{4} \left[\frac{i(t)}{z(t)} \right]^2 \tag{2}$$

where, N denotes the number of turns of the electromagnet coil; $z(t)$ is the suspension air gap; $i(t)$ represents the current of the solenoid coil.

According to Newton's law and current saturation, we have:

$$m\ddot{z} = -\frac{\mu_0 A N^2}{4} \left\{ \left[\frac{\mathrm{sat}_{imax}[i(t)]}{z(t)} \right]^2 \right\} + mg \tag{3}$$

where, m is the nominal mass of vehicle body; g is the acceleration of gravity; $\mathrm{sat}_{imax}(i)$ is defined as:

$$\mathrm{sat}_{imax}(i) \triangleq \begin{cases} i_{max}, & i > i_{max}, \\ i(t), & 0 \leq i \leq i_{max} \end{cases} \tag{4}$$

The saturated nonlinearity can be removed by a bilateral constraint, and the dynamic equation is transformed as follows:

$$m\ddot{z} = -\frac{\mu_0 A N^2}{4} \left[\frac{i(t)}{z(t)} \right]^2 + mg, \quad \text{subject } 0 \leq i(t) \leq i_{max} \tag{5}$$

The control objective of the maglev vehicle is to keep the levitation airgap $z(t)$ at the desired airgap z_d regardless of the time-varying mass in the process of running. In order to quantitatively describe the control objective, the error signal is defined as follows:

$$e(t) = z(t) - z_d; \quad \dot{e}(t) = \dot{z}(t) - \dot{z}_d; \quad \ddot{e}(t) = \ddot{z}(t) - \ddot{z}_d \tag{6}$$

where, z_d is the desired airgap, $\dot{z}_d$ and $\ddot{z}_d$ are the first and second order derivatives of the desired airgap, respectively. If z_d is constant, then $\dot{z}_d$ and $\ddot{z}_d$ are equal to zero.

Based on (5) and (6), the error system can be derived as follows:

$$m\ddot{e} = -\frac{\mu_0 A N^2}{4} \left[\frac{i(t)}{z(t)} \right]^2 + m(g - \ddot{z}_d) \tag{7}$$

3 Controller Design and Stability Analysis

Based on the constructed dynamic model of maglev vehicle, this section will design a neural network-based adaptive controller to achieve stable levitation and accurately estimate the time-varying mass simultaneously.

3.1   Controller Design

To achieve the control objective, a positive definite function $E(t)$, based on the error signals, depicts as:

$$E(t) = \frac{1}{2}k_p e^2 + \frac{1}{2}m\dot{e}^2 \tag{8}$$

whose derivative, by using (7), can be calculated as:

$$\begin{aligned}
\dot{E}(t) &= k_p e\dot{e} + \dot{e}(m\ddot{e}) \\
&= k_p e\dot{e} + \dot{e}\left[-(\mu_0 AN^2/4)\left[i(t)^2/z(t)^2\right] + m(g - \ddot{z}_d)\right]
\end{aligned} \tag{9}$$

where, $k_p \in \mathbb{R}^+$ is a positive control parameter.

As the airgap $z(t)$ can be measured by the sensor in real time, if nominal mass kept unchanged, a nonlinear controller could be developed as follows:

$$i(t) = \sqrt{\frac{4z^2\left[k_p e + k_d \arctan(\dot{e}) + mg - m\ddot{z}_d\right]}{\mu_0 AN^2}} \tag{10}$$

where, $k_d \in \mathbb{R}^+$ is another positive control parameter.

Theorem 1. If nominal mass m kept unchanged, the proposed nonlinear levitation controller (10) can achieve asymptotic stability of the closed-loop system in the sense that

$$\lim_{t\to\infty} e(t) = 0, \ \lim_{t\to\infty} \dot{e}(t) = 0 \tag{11}$$

Proof. To prove the Theorem 1, the positive definite Lyapunov candidate function is selected as (8). Then, differentiating (8) and inserting (10) yields

$$\dot{E}(t) = -k_d \dot{e} \arctan(\dot{e}) \leq 0 \tag{12}$$

The $\dot{E}(t)$ is negative semi-definite, so the system with the presented control law is Lyapunov stable, and we can obtain

$$E(t) \in L_\infty \Rightarrow \dot{z}, e, z, \dot{e} \in L_\infty \tag{13}$$

LaSalle's invariance theorem is utilized to prove the asymptotic stability of the closed-loop system. A set S is defined as:

$$S \triangleq \left\{(e, \dot{e})/\dot{E} = 0\right\} \tag{14}$$

The largest invariant set M contained in S can be calculated as:

$$\dot{E}(t) = 0 \Rightarrow \dot{e} = 0, \dot{z} = 0, \ddot{z} = 0 \tag{15}$$

For maglev vehicles, the desired airgap z_d is always constant, so $\ddot{z}_d$ are equal to zero. Insert (22) into (5) to obtain:

$$k_p e + k_d \arctan(\dot{e}) = 0 \tag{16}$$

Since $\dot{e} = 0$, the e shall be equal to zero. Therefore, it can be concluded that M contains only the equilibrium point $[e, \dot{e}]^T = [0, 0]^T$.

According to LaSalle's invariance theorem, we can conclude that the system is asymptotic stability.

However, the nominal mass m is time-varying or even unknown. To address this problem, a modified control law is developed as follows:

$$i(t) = \sqrt{\frac{4z^2 \left[k_p e + k_d \arctan(\dot{e}) + \hat{m}(E)g - \hat{m}(E)\ddot{z}_d \right]}{\mu_0 A N^2}} \tag{17}$$

where, $\hat{m}(E)$ denotes the online identification for m utilizing radial basis function (RBF) neural network.

The RBF neural network algorithm can be depicted as follows.

$$r_j(E - P_i) = \exp\left(-\|E - P_i\|^2 / 2\sigma_j^2 \right), \; i = 1, 2 \; and \; j = 1, 2, \cdots, n \tag{18}$$

$$m = W^T R(E) + \varepsilon \tag{19}$$

where, E represents input of the neural network; j denotes the jth node of the hidden layer of the neural network; $R = [r_1, r_2, \cdots, r_n]$ is output of the gaussian function. $W = [w_1, w_2, \cdots, w_n]$ denotes the weight matrix of neural network; ε is network approximation error, $\varepsilon \leq \varepsilon_M$; $P_i = [p_{i1}, p_{i2}, \cdots, p_{in}]$ is the coordinate matrix of the center point of the hidden layer Gaussian function.

The proposed control law chooses $E = [e_1, e_2]^T = [e, \dot{e}]^T$ as the input of neural network. The online identification output can be given as:

$$\hat{m}(E) = \hat{W}^T R(E) \text{ or } \hat{m}(E) = \sum_{i=1}^{2} \sum_{j=1}^{n} w_j r_j(e_i - p_{ij}) \tag{20}$$

where, $\hat{W}$ denotes the estimation value of the ideal weight matrix W^*.

The adaptive law of the $\hat{W}$ can be designed as follows.

$$\dot{\hat{W}} = \gamma g \dot{e} R(E) \tag{21}$$

where, $\gamma \geq 0$; g denotes acceleration of gravity.

3.2 Stability Analysis

To begin with, by substituting (17) into (5), the error dynamics can be calculated as:

$$\ddot{e} = -\frac{k_p}{m}e - \frac{k_d}{m}\arctan(\dot{e}) - \frac{g}{m}(\hat{m}(\boldsymbol{E}) - m) \qquad (22)$$

Before stability analysis, an assumption is made here.

Assumption 1. A sufficient number of hidden nodes are included into the neural network to obtain sufficient approximation accuracy, so that there exists an optimal weight matrix as follows:

$$\boldsymbol{W}^* = \arg\min_{\boldsymbol{W}\in\Omega}[\sup|\hat{m}(\boldsymbol{E}) - m|] \qquad (23)$$

The mass approximation error is defined as:

$$\eta_t = \hat{m}(\boldsymbol{E}|\boldsymbol{W}^*) - m \qquad (24)$$

Hence, (22) can be rewritten as follows:

$$\ddot{e} = -\frac{k_p}{m}e - \frac{k_d}{m}\arctan(\dot{e}) - \frac{g}{m}\left\{[(\hat{m}(\boldsymbol{E}|\hat{\boldsymbol{W}}) - \hat{m}(\boldsymbol{E}|\boldsymbol{W}^*)] + \eta_t\right\} \qquad (25)$$

By substituting (20) into (25), we can get:

$$\ddot{e} = -\frac{k_p}{m}e - \frac{k_d}{m}\arctan(\dot{e}) - \frac{g}{m}\left[(\hat{\boldsymbol{W}} - \boldsymbol{W}^*)^T \boldsymbol{R}(\boldsymbol{E}) + \eta_t\right] \qquad (26)$$

To analyze the system stability with proposed neural network-based adaptive controller, we construct the following Lyapunov candidate function:

$$V = \frac{k_p}{2m}e^2 + \frac{1}{2}\dot{e}^2 + \frac{1}{2\gamma m}\left[(\hat{\boldsymbol{W}} - \boldsymbol{W}^*)^T (\hat{\boldsymbol{W}} - \boldsymbol{W}^*)\right] \qquad (27)$$

where, $\gamma \in \mathbb{R}^+$ is a positive constant. V is positive definite.

Then, we have

$$\begin{aligned}
\dot{V} &= \frac{k_p}{m}e\dot{e} + \dot{e}\ddot{e} + \frac{1}{\gamma m}(\hat{\boldsymbol{W}} - \boldsymbol{W}^*)^T \dot{\hat{\boldsymbol{W}}} \\
&= \frac{k_p}{m}e\dot{e} + \dot{e}\left\{-\frac{k_p}{m}e - \frac{k_d}{m}\arctan(\dot{e}) - \frac{g}{m}\left[(\hat{\boldsymbol{W}} - \boldsymbol{W}^*)^T \boldsymbol{R}(\boldsymbol{E}) + \eta_t\right]\right\} + \frac{1}{\gamma m}(\hat{\boldsymbol{W}} - \boldsymbol{W}^*)^T \dot{\hat{\boldsymbol{W}}} \\
&= -\frac{k_d}{m}\dot{e}\arctan(\dot{e}) - \frac{\dot{e}g}{m}\left[(\hat{\boldsymbol{W}} - \boldsymbol{W}^*)^T \boldsymbol{R}(\boldsymbol{E}) + \eta_t\right] + \frac{1}{\gamma m}(\hat{\boldsymbol{W}} - \boldsymbol{W}^*)^T (\gamma \dot{e}g \boldsymbol{R}(\boldsymbol{E})) \\
&= -\frac{k_d}{m}\dot{e}\arctan(\dot{e}) - \frac{\dot{e}g}{m}\eta_t
\end{aligned}$$

$$\qquad (28)$$

Since $-\frac{k_d}{m}\dot{e}\arctan(\dot{e}) \le 0$, we can design the RBF neural network to make the approximation error small enough, so that $\dot{V} \le 0$.

4 Numerical Simulation

In this section, a series of numerical simulations are provided to demonstrate the effectiveness of the proposed controller for the MLS with the time-varying mass. The system parameters of the maglev vehicle are listed in Table 1.

Table 1. Parameter values for maglev system.

Parameters	Value	Parameters	Value
m	600 kg	z_0	0.016 m
N	340	z_d	0.008 m
μ_0	$4\pi \times 10^{-7}\ H \cdot m^{-1}$	A	0.02 m^2

The control gains of the neural network-based adaptive (NNBA) control approach are selected as $k_p = 9 \times 10^6$; $k_d = 3 \times 10^5$. The structure of RBF neural network is selected as 2-5-1. $\gamma = 3 \times 10^6$, $g = 9.8$, $\sigma = 0.5$ The initial value of the network weight w is set to zero. The center points P is set as follows:

$$P = \begin{bmatrix} -0.02 & -0.01 & 0 & 0.01 & 0.02 \\ -0.02 & -0.01 & 0 & 0.01 & 0.02 \end{bmatrix}$$

In addition, the basic controller without mass identification is utilized for comparison, whose k_p and k_d remain the same.

Simulation studies are performed for the three cases.

Case 1: $m = 600$.

Case 2: $m = 600 + \sum\limits_{i=1}^{4} 200 step(t - i)$.

Case 3: $m = 600 + 200\sin(5t)$.

The simulation results of case 1 are illustrated in Figs. 3, 4 and 5. We can learn that if the mass information is unknown, static error of 0.654 mm is existing in the system. However, the NNBA control law can remove the static error and estimate the mass accurately.

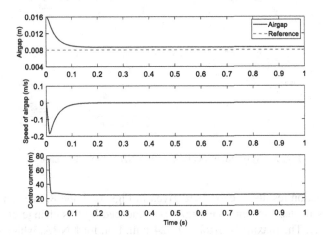

Fig. 3. Simulation results: basic control in case 1.

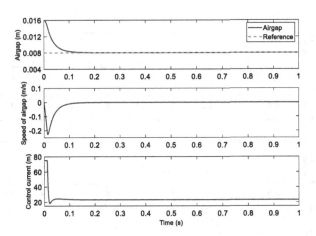

Fig. 4. Simulation results: NNBA control in case 1.

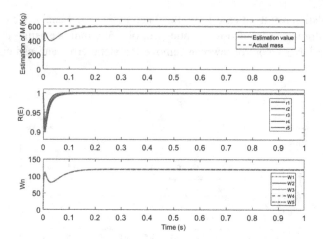

Fig. 5. Mass identification with NNBA control in case 1.

The simulation results of case 2 are given in Figs. 6, 7 and 8. We can see that the mass changes in step type and the airgap is influenced by this change obviously with basic controller. The maximum error is 1.524 mm. But, for NNBA, which can identify the mass change quickly and accurately, can keep the levitation airgap $z(t)$ at the desired airgap z_d regardless of the time-varying mass.

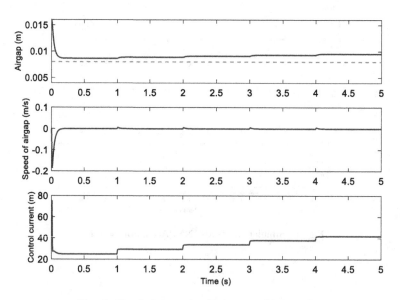

Fig. 6. Simulation results: basic control in case 2.

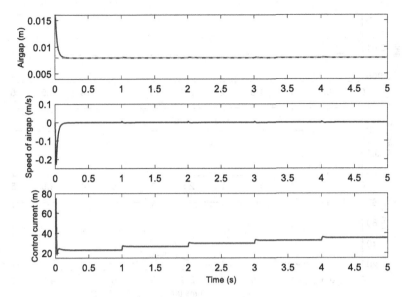

Fig. 7. Simulation results: NNBA control in case 2.

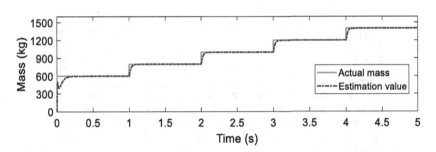

Fig. 8. Time-varying mass identification with NNBA control in case 2.

The simulation results of case 3 are provided in Figs. 9, 10 and 11. We can learn that the mass changes in sine type, which cause the airgap to fluctuate in the type of sine. The maximum error is 0.924 mm. The NNBA, however, can quickly and accurately estimate the time-varying mass and ensure the system error converges to zero.

In summary, the simulation results demonstrated that with the proposed neural network-based adaptive control approach, the overall control performance is well guaranteed even the mass changes sharply, implying that the stable levitation task of the MLS can be well achieved, and the time-varying mass can be quickly and accurately estimated during the levitation process.

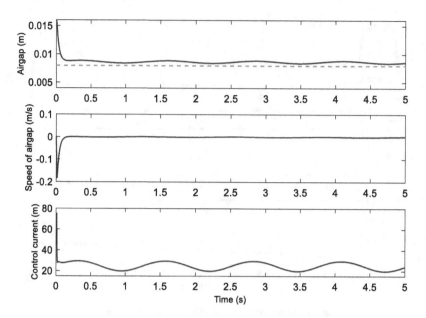

Fig. 9. Simulation results: basic control in case 3.

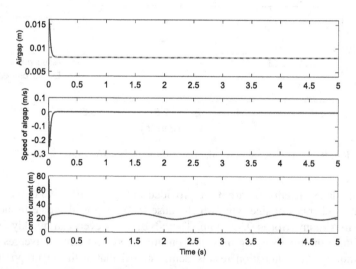

Fig. 10. Simulation results: NNBA control in case 3.

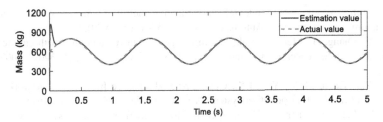

Fig. 11. Time-varying mass identification with NNBA control in case 3.

5 Conclusion

In this paper, to accomplish the stable levitation control objective, a neural network-based adaptive control approach for MLS of EMS Type maglev vehicles. The minimum levitation dynamic model of the maglev vehicle was presented. Then, a basic nonlinear control law was developed, and the asymptotic stability is proven by Lyapunov technology and LaSalle's invariance theorem. The RBF neural network was utilized to estimate the unknown or time-varying mass of vehicle body. A Lyapunov-based stability analysis is provided for the NNBA controller. The simulation results indicated that the adaptive NNBA control scheme can be more effective than the basic nonlinear control scheme. The time-varying mass can be effectively identified with our proposed approach, and the system error can converge to zero. Our future works will focus on the adjusting rules of control parameters of NNBA controller, so that it can be more convenient to apply the control strategy to the full-scale maglev train.

Acknowledgement. This work was supported in part by the National Key Technology R&D Program of the 13th Five-year Plan (2016YFB1200601), by the National Natural Science Foundation of China under Grant 51905380, by China Postdoctoral Science Foundation under Grant 2019M651582.

References

1. Lee, H.W., Kim, K.C., Lee, J.: Review of maglev train technologies. IEEE Trans. Magn. **42**(7), 1917–1925 (2006)
2. Sun, Y., Xu, J., Qiang, H., Chen, C., Lin, G.: Fuzzy H∞ robust control for magnetic levitation system of maglev vehicles based on T-S fuzzy model: design and experiments. J. Intell. Fuzzy Syst. **36**(2), 911–922 (2019)
3. Thornton, R.D.: Efficient and affordable maglev opportunities in the United States. Proc. IEEE **97**(11), 1901–1921 (2009)
4. Boldea, I., Tutelea, L., Xu, W., et al.: Linear electric machines, drives, and MAGLEVs: an overview. IEEE Trans. Ind. Electron. **65**(9), 7504–7515 (2017)
5. MacLeod, C., Goodall, R.M.: Frequency shaping LQ control of maglev suspension systems for optimal performance with deterministic and stochastic inputs. IEE Proc.-Control Theory Appl. **143**(1), 25–30 (1996)
6. Sinha, P.K., Hadjiski, L.M., Zhou, F.B., et al.: Electromagnetic suspension: new results using neural networks. IEEE Trans. Magn. **29**(6), 2971–2973 (1993)

7. Sinha, P.K., Pechev, A.N.: Model reference adaptive control of a maglev system with stable maximum descent criterion. Automatica **35**(8), 1457–1465 (1999)
8. Sun, Y., Qiang, H., Mei, X., et al.: Modified repetitive learning control with unidirectional control input for uncertain nonlinear systems. Neural Comput. Appl. **30**(6), 2003–2012 (2017)
9. Wai, R.J., Chen, M.W., Yao, J.X.: Observer-based adaptive fuzzy-neural-network control for hybrid maglev transportation system. Neurocomputing **175**, 10–24 (2016)
10. Sun, N., Fang, Y., Chen, H.: Tracking control for magnetic-suspension systems with online unknown mass identification. Control Eng. Pract. **58**, 242–253 (2017)
11. Morales, R., Feliu, V., Sira-Ramirez, H.: Nonlinear control for magnetic levitation systems based on fast online algebraic identification of the input gain. IEEE Trans. Control Syst. Technol. **19**(4), 757–771 (2011)
12. Zhang, M., Jing, X.: A bioinspired dynamics-based adaptive fuzzy SMC method for half-car active suspension systems with input dead zones and saturations. IEEE Trans. Cybern. (2020). https://doi.org/10.1109/tcyb.2020.2972322
13. Sun, Y., Xu, J., Qiang, H., Lin, G.: Adaptive neural-fuzzy robust position control scheme for maglev train systems with experimental verification. IEEE Trans. Ind. Electron. **66**(11), 8589–8599 (2019)
14. Kusagawa, S., Baba, J., Shutoh, K., et al.: Multipurpose design optimization of EMS-type magnetically levitated vehicle based on genetic algorithm. IEEE Trans. Appl. Supercond. **14**(2), 1922–1925 (2004)
15. Liu, C., Rong, G.: SVM α order inverse system decoupling time-varying sliding mode control of double suspension systems of machining center. China Mech. Eng. **26**(5), 668–674 (2015)
16. Qiang, H., Li, W., Sun, Y., et al.: Levitation chassis dynamic analysis and robust position control for maglev vehicles under nonlinear periodic disturbance. J. VibroEng. **19**(2), 1273–1286 (2017)

A Novel Collision-Avoidance TDMA MAC Protocol for Fog-Assisted VCNs

Bingyi Liu[1]([✉]), Jing Qin[1], Dongxiao Deng[1], Ze Wang[1], Shengwu Xiong[1], and Luyao Ye[2]

[1] Wuhan University of Technology, Wuhan 430070, China
{byliu,qinjing,dxdeng,cs_wangze,xiongsw}@whut.edu.cn
[2] City University of Hong Kong, Kowloon 999077, Hong Kong
luyaoye2-c@my.cityu.edu.hk

Abstract. In recent years, the advances of vehicular communication networks (VCNs) have greatly promoted the development of intelligent transportation systems (ITSs). Moreover, by the aid of the vehicular communication and computing technologies, the driving safety and traffic efficiency can be substantially improved. However, the contention-based nature and inability to handle hidden-terminal problems of existed vehicle-to-vehicle (V2V) and vehicle-to-infrastructure (V2I) communications may incur high packet collision probability. To solve this problem, we develop a fog-assisted communication framework for message dissemination in VCNs. Specifically, we propose a centralized time division multiple access (TDMA)-based multiple access control (MAC) protocol that combines the advantages of fog computing to predict and avoid the packet collisions in advance, so that the channel utilization and packet transmission can be improved significantly. Simulation results confirm the efficiency of the proposed fog-oriented TDMA-based MAC protocol in urban scenarios.

Keywords: Vehicular communication networks (VCNs) · Intelligent Transportation Systems (ITSs) · Medium Access Control (MAC) · Fog computing · Collisions avoidance

1 Introduction

The Intelligent Transportation Systems (ITSs) are currently undergoing key technological transformation, as more and more vehicles are connected to each other and to the Internet. In order to deal with the potential risks and traffic anomalies in the increasing complex traffic situations, vehicles need to cooperate with each other, rather than make decisions on their own. Obviously, to achieve

This work was supported by National Natural Science Foundation of China (No. 61802288), Research Funding Project of Qingdao Research Institute of WHUT (No. 2019A05), Major Technological Innovation Projects in Hubei Province (No. 2019AAA024), Hong Kong Research Grant Council under project GRF (No. 11216618).

such objectives, it is expected to design reliable and efficient communication frameworks and protocols which can ensure the freshness of information and the high throughput of data transmission. Nevertheless, since vehicles share a common wireless channel, an inappropriate access of the wireless channel may result in transmission collisions. In order to provide an efficient communication channel, several Time Division Multiple Access (TDMA) based Medium Access Control (MAC) protocols are proposed, which can ensure data transmission with ultra-low delays through the time-slotted access scheme.

However, packet collisions still occur frequently in the situations with high traffic density. There are mainly two types of collisions: access collisions and encounter collisions. An access collision happens when two or more vehicles within two hops of each other attempt to acquire the same available time slot. On the other hand, an encounter collision happens when two or more vehicles acquiring the same time slot become members of the same two-hop set due to vehicle mobility. In recent works, the centralized TDMA-based scheduling protocol can significantly avoid the two types of collisions by efficient time-slots management. For instance, [1] proposed a Collision-Predicted TDMA MAC protocol in which the roadside unit (RSU) allocates the time slots to vehicles to avoid access collisions, and predicts the encounter collisions in the upcoming time slots, and then adjusts the time slots allocation to avoid the encounter collisions. Although the RSU-assisted centralized TDMA-based scheduling can effectively avoid the packet collisions, there are still several issues that have not been fully addressed. First, the infrastructures (such as RSUs) deployment and maintenance are quite costly and labor-consuming. Second, in urban roads, low-latency communication between vehicles and infrastructures is a challenging issue due to the roadside obstacles, rapid topology change and unbalanced traffic.

Nowadays, due to flexibility of deployment and its ability to provide low-latency and high-efficient data traffic, fog computing is becoming more and more attractive in vehicle communication field. In study [2], the authors make an initial effort on proposing a fog computing empowered architecture together for data dissemination in software defined heterogeneous vehicular ad-hoc networks (VANETs). The authors in [3] propose a two-layer vehicular fog computing (VFC) architecture to explore the synergistic effect of the cloud, the static fog and the mobile fog on processing time-critical tasks in Internet of Vehicles (IoV). In this study, we consider a novel time slot allocation method based on fog computing, which utilizes buses as fog nodes instead of roadside infrastructures to provide time slots management and reduces the possibility of encounter collisions. Benefiting from collaboration among cloud, fog nodes and ordinary vehicles, city-level real-time network can be constructed to obtain local and global information for packet collision prediction and avoidance. Our main contributions in this paper can be summarized as follows.

(1) We develop a fog-assisted communication framework for message dissemination in VCNs. Buses are considered as fog nodes to collect local and global information from the cloud and surrounding vehicles.

(2) We propose a centralized TDMA-based MAC protocol that combines the advantages of fog computing to predict and avoid the packet collisions in advance.

(3) Extensive simulation experiments confirm the efficiency of the proposed protocol. The collisions can be reduced significantly under different traffic load and a higher packet transmission rate can be achieved.

The remainder of this paper is organized as follows. We first discuss the related work in Sect. 2. Section 3 describes the system model and problem formulation. In Sect. 4, we elaborate the proposed fog-assisted TDMA-based MAC protocol. We conduct simulations to validate the protocol in Sect. 5, before concluding the paper in Sect. 6.

2 Related Work

In this section, we discuss related work about TDMA-based message dissemination methods and fog-oriented VANETs.

Many TDMA-based MAC protocols have been developed for VANETs in recent years. DMCMAC [4] proposes a method that the number of time slots changed adaptively with traffic densities to allow the vehicles to broadcast messages. MoMAC [5] assigns time slots to vehicles according to the underlying road topology and lane distribution on roads with the consideration of vehicles' mobilities. VeMAC [6] assigns disjoint sets of time slots to RSUs and vehicles moving in opposite directions. In this method, three sets of slots (L, R and F) are related to the vehicles moving in left, right and the RSUs respectively, hence can decrease the rate of transmission collisions on the control channel which caused by vehicles mobility. Based on VeMAC protocol, CFR MAC [7] considers the mobility features of VANETs and assigns disjoint sets of time slots to vehicles according to their driving speed. This mechanism proposes each time slots set of L, R and F can be further divided into three subsets associated to three speed intervals respectively: High, Medium and Low.

Some existing researches mainly focus on predicting packet collisions in advance based on traffic information to eliminate collisions. The authors in [8] utilize variable transmission power to enhance message delivery ratio and reduce the transmission collisions, and pre-detect and eliminate encounter collisions by cooperation among the vehicles. SdnMAC [9] designs a novel roadside openflow switch (ROFS) which is controlled by the openflow controller as the roadside unit, and the controller schedules the cooperative sharing of time slot information among ROFSes. As each ROFS is aware of the time slots information of adjacent nodes, pre-warning of encounter collisions can be provided to vehicles. Although sdnMAC can adapt to the topology change and varying density of vehicles, the centralized algorithms suffer from poor scalability since the sparse deployment and communication coverage of RSUs. In [10], PTMAC protocol is proposed which utilizes vehicles on the road as Intermediate Vehicles to predict whether vehicles using the same time slot outside the two hop range will encounter. However, the use of intermediate vehicles needs extra coordinating overhead and transmission delays.

Recently, much attention has been paid to fog-oriented VANETs. Ullah *et al.* [11] propose a emergency messages dissemination scheme which utilize fog nodes as servers to reduce the accessibility delays and congestion. Pereira *et al.* [12] propose a generic architecture for the deployment of fog computing and services in VANETs environment, which can provide reliable information in a considerable shorter period of time and reduce the amount of traffic over the VANETs backhaul infrastructure. Xiao *et al.* [13] present a hierarchical architecture, which integrates the paradigm of both fog computing and the software defined networking (SDN), and maximizes the bandwidth efficiency by coordinating the service in both the fog layer and the cloud layer in a novel problem which called Cooperative Service in Vehicular Fog Computing (CS-VFC).

Although the aforementioned researches on TDMA-based MAC protocols and fog-oriented VANETs are important to improve communication efficiency, few of which have combined these two research fields and utilize the advanced capability of fog computing for time slots management in TDMA-based communication. In this study, we propose a fog-assisted TDMA-based MAC protocol that utilizes buses as the fog nodes to predict collisions and allocate time slots to eliminate collisions. To the best of our knowledge, this is a first attempt that a TDMA-based MAC protocol in VCN is designed with the consideration of abilities from fog computing.

3   Architecture and Problem Formulation

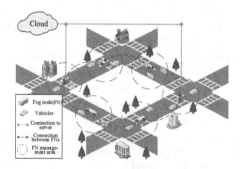

Fig. 1. Network architecture.

Here, we elaborate on the proposed fog-assisted network architecture and the definition of fog areas.

3.1   Architecture

Figure 1 shows the proposed fog-assisted wireless networks which consist of three tiers of nodes. The low-tier nodes are ordinary vehicles, which collect instantaneous traffic information and broadcast the traffic information to mid-tier nodes.

The mid-tier nodes, namely, Fog Nodes (FNs) that are responsible for collecting local information about surrounding vehicles and managing channel resources in the management area. The high-tier node is the Cloud, which can provide global traffic information and guidance. We assume each vehicle is equipped with a global positioning system (GPS) device and broadcasts a message at every frame, includes its traffic status, such as location, speed, and moving direction.

The general message dissemination process under the proposed architecture is as follows. At the beginning, ordinary vehicles broadcast their basic information to neighboring fog nodes. Then, the fog nodes should broadcast the basic information and neighborhood tables. Meanwhile, the fog nodes can get global information from the cloud and broadcast control message to ordinary vehicles in the management area. The fog nodes should also dynamically report their basic information to the cloud, such as geographic coordinates, access delay and bandwidth.

In order to manage the channel resources efficiently, buses are selected as fog nodes and form the fog area to guarantee the broadcast coverage in target area. To this end, the road is divided into multiple fog areas with an equal length of L_f and fog area is further divided into $\frac{L_f}{R}$ sub-areas. Let L_f denotes the integral multiple of the communication range R of vehicles. The fog nodes are uniformly distributed along the road and each sub-area has a fog node as a manager to manage ordinary vehicles. In work [14], the authors have theoretically verified that the number of hops between two buses is less than or equal to 1. Also, the real traffic data in [15] shows that the bus density of urban cities in daytime is 3–30 buses per kilometer and most of the vehicles could connect with a bus node in their communication range. We provide an example of fog area as shown in Fig. 2 which has six sub-areas labeled as F1–F6. Without loss of generality, the ordinary vehicles in the fog area are called fog member vehicles (FMVs), and buses are called fog nodes (FNs) while the vehicles out of the fog area are dissociate vehicles (DVs).

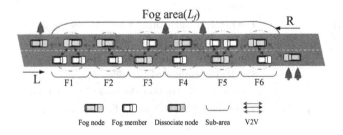

Fig. 2. Fog area.

3.2   Problem Formulation

We set up the same rules as in [16] to regulate the time sequence of FMVs' beacons.

Constraint 1: To avoid beacon collisions, all FMVs within the two-hop transmission range are allocated with non-overlapping time slots, i.e.,

$$\forall v_j \in N_{cch}^2(v_i), TS(v_j) \neq TS(v_i) \tag{1}$$

Constraint 2: To maximize the channel utilization, any two FMVs out of each other's two-hop communication range could be potentially allocated with the same time slot, i.e.,

$$\exists v_j \notin N_{cch}^2(v_i), TS(v_j) = TS(v_i) \tag{2}$$

Constraint 3: To avoid channel congestion, beacon bandwidth should not exceed the time slots that the system can totally support, i.e.,

$$\sum_i F(v_i) \leq N_{slot}, v_i \in Area(F_n, F_{n+1}, F_{n+2}), n = 1, 2, 3, 4; \tag{3}$$

The symbols in this section are summarized in Table 1. Our target is to maximize the number of vehicles that can get the channel resources to guarantee the basic safety message dissemination, and maximize successful safety packet receptions $R(v_i)$ (without collisions) of each FMVs. This problem can be formalized as follows.

$$max \sum_{i=1}^{n} F(v_i) \times R(v_i) \tag{4}$$

Table 1. Summary of the notations

Symbol	Description
V	The set of ordinary vehicles in the network
v_i	Vehicle i in set V
$N_{cch}(v_i)$	The one-hop neighboring set of the vehicle v_i
$N_{cch}^2(v_i)$	The two-hop neighboring set of the vehicle v_i
N_{slot}	The number of time slots in per frame
$TS(v_i)$	Slot ID currently used by vehicle v_i
$F(v_i)$	A value 1 indicates that the vehicle v_i acquired a time slot
$Area(F_n, F_{n+1}, F_{n+2})$	Three consecutive sub-areas F_n, F_{n+1}, F_{n+2}

4   Protocol Design

In this section, we first use the simulated annealing algorithm to solve the problem in Eq. (4). Then we propose a novel centralized TDMA protocol which solves encounter collisions by the collaboration of fog computing. It shall be noted that the access collisions can be fully avoided with the centralized TDMA protocols. Thus, unless explicitly stated, the collisions we mentioned mean encounter collisions in the remaining of this paper. In our framework, the fog-assisted TDMA protocol is implemented by three steps. First, a sub-area switching should be detected by FNs based on FMVs' information since collisions may happen after FMVs switch the sub-area. Then, the collisions among FMVs need to be predicted based on real-time traffic information. Finally, a time slots rescheduling approach is needed to eliminate the upcoming collisions. We provide the detailed descriptions of these steps in the following parts of this section.

4.1 Simulated Annealing Algorithm Design

As shown in Algorithm 1, the simulated annealing algorithm is actually a greedy algorithm, but its search process introduces random factors. When updating the feasible solution iteratively, simulated annealing algorithm accepts a solution that is worse than the current solution with a certain probability. Therefore, it may jump out of current local optimal solution and approach the global optimal solution. We simplify the initial solution to x and the objective function to $f(x)$, and introduce the simulated annealing algorithm to find the optimal solution.

Algorithm 1. Simulated annealing algorithm for optimal solution

---

Input: An initial solution x; initial objective function $f(x)$;
Output: The optimal solution
1: Perturbation produces new solutions x_{new}, and calculate the objective function $f(x_{new})$
2: Calculation $temp = f(x_{new}) - f(x)$
3: **if** $temp < 0$ **then**
4:     Accept new solution x_{new}
5: **else**
6:     Calculate probability p
7: **end if**
8: **if** $p > 1$ **then**
9:     Accept new solution x_{new}
10: **else**
11:     **if** Number of iterations is reached || the termination condition is met **then**
12:         return the optimal solution
13:     **else**
14:         go to line 2
15:     **end if**
16: **end if**

4.2 Detection of Sub-area Switching

In this part, some specifications and definitions are made to facilitate our further discussions.

Definition of Sub-area Edge. Due to the difference of vehicles' velocity in a time period, FMVs have different sub-area edge values x which is defined as the distance between the border of sub-area and the FMV. If the FMV is driving to the right, this boundary is referred to the right boundary of sub-area. For the convenient calculation of the sub-area edge values x, speeds of the FMVs are divided into multiple intervals. The sub-area edge value x is calculated as follows:

$$x = V_{max} \times T \qquad (5)$$

where V_{max} is the maximum value in each speed interval, and T stands for a short duration time.

Definition of Time Slot Usage Table. Then we define the time slot usage table (TSUT) as shown in Fig. 3 and S_{ID} indicates the number of the time slot. According to *Constraint* 2, any two FMVs out of each other's two-hop communication range could be allocated with the same time slot, thus the last two sections in TSUT distinguished by different colors record the information of vehicles in the two sub-areas. The information includes $Flag_1$ which indicates whether the time slot is used by the FMVs, vehicle ID V_{ID}, driving direction Dir and the vehicle location in sub-area $SubArea$. Each fog node maintains a TSUT. Once the TSUT is updated, fog nodes send the latest table to the neighboring fog nodes.

| S_{ID} | Flag1 | V_{ID} | Dir | SubArea | Flag2 | V_{ID} | Dir | SubArea |

Fig. 3. Time slot usage table (TSUT).

Once a FMV enters the sub-area edge, the neighboring FN will check the TSUT whether there is another FMV sharing the same time slot. We call this neighboring FN as coordination FN (*coorFN*). The FMV entering the sub-area edge is considered as a dangerous vehicle (*dangeFMV*). We say that there is a potential collision between *dangeFMV* and another FMV if they share a time slot. The details of potential collisions prediction and elimination are described in the next section.

4.3   Potential Collisions Prediction and Elimination

Here, an example is given for describing the potential collisions prediction. When a FMV A (i.e., *dangeFMV*) gets close to the sub-area edge and uses the same time slot with FMV B, the neighboring FN (i.e., *coorFN*) of A needs to predict whether the potential collision between A and B will happen in the future. The prediction can be made based on the kinetic status of vehicles, such as the locations, speeds and moving directions of A and B. First the *coorFN* gets the latest information of B from neighboring FNs since B is out of the communication range of the $coorFN$. Then the $coorFN$ will make the potential collision prediction for $dangeFMV$ A.

In the case when FMVs A and B drive in the same direction, they are likely to catch up with each other if the one behind has a faster speed. Assuming A locates behind B, potential collision will happen if the distance between them can be reduced to $2L$ in a short time, where L represents the length of a sub-area. This can be shown in the following equations:

$$\begin{cases} (V_a - V_b) \times T \geqslant D_{AB} - 2L, \ (if \quad V_b < V_a) \\ T = min\{T_f, \frac{D_{fog}}{V_b}\} \end{cases} \quad (6)$$

where V_a and V_b denote the speeds of A and B respectively. $\frac{D_{fog}}{V_b}$ denotes the time required for B to drive out of the fog area, and D_{AB} is the current distance between A and B. T_f denotes a short time that enables the $dangeFMV$ to change its time slot with high success probability. Since T is a short period of time, we regard V_a and V_b as a constant value within T. If Eq. (6) is satisfied, this potential collision has to be eliminated. Otherwise, we can ignore this potential collision.

In another case when A and B drive in the opposite direction, the potential collisions between them can be ignored when they are running farther away from each other. However, a potential collision may happen between A and B that they are running toward each other. If A and B satisfy the following condition, this potential collision should be eliminated timely.

$$(V_a + V_b) \times T \geqslant D_{AB} - 2L \quad (7)$$

Next, we introduce how to eliminate the upcoming potential collisions. First, we give the priority for time slots according to the USTB so that FNs can choose the higher priority time slots to eliminate potential collisions. The time slots are prioritized from two aspects: (1) The high priority time slots are the idle time slots which are not used by any FMVs; (2) The reduced priority time slots are the occupied time slots, and the priority decreases with the decrease of distance between the current vehicle and the vehicle that occupies the time slot, which means the time slots that are occupied by the further vehicles have a higher priority for the current FMVs. Then the $dangeFMV$ is selected as the vehicle to switch time slot since it can communicate with $coorFN$ directly without further packet forwarding. Finally, the $coorFN$ selects a new time slot with the high priority for the $dangeFMV$ based on the TSUT and sends a packet including the ID of the new time slot. The $dangeFMV$ will use the new time slots to send message in upcoming frames.

5   Performance Evaluation

In this section, we first describe the experiments settings, then evaluate the performance of our proposed protocol.

5.1   Simulation Settings

In the experiments, we choose an open-source vehicular network simulator Veins which combines OMNET++ for discrete event-based network simulation and Simulation of Urban MObility (SUMO) for the generation of real-world mobility models, including the road map, traffic lights, and vehicle's moving pattern. For the traffic scenario, we deploy vehicles on a bidirectional city road with four lanes in either direction. The buses are uniformly distributed along the road and each sub-area has a bus as a manager to manage ordinary vehicles. Ordinary vehicles are running at different speeds and a vehicle can catch up with other vehicles if its speed is faster. The simulation parameters are specified in Table 2. The transmitting power is set to meet the requirement of the communication range with R = 300 m for all vehicles. The average speed of ordinary vehicles is set to 30 m/s. In this paper, we compare the performance of our scheme with A-VeMAC [17], which the frame partitioning is not equal and the frame can adaptively vary with the vehicle traffic conditions. We evaluate the performance of our proposed protocols and A-VeMAC in terms of number of collisions and Packet Transmission Rate (PTR). The number of collisions is the sum of the number of encounter collisions and the number of access collisions. The packet transmission rate is defined as the ratio of the successful transmission numbers to the total transmission numbers.

Table 2. Simulation parameters

Parameter	Value
Road length	2000 m
Number of lines (each direction)	4
Traffic load	0.2, 0.4, 0.6, 0.8, 1.0
Communication range	300 m
Speed mean value	30 m/s
Frame duration	100 ms
Slots/frame	70, 75, 80, 85, 90
Simulation time	100 s

5.2 Performance Evaluation

Impact of Traffic Load on Protocol Performance. We introduce a parameter called traffic load which is defined in [17]. Figure 4 represents the total number of collisions and the packet transmission rate with traffic load 0.2, 0.4, 0.6, 0.8 and 1.0, respectively. As shown in Fig. 4(a), the total number of collisions of our scheme is much lower than the A-VeMAC since our scheme predicts collisions and eliminates them in a timely manner. Figure 4(b) shows the packet transmission rate of the two schemes. We can see that the packet transmission rate of our scheme is higher than A-VeMAC, and the packet transmission rate of our scheme is not greatly affected by different values of traffic load and keeps a steady and high level. In our proposed scheme, fog nodes eliminate collisions in advance, thus the packet transmission rate of our scheme is relative higher than that of A-VeMAC.

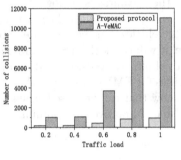
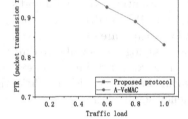

(a) Number of collisions   (b) Packet transmission rate

Fig. 4. Performance with different traffic load

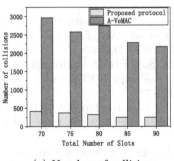
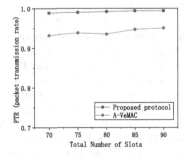

(a) Number of collisions   (b) Packet transmission rate

Fig. 5. Performance with different number of slots

Impact of Time Slots Number on Protocol Performance. In this part, we evaluate the impact of the time slots number on the effectiveness of our scheme. The number of time slots in a frame is changed from 70 to 90 and its impact on the performance is investigated. Figure 5 shows the collision number and the packet transmission rate with different time slots. Obviously, as the number of time slots increases, the number of collisions in both schemes decreases in Fig. 5(a), but the number of collisions of our scheme is much lower than A-VeMAC. Figure 5(b) shows that the packet transmission rate of our scheme and A-VeMAC are affected slightly by the number of time slots, and our scheme can achieve a higher packet transmission rate.

6   Conclusion

In this paper, we have proposed a fog-assisted TDMA-based MAC protocol to effectively eliminate collisions by exploiting the advantages of fog computing technologies. In our protocol, the buses act as fog nodes and collect local and globe information from cloud and surrounding vehicles. Once a fog node detects that the FMVs are driving into sub-area edge, it will check the TSUT and detect if FMVs with shared time slots will collide. To eliminate collisions and minimize packet loss, fog nodes will give different priorities for FMVs to acquire their time slots. The extensive simulation experiments show that the proposed scheme not only can eliminate collisions efficiently but also can improve the communication efficiency significantly.

References

1. Lin, L., Hu, B.J., Wei, Z., Wu, C.: A collision-predicted TDMA MAC protocol in centralized vehicular network. In 2017 IEEE 17th International Conference on Communication Technology (ICCT), pp. 593–597. IEEE (2017). https://doi.org/10.1109/ICCT.2017.8359705
2. Liu, K., Xiao, K., Dai, P., Lee, V., Guo, S., Cao, J.: Fog computing empowered data dissemination in software defined heterogeneous VANETs. IEEE Trans. Mob. Comput. (2020). https://doi.org/10.1109/TMC.2020.2997460
3. Liu, C., Liu, K., Guo, S., Xie, R., Lee, V.C.S., Son, S.H.: Adaptive offloading for time-critical tasks in heterogeneous internet of vehicles. IEEE Internet Things J. (2020). https://doi.org/10.1109/JIOT.2020.2997720
4. Lin, Z., Tang, Y.: Distributed multi-channel MAC protocol for VANET: an adaptive frame structure scheme. IEEE Access **7**, 12868–12878 (2019). https://doi.org/10.1109/ACCESS.2019.2892820
5. Lyu, F., et al.: MoMAC: mobility-aware and collision-avoidance MAC for safety applications in VANETs. IEEE Trans. Veh. Technol. **67**(11), 10590–10602 (2018). https://doi.org/10.1109/TVT.2018.2866496
6. Omar, H.A., Zhuang, W., Li, L.: VeMAC: a TDMA-based MAC protocol for reliable broadcast in VANETs. IEEE Trans. Mob. Comput. **12**(9), 1724–1736 (2012). https://doi.org/10.1109/TMC.2012.142

7. Zou, R., Liu, Z., Zhang, L., Kamil, M.: A near collision free reservation based MAC protocol for VANETs. In2014 IEEE Wireless Communications and Networking Conference (WCNC), pp. 1538–1543. IEEE (2014). https://doi.org/10.1109/WCNC.2014.6952438

8. Li, S., Liu, Y., Wang, J.: An efficient broadcast scheme for safety-related services in distributed TDMA-based VANETs. IEEE Commun. Lett. **23**(8), 1432–1436 (2019). https://doi.org/10.1109/LCOMM.2019.2923412

9. Luo, G., Li, J., Zhang, L., Yuan, Q., Liu, Z., Yang, F.: SdnMAC: a software-defined network inspired MAC protocol for cooperative safety in VANETs. IEEE Trans. Intell. Transp. Syst. **19**(6), 2011–2024 (2018). https://doi.org/10.1109/TITS.2017.2736887

10. Jiang, X., Du, D.H.: PTMAC: a prediction-based TDMA MAC protocol for reducing packet collisions in VANET. IEEE Trans. Veh. Technol. **65**(11), 9209–9223 (2016). https://doi.org/10.1109/TVT.2016.2519442

11. Ullah, A., Yaqoob, S., Imran, M., Ning, H.: Emergency message dissemination schemes based on congestion avoidance in VANET and vehicular FoG computing. IEEE Access **7**, 1570–1585 (2018). https://doi.org/10.1109/ACCESS.2018.2887075

12. Pereira, J., Ricardo, L., Luís, M., Senna, C., Sargento, S.: Assessing the reliability of fog computing for smart mobility applications in VANETs. Future Gener. Comput. Syst. **94**, 317–332 (2019). https://doi.org/10.1016/j.future.2018.11.043

13. Xiao, K., et al.: A fog computing paradigm for efficient information services in VANET. In: 2019 IEEE Wireless Communications and Networking Conference (WCNC), pp. 1–7. IEEE (2019). https://doi.org/10.1109/WCNC.2019.8885810

14. Liu, B., Jia, D., Wang, J., Lu, K., Wu, L.: Cloud-assisted safety message dissemination in VANET-cellular heterogeneous wireless network. IEEE Syst. J. **11**(1), 128–139 (2015). https://doi.org/10.1109/JSYST.2015.2451156

15. Zeng, L., et al.: A bus-oriented mobile FCNs infrastructure and intra-cluster BSM transmission mechanism. IEEE Access **7**, 24308–24320 (2019). https://doi.org/10.1109/ACCESS.2019.2900392

16. Liu, B., et al.: Infrastructure-assisted message dissemination for supporting heterogeneous driving patterns. IEEE Trans. Intell. Transp. Syst. **18**(10), 2865–2876 (2017). https://doi.org/10.1109/TITS.2017.2661962

17. Chen, P., Zheng, J., Wu, Y.: A-VeMAC: an adaptive vehicular MAC protocol for vehicular ad hoc networks. In2017 IEEE International Conference on Communications (ICC), pp. 1–6. IEEE (2017). https://doi.org/10.1109/ICC.2017.7997358

Design Optimization of Plate-Fin Heat Exchanger Using Sine Cosine Algorithm

Tianyu Hu[1], Lidong Zhang[1(✉)], Zhile Yang[2(✉)], Yuanjun Guo[2], and Haiping Ma[3]

[1] Northeast Electric Power University, Jilin 132012, China
nedu1015@aliyun.com
[2] Shenzhen Institute of Advanced Technology, Chinese Academy of Sciences, Shenzhen 518055, China
zl.yang@siat.ac.cn
[3] Shaoxing University, Shaoxing 312099, China

Abstract. Plate-fin heat exchanger (PFHE) is a popular and featured compact heat exchanger, which transfers heat at relatively low-temperature differences and has a high heat transfer surface area to volume ratio. Due to the compact size and light weight, it has been widely used in various engineering areas such as energy, transportation and aerospace. The optimal design of PFHE aims to minimize the economic cost or maximize the efficiency, formulating a mixed integer non-linear optimization problem which challenges the optimization tools. In this paper, the number of entropy generation units is formulated as the objective functions. Given the strong non-linear behaviors, a recent proposed meta-heuristic algorithm named sine cosine algorithm (SCA) is adopted to solve the problem of the design and optimization of the PFHE and compared with the results of several classical and new algorithms. Test results show that the sine cosine algorithm is the a competitive solver for PFHE optimal design problems.

Keywords: Plate-fin heat exchanger · Optimization design · Sine cosine algorithm

1 Introduction

Energy use is one of the most important issues in the world, in particular under the unprecedented global warming scenarios during recent years. Improving energy conservation and efficiency not only reduces the world's energy consumption but also directly reduce the economic cost. Heat exchanger is a featured heat exchange equipment, which can be used for the cooling or heating process of fluid. According to their structure, common heat exchangers can be divided into the Plate heat exchanger, the fixed tube-sheet exchanger, the Shell and tube heat exchanger (STHE), and the floating heat exchanger, etc.

The PFHE is mainly made of aluminum with a compact structure and lightweight. It is often used in energy, chemical, aerospace, railway and other

H. Zhang et al. (Eds.): NCAA 2020, CCIS 1265, pp. 408–419, 2020.
https://doi.org/10.1007/978-981-15-7670-6_34

engineering areas. The design of PFHE directly influences the efficiency of energy utilization and economic benefit. Therefore, the design of the PFHE optimization problem has become a hot issue. Common decision variables of PFHE optimal design problem include the geometry size, the density, the number of layers and the shape of the fin, etc. These design variables are of continual and integer forms, leading to the problem become a nonlinear optimization task. In previous researches, PFHE design mainly relies on the experience and intuition of the designers, providing significant potentials to improve the efficiency by intelligent algorithms.

In the recent years, a number of studies have been addressed in terms of the design optimization of PFHE. Sanaya et al. [15] used non-dominated sorting based algorithm (NSGA II) and took the heat exchanger of maximum effectiveness (heat recovery) and the minimum total cost as the target, obtaining a set of Pareto optimal solution. In 2013, Rao et al. [13] used a modified teaching-learning-based algorithm (TLBO) to optimize STHE and PFHE with maximum heat exchanger efficiency and minimum cost as the optimization objectives. It is proved that TLBO is effective in heat exchanger design optimization compared with genetic algorithm (GA). In 2011, Rao et al. [12] used TLBO to optimize the PFHE and minimize the total number of entropy generation units. In 2014, Petal et al. [11] proposed a multi-objective TLBO (MO-ITLBO) algorithm considering annual cost, the total weight of the heat exchanger as well as total pressure drop as the optimization goal, proves the validity of the MO-ITLBO. In 2018, Zarea et al. [19] proposed a bee algorithm hybrid with particle swarm optimization (BAHPSO) to optimize the PFHE with maximum effectiveness and the minimum total annual cost (TAC). It is proved that BAHPSO is more accurate than other algorithms.

Though a number of methods have been adopted, the complex optimal designing problems of different types of heat exchanger still call for powerful computational methods. The sine cosine algorithm (SCA) was proposed by Mirjalili et al. [9] and have been utilized in solving various engineering optimization problems [1,3]. In the design problem of PFHE, the search characteristic of SCA algorithm may make it more potential. The work in this article is more about design optimization problems and getting the only optimal design. Repetitive optimization problems are different from repetitive optimization problems, which focus more on stable performance [18]. In this paper, it will be used in the optimal designing problem of PFHE. The rest of this paper is organized as follows: Sect. 2 describes the design content of PFHE and the thermodynamic model; Sect. 3 introduces the sine cosine algorithm and demonstrates the operation principle. In Sect. 4, numerical results are illustrates and compared. Finally, Sect. 5 concludes the paper.

2   Problem Formulation

Plate-fin is one of most commonly used heat exchanger type. It uses simple but efficient structure to get the heat exchanged in an well planned scenarios.

Nomenclature			
ϵ	Effectiveness	m	Mass flow rate (kg/s)
A	Heat exchanger surface area (m^2)	max	Maximum
Aff	Free flow area (m^2)	min	Minimum
c	Cold stream	N	Number of fin layers
Cp	Specific heat (J/kg K)	n	Fin frequency
Dh	Hydraulic diameter (m)	NTU	Number of transfer units
e	Euler's number, e = 2.71828...	Pr	Prandtl number
H	Height of the fin (m)	Q	Heat duty (W)
h	Hot stream	Re	Reynolds number
i	Inlet	$Rcte$	Specific gas constant (J/kg K)
L	Heat exchanger length (m)	s	Fin spacing (m)
l	Lance length of the fin (m)	t	Fin thickness (m)

Figure 1 shows a multi-stream PFHE. The three colored arrows in the figure represent three fluids. Figure 2 shows a detailed structure diagram of the plate-fin heat exchanger. The PFHE in Fig. 2 demonstrates that for the rectangular flat fins, dimensions of the fins, thickness and number of fins, as well as the number of layers of plates are to be designed and should be determined in the optimal design problem formulation.

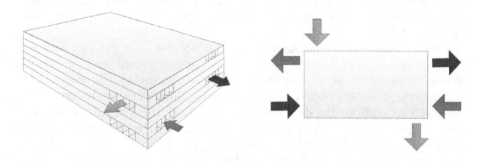

Fig. 1. Multistream plate-fin heat exchanger.

For the cross-flow PFHE, the formula of heat transfer can be represented as,

$$Q = \epsilon C_{min} (T_{hi} - T_{ci}) \tag{1}$$

where C_{min} is minimum heat capacity rate, $C_{min} = m * C_p$, T_{hi}, T_{ci} are the hot fluid and cool fluid inlet temperature. The effectiveness ϵ is obtained from [5] as shown below,

$$\epsilon = 1 - e^{\left(\left(\frac{1}{C_r} \right) NTU^{0.22} \left(e^{\left(-C_r NTU^{0.78} \right)} - 1 \right) \right)} \tag{2}$$

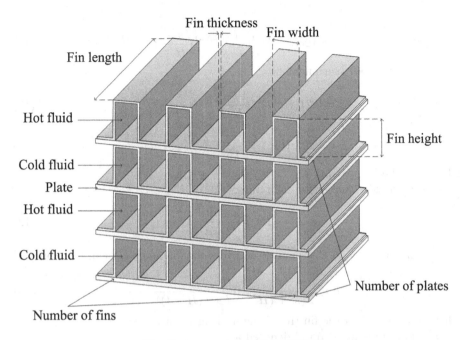

Fig. 2. Arrangement of the PFHE.

where NTU for heat transfer unit number, specific heat ratio $C_r = C_{min}/C_{max}$, e is the Euler's number, NTU from [5,16] as shown below,

$$\frac{1}{NTU} = C_{min} \left(\frac{A_{ffh}}{j_h Cp_h Pr_h^{-0.667} m_h A_h} + \frac{A_{ffc}}{j_c Cp_c Pr_c^{-0.667} m_c A_c} \right) \quad (3)$$

where Aff for the free flow area (m^2), and m is the mass flow rate (kg/s).

The free flow area of hot and cold fluids are as below,

$$Aff_h = (H_t - t_h)(1 - n_h t_h) L_c N_h \quad (4)$$

$$Aff_c = (H_c - t_c)(1 - n_c t_c) L_h N_c \quad (5)$$

where N is the number of fin layers, and $N_c = N_h + 1$.

The heat transfer areas of hot and cold fluid in the PFHE id denoted as below,

$$A_h = L_h L_c N_h (1 + (2n_h (H_h - t_h))) \quad (6)$$

$$A_c = L_h L_c N_h (1 + (2n_c (H_c - t_c))) \quad (7)$$

where j, f is the Colburn factor and Fanning friction factor [6]. For laminar flow, $(Re \le 1500)$ is denoted as shown below,

$$j = 0.53(Re)^{-0.5} (\frac{l}{D_h})^{-0.15} (\frac{s}{H} - t)^{-0.14} \quad (8)$$

$$f = 8.12(Re)^{-0.74}(\frac{l}{D_h})^{-0.41}(\frac{s}{H} - t)^{-0.02} \tag{9}$$

For $(Re > 1500)$ (turbulent flow),

$$j = 0.21(Re)^{-0.4}(\frac{l}{D_h})^{-0.24}(\frac{s}{H} - t)^{0.02} \tag{10}$$

$$f = 1.12(Re)^{-0.36}(\frac{l}{D_h})^{-0.65}(\frac{s}{H} - t)^{0.17} \tag{11}$$

where s is the fin spacing, $s = (1/n - t)$.

The Reynolds number for each fluid,

$$Re = \frac{m \cdot dh}{A_{ff}\mu} \tag{12}$$

For the fin shape in this paper, the hydraulic diameter Dh can be expressed as:

$$dh = \frac{4s(H - t)l}{2(sl + (H - t)l + t(H - t)) + ts} \tag{13}$$

where s is spacing of the fin (m), l the fin length (m).

The fluid pressure drop id denoted as:

$$\Delta P = \frac{2fL\left(\frac{m}{A_{ff}}\right)^2}{\rho Dh} \tag{14}$$

where ρ is the density (kg/m^3), Dh is the hydraulic diameter (m). Finally, the entropy generation units [2],

$$Ns = (1 - \epsilon)\left[\frac{(T_{ci} - T_{hi})^2}{T_{ci}T_{hi}}\right] + \left(\frac{Rcte_h}{Cp_h}\right)\left(\frac{\Delta P_h}{P_{hi}}\right) + \left(\frac{Rcte_c}{Cp_c}\right)\left(\frac{\Delta P_c}{P_{ci}}\right) \tag{15}$$

Ns is the optimized objective function in this paper. We can express the optimization problem as follows,

$$Minimise f(\chi) = Ns \tag{16}$$

The optimization variables in this paper are subject to the following constraints (length unit: mm):

$$\left. \begin{array}{l} g_1(\chi) \Rightarrow 100 \leq L_h \leq 1000 \\ g_2(\chi) \Rightarrow 100 \leq L_c \leq 1000 \\ g_3(\chi) \Rightarrow 2 \leq H \leq 10 \\ g_4(\chi) \Rightarrow 100 \leq n \leq 1000 \\ g_5(\chi) \Rightarrow 0.1 \leq t \leq 0.2 \\ g_6(\chi) \Rightarrow 1 \leq l \leq 10 \\ g_7(\chi) \Rightarrow 1 \leq N_h \leq 10 \\ g_8(\chi) \Rightarrow \xi(\chi) - Q = 0 \end{array} \right\} \tag{17}$$

where $g_8(X)$ is the heat transfer rate constraint of the heat exchanger, and $\xi(X)$ can be calculated by Eq. 1. In this paper, the size of Q is set as 160 kW. The design of a gas - to - air crossflow PFHE is in need, and the size is 1 m by 1 m associated with the size requirements shown in Eq. 17.

The data for the two fluids are given in Table 1.

Table 1. Two fluid data.

Parameters	Hot fluid	Cold fluid
Mass flow rate, m(kg $\cdot$s^{-1})	0.8962	0.8296
Inlet temperature, T$_i$(K)	513	277
Inlet pressure, P$_i$(Pa)	100000	100000
Specific heat, Cp(J $\cdot$kg$^{-1}\cdot$K^{-1})	1017.7	1011.8
Density, ρ(kg $\cdot$ m^{-3})	0.8196	0.9385
Dynamic viscosity, μ(N $\cdot$ s $\cdot$ m^{-2})	241.0	218.2
Prandtl number, Pr	0.6878	0.6954
Heat duty of the exchanger, Q(kW)	160	

3   Sine Cosine Algorithm Technique

The SCA was put forward by Mirjalili et al. [9] in 2016. Generally speaking, the idea of the algorithm mainly comes from several aspects includes the fundamental scheme of evolutionary techniques and the evolving process of mathematical vibration effects. Swarm intelligence techniques are often inspired by the behavior of natural lives, such as birds, insects and mammals. Techniques from human social behaviors, such as human teaching and learning behaviors, the rules of human race and practice in industrial production, have all be considered. There are also algorithms inspired from physics, such as black holes, gravity and metal-making.

Most algorithms operates the formation of a collection of random solution. After that, the set of random solutions will be improved according to the formula defined by the algorithm, and the objective function are continuously calculated. These directions are fairly random determined by the evolutionary logic, and the probability of finding the optimal solution can be improved by the continuous calculation. In SCA, this scheme is different. The value of solution is constantly changing by adding more stochastic items, mimics the value vibration as in the sine/cosine wave. The core evolutionary logic of SCA can be represented as below,

$$X_i^{t+1} = \begin{cases} X_i^t + r_1 \times sin(r_2) \times \left| r_3 P_t^i - X_i^t \right|, & r_4 < 0.5 \\ X_i^t + r_1 \times cos(r_2) \times \left| r_3 P_t^i - X_{i_i}^t \right|, & r_4 \geq 0.5 \end{cases} \quad (18)$$

where X is the position, t is the $t-th$ iteration, P is final position, r_1, r_2 and r_3 are random numbers. Considering that there are sines and cosines (Fig. 3) in the formula, the algorithm is named sines and cosine algorithm. To equilibrium the algorithm convergence and exploitation performance, r_1 takes an adaptive change:

$$r_1 = a - t\frac{a}{T} \tag{19}$$

T in the formula is the maximum number of iterations, a is a constant.

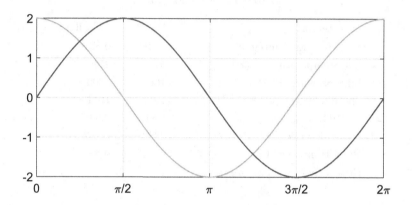

Fig. 3. One cycle of sine and cosine function image, range of $[-2, 2]$.

In the four parameters of SCA, r_1, r_2, r_3, r_4, r_1 determines the direction of movement, r_2 determines the distance of movement, r_3 determines the direction whether to move inward or outward, bringing randomness to the position, and r_4 determines whether to move by sine or cosine manner. The calculation principle of SCA as shown in Fig. 4

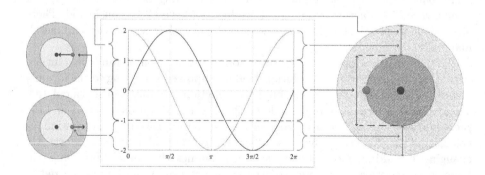

Fig. 4. A diagram of basic computing principles for SCA. (Color figure online)

The green dots represent the position of the solution in this calculation and the blue dots represent the position of the optimal solution. As can be seen from the Fig. 4, when the position of sine or cosine is at $[-1, 1]$, the algorithm will search for the optimal solution in the determined range. In the other space of $(-2, 2)$, the algorithm will search for a new range. By adopting this manner, SCA can avoid falling into local optimality, and the solution is searched around the optimal solution obtained so far. Therefore, SCA possesses high possibility in obtaining the optimal solution for any problems. In this paper, the SCA method is adopted for solving optimal design of PFHE. The flowchart of PFHE design optimization problem solved by SCA is shown in Fig. 5.

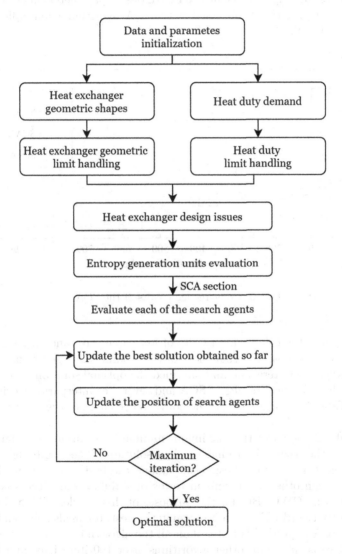

Fig. 5. PFHE design problem flowchart.

4   Results and Discussion

In addition to using SCA for calculation, real coding genetic algorithm (rcGA) [4], differential evolution (DE) [17], particle swarm optimization (PSO) [7], TLBO [14], moth-flame optimization (MFO) [8] and grey wolf optimizer (GWO) [10] have all been adopted for application. In the numerical study, we set the parameter number of iterations Gm to be 1000 and the population size Np to be 50.

The core of this paper is solving the PFHE design parameters under minimum Ns. The convergence results obtained by the SCA algorithm of four different runs are shown in Fig. 6.

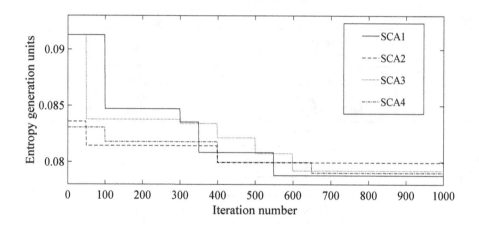

Fig. 6. Convergence results of the SCA.

In the figure, the calculation results of four times are randomly selected for drawing. It can be observed that most of the results converged from about 0.1 to 0.0825 within 100 iteration, and obtained the optimal solution in around 650 iteration. The statistical results of SCA and several counterparts algorithms are shown in Table 2, associated with the convergence curve of the algorithm given in Fig. 7.

The Table 2 compares the maximum, minimum, mean and standard deviation (Std) of the results for 10 independent solutions. The results indicate that the SCA algorithm gets the best result among the best values. It is also more competitive than other algorithms in terms of stability, ranks the second only outperformed by GWO. But for the purposes of this article, SCA had the best ability to solve the PFHE design problem in this test. It can also be seen from the Fig. 6 that SCA algorithm is faster than all the algorithms in the first 20 times, and the curve is similar to other algorithms after 100 iterations, and becomes the optimal solution of all the algorithms after 550 iterations. At 1000 iterations, the result is very close to TLBO but far better than other algorithms.

Table 2. The calculation result of the algorithm.

Algorithms	Entropy generation units			
	Best	Worst	Mean	Std(%)
rcGA [4]	0.084127	0.090710	0.086948	0.269
DE [17]	0.082986	0.106877	0.093368	0.834
PSO [7]	0.081034	0.086455	0.084047	0.200
TLBO [14]	0.078529	0.081652	0.080087	0.104
MFO [8]	0.080117	0.084737	0.082151	0.126
GWO [10]	0.078483	0.079229	0.078927	0.023
SCA	**0.078359**	0.084074	0.080079	0.149

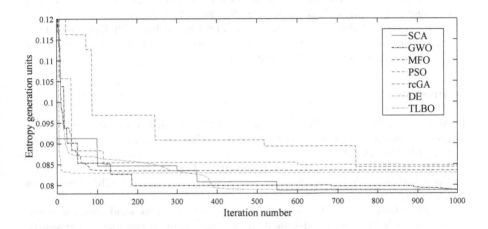

Fig. 7. Convergence curves for all algorithms.

Finally, according to the optimal solution obtained by SCA, we can obtain the optimal solution of PFHE design problem under the given scenario, which is shown in the Table 3.

Table 3. The optimum solution obtained by SCA

L_h, m	L_c, m	H, mm	t, mm	n, fins/m	l, mm	N_h	Ns	Q, kW
1	1	10	0.1	127.697	10	10	0.07445	162

5   Conclusions

In this paper, The design optimization problem for the PFHE is solved using SCA algorithms under given working conditions and designed constraints. Some other algorithms have been adopted to compared the computational results. The results indicate that the SCA has great potential in the design optimization of PFHE, and has achieved excellent results in obtaining the optimal solution. Compared with other algorithms, the randomness of SCA leads to a faster convergence speed and higher exploitation performance. In a result, SCA method has potential in solving the complex internal structural optimization design, including PFHE and other heat exchangers. It can also be used for solving further tasks such as multi-objective design optimization problems and repetitive optimization problems of heat exchangers under various structures and scenarios.

Acknowledgement. This research work is supported by the Zhejiang Provincial Natural Science Foundation of China under Grant No. LY19F030011, National Natural Science Foundation of Guangdong (No. 2018A030310671, 2016A030313177), Guangdong Frontier and Key Technological Innovation (No. 2017B090910013), Science and Technology Innovation Commission of Shenzhen (ZDSYS20190902093209795), and Outstanding Young Researcher Innovation Fund of Shenzhen Institute of Advanced Technology, Chinese Academy of Sciences (201822).

References

1. Attia, A.F., El Sehiemy, R.A., Hasanien, H.M.: Optimal power flow solution in power systems using a novel sine-cosine algorithm. Int. J. Electr. Power Energy Syst. **99**, 331–343 (2018)
2. Bejan, A.: Entropy Generation Minimization: The Method of Thermodynamic Optimization of Finite-size Systems and Finite-time Processes. CRC Press, New York (2013)
3. Fu, W., Wang, K., Li, C., Tan, J.: Multi-step short-term wind speed forecasting approach based on multi-scale dominant ingredient chaotic analysis, improved hybrid gwo-sca optimization and elm. Energy Convers. Manag. **187**, 356–377 (2019)
4. Herrera, F., Lozano, M., Verdegay, J.L.: Tackling real-coded genetic algorithms: operators and tools for behavioural analysis. Artif. Intell. Rev. **12**(4), 265–319 (1998)
5. Incropera, F.P., Lavine, A.S., Bergman, T.L., DeWitt, D.P.: Fundamentals of Heat and Mass Transfer. Wiley, Hoboken (2007)
6. Joshi, H.M., Webb, R.L.: Heat transfer and friction in the offset stripfin heat exchanger. Int. J. Heat Mass Transf. **30**(1), 69–84 (1987)
7. Kennedy, J., Eberhart, R.: Particle swarm optimization. In: Proceedings of ICNN 1995-International Conference on Neural Networks, vol. 4, pp. 1942–1948. IEEE (1995)
8. Mirjalili, S.: Moth-flame optimization algorithm: a novel nature-inspired heuristic paradigm. Knowl.-Based Syst. **89**, 228–249 (2015)
9. Mirjalili, S.: SCA: a sine cosine algorithm for solving optimization problems. Knowl.-Based Syst. **96**, 120–133 (2016)
10. Mirjalili, S., Mirjalili, S.M., Lewis, A.: Grey wolf optimizer. Adv. Eng. Softw. **69**, 46–61 (2014)
11. Patel, V., Savsani, V.: Optimization of a plate-fin heat exchanger design through an improved multi-objective teaching-learning based optimization (MO-ITLBO) algorithm. Chem. Eng. Res. Des. **92**(11), 2371–2382 (2014)

12. Rao, R.V., Patel, V.: Thermodynamic optimization of plate-fin heat exchanger using teaching-learning-based optimization (TLBO) algorithm. Optimization **10**, 11–12 (2011)
13. Rao, R.V., Patel, V.: Multi-objective optimization of heat exchangers using a modified teaching-learning-based optimization algorithm. Appl. Math. Model. **37**(3), 1147–1162 (2013)
14. Rao, R.V., Savsani, V.J., Vakharia, D.: Teaching-learning-based optimization: an optimization method for continuous non-linear large scale problems. Inf. Sci. **183**(1), 1–15 (2012)
15. Sanaye, S., Hajabdollahi, H.: Multi-objective optimization of shell and tube heat exchangers. Appl. Therm. Eng. **30**(14–15), 1937–1945 (2010)
16. Shah, R.K., Sekulic, D.P.: Fundamentals of Heat Exchanger Design. Wiley, Hoboken (2003)
17. Storn, R., Price, K.: Differential evolution-a simple and efficient heuristic for global optimization over continuous spaces. J. Global Optim. **11**(4), 341–359 (1997)
18. Yuen, S.Y., Lou, Y., Zhang, X.: Selecting evolutionary algorithms for black box design optimization problems. Soft. Comput. **23**(15), 6511–6531 (2018). https://doi.org/10.1007/s00500-018-3302-y
19. Zarea, H., Kashkooli, F.M., Soltani, M., Rezaeian, M.: A novel single and multi-objective optimization approach based on bees algorithm hybrid with particle swarm optimization (bahpso): Application to thermal-economic design of plate fin heat exchangers. Int. J. Therm. Sci. **129**, 552–564 (2018)

Mr-ResNeXt: A Multi-resolution Network Architecture for Detection of Obstructive Sleep Apnea

Qiren Chen[1], Huijun Yue[2], Xiongwen Pang[1(✉)], Wenbin Lei[2(✉)], Gansen Zhao[1], Enhong Liao[3], and Yongquan Wang[2]

[1] School of Computer Science, South China Normal University,
Guangzhou, China
1248105154@qq.com, augepang@163.com,
gzhao@m.scnu.edu.cn
[2] Otorhinolaryngology Hospital, The First Affiliated Hospital,
Sun Yat-sen University, Guangzhou, China
yhjent@163.com, leiwb2013@aliyun.com
[3] Guangdong Engineering Polytechnic, Guangzhou, China
315895436@qq.com

Abstract. Obstructive sleep apnea (OSA) is the most common sleep related breathing disorder causing sleepiness and several chronic medical conditions, for which the gold-standard diagnostic test is polysomnography (PSG), however the analysis of PSG is a time-consuming and labor-intensive procedure. To address these issues, we use deep learning as a new method to detect sleep respiratory events which can provide effective and accurate OSA diagnosis. We present a network named Mr-ResNeXt improved from ResNeXt, in which the 3×3 filters was replaced by a new block containing multi-level group convolution. The first level group convolution is used to exchange information between groups and the second level group convolution contains filters of different sizes which are used to extract features of different resolutions. All group convolutions involve residual-like connections. All the above changes help to extract multi-resolution image features more easily. Firstly, the experimental results show that our network can achieve a nearly 3% improvement (from 91.02% to 94.23%) comparing with the excellent networks in this field. Secondly, our network achieves nearly 1% accuracy improvement while comparing with ResNet and ResNetXt. Thus, we confirm that our method can improve the efficiency of detecting respiratory events.

Keywords: Obstructive sleep apnea · Respiratory event · Biomedical image classification · Mr-ResNeXt

1 Introduction

Obstructive sleep apnea (OSA) is a sleep disordered breathing characterized by upper airway obstruction during sleep resulting in chronic intermittent hypoxia (CIH) and fragmented sleep. Importantly, OSA is associated with several chronic medical

H. Zhang et al. (Eds.): NCAA 2020, CCIS 1265, pp. 420–432, 2020.
https://doi.org/10.1007/978-981-15-7670-6_35

conditions that contribute to poor quality of life [1]. The PSG is the gold-standard diagnostic test for OSA, which contains multiple sensors with simultaneous recording of airflow, blood oxygen levels, electrocardiogram (ECG), electroencephalogram (EEG), body position, and so on. The standard metric used to characterize OSA remains the apnea-hypopnea-index (AHI), which is defined as the number of sleep respiratory events (apneas and hypopneas) per hour during total sleep time.

There are two main sleep respiratory events associated with OSA [2] in Fig. 1. On the left, the normal waveform appears regularly, the middle one is hypopnea waveform which flow signal drops by 30% to 90%, and the apnea waveform which flow signal drops more than 90% is on the far right.

Because of the massive data, detecting sleep respiratory events is a time-consuming and labor-intensive procedure, some new schemes should be introduced to replace traditional PSG analysis. Deep learning solutions came into being.

To extract the features of pictures more effectively, we proposed Mr-ResNeXt, which is an improved network structure based on ResNeXt. We did a lot of comparative experiments in Sect. 4 to verify the effectiveness of Mr-ResNeXt, and the experimental results have confirmed the superiority of our proposed network.

The rest of this paper is arranged as follows: Sect. 2 mainly introduces the related research in this field. The dataset used in this paper and the proposed method are introduced in Sect. 3. Section 4 evaluates the results of this method and Sect. 5 is the conclusion and future work.

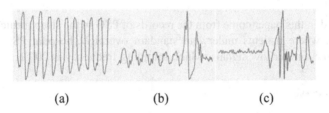

(a)   (b)   (c)

Fig. 1. (a) Normal airflow waveform, (b) hypopnea, (c) apnea

2   Related Work

Airflow data is the main factor in determining OSA, and it is also the most direct reflection of the severity of the respiratory event. Sleep Apnea Event Detection from Nasal Airflow Using CNN was analyzed by Haidar et al. [3], the dataset contained 2057 participants in sleep experiments between 2010 and 2011, which come from the MESA sleep study [4]. The paper proposed two classification methods, one was to use CNN for feature extraction, the other is to perform feature extraction first, and then use SVM for classification. The accuracy of the classification in CNN can reach 74.70%, while it reached 72% using the SVM method.

Almuhammadi et al. [5] proposed an efficient OSA classification based on EEG signals. This method is divided into four stages: filtering noise, bandwidth decomposition, feature extraction, and classification.

A multi-scale parallel convolutional neural network for automatic sleep apnea detection using single-channel EEG signals was proposed by Jiang et al. [6]. The innovation of this paper lies in the multi-scale Parallel Convolutional Neural Network.

Banluesombatkul et al. [7] proposed two classification methods. The dataset was taken from the MrOS sleep study (Visit 1) database [8]. The deep learning scheme combines multiple models, including one-dimensional CNN, LSTM and fully connected network.

Song et al. [9] proposed a method based on the recognition of the hidden Markov model (HMM) for OSA [10]. The results showed that classifying OSA by CNN and HMM can improve classification accuracy.

Vaquerizo et al. [11] proposed using CNN to automatically detect OSA events from SpO2 raw data. By using CHAT-baseline dataset, which contained 453 SpO2 recordings, the experiment achieved a classification accuracy of 88%.

Korkalainen et al. [12] used deep learning to solve the problem of manual scoring of sleep stages. For clinical data, the accuracy of the model through a single EEG signal and two channels (EEG + EOG) is 82% and 83% respectively. He concluded that deep learning can automatically detect sleep staging with an higher accuracy in patients with suspected OSA.

3 Methods

3.1 Data

The data used in this paper come from the records of PSG in the sleep medical center of the hospital, where patients underwent standard overnight hospital PSG. PSG data contained time and airflow signals, which sampling frequency is 200 Hz.

3.2 Preprocessing

Converting the original time series data into a spectrogram so that we can take full advantage of some classic neural networks to solve the problem. The original data need to be processed before the convert that every 8 h data was splited into 30-s segments, each segment has a corresponding label. As shown in Fig. 2, the blue box represents a 30 s airflow signal and the red box represents a respiratory event, the left side of the red box indicates the start time of the event and the right side indicates the end time of the event. In each 30 s, we determined a respiratory event based on whether it reached 90% or more of a standard duration. It is marked as 1 when hypopnea occurs while as 2 if apnea occurs.

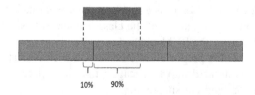

Fig. 2. OSA time percentage chart (Color figure online)

After the operation above, we can convert the segmented data into spectrograms. Short-time Fourier transform (STFT) [13], as the conversion method, is a general tool for time-frequency signal processing. The STFT can be described as Eq. 1, where $z(u)$ is input signal at time n and the g(u-t) is the window function.

$$STFT_z = \int_{-\infty}^{+\infty} [z(u)g(u-t)]e^{-j2\pi fu} du \tag{1}$$

In this way, the original input signal will be converted into a spectrogram and our dataset can be generated.

3.3 Problem Description

Figure 1 shows a standard airflow signal waveform, from which we can easily identify the type of airflow signal. But more than 60,000 airflow signal spectrograms in our dataset, a large number of it is not so easily identified. Figure 3 shows a set of non-standard airflow waveforms. The two pictures above are normal airflow waveforms while the below ones are abnormal, which are easy to confuse. This will increase the difficulty of picture classification after converting into spectrograms, so we need a network with stronger feature extraction capabilities.

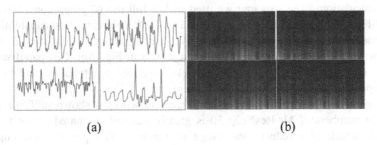

(a)   (b)

Fig. 3. (a) Irregular waveforms of airflow signal. (b) The spectrograms corresponding to (a)

3.4 Mr-ResNeXt: Multi-resolution ResNeXt

ResNet (Residual Network) [14] can solve the problem of "accuracy decreases as the network deepens", which was proposed by Kaiming He et al.

ResNeXt [15] is an upgraded version of ResNet, of which the core innovation is to propose aggregated transformations. A three-layer convolutional block of the original ResNet is replaced with a parallel stacked block of the same topology.

According to the idea of octave convolution [16], the image can be decomposed into low-frequency features and high-frequency features. One part of the convolution is focused on extracting low-frequency features, while the other is focused on extracting high-frequency features. Since the picture can be divided into high-frequency and low-frequency parts, it should also be divided into high-resolution parts and low-resolution parts [17]. The high-resolution features can be captured by large filters while low-resolution features can be captured by small ones.

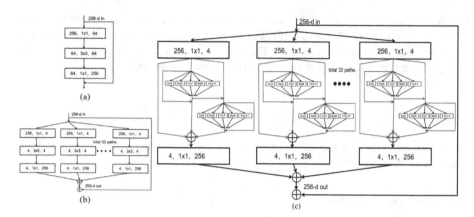

Fig. 4. (a) A block of ResNet. (b) A block of ResNeXt. (c) A block of Mr-ResNeXt

Group convolution [18] can increase the diagonal correlation between filters and reduce the training parameters. In our network, the group convolution is fully used. ResNeXt itself can be understood as a group convolution. The internal 3 × 3 filters have limitations in extracting features. Thus, it is possible to replace the 3 × 3 filters with a structure with more powerful feature extraction capabilities while maintaining similar calculations. The structure we used makes full use of group convolution and residual-like connections at both levels to maximize feature extraction capabilities and reduce the number of network parameters. According to the characteristics above, we named it Mr-ResNeXt, which architecture is shown in Fig. 4(c). Compared with the structure of ResNeXt (Fig. 4 (b)), these blocks have the same topology and share the same hyper-parameters (width and filter sizes).

The parameter comparison between different networks is shown in Table 1. The parameter numbers of Mr-ResNeXt-50 is greatly reduced compared to the first two networks, which also reduces the usage of memory. The performance improved because we use group convolution and shortcut connection in multiple places.

One disadvantages of PSG analysis is time-consuming. By full use of the group convolution, the number of parameters has dropped significantly compared to previous networks, which greatly improved the training speed of our network. The time spent by three networks training 20 epochs is shown in Table 1.

Table 1. Parameter comparison between different networks

	ResNet-50	ResNeXt-50	Mr-ResNeXt-50
# params.	25.5×10^6	25.0×10^6	9.3×10^6
FLOPs	4.1×10^9	4.2×10^9	1.6×10^9
Time(s)	33821	33711	32104

Since our network is composed of parallel and similar topologies, and each branch has the same spatial size, we consider using a more general function to explain it, whose corresponding formula is as follows:

$$f(x) = \sum_{k=1}^{C} R_k(x) \qquad (2)$$

Where $R_k(x)$ is a multi-resolution block and k represents the kth branch. C is the cardinality proposed by ResNeXt, which value in Mr-ResNeXt is 32.

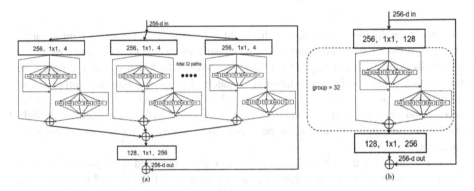

Fig. 5. Equivalent building blocks of Mr-ResNeXt

We can build equivalent blocks of our network, which is shown in Fig. 5. The difference between Fig. 4(c) and Fig. 5(a) is that the later implemented as early concatenation. Figure 5(a) is similar to the Inception-ResNet [19] block in that it involves branching and concatenating in the residual function. Figure 5(b) uses the notation of group convolutions, in which all the first and the last 1 × 1 layers were replaced with a single, wider layer.

3.5 Multi-resolution Block

We named the block in Fig. 6 a multi-resolution block, which is located between the upper and lower 1 × 1 convolutions. The output of the first 1 × 1 convolutional layer is evenly divided into 3 feature map subsets. Except for the first branch, group convolutions are added to each branch and shortcut connection is added to each group convolution. There is also a relationship between the second and the third branch, which promote information exchange between them. In this way, each branch can obtain important information from other branches.

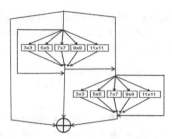

Fig. 6. Multi-resolution block

As shown in Fig. 6, we group the input information and establish information channels between different groups. This processing method is beneficial to extract the global and local information, and the receptive field of the block is added through the information exchange between different channels. Group convolutions composed of multiple convolution kernels (3×3, 5×5, 7×7, 9×9, 11×11) are used on the second and the third branch, which can not only easily extract the features of high-resolution part and low-resolution part, but also effectively reduce the number of parameters.

The input of the three branches from left to right is recorded as x_i, where $i \in \{1, 2, 3\}$, the corresponding output is recorded as y_i, where $i \in \{1, 2, 3\}$, and g_j represents a certain branch of the group convolution. Thus, the multi-resolution block can be expressed as:

$$y_i \begin{cases} x_i & i = 1 \\ \sum_{j=1}^{D} g_j(x_i) + x_i & i = 2 \\ \sum_{j=1}^{D} g_j(x_i + y_{i-1}) + (x_i + y_{i-1}) & i = 3 \end{cases} \tag{3}$$

To integrate the information of each branch, we concatenate all branches. This split and concatenation strategy make the convolution processing features more effectively. Multi-resolution blocks are inspired by residual blocks and group convolution, which can reduce parameters while ensuring performance. ResNeXt also adopts this idea, and we extend it.

4 Experiments

Qualitative Analysis of Normal and Abnormal
Our first task is to determine whether there is a respiratory event, the Mr-ResNeXt was analyzed by feature extraction capability and comprehensive performance.

Feature Extraction Capability
One of the performance evaluation metrics of a network is the feature extraction capability. The difficulty mentioned in Sect. 3.3 is that the spectrogram caused by irregular airflow signals is not easy to identify, so we use graphs to analyze which network has the stronger feature extraction capability in this section.

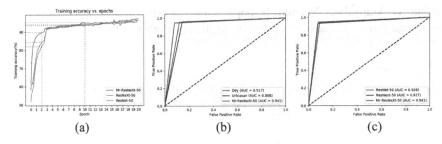

(a)   (b)   (c)

Fig. 7. (a) Training curves on our data, (b) (c) comparison of ROC curves and AUC value (Color figure online)

Figure 7(a) shows the curve of accuracy vs. epochs. Two epochs are randomly selected for comparison, and marked with red dashed lines in the figure. When training to the second epoch, the accuracy of Mr-ResNeXt-50 is 5% higher than the other two. During the entire training process, the blue line was above the other two lines. Therefore, the feature extraction ability of Mr-ResNeXt-50 is superior to the other two.

Comprehensive Performance Comparison

Mr-ResNeXt needs to be compared with the previous researches. We choose two excellent networks as the experimental subjects. From Table 2, we can see that the performance of Mr-ResNeXt-50 is improved by 3% compared with the other two.

Table 2. Performance comparison between different networks

	Acc	Precision	Recall	F1
Dey	91.02%	84.42%	93.40%	89.58%
Urtnasan	89.78%	81.94%	93.88%	88.36%
Mr-ResNeXt-50	94.23%	92.02%	94.99%	91.91%

ROC curves and AUC [20] are universal evaluation metrics. The result is shown in Fig. 7(b), from which it can be seen that the area represented by the green line is the largest and the value of AUC is the largest too, so its performance is the best.

Experiments to verify generalization are necessary. From Table 3, Mr-ResNeXt-50 has improved the performance compared to the other two. So we confirmed that Mr-ResNeXt-50 indeed improve performance. The ROC curve is shown in Fig. 7(c).

Table 3. Performance comparison between different networks

	Acc	Precision	Recall	F1
ResNet-50	92.64%	87.85%	94.93%	91.25%
ResNeXt-50	92.83%	88.06%	93.82%	91.87%
Mr-ResNeXt-50	94.23%	92.02%	94.99%	91.91%

Qualitative Analysis of Normal, Hypopnea and Apnea

Our second task is to classify the type of respiratory events of the subjects.

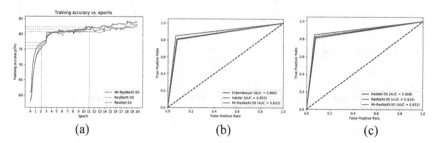

Fig. 8. (a) Training curves on our data, (b) (c) comparison of ROC curves and AUC value (Color figure online)

Feature Extraction Capability

Figure 8(a) shows the curves of accuracy vs. epochs. In the second epoch, the accuracy of Mr-ResNeXt-50 is higher than the other two. The accuracy of the three networks in the middle part (4th to 12th epoch) is very close, but after 12th epochs, the blue line starts to move away from the other two lines.

Comprehensive Performance Comparison

We need to compare Mr-ResNeXt with previous researches and compare Mr-ResNeXt-50 with ResNet-50 and ResNeXt-50. Table 4 shows that when the number of categories increased, Mr-ResNeXt also showed better performance, so it can be extended to multi-class networks.

Table 4. Performance comparison between different networks

	Acc	Precision	Recall	F1
Erdenebayar	80.06%	80.32%	79.98%	79.68%
Haidar	80.28%	79.90%	80.05%	79.36%
Mr-ResNeXt-50	83.26%	82.17%	81.32%	81.25%
	Acc	Precision	Recall	F1
ResNet-50	80.95%	80.25%	80.50%	79.76%
ResNeXt-50	81.81%	80.73%	80.53%	79.63%
Mr-ResNeXt-50	83.26%	82.17%	81.32%	81.25%

ROC curve and AUC are universal evaluation metrics in binary classification problems but do not support multi-classification problems. We convert multi-classification into binary classification in each category for computing the average ROC curves and AUC values [21]. The comparison results of various classification methods are shown in Fig. 8(b) and Fig. 8(c).

Group and Depth

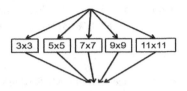

Fig. 9. Group

In this section, we verify the influence of group on network performance. The structure shown in Fig. 9 is called "group". We verify the performance of the network by changing the size of the group and the depth of the network. Table 6 shows the effect of group size on performance. According to Table 5 (a), when the group is equal to 4, the classification accuracy of Mr-ResNeXt-50 is similar to ResNeXt-50, but only slightly improved. According to Table 5 (b), when the group is equal to 5, it can be seen that the model with the Multi-resolution block has a more than 1% improvement in classification accuracy compared to ResNeXt-50.

To verify the effect of group size on performance, we conducted a third comparison experiment by setting the group to 6. According to Table 5 (c), when the group is equal to 6, the performance starts to decrease. So we can know that the performance was not correlated positively with group number, when the group is equal to 5, the network achieves the best performance.

Table 5. Performance comparison between different groups of Mr-ResNeXt-50

(a)				
	Acc	Precision	Recall	F1
Mr-ResNeXt-50 (Group = 4)	82.36%	80.37%	81.56%	80.05%
ResNeXt-50	81.81%	80.73%	80.53%	79.63%
(b)				
	Acc	Precision	Recall	F1
Mr-ResNeXt-50 (Group = 5)	83.26%	82.17%	81.32%	81.25%
ResNeXt-50	81.81%	80.73%	80.53%	79.63%
(c)				
	Acc	Precision	Recall	F1
Mr-ResNeXt-50 (Group = 6)	81.28%	79.20%	78.82%	80.12%
ResNeXt-50	81.81%	80.73%	80.53%	79.63%

As we known that the performance is best when the group is equal to 5, the impact of network depth on performance will be discussed below between Mr-ResNeXt-101 and ResNeXt-101. Results are shown in Table 6.

Table 6. Performance comparison between different depths

	Acc	Precision	Recall	F1
Mr-ResNeXt-101 (Group = 5)	83.56%	82.67%	81.52%	81.55%
ResNeXt-101	82.01%	81.03%	80.83%	80.12%

Seen from Table 6, the performance of both Mr-ResNeXt and ResNeXt improve slightly, not obviously, with the depth of the network increasing. While deeper networks increase the number of parameters, which affect the speed of training, thus Mr-ResNeXt-50 will be a better choice for fast diagnosis in practical applications.

In summary, both the number of groups and the depth can bring performance improvements to the network. Finally, we chose Mr-ResNeXt-50 with similar performance but more time-saving from the perspective of practical applications.

From the experiments above, we can know the reasons why Mr-ResNeXt has advantages in classification accuracy and training speed:

- From a global perspective: our network is composed of multi-level residual connections and group convolutions, both of which make the network learning ability more powerful.
- From a local perspective: Mr-ResNeXt has taken local informations of the picture into account and process them with filters of different sizes.

5   Conclusion and Future Work

In this work, we proposed Mr-ResNeXt, an optimization improvement based on ResNeXt. The most distinctive characteristics of Mr-ResNeXt were as follows:

- Mr-ResNeXt inherits the excellent classification performance of ResNeXt.
- Mr-ResNeXt is suitable for capturing different resolution information of pictures.
- The number of parameters in Mr-ResNeXt is less than ResNeXt (Table 1).
- The training speed of Mr-ResNeXt is faster than ResNeXt.
- The performance of Mr-ResNeXt does not increase with the number of groups, which is best when the group is equal to 5 (Table 5).
- The performance of Mr-ResNeXt is stronger than the methods for OSA diagnosis proposed in other papers.

Although Mr-ResNeXt has achieved better performance than previous networks in the detection of respiratory events, a higher classification accuracy is needed. In future, it is necessary to develop a network with stronger feature extraction capabilities to meet the clinical demand.

References

1. Jordan, G., Vgontzas, A.N., et al.: Obstructive sleep apnea and the metabolic syndrome: the road to clinically-meaningful phenotyping, improved prognosis, and personalized treatment. Sleep Med. Rev. **42**, 211–219 (2018)
2. Berry, R.B., Budhiraja, R., Gottlieb, D.J., et al.: Rules for scoring respiratory events in sleep: update of the 2007 AASM manual for the scoring of sleep and associated events. J. Clin. Sleep Med. **8**(5), 597–619 (2012)
3. Haidar, R., Koprinska, I., Jeffries, B.: Sleep apnea event detection from nasal airflow using convolutional neural networks. In: Liu, D., Xie, S., Li, Y., Zhao, D., El-Alfy, E.S.M. (eds.) ICONIP 2017. LNCS, vol. 10638, pp. 819–827. Springer, Cham (2017). https://doi.org/10.1007/978-3-319-70139-4_83
4. Dean, D.A., Goldberger, A.L., Mueller, R., et al.: Scaling up scientific discovery in sleep medicine: the national sleep research resource. Sleep **39**(5), 1151–1164 (2016)
5. Almuhammadi, W.S., Aboalayon, K.A.I., Faezipour, M.: Efficient obstructive sleep apnea classification based on EEG Signals. In: 11th IEEE Long Island Systems, Applications and Technology Conference (LISAT). IEEE (2015)
6. Jiang, D., Ma, Y., Wang, Y.: A multi-scale parallel convolution neural network for automatic sleep apnea detection using single-channel EEG signals. In: 2018 11th International Congress on Image and Signal Processing, BioMedical Engineering and Informatics (CISP-BMEI), pp. 1–5. IEEE (2018)
7. Banluesombatkul, N., Rakthanmanon, T., Wilaiprasitporn, T.: Single channel ECG for obstructive sleep apnea severity detection using a deep learning approach (2018)
8. Blank, J.B., Cawthon, P.M., Carrion-Petersen, M.L., et al.: Overview of recruitment for the osteoporotic fractures in men study (MrOS). Contemp. Clin. Trials **26**(5), 557–568 (2005)
9. Song, C., Liu, K., Zhang, X., et al.: An obstructive sleep apnea detection approach using a discriminative hidden markov model from ECG signals. IEEE Trans. Biomed. Eng. **63**(7), 1532–1542 (2016)
10. Marcos, J.V., Hornero, R., Álvarez, D., et al.: Automated detection of obstructive sleep apnoea syndrome from oxygen saturation recordings using linear discriminant analysis. Med. Biol. Eng. Comput. **48**(9), 895–902 (2010)
11. Vaquerizo-Villar, F., Álvarez, D., Kheirandish-Gozal, L., et al.: Convolutional Neural Networks to Detect Pediatric Apnea-Hypopnea Events from Oximetry. In: 2019 41st Annual International Conference of the IEEE Engineering in Medicine and Biology Society (EMBC), pp. 3555–3558. IEEE (2019)
12. Korkalainen, H., Aakko, J., Nikkonen, S., et al.: Accurate deep learning-based sleep staging in a clinical population with suspected obstructive sleep apnea. IEEE J. Biomed. Health Inform. **24**, 2073–2081 (2019)
13. Kıymık, M.K., Güler, İ., Dizibüyük, A., et al.: Comparison of STFT and wavelet transform methods in determining epileptic seizure activity in EEG signals for real-time application. Comput. Biol. Med. **35**(7), 603–616 (2005)
14. He, K., Zhang, X., Ren, S., et al.: Deep residual learning for image recognition. In: Proceedings of the IEEE Conference on Computer Vision and Pattern Recognition, pp. 770–778 (2016)
15. Xie, S., Girshick, R., Dollár, P., et al.: Aggregated residual transformations for deep neural networks. In: Proceedings of the IEEE Conference on Computer Vision and Pattern Recognition, pp. 1492–1500 (2017)

16. Chen, Y., Fan, H., Xu, B., et al.: Drop an octave: reducing spatial redundancy in convolutional neural networks with octave convolution. In: Proceedings of the IEEE International Conference on Computer Vision, pp. 3435–3444 (2019)
17. Tan, M., Le, Q.V.: Mixconv: mixed depthwise convolutional kernels. CoRR, abs/1907.09595 (2019)
18. Rieffel, M.A.: Lie group convolution algebras as deformation quantizations of linear poisson structures. Am. J. Math. **112**(4), 657–685 (1990)
19. Szegedy, C., Ioffe, S., Vanhoucke, V.: Inceptionv4, inception-ResNet and the impact of residual connections on learning. In: ICLR Workshop (2016)
20. Fan, J., Upadhye, S., Worster, A.: Understanding receiver operating characteristic (ROC) curves. Can. J. Emerg. Med. **8**(1), 19–20 (2006)
21. Rice, M.E., Harris, G.T.: Comparing effect sizes in follow-up studies: ROC area, cohen's d, and r. Law Hum Behav. **29**(5), 615–620 (2005)

Scalable Multi-agent Reinforcement Learning Architecture for Semi-MDP Real-Time Strategy Games

Zhentao Wang$^{(\boxtimes)}$, Weiwei Wu$^{(\boxtimes)}$, and Ziyao Huang

School of Cyber Science and Engineering, Southeast University, Nanjing, China
jsyxl994@hotmail.com, weiweiwu@seu.edu.cn,
seuhuangziyao@outlook.com

Abstract. Action-value has been widely used in multi-agent reinforcement learning. However, action-value is hard to be adapted to scenarios such as real-time strategy games where the number of agents can vary from time to time. In this paper, we explore approaches of avoiding the action-value in systems in order to make multi-agent architectures more scalable. We present a general architecture for real-time strategy games and design the global reward function which can fit into it. In addition, in our architecture, we also propose the algorithm without human knowledge which can work for Semi Markov Decision Processes where rewards cannot be received until actions last for a while. To evaluate the performance of our approach, experiments with respect to micromanagement are carried out on a simplified real-time strategy game called MicroRTS. The result shows that the trained artificial intelligence is highly competitive against strong baseline robots.

Keywords: Markov Decision Process · Multi-agent systems · Reinforcement learning · Real-Time strategy games · Deep learning · Game theory

1 Introduction

Reinforcement learning (RL), which is a subfield of machine learning, is about learning what to do by try and error so as to maximize a numerical reward signal [1]. Reinforcement learning along with deep neural network has achieved great success universally these years, especially in game areas [2–4]. Encouraged by this, researchers have been further studying on multi-agent reinforcement learning (MARL) algorithms and use games for performance tests. However, solving multi-agent real-time planning problems such as real-time strategy (RTS) games is of great challenge by MARL approaches. The reason is that, unlike single agent games like Atari [5], in RTS games: 1) the number of units one player can control can be quite large and changeable, which indicates that with the increasing units, the state-action spaces grow exponentially; 2) The game is real-time, which means time is limited (usually within a semi-second) for players to make decisions to deploy units at every timestep. 3) Units in RTS games are generally heterogenetic; 4) There exists both competition and cooperation for units.

© Springer Nature Singapore Pte Ltd. 2020
H. Zhang et al. (Eds.): NCAA 2020, CCIS 1265, pp. 433–446, 2020.
https://doi.org/10.1007/978-981-15-7670-6_36

The four features mentioned above lead to the difficulty of training an efficient RL artificial intelligence (AI) for RTS games.

Existing algorithms designed for RTS games are mostly script-based or search-based. Some of these algorithms typically apply domain knowledge to RTS AI [6–9], the others search the inner game model to look ahead and use a heuristic function to evaluate the simulating results [10, 11]. These AIs usually work well, however, to some extent, the performance depends on how "smart" their designers are, not how smart themselves are. On the contrary, several MARL approaches take little domain knowledge from human, but strong prerequisite such as homogeneous of units or fixed number of units are demanded for the architecture and training algorithms [12].

Thus, it calls for a more flexible multi-agent reinforcement learning architecture to be able to scale to complex scenarios. In this work, we mainly focus our research on a simplified RTS game environment called MicroRTS (we also make it gym-like [13] for RL training). We present a multi-agent reinforcement learning architecture that shares one global critic and multiple switchable actors, which is scalable and general for RTS games. We also give our Semi Markov Decision Process (SMDP) algorithm for this architecture. Then, we train our AI in a self-play manner in MicroRTS environment. Finally, to measure the performance, the trained AI are compared with the traditional script-based and search-based baseline algorithms by combating with each other. Result shows that our self-played AI, as the first AI for MicroRTS game without any human knowledge, outperforms these baseline AIs in micromanagement scenarios.

2   Related Work

In this section, we mainly review relevant works on search-based and MARL methods and respectively take these two kinds into consideration.

2.1   Search-Based Methods

Many AI methods in RTS games are hard-coded by game experts, which may fail to work because other players can easily take advantages of these rule-based methods. Researches on RTS games start from search-based algorithms. NaiveMCTS makes effort to solve the Combinatorial Multi-Armed Bandit (CMAB) problems in RTS games by two Monte Carlo Tree Search (MCTS) modules taking turns to interact with the environment [11]. From another perspective to solve CMAB problem, portfolio greedy search method searches scripts rather than actions in multi-unit combat scenarios [7]. Moreover, Strategy Creation via Voting (SCV) algorithm uses a voting method to generate a large set of novel strategies from existing expert-based ones and use them in games [14]. All these search-based methods mentioned above need inner models of games to do some simulations so as to get evaluations of states required by their algorithms. However, in practice, it is impossible to acquire the model when put in the environment with which we are not familiar. To handle this situation, in this work, we present a model-free MARL approach.

2.2 Multi-agent Reinforcement Learning Methods

Although independent deep Q-networks (DQN) makes big achievements on Atari games [5], its performance on multi-agent systems is not quite competitive. Altering the Actor-critic architecture [15], counterfactual multi-agent policy gradient (COMA) presents a centralised critic for estimation of a counterfactual advantage for decentralised policies [16]. Like COMA, multi-agent deep deterministic policy gradient (MADDPG) adds observations and actions of all agents in centralised critic to improve cooperation and competition [17], which is infeasible when impossible to have the model. Moreover, Multi-agent bidirectionally-coordinated network (BiCNet) presents a vectorized actor-critic framework to learn better coordination strategies [18]. However, none of These approaches can handle situations where the number of the units will vary in one game. The main drawback may lie in the poor scalability of the action-value function which is also called Q-function. Most recently, mean field multi-agent reinforcement learning (MFRL) is proposed to tackle the multi-agent reinforcement learning problems when a large and variable number of agents co-exist [12]. Unfortunately, MFRL needs all agents to be homogenous, which is a strong limitation to apply to other scenarios. In this paper, we extend the work above and try to explore a more general architecture for RTS games.

3 Preliminary

In this section, we will mainly introduce the background of the basic theories about MARL and related academic fields.

3.1 Stochastic Game and Reinforcement Learning

A stochastic game is a model for dynamic games repeatedly played by one or more players with uncertainty of the next game state [19]. This model can be generally adapted to the MARL by adding extra ingredients. An N-agent stochastic game is a Markov Decision Process (MDPs) defined by a tuple

$$\Gamma \triangleq \langle \mathcal{S}, \mathcal{A}^1, \ldots, \mathcal{A}^N, \mathcal{U}, p, r^1, \ldots, r^N, N, \gamma \rangle$$

where $\mathcal{S}$ denotes the state space; $\mathcal{A}^i$ is the action space of agent $i \in \{1, \ldots, N\}$; $\mathcal{U} = \mathcal{A}^1 \times \cdots \times \mathcal{A}^N$ is the joint actions space; p denotes the transition function: $\mathcal{S} \times \mathcal{U} \times \mathcal{S} \to [0, 1]$; The individual payoff function r_i gives utilities to each players participated in the game, which is the immediate reward in RL: $\mathcal{S} \times \mathcal{A}^i \to \mathbb{R}$, $i \in \{1, \ldots, N\}$; The constant $\gamma \in [0, 1)$ denotes the reward discount, which can serve as the terminator in the infinite-time-step scenarios (Fig. 1).

Unlike independent RL, there are more than one agent interacting with the environment simultaneously in MARL. Given the state, each agent takes one action every timestep according to their policy $\pi^i: \mathcal{S} \to \Omega(\mathcal{A}^i)$, in which Ω is the probability over agent i's action space $\mathcal{A}^i$. After taking in the joint action of all the agent, the

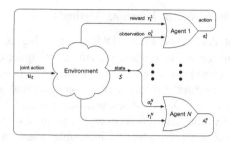

Fig. 1. Interaction between the agents and the environment in multi-agent scenario

environment produces the next state and gives the feedbacks for each agent. Each agent's cumulative rewards G_t^i in one episode can be written as:

$$G_t^i \triangleq \sum_{k=0}^{\infty} \gamma^k r_{t+k+1}^i \tag{1}$$

Thus, the state-value function and the action-value for policy π can be respectively defined as:

$$\begin{aligned} v_\pi^i(s) &\triangleq \mathbb{E}_\pi[G_t|s], \text{ and} \\ q_\pi^i(s, a_t^i) &\triangleq \mathbb{E}_\pi[G_t|s, a_t^i] = \mathbb{E}_p[r_t^i + v_\pi^i(s')] \end{aligned} \tag{2}$$

where s' is the next state transferred from s according to the probability p at time t.

3.2   Policy Gradient Theorem and Actor-Critic Architecture

Policy Gradient (PG) method is one branch of reinforcement learning. By this method, the policy can be parameterized as $\pi_\theta(a|s)$ if π is differentiable with respect to its parameters θ [1]. More specifically, the goal of PG method is to figure out the optimal stochastic policy $\pi_\theta^*: \mathcal{S} \times \mathcal{A} \to [0, 1]$ which maximizes the performance measure, $J(\theta) \triangleq v_{\pi_\theta}(s_0)$, and its gradient is proved to be:

$$\nabla_\theta J(\theta) = \mathbb{E}_{\pi,p}[G_t \nabla_\theta \log \pi_\theta(a|s)] \tag{3}$$

Equation (3) is the core formula of REINFROCE algorithm [20]. Similarly, using a parameterized baseline $\hat{v}_w(s)$ to significantly reduce the variance of the gradient, PG methods have been applied to modern actor-critic architecture:

$$\nabla_\theta J(\theta) = \mathbb{E}_{\pi,p}[(G_t - \hat{v}_w(s))\nabla_\theta \log \pi_\theta(a|s)] \tag{4}$$

where $\widehat{v}_w(s)$ can be viewed as the critic and $\pi_\theta(a|s)$ is called the actor. Sometimes, the critic also can be replaced by the action-value function $q_w(s,a)$, and the term $G_t - \widehat{v}_w(s)$ can be changed to:

$$Adv = q_w(s,\, a) - \widehat{v}(s) \tag{5}$$

4   Methods

We mainly discuss our methods in three aspects in this section. Firstly, we will introduce our scalable architecture. Secondly, we design the reward function to fit into this architecture. Finally, we present our algorithm in the architecture, particularly for RTS games.

4.1   The Architecture

The proposed MARL architecture follows the paradigm of centralized training and decentralized executing [21], which can be integrated into the actor-critic methods, where units (both homogeneous and heterogenetic) play the role of the actor and an evaluation function play the role of the critic. Two procedures iteratively go through during the whole learning process, which are executing and training. The critic is only used in the training. During the executing phase, each actor follows its own policy π_i to simultaneously interact with the environment and save samples respectively in the memory buffer. While in the training phase, the critic reads samples from the memory and train the network (Fig. 2).

Many deep learning architectures related to MARL failed to handle the situation where agents are heterogenetic, or the number of the units varies from time to time in one game episode. This may root in the intractable problem of the poor scalability of the widely-used parameterized action-value function. Because $Q(s,a)$ needs the action of agents, but units will be created or killed, which causes the input module of the Q network collapse. Another reason why those approach hardly work in RTS games may be that, different kinds of units have different action spaces, which requires multiple Q functions, resulting in redundant trainings.

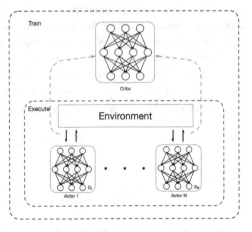

Fig. 2. Learning architecture

Therefore, to avoid adding actions in deep neural network, in our architecture, we just use the state-value function $V(s)$ as the critic, and alter the Eq. (5) by

$$\delta = r_t + \gamma v(s') - v(s) \tag{6}$$

in which δ is called Temporal-Difference (TD) error. It is worth to note that, unlike other approaches which uses multiple critics for actors, we simply use one and the only global state-value critic to evaluate how good states are. This global state-value function is shared by all actors during training, which can be generally adapted to both cooperation and competition scenarios. It could work because units from one player usually make actions to stay at states which have the higher value. Apart from this, we regard instances of different types of agents as the actors. Individual units on the map in RTS games could be viewed as instances or actors. In our setting, actors of the same type share experiences and parameters to accelerate the training process.

To make the architecture complete, as Fig. 3 shows, we also design the deep neural network framework. The encoded spatial features first go through the shared layers for both critic and actor network, then the output flows to the heads of critic and actor respectively. The parameters of all types of units are initiated as different modules ahead in actor network. In order to avoid all units following the same trajectory, we concatenate the encoding of the vary unit's information (including their location and attributes) on the game map into the shared outputs upstream, and the actor network activate the module according to the unit type.

4.2   Reward Design

Not only RL agents learn from rewards, but also the ultimate target of RL is to maximize the expected cumulative rewards. It is of great importance to set up an appropriate reward function for learning in multi-agent systems. Inspired by BiCNet [18], we establish our global reward function as follows.

Using the notations in Sect. 3.1, we define the hit points (hp) of all units of player p at timestep t as $HP_t^p \triangleq \sum_i hp_t^i$, where hp_t^i stands for the hp of agent i; Likewise, on the other hand, $HP_t^{-p} \triangleq \sum_{-i} hp_t^{-i}$ represents for the opponent's situation. The global reward function for player p can be designed by:

$$R_{\Delta t}^p = \Delta HP_{\Delta t}^p - \Delta HP_{\Delta t}^{-p} \tag{7}$$

where

$$\Delta HP_{\Delta t}^p = \left(HP_t^p - HP_{t'}^p \right) - \left(HP_t^{-p} - HP_{t'}^{-p} \right) \tag{8}$$

Equation (8) indicates that sampling starts at time t' and ends at time t because the action last for Δt. From Eq. (7), we can easily get $R_{\Delta t}^p + R_{\Delta t}^{-p} = 0$, which means this reward function leads to a two-player zero-sum game.

We take this global reward function rather than other shaped reward or individual reward for several considerations: (1) The function is simple and general for RTS multi-agent systems. In addition, it is very flexible for that we can easily change rewards for all units by simply changing Eq. (8); (2) It meets the requirement of maximizing the team's utility. In other words, the goal of player p's units is to increase team's common interests rather than themselves', which allows units to sacrifice for team's benefit. (3) In our architecture, this function can be adapted to both cooperation and competition scenarios. The reward generated by environment is shared by all units from one side. Correspondingly, we also use one critic shared by all actors, which naturally combines the reward function and the architecture together.

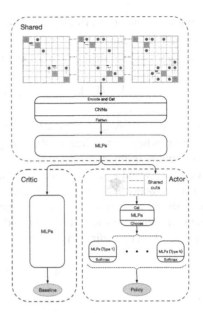

Fig. 3. Network architecture of our approach

4.3 Policy Gradient for Semi-MDP

Markov Decision Process (MDP) is a classic mathematical model for reinforcement learning. In this model, actions are taken in consecutive timesteps with immediate rewards coming afterwards. In fact, situation is much more complicated where rewards are received when actions have been last for a while. This closed-loop policies for taking action over a period of time is called *options* [23], which extends MDP to Semi-MDP (SMDP) (Fig. 4).

The feature mentioned above regarding SMDP implies that not in every single timestep will the unit receive a meaningful reward from environment and different actions takes different time to finish, which indicates that units may asynchronously take their actions and receive their rewards. To tackle this problem, we propose Semi-MARL Policy Gradient (SPG) for multi-agent learning in SMDP.

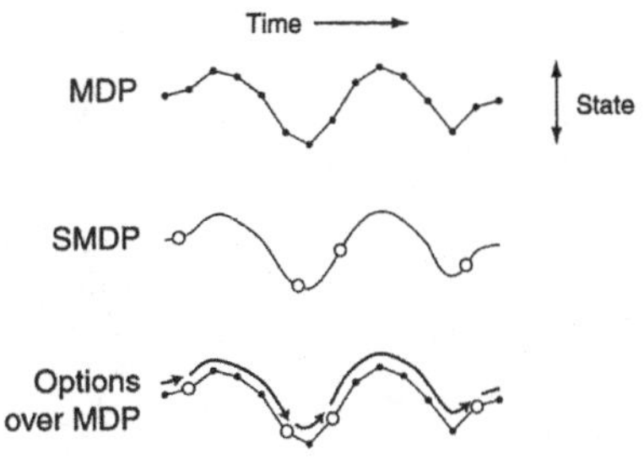

Fig. 4. Difference between MDP and SMDP [23]

Using the notation previously defined, the policies or the options of M heterogenetic agents in game can be parameterized by $\Theta = \{\theta^1, \cdots, \theta^M\}$, $\Pi = \{\pi^1, \cdots, \pi^M | \pi^i = \pi_{\theta^i}\}$, and the gradient for agent i in SMDP model can be written as:

$$\nabla_{\theta^i} J(\theta^i) = \mathbb{E}_{a \sim \pi_{\theta^i},\, s \sim p} \left[\delta_{t \to t'} \nabla_{\theta^i} \log \pi_{\theta^i} (a_t^i | s_t) \right] \tag{9}$$

Here we call $\delta_{t \to t'} = r^i(s_t, a_t^i) + \gamma^{t'} v(s_{t'}) - v(s_t)$ as *Semi-Temporal-Difference* (STD) error, where $t' > t$ denotes that sampling begins at time t and terminates at t'.

Furthermore, samples are recorded as tuples of $\langle k, s, a, r, s' \rangle$, where k denotes the type of units, s and s' denotes the beginning state and ending state respectively, a represents the action taken at the state, and r is the delayed reward received when doing action at state s and finished at state s'. During training, as Fig. 3 shows, samples are sent into different actors according to k and the loss of the centralized state-value critic parametrized by ω is defined as:

$$\mathcal{L}(\omega) = (y - v_\omega(s))^2 \tag{10}$$

where

$$y = \begin{cases} r + \gamma^k v_\omega(s') & \text{for terminal} \quad s' \\ r & \text{for non terminal} \quad s' \end{cases} \tag{11}$$

If ω is differentiable, we can get the following term for optimizers to update:

$$\nabla_\omega \mathcal{L}(\omega) = (y - v_\omega(s)) \nabla_\omega v_\omega(s') \tag{12}$$

As for the update for actor i, the optimizing target is to maximize the performance $J(\theta^i)$ or minimize the negative performance, and differentiating it gives

$$\nabla_{\theta^i} \mathcal{J}(\theta^i) = -\nabla_{\theta}^i J(\theta^i) = -\delta \nabla_{\theta^i} \log \pi_{\theta^i}(a|s) \tag{13}$$

Algorithm 1 Semi-MARL Policy Gradient for RTS Games

Initialize critic parameters ω
Initialize actor parameters $\theta^1, \cdots, \theta^M$
Initialize counter $T = 0$
Input: Initial state s

procedure ASSIGN ACTION(State s, Player p)
    Sample actions for free units player p has at state s based on Π
    return sampled actions

procedure SELF PLAY
    while $T < T_{max}$ **do**
        $\mathcal{A}_1 \leftarrow$ **ASSIGN ACTION**$(s, 1)$
        $\mathcal{A}_2 \leftarrow$ **ASSIGN ACTION**$(s, 2)$
        Take actions $\mathcal{A} = (\mathcal{A}_1, \mathcal{A}_2)$ and wait for actions to finish
        Receive next state s'
        Receive rewards **r** asynchronously and records samples $\triangleright$ In backend
        Divide samples into different parts according to unit type
        Use Eq.(12) and Eq.(13) to update ω and Θ
        $s \leftarrow s'$
        $T \leftarrow T + 1$

---

5 Experiments

5.1 Environment

We choose MicroRTS as the main body of the study. MicroRTS is a mini real-time strategy game implemented in Java. To make the environment work for reinforcement learning, we add the interfaces for Python so as to wrap the environment like gym [13]. The motivation of MicroRTS is to strip the RTS logic from large games like StarCraft to improve the performance of the system and reduce the work of engineering so that experiments can be quickly conducted on it.

The environment of the game is shown in Fig. 5, where the entire battlefield is divided into $m \times n$ grids, each of which can only be occupied by one unit every timestep. There are seven different types of units, which are respectively Base, Worker, Barracks, Heavy, Light, Ranged, Resource. Different types of units have different attributes and behaviors: Base can only produce Worker; Worker can collect resources for production and construction tasks; Barracks are responsible for the production of combat types of units; Heavy and light are melee

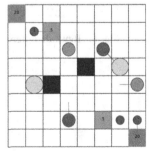

Fig. 5. Screenshot of MicroRTS

units (they can only attack the enemy directly adjacent to them); Ranged is a ranged attack unit; Resource is a neutral unit that is consumed by construction and production tasks. The game is two-player and different side of the units is of different colors. In this game, each player's goal is to destroy all enemy units in a limited time.

Like other RTS games, MicroRTS also has the following challenges:

- Long-term and short-term planning: similar to the real world, many cause-and-effect relationships are not clear at a glance; Decisions made earlier in the game may take a few steps to show effects.
- Real-time: unlike chess in which two players take turns to play, MicroRTS requires the player to make decisions on all units of his in 100 ms.
- Large action space: as the number of units increases, the number of action combinations will increase exponentially.

5.2 Experiment Setup

In our experiment, we mainly evaluate micro-management of our AI which is trained by the approach we proposed. Experiments are conducted in the following settings:

- 1 VS 1: The map size is 4×4, each player only has one Light to fight for them. The max cycle of game is 1000 timesteps. No war fog on the battlefield (perfect information).
- 2 VS 2: The map size is 4×4, each player has one Light and one Heavy. The max cycle of game is 1000 timesteps. No war fog on the battlefield.
- 6 VS 6: The map size is 6×6, each player can control six Lights and totally 12 units. The max cycle of game is 1500 timesteps. No war fog on the battlefield.

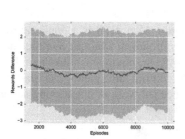

Fig. 6. Rewards difference

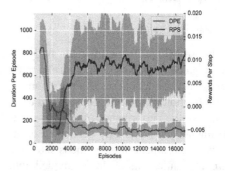

Fig. 7. Rewards and duration curve (6 VS 6)

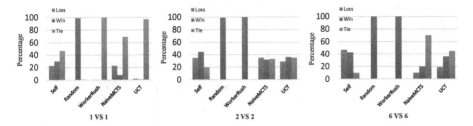

Fig. 8. Win, loss, tie rates of 1 VS 1(left), 2 VS 2 (middle) and 6 VS 6 (right) against the baseline AIs (Random, WorkerRush, NaiveMCTS and UCT) respectively

All experiments above are carried out in self-play manner with multiple threads and we train a single network both for player 1 and player 2. We use no abstract actions with human knowledge, but we use translated actions. Translated action functions like that if unit take "left probe" and there is an enemy, then the action will be translated to "attack left"; if there is an empty box, then the action will be translated to "move left". After training, four baseline AIs are selected to combat with our AI to measure the performance. The four baseline AIs we choose are Random, UCT with UCB1 [24], NaiveMCTS, and WorkerRush [25]. NaiveMCTS and UCT are search-based and WorkerRush is script-based. Random is included to test the generalization ability of our neural network since Random robot can stochastically bring various states, which is necessary because some AIs can defeat skillful opponents but fail to win the weaker one they have never seen before. Script-based robot WorkerRush is chosen as it is known to perform well in MicroRTS. UCT is selected because it is the classic and the traditional search-based MCTS algorithm. Additionally, we choose NaiveMCTS because it performs significantly better than other search-based algorithms as the number of units increases [11]. These search-based AI mentioned above are quite strong in small scenarios.

5.3 Experiment Results

During training, we log the rewards difference between player 1 and player 2 every episode to see if anyone takes advantage of the other or gives up. As Fig. 6 shows, the curve representing rewards difference is around zero over time. It indicates that both players parameterized in one network in our self-play settings struggle to get more rewards. Figure 7 shows the 6 VS 6 curve of the duration of battle and the average rewards received for both two players over 16000 episodes. We can infer that, both two players we trained tend to win games as fast as possible and get as much rewards as they can.

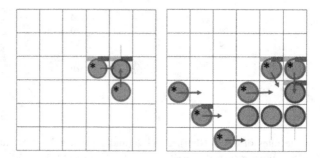

Fig. 9. Learning to cooperate after training in 6 VS 6 settings. Units with blue profile are ours (also marked with '*') and the ones with red profile are enemies'. The red arrows extend from units indicate trends of the next move. (Color figure online)

After training, we select our best model to combat with the baseline AIs and record the average rate of loss, win and tie in 100 games respectively. If one of the two players eliminates all units of the other, then he wins and if no one defeats his opponent in a limited time, then they tie. Figure 8 shows the result of our AI's performance. As we can see, in the 1 VS 1 setting, the result of battles against other AIs is mostly tie except Random. However, in the setting of 2 VS 2 which needs units to cooperate with each other, our AI can completely defeat WorkerRush which is a strong script-based AI. Furthermore, in the 6 VS 6 experiment, the AI's winning rate surpasses the search-based AIs (NaiveMCTS and UCT) for the first time. The reason why our AI does better in larger scenarios (2 vs 2 and 6 vs 6) may be that the search-based algorithm only has 100 ms (because of real-time strategy) for them to search the game. In all, our AI is highly competitive to these baseline AIs.

To visualize whether our AI learns to cooperate with each other, we capture the frames during game. As Fig. 9 shows, the AI is able to chase the enemy and make a joint attack on it (left). Furthermore, it also learns to siege the enemies in the corner to eliminate them (right).

6   Conclusion and Future Work

In this paper we firstly proposed a general and scalable reinforcement learning architecture for multi-agent systems and make it work for semi-MDP model. We then design the reward function for this architecture and present our algorithms. We mainly use MicroRTS environment to evaluate the micro-management performance of our algorithm. The result shows that the AI trained by our approach is highly competitive. For the future work, we are going to test the full game performance and we observe that the neural network might be the bottleneck of our algorithm and further experiments about how networks should be better built and the better tuning of the network hyperparameter is worth to be carried out.

Acknowledgement. This work is supported by the national key research and development program of China under grant No. 2019YFB2102200, National Natural Science Foundation of China under Grant No. 61672154, No. 61972086, and the Fund of ZTE Corporation under Grant No. HC-CN-20190826009.

References

1. Sutton, R.S., Barto, A.G.: Reinforcement Learning: An Introduction. MIT Press, Cambridge (2018)
2. Silver, D., et al.: Mastering the game of go without human knowledge. Nature **550**(7676), 354–359 (2017)
3. Vinyals, O., et al.: Grandmaster level in StarCraft II using multi-agent reinforcement learning. Nature **575**(7782), 350–354 (2019)
4. Berner, C., et al.: Dota 2 with large scale deep reinforcement learning (2019)
5. Mnih, V., et al.: Playing atari with deep reinforcement learning (2013)
6. Ontanón, S., Buro, M.: Adversarial hierarchical-task network planning for complex real-time games. In: Twenty-Fourth International Joint Conference on Artificial Intelligence (2015)
7. Churchill, D., Buro, M.: Portfolio greedy search and simulation for large-scale combat in StarCraft. In: 2013 IEEE Conference on Computational Intelligence in Games (CIG). IEEE (2013)
8. Churchill, D., Saffidine, A., Buro, M.: Fast heuristic search for RTS game combat scenarios. In: Eighth Artificial Intelligence and Interactive Digital Entertainment Conference (2012)
9. Marino, J.R.H., et al.: Evolving action abstractions for real-time planning in extensive-form games. In: Proceedings of the AAAI Conference on Artificial Intelligence, vol. 33 (2019)
10. Barriga, N.A., Stanescu, M., Buro, M.: Combining strategic learning with tactical search in real-time strategy games. In: Thirteenth Artificial Intelligence and Interactive Digital Entertainment Conference (2017)
11. Ontanón, S.: The combinatorial multi-armed bandit problem and its application to real-time strategy games. In: Ninth Artificial Intelligence and Interactive Digital Entertainment Conference (2013)
12. Yang, Y., et al.: Mean field multi-agent reinforcement learning (2018)
13. Brockman, G., et al.: Openai gym. arXiv preprint arXiv:1606.01540 (2016)
14. Silva, C.R., et al.: Strategy generation for multi-unit real-time games via voting. IEEE Trans. Games **11**, 426–435 (2018)
15. Mnih, V., et al.: Asynchronous methods for deep reinforcement learning. In: International Conference on Machine Learning (2016)
16. Foerster, J.N., et al.: Counterfactual multi-agent policy gradients. In: Thirty-Second AAAI Conference on Artificial Intelligence (2018)
17. Lowe, R., et al.: Multi-agent actor-critic for mixed cooperative-competitive environments. In: Advances in Neural Information Processing Systems (2017)
18. Peng, P., et al.: Multiagent bidirectionally-coordinated nets: emergence of human-level coordination in learning to play starcraft combat games (2017)
19. Shapley, L.S.: Stochastic games. Proc. Nat. Acad. Sci. **39**(10), 1095–1100 (1953)
20. Williams, R.J.: Simple statistical gradient-following algorithms for connectionist reinforcement learning. Mach. Learn. **8**(3-4), 229–256 (1992)
21. Oliehoek, F.A., Spaan, M.T.J., Vlassis, N.: Optimal and approximate Q-value functions for decentralized POMDPs. J. Artif. Intell. Res. **32**, 289–353 (2008)

22. Kraemer, L., Banerjee, B.: Multi-agent reinforcement learning as a rehearsal for decentralized planning. Neurocomputing **190**, 82–94 (2016)
23. Sutton, R.S., Precup, D., Singh, S.: Between MDPs and semi-MDPs: a framework for temporal abstraction in reinforcement learning. Artif. Intell. **112**(1-2), 181–211 (1999)
24. Kocsis, L., Szepesvári, C.: Bandit based Monte-Carlo planning. In: Fürnkranz, J., Scheffer, T., Spiliopoulou, M. (eds.) ECML 2006. LNCS (LNAI), vol. 4212, pp. 282–293. Springer, Heidelberg (2006). https://doi.org/10.1007/11871842_29
25. Ontañón, S., et al.: The first microrts artificial intelligence competition. AI Mag. **39**(1), 75–83 (2018)

Stacked Deep Learning Structure with Bidirectional Long-Short Term Memory for Stock Market Prediction

Ying Xu[1(✉)], Lymeng Chhim[1,2], Bingxin Zheng[1,2],
and Yusuke Nojima[2]

[1] College of Computer Science and Electronics Engineering,
Hunan University, Changsha, China
hnxy@hnu.edu.cn
[2] Graduate School of Engineering,
Osaka Prefecture University, Osaka 599-8531, Japan

Abstract. The rapid growth of deep learning research has introduced numerous methods to solve real-world applications. In the financial market, the stock price prediction is one of the most challenging topics. This paper presents design and implementation of a stacked system to predict the stock price of the next day. This approach is a method that considers the historical data of the real stock prices from Yahoo Finance. This model uses the wavelet transform technique to reduce the noise of market data, and stacked auto-encoder to filter unimportant features from preprocessed data. Finally, recurrent neural network and bidirectional long-short term memory are used to predict the future stock price. We evaluate our model by analyzing the performance of different models with time series evaluation criteria.

Keywords: *Terms*—LSTM · RNN · Stacked auto-encoder · Stock prediction · Wavelet transform

1 Introduction

Financial investment is one of the fields whose nature keeps changing dramatically in the stochastic environment. The predictability of this market is almost unforeseeable. Yet, many existing algorithms attempted to optimize portfolio and forecasting the future stock price [1]. Stock price forecasting can be expressed as the estimation of future stock market prices by exploring historical and relevant data. It can be categorized into short term, medium term, and long term prediction. Due to its noise, volatile and chaotic characters, stock prediction is considered as one of the most challenging subjects in the time series prediction [2].

Stock price prediction is the act of attempting to estimate the future stock value in the stock exchange. The benefit of successive estimation could yield significant profit for traders and investors. According to the efficient market hypothesis, stock prices reflect all currently available information including the historical prices. Thus, it is fundamentally unpredictable. On the other hand, investors and financial academics with countless methods and technology tend to disagree with this theory.

© Springer Nature Singapore Pte Ltd. 2020
H. Zhang et al. (Eds.): NCAA 2020, CCIS 1265, pp. 447–460, 2020.
https://doi.org/10.1007/978-981-15-7670-6_37

Stock forecasting, which is a time series problem, can be analyzed by identifying the patterns, trends, and cycles in the price data. The methods for predicting the stock usually fall into three categories, including the fundamental analysis, the technical analysis and technological methods.

The fundamental analysis and the technical analysis are two commonly used methods by professional investors to analyze the stock for their portfolio. The fundamental analysis is the traditional method of attempting to evaluate a company's intrinsic value. It is constructed on the principle that a company needs capital to make progress and if a company performs well, additional capital should be invested and then a surge in the stock price is resulted. The technical analysis is a set of trading tools used to estimate the stock movement based on the historical price and volume. This technique seeks to forecast the future price of a stock based solely on the trends of the past price. Numerous patterns are employed in the form of time series analysis.

The technological method incorporates the machine learning capability to capture the nonlinear relationship with relevant factors. These soft computing techniques are widely accepted to forecast financial times with or without prior knowledge of input data statistical distributions [3]. Artificial neural networks (ANN) is one of the techniques that have been widely applied in forecasting time series, since they are data-oriented, and self-adapted behaviors [4]. Other approaches investigated on mining important sentimental features from the textual data that are extracted from online news platform and social media [5].

The motivation in this work is to design an efficient time series predictive model based on a deep learning framework for stock forecasting. This work has the following three main contributions. Firstly, the maximal overlap discrete wavelet transform is applied to de-noise the stock data. Secondly the stacked auto-encoders are used to generate deep features from technical indicators as multi-variances. Thirdly, the bidirectional long-short term memory (LSTM) is designed to forecast future stock prices based on the processed stock data.

The rest of this paper is organized as follows. In Sect. 2, we provide a literature review on related works in the field of stock forecasting. Section 3 illustrates our proposed model with the associated theoretical foundation. Section 4 is the description of experimental design and results. The final section concludes our study.

2   Related Works

Prediction of the stock price has attracted more attention from investors to gain higher returns and this has led researchers to propose various predicting models. The time series models are divided into two classes of algorithms including the linear (statistical) model and the non-linear (AI based) model [6, 7]. Linear based models for prediction have developed in the literature a long time ago and they are still used in practice [7]. Some time series linear models have been proposed such as autoregressive (AR) [8], autoregressive moving average (ARMA) [9], autoregressive integrated moving average

(ARIMA) [10], and other variations of these models. The early study on time series models [11, 12] inspired Box and Jenkins to work on prediction theory. Box and Jenkins [13] work is considered as an important contribution in time series analysis in forecasting, mainly autoregressive moving average (ARMA) model. The problem for linear models is that they only consider time series variables. Since stock data is non-stationary or non-linear in nature, it is difficult for linear models to capture the dynamic patterns of the data.

In order to overcome the limitation of linear models, AI-based models have been adopted. These non-linear models include support vector machine (SVM), support vector regression (SVR), and deep learning algorithms [7]. A deep learning framework has provided a promising result in mapping non-linear function for stock prediction [14]. These frameworks include artificial neural network (ANN), multi-layer perceptron (MLP), recursive neural network (RNN), long short-term memory (LSTM), and con-volutional neural network (CNN).

Reference [15] employed artificial neural network to forecast daily direction of Standard & Poor's 500 (S&P 500). S&P500 is the indicator for 500 top traded com-panies in the U.S. market. They used 27 financial and economic factors as inputs of the neural network. The designed ANN has outperformed logit model for correct fore-casting and buy-and-hold strategy for trading profit. It is also pointed out that exchange rates of US dollar significantly influence the predictability of the model, and some input vectors may confuse the model and decrease the performance of ANN. Before applying ANN classification procedure, authors in [16] used three different dimensionality reduction techniques, including the principal component analysis (PCA), the fuzzy robust principal component analysis (FRPCA), and the kernel-based principal com-ponent analysis (KPCA), to restructure the whole data using 60 financial and economic variables. The ANN-PCA mining model gives a slightly higher prediction accuracy with the simplest procedure.

To fix the vanishing gradient problem in neural networks, a RNN algorithm called LSTM has been proposed in [17]. LSTM introduces a unit of computational memory cell that can remember values over arbitrary time intervals. With these memory cells, LSTM is able to obtain dynamic structures of the stock data over time with high prediction accuracy [18]. LSTM becomes a popular model in the recent study of forecasting in the stock market by enhancing with data preprocessing method [6, 18–21]. To produce deep features for one-step-ahead stock price prediction, a deep learning-based forecasting scheme [19] has been designed which combines LSTM with the stacked auto-encoders architecture. Discrete wavelet transform was used to reduce the noise of the input data which includes the daily OHLC (Open, High, Low, Close) variables, the technical indicators and the macroeconomic variables. In contrast to previous approaches, authors [6] applied three independent deep learning architectures, i.e. RNN, LSTM, and CNN, to capture hidden patterns in the data. The sliding window approach is used in their work for the future short-term prediction. It shows that the

CNN model provided a more accurate prediction by only using the current sliding window data without previous prediction information. In [20], CNN is combined with LSTM to analyze the quantitative strategy in stock markets. CNN is used to extract the useful features from the series data, and LSTM is used to predict the future stock price.

3 Our Proposed Hybrid Prediction Model

According to a comparative survey [22], both ANNs and wavelet transforms have a better performance in many cases of financial forecasting in single and hybrid models. To produce deep and static features for the next-day stock price prediction, we propose a hybrid deep learning model, namely WAE-BLSTM, which hybridizes the wavelet transform, the stacked auto-encoders and the bidirectional long-short term memory.

3.1 The Wavelet Transform

The main objective of de-noising is to remove the noise while retaining the fundamental features of the data. Wavelet transform is applied in our work due to its good performance on vagaries financial time series data [23]. This transformation is a mathematical signal analysis process when the signal frequency varies over time. Similar to the discrete wavelet transform (DWT), Maximal Overlap Discrete Wavelet Transform (MODWT) [24] is a linear filtering operation. MODWT transforms series data into coefficients related to variations over a set of scales. MODWT has been proven to be more effective than DWT for time series data such as the rainfall and Mackey-Glass time series prediction [25].

The MODWT wavelet ($\widetilde{W}$) and scaling ($\widetilde{V}$) coefficients are defined as follows:

$$\widetilde{W}_{j,t} = \sum_{l=0}^{L_j-1} \widetilde{h}_{j,l} X_{t-l \bmod N} \tag{1}$$

$$\widetilde{V}_{j,t} = \sum_{l=0}^{L_j-1} \widetilde{g}_{j,l} X_{t-l \bmod N} \tag{2}$$

where time $t = 0, \ldots, N-1$ and N is the sample sizes. For the jth level of decomposition, h is the wavelet high-pass filter and g is the scaling low-pass filter. In our proposed hybrid deep learning algorithm *WAE-BLSTM*, MODWT is used by applying the Haar wavelet as the bias function and the transformation with four levels of decomposition (Fig. 1).

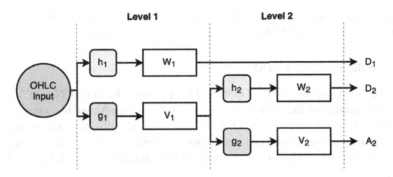

Fig. 1. Example of discrete wavelet transform with 2 levels of coefficient. D is the detail coefficients arrays and A is the approximation coefficients array.

3.2   The Stacked Auto-Encoder

A stacked auto-encoder is a kind of neural network consisting of multiple layers of sparse auto-encoders in which the outputs of each layer are wired to the inputs of the next layer. A single auto-encoder is a supervised learning structure that contains three layers: the input layer, the hidden layer, and the output layer.

Encoder maps the input data into hidden representation, whereas decoder reconstructs the input data from the hidden layer. The process of encoder and decoder can be defined as follows.

$$h_n = f(W_1 x_n + b_1) \tag{3}$$

$$x'_n = g(W_2 h_n + b_2) \tag{4}$$

where f is the encoding function and g is the decoding function. W and b are the weight matrix and the bias vector, respectively. h_n is the encoded value, x'_n is the decoded value, x_n is the original value, b_1 is the bias of the encoding function, b_2 is the bias of the decoding function.

The single layer auto-encoder is trained to obtain the feature vector and to minimize the error between the input vector and the reconstruction vector. The optimization function for minimizing the error can be defined in (5).

$$\arg\min_{W,b}[J] = \arg\min_{W,b} \frac{1}{n} \sum_{i=1}^{n} L(x^i, x'^i) \tag{5}$$

where L is the loss function $L(x, x') = \|x - x'\|^2$.

A stacked auto-encoder is constructed by assembling a sequence of single-layer auto-encoders. The stacked auto-encoders are arranged serially such that the output of the first layer is set as the input to the second and so on. Each auto-encoder layer is trained greedily. Hidden features from the first auto-encoder are used as the input to the second auto-encoder layer. The process is repeated for all subsequent layers. Finally, the deep feature vector from the last encoding layer acts as the final encoded feature vector in our WAE-BLSTM based on the stacked auto-encoders.

3.3   The Bidirectional LSTM

At the beginning of the 1980s, Hopfield [26] proposed Recurrent Neural Networks (RNN) by introducing the neuron as a delocalized content-addressable memory in the asynchronous neural network model. RNN is a class of artificial neural network that takes advantage of the sequential connections of nodes to form a directed graph. It means RNN has the ability to process time-based dynamic behavior for a sequence of values.

LSTM is a type of RNN's architecture that was introduced in 1997 [17]. It tackles the vanishing gradients problem which is practically impossible for a simple RNN to learn from long term events. A common LSTM model is composed of a memory cell, an input gate, an output gate and a forget gate. LSTM uses the hidden activation functions between each gate and the rest of its layer. The cell state is protected by the hidden activation. The following equations describe the behavior of the LSTM model:

$$f_t = \sigma\left(W_{xf}\, x_t + W_{hf}\, h_{t-1} + b_f\right) \tag{6}$$

$$i_t = \sigma(W_{xi}\, x_t + W_{hi}\, h_{t-1} + b_i) \tag{7}$$

$$o_t = \sigma(W_{co}\, c_t + W_{ho}\, h_{t-1} + b_o) \tag{8}$$

$$\widetilde{c}_t = tanh(W_{xc}\, x_t + W_{hc}\, h_{t-1} + b_c) \tag{9}$$

$$c_t = f_t \circ c_{t-1} + i_t \circ \widetilde{c}_t \tag{10}$$

$$h_t = o_t \sigma(c_t) \tag{11}$$

where f_t, i_t, and o_t are the activations of the input, forget, and output gates at time-step t, which control how much of the input and the previous state will be considered and how much of the cell state will be included in the hidden activation of the network. $\widetilde{c}_t$, c_t, x_t and h_t are the candidate state, the protected cell state vector, the input vector, and the output vector of the LSTM unit. W and b are the weight matrices and the bias vectors which need to be learned during the training.

The bidirectional module over the RNN was developed by Schuster and Paliwal [27] to increase the amount of information available in the network. Traditional LSTM learns the representation from the past data, while bidirectional long short-term memory (BLSTM) can get information from the past and the future simultaneously. Two hidden layers of opposite direction connected to the same output are shown in Fig. 2.

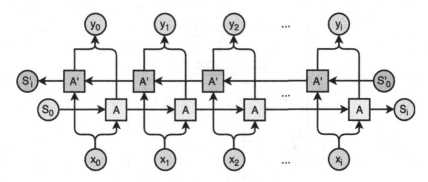

Fig. 2. Bidirectional neural network.

3.4 Our Proposed Hybrid Model WAE-BLSTM

Our proposed hybrid deep learning model includes three components: (1) Decomposing stock data to eliminate the noise by applying the maximal overlap discrete wavelet transform; (2) Stacked auto-encoders are used to reduce the dimensionality of variables; (3) Predicting the next day stock price by using bidirectional long short-term memory BLSTM. Figure 3 illustrates the work flow of our proposed model architecture with flowcharts.

Our model uses de-noising and calculating technical indicators from the stock input signals: open, high, low, close (OHLC). The structural elements of de-noising processing consist of the type of wavelet, level of de-noise, threshold strategy and reconstruction method. Here, for each OHLC signals, we use Haar wavelets with level four when decomposing by multilevel transform. The coefficients arrays are calculated by using Eq. (1) and (2). All levels of detail coefficients arrays go through the soft threshold processing method. Then, we reconstruct each signal by using the sum of detail coefficients arrays and approximate coefficients array.

$$SMA_t = \frac{C_t + C_{t-1} + \ldots + C_{t-n+1}}{n} \tag{12}$$

Despite the de-noised signals, we also calculate key technical indicators from the OHLC data. The survey [3] highlighted the importance of technical indicators and theirs popularity for using input variables. We employ eight technical indicators including the daily trading volume, the simple moving average (SMA), the exponential moving average (EMA), the moving average convergence divergence (MACD), the relative strength index (RSI), the stochastic oscillator (%K, %D), the commodity channel index (CCI), and Williams %R. These indicators can be defined as follows.

$$EMA_t = \frac{(n)(C_t) + (n-1)(C_{t-1}) + \cdots + C_{t-n+1}}{n + (n-1) + \cdots + 1} \tag{13}$$

$$MACD(n)_t = EMA12_t - EMA26_t \tag{14}$$

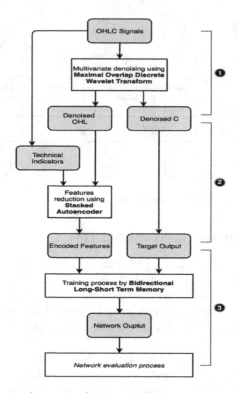

Fig. 3. Flowchart of our three-stage framework. The process is starting by (1) applying denoising method to OHLC data, (2) reducing the input vectors with stack auto-encoders, then (3) training the BLSTM network to predict stock data.

$$RSI_t = 100 - \frac{100}{1 + \frac{\left(\frac{\sum_{i=0}^{n-1} UP_{t-i}}{n}\right)}{\frac{\sum_{i=0}^{n-1} DOWN_{t-i}}{n}}} \tag{15}$$

$$\%K_t = \frac{C_t - LL_{t-n}}{HH_{t-n} - LL_{t-n}} \times 100 \tag{16}$$

$$\%D_t = \frac{\sum_{i=0}^{n-1} K_t}{n} \tag{17}$$

$$CCI_t = \frac{M_t - SMA_t}{0.015D_t} \tag{18}$$

$$Williams\, \%R_t = \frac{HH_{t-n} - C_t}{HH_{t-n} - LL_{t-n}} \times 100 \tag{19}$$

where $M_t = (H_t + L_t + C_t)/3$ is the typical price. C_t, H_t, and L_t are the closing, high, and low price at time t, respectively. HH_{t-n} and LL_{t-n} represent the highest high, and lowest low of the past $t - n$ days.

All technical indicators and de-noised OHL signals account for thirteen variables. The stacked auto-encoder compresses these variables into five encoded features by using the five-layer neural network with 40, 30, 5, 30, and 40 hidden units. The first two layers are the encoder to product encoded features for the third layer, whereas the last two layers provide the training for the decoder. The network is trained with 200 epochs with Adam optimizer [30] and the mean square error as the loss function. The optimization function is defined in Eq. (5).

The final step is to train the bidirectional LSTM as the predictive model. The architecture of bidirectional LSTM contains a number of hidden layers and a number of units in each hidden layer. To test the significance of the workflow, we only apply one hidden layer with 24 units in our proposed hybrid model WAE-BLSTM.

4 Experiments

4.1 Experimental Setup

Our sample data are from Yahoo finance. The data contains historical prices of the Standard & Poor 500 Index (S&P500) from 01/01/2010 to 31/12/2017 (eight years period). It is one of the important benchmark datasets for stock prediction. The financial time series data was divided into the training and testing subsets, the proportion is 80% and 20% of the time series data, respectively. The number of epochs is 200 and the batch size is 32. We implement the proposed stock prediction framework in Python with Keras [28], an open source API for deep learning, and TensorFlow as our deep learning platform.

The prediction procedure is divided into two parts. The first part is the training part, where the past two years' data is used to train the model. The second is the testing part, where we use the next six months' data to test the performance of each model as shown in Fig. 4.

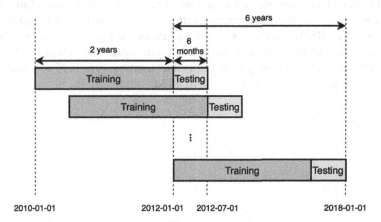

Fig. 4. Training and testing dataset split.

To measure the predictive accuracy of our proposed model, we adapted three performance measures described in [29]. These performance indicators are the mean absolute error (MAE), the root mean square error (RMSE), and the coefficient of determination (R^2 or R squared). The mathematical definitions of these indicators as follows:

$$MAE = \frac{1}{n}\sum_{t}^{n-1}|y_t - \widehat{y}_t| \qquad (20)$$

$$RMSE = \sqrt{\frac{1}{n}\sum_{t}^{n-1}(y_t - \widehat{y}_t)^2} \qquad (21)$$

$$R^2 = 1 - \frac{\sum_{t}^{n-1}(y_t - \widehat{y}_t)^2}{\sum_{t}^{n-1}(y_t - \overline{y})^2} \qquad (22)$$

where y and y_t represent the actual and predicted value at time t for $0 \leq t < n$, respectively, and $\overline{y} = \sum_{t}^{n-1} y_t/n$. For better prediction performance, MAE and RMSE should be close to zero. R^2 is the proportion of the variance in the dependent variable that is predictable from the independent variables. It provides a measure of how well the model can predict the stock price. For better predictions approximate to the real data points, the value of R^2 should be close to one.

4.2   Experimental Results

For comparison, we test our proposed WAE-BLSTM model compared with four different models, including LSTM, BLSTM, W-LSTM, W-BLSTM. Figure 5 presents yearly price prediction of five models in comparison with the actual price of the S&P500 Index. It can be seen that most models can predict the actual price when the price trend is in less volatile periods. LSTM and BLSTM have a larger variation from the actual price than others during the bull and bear market points. WAE_BLSTM and W_LSTM have less fluctuated trends and are closer to the actual price (Table 1).

Figure 6 compares the experimental results of these five models with respect to the values of MAE, RMSE, and R^2, respectively. From the figure, we can see that our proposed WAE-BLSTM is better than the other four algorithms in terms of MAE and RMSE. For the coefficients R^2, our model is fluctuated between 0.97 and 0.82, which are slightly smaller than the values of W-BLSTM.

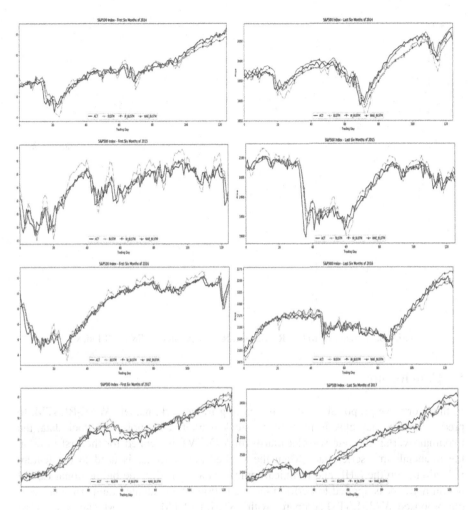

Fig. 5. The comparison of predictions of five models on the S&P500 Index data (2014–2017).

Table 1. Average accuracy on the test dataset.

Model	MAE	RMSE	R^2
WAE-BLSTM	**0.0211**	**0.0272**	0.8934
W-BLSTM	0.0238	0.0309	**0.9125**
W-LSTM	0.0351	0.0423	0.8524
BLSTM	0.0293	0.0375	0.8770
LSTM	0.0334	0.0419	0.8538

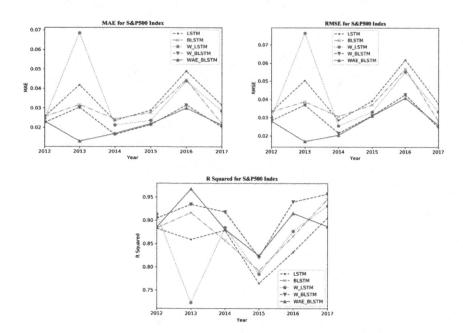

Fig. 6. Results of MAE, RMSE and R2 for predicting S&P500 Index.

5 Conclusions

In this work, we proposed a hybrid forecasting framework, namely WAE-BLSTM, to predict the next day closing price of the stock market. To denoise the stock data, the maximal overlap discrete wavelet transform MODWT is applied for the first time on the financial time series data. Then the stacked auto-encoder is used to the feature reduction from the OHL and technical indicator data. Finally, the bidirectional LSTM neural network is applied to perform the prediction. We evaluated the performance of our proposed WAE-LSTM compared with LSTM, BLSTM, W-LSTM, and W-BLSTM on three indicators including MAE, RMSE, and R^2. The results showed that WAE-BLSTM can obtain the best prediction accuracy among the five algorithms in terms of MAE and RMSE. WAE-BLSTM performs slightly worse than that of W-BLSTM in terms of R^2. Since our proposed WAE-BLSTM framework is based on three distinct parts which can be improved individually in our future work. In addition, existing models can also be the ensemble to produce more complex prediction algorithm based on ensemble learning in the future.

References

1. Pareek, M.K., Thakkar, P.: Surveying stock market portfolio optimization techniques. In: 2015 5th Nirma University International Conference on Engineering (NUiCONE), pp. 1–5 (2015)

2. Wang, B.H., Huang, H.J., Wang, X.L.: A novel text mining approach to financial time series forecasting. Neurocomputing **83**, 136–145 (2012)
3. Atsalakis, G.S., Valavanis, K.P.: Surveying stock market forecasting techniques – part II: soft computing methods. Expert Syst. Appl. **36**(3), 5932–5941 (2009)
4. Cavalcante, R.C., Brasileiro, R.C., Souza, V.L.F., Nobrega, J.P., Oliveira, A.L.I.: Computational intelligence and financial markets: a survey and future directions. Expert Syst. Appl. **55**, 194–211 (2016)
5. Groth, S.S., Muntermann, J.: An intraday market risk management approach based on textual analysis. Decis. Support Syst. **50**(4), 680–691 (2011)
6. Selvin, S., Vinayakumar, R., Gopalakrishnan, E.A., Menon, V.K., Soman, K.P.: Stock price prediction using LSTM, RNN and CNN-sliding window model. In: 2017 International Conference on Advances in Computing, Communications and Informatics (ICACCI), pp. 1643–1647 (2017)
7. Rather, A.M., Sastry, V.N., Agarwal, A.: Stock market prediction and Portfolio selection models: a survey. OPSEARCH **54**(3), 558–579 (2017). https://doi.org/10.1007/s12597-016-0289-y
8. Marcek, D.: Stock price forecasting: autoregressive modelling and fuzzy neural network. Math. Soft Comput. **7**(2), 139–148 (2000)
9. Saxena, H., Anurag, A.V., Chirayath, N., Bendale, R., Kaul, S.: Stock prediction using ARMA. Int. J. Eng. Manag. Res. **8**(2), 1–4 (2018)
10. Adebiyi, A.A., Adewumi, A.O., Ayo, C.K.: Stock price prediction using the ARIMA model. In: Proceedings - UKSim-AMSS 16th International Conference on Computer Modelling Simulation, UKSim 2014, June, pp. 106–112 (2014)
11. Yule, G.U.: Why do we Sometimes get nonsense-correlations between time-series?–A study in sampling and the nature of time-series. J. R. Stat. Soc. **89**(1), 1 (1926)
12. Wold, H.: A study in analysis of stationary time series. J. R. Stat. Soc. **102**(2), 295–298 (1939)
13. Box, G.E.P., Jenkins, G.J.: Time Series Analysis: Forecasting and Control. Holden Day, San Francisco (1970)
14. Heaton, J.B., Polson, N.G., Witte, J.H.: Deep learning in Finance. arXiv Prepr. arXiv:1602.06561, pp. 1–20 (2016)
15. Niaki, S.T.A., Hoseinzade, S.: Forecasting S&P 500 index using artificial neural networks and design of experiments. J. Ind. Eng. Int. **9**(1), 1 (2013)
16. Zhong, X., Enke, D.: Forecasting daily stock market return using dimensionality reduction. Expert Syst. Appl. **67**, 126–139 (2017)
17. Hochreiter, S., Schmidhuber, J.: Long short-term memory. Neural Comput. **9**(8), 1–32 (1997)
18. Murtaza, R., Harshal, P., Shraddha, V.: Predicting stock prices using LSTM. Int. J. Sci. Res. **6**(4), 2319–7064 (2017)
19. Bao, W., Yue, J., Rao, Y.: A deep learning framework for financial time series using stacked autoencoders and long-short term memory. PLoS ONE **12**(7), e0180944 (2017)
20. Liu, S., Zhang, C., Ma, J.: CNN-LSTM neural network model for quantitative strategy analysis in stock markets. In: Liu, D., Xie, S., Li, Y., Zhao, D., El-Alfy, E.S. (eds.) ICONIP 2017. LNCS, vol. 10635, pp. 198–206. Springer, Cham (2017). https://doi.org/10.1007/978-3-319-70096-0_21
21. Yu, S., Li, Z.: Forecasting stock price index volatility with LSTM deep neural network. In: Tavana, M., Patnaik, S. (eds.) Recent Developments in Data Science and Business Analytics. SPBE, pp. 265–272. Springer, Cham (2018). https://doi.org/10.1007/978-3-319-72745-5_29

22. Bahrammirzaee, A.: A comparative survey of artificial intelligence applications in finance: artificial neural networks, expert system and hybrid intelligent systems. Neural Comput. Appl. **19**(8), 1165–1195 (2010)
23. Ramsey, J.B.: The contribution of wavelets to the analysis of economic and financial data. Philos. Trans. Math. Phys. Eng. Sci. **357**(1760), 2593–2606 (1999)
24. Cornish, C.R., Bretherton, C.S., Percival, D.B.: Maximal overlap wavelet statistical analysis with application to atmospheric turbulence. Boundary-Layer Meteorol. **119**, 339–374 (2005)
25. Zhu, L., Wang, Y., Fan, Q.: MODWT-ARMA model for time series prediction. Appl. Math. Model. **38**, 1859–1865 (2014)
26. Hopfield, J.J.: Neural networks and physical systems with emergent collective computational abilities. Proc. Natl. Acad. Sci. U.S.A. **79**(8), 2554–2558 (1982)
27. Schuster, M., Paliwal, K.K.: Bidirectional recurrent neural networks. IEEE Trans. Signal Process. **45**(11), 2673–2681 (1997)
28. Chollet, F., et al.: "Keras." Github (2015)
29. Althelaya, K.A., El-Alfy, E.-S.M., Mohammed, S.: Evaluation of bidirectional LSTM for short-and long-term stock market prediction. In: 2018 9th International Conference on Information and Communication Systems (ICICS), pp. 151–156 (2018)
30. Kingma, D., Ba, J.: Adam: a method for stochastic optimization. In: International Conference on Learning Representations (2014)

Multi-period Distributed Delay-Sensitive Tasks Offloading in a Two-Layer Vehicular Fog Computing Architecture

Yi Zhou, Kai Liu$^{(\boxtimes)}$, Xincao Xu, Chunhui Liu, Liang Feng, and Chao Chen

College of Computer Science, Chongqing University, Chongqing 400040, China
{zyzhouyi,liukai0807,near,chhliu0302,liangf,cschaochen}@cqu.edu.cn

Abstract. Vehicular fog computing (VFC) is a new paradigm to extend the fog computing to conventional vehicular networks. Nevertheless, it is challenging to process delay-sensitive task offloading in VFC due to high vehicular mobility, intermittent wireless connection and limited computation resource. In this paper, we first propose a distributed VFC architecture, which aggregates available resources (i.e., communication, computation and storage resources) of infrastructures and vehicles. By considering vehicular mobility, lifetimes of tasks and capabilities of fog nodes, we formulate a multi-period distributed task offloading (MPDTO) problem, which aims at maximizing the system service ratio by offloading tasks to the suitable fog nodes at suitable periods. Then, we prove that the MPDTO problem is NP-hard. Subsequently, an Iterative Distributed Algorithm Based on Dynamic Programming (IDA_DP) is proposed, by which each fog node selects the appropriate tasks based on dynamic programming algorithm and each client vehicle determines the target fog node for its tasks according to the response delay. Finally, we build the simulation model and give a comprehensive performance evaluation, which demonstrate that IDA_DP can obtain the approximate optimal solution with low computational cost.

Keywords: Vehicular Fog Computing (VFC) · Distributed scheduling · Task offloading · Delay-sensitive

1 Introduction

In vehicular networks, there are a variety of computation-intensive and delay-sensitive applications, such as ultra-high definition video streams [1], autonomous driving [2] and real-time traffic information management [3], which involve massive data transmission and complex task processing [4,5]. Due to the limitation of computation and storage capacities of vehicles, the contradiction between resource-requirement of applications and resource limitation of vehicles poses significant challenges on task offloading.

Mobile Edge Computing (MEC) [6,7] and Fog Computing (FC) [8,9] are envisioned as promising approaches to address such challenge by offloading

© Springer Nature Singapore Pte Ltd. 2020
H. Zhang et al. (Eds.): NCAA 2020, CCIS 1265, pp. 461–474, 2020.
https://doi.org/10.1007/978-981-15-7670-6_38

the computation-intensive tasks to the edge infrastructures via Vehicle-to-Infrastructure (V2I) communication [10]. The general idea is to migrate the resources from the centralized servers to the edge of network and provide computing services in the proximity of vehicles, which can improve resource utilization of edge devices and facilitate the processing of tasks generated by vehicles [11]. Due to the high cost of deploying infrastructures, such as the roadside units (RSUs) and Basic Stations (BSs), the paradigm of Vehicular Fog Computing (VFC) [12] has been envisioned as an promising approach in vehicular networks, which considers vehicles as the fog nodes by exploiting their communication, computation and storage resources to provide task processing services. Zhu et al. [13] proposed a VFC architecture and an innovative task allocation scheme to optimize offloading latency and workload. Zhang et al. [14] proposed a VFC system by exploiting parked vehicles to extend the capabilities of fog nodes, which can assist the delay-sensitive computing services. Liu et al. [15] proposed a two-layer VFC architecture to explore the synergistic effect of the cloud and the fog nodes on processing time-critical tasks in vehicular networks. However, these studies are based on centralized scheduling, which requires much communication overhead to collect global information. With an increasing of network scale and data volume, it is difficult to obtain global information with the constraints of low latency and finite resources. To overcome this challenge, this work is dedicated to designing a distributed VFC architecture and offloading scheme for individual fog node to improve the system efficiency.

Specifically, we present a distributed VFC architecture with two layers, namely, the fog layer and the client layer. Each node in the fog layer schedules the offloading requests submitted by the vehicles in the client layer. However, due to the limited resources and heterogeneous capabilities of fog nodes, it is critical to have the coordination of different fog nodes. Thus, we formulate a multi-period distributed task offloading (MPDTO) problem, aiming at maximizing the system service ratio by best coordinating of task offloading and exploiting the resources of fog nodes in a distributed way. Then, we prove that the MPDTO is NP-hard by constructing a polynomial-time reduction from the maximum cardinality bin packing (MCBP) problem. Subsequently, we design a heuristic distributed algorithm based on dynamic programming to reduce computational overhead.

The main contributions of this paper are listed as follows:

(1) We construct a distributed VFC architecture with fog layer and client layer, in which vehicles that generate tasks are called clients and the others with available resources are regarded as fog nodes. Furthermore, the infrastructures (i.e., RSUs and BSs) are also regarded as fog nodes. In this architecture, each fog node plays two roles: the scheduler and the processor, which are responsible for making offloading decisions and processing tasks, respectively.

(2) We formulate an MPDTO problem, which aims at maximizing the system service ratio by considering multiple critical factors, such as heterogeneous resources of fog nodes, lifetime of tasks and vehicular mobility. Then, we prove that the problem is NP-hard.

(3) We propose a Iterative Distributed Algorithm Based on Dynamic Programming (IDA_DP), which is decomposed into two subprocess in each period and schedules tasks offloading by coordinating fog nodes and clients. Specifically, a three-dimensional table is constructed to record the maximum number of selected tasks and is updated iteratively based on a designed update policy. According to the obtained scheduling strategy, an exploitation method is executed at each client vehicle to select the final fog node based on the minimum response delay.
(4) We build the simulation model and give a comprehensive performance evaluation. The simulation results demonstrate the effectiveness and the scalability of the proposed algorithm.

The remainder of this paper is structured as follows. Section 2 presents the two-layer VFC architecture. Section 3 formulates the MPDTO problem and prove its NP-hardness. A heuristic distributed algorithm is proposed in Sect. 4. Section 5 gives the performance evaluation and Sect. 6 summarizes this work.

2   System Model

A typical system of distributed VFC architecture with the client layer and the fog layer is presented in Fig. 1. In this system, smart vehicles can play two roles: client and fog node, which are equipped with multiple sensors (e.g., GPS, radars, cameras), and have certain computation, storage and communication capacities. As a client, vehicle may generate some delay-sensitive and computation-intensive tasks, which require specified resources from fog nodes. As a fog node, vehicle has extra resources to provide diverse services to client vehicles. Furthermore, static infrastructures can also be served as fog nodes to form the heterogeneous fog environment.

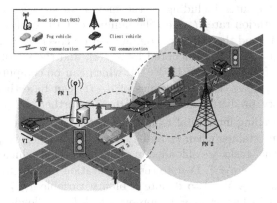

Fig. 1. Distributed VFC architecture.

In the concerned service scenario, client vehicles collect data from various sensors and generate some latency-sensitive and computation-intensive tasks.

With the increasing of data volume, the great demands on low-latency response and the finite resources of mobile terminal make all the tasks unable to be completed locally on time. Hence, fog nodes located in proximity to client vehicles can provide services to reduce the response delay and alleviate such limitations. As shown in Fig. 1, the communication range (denoted by the circles) of a fog node covers a region of the city and can involve several intersections. Each client may drive in the regions of multiple fog nodes at the same time, and its offloading requirements will be submitted to these fog nodes. A fog node can offer effective services to multiple client vehicles in its communication range as long as it can meet resource requirements of all the tasks. However, each task can only select one fog node to offload. It is urgent to design a task offloading algorithm implemented at each fog nodes in the distributed VFC architecture to maximizing the system service ratio.

3   Problem Formulation

3.1   Notations

Consider a set of $V = \{v_1, v_2, \cdots, v_{|V|}\}$ as client vehicles, where each client has a set of computation-intensive and delay-sensitive tasks to be completed. Denote $B = \{b_1, b_2, \cdots b_{|B|}\}$ as the set of tasks generated by all the client vehicles. The binary matrix $\mathbf{A} \in \{0,1\}^{|V| \times |B|}$ represents the relationship between clients and tasks, where $a_{ik} = 1$ represents b_i is generated by v_k; otherwise, $a_{ik} = 0$. For each task $b_i \in B$, the generation time, the deadline, the data size and the required number of CPU cycles are denoted by st_i, dt_i, s_i and c_i, respectively. The interval $[st_i, dt_i]$ represents the lifetime of b_i, denoted by LP_i, which indicates b_i need to be completed during this interval.

The set of fog nodes is denoted by $F = \{f_1, f_2, \cdots, f_{|F|}\}$. For each $f_j \in F$, there are four attributes including of the maximum storage capacity S_j, the maximum computation rate R_j (i.e., the number of CPU cycles per unit time), the bandwidth W_j and the radius of communication coverage D_j. At time t, the distance between f_j and v_k is denoted by $d_{j,k}(t)$, and the set of client vehicles in the coverage of f_j is denoted by $V_{f_j}(t)$, which can be computed by $V_{f_j}(t) = \{v_k | d_{j,k}(t) \leq D_j, \forall v_k \in V\}$. The values of $d_{j,k}(t)$ can be estimated based on existing trajectory prediction techniques [16,17]. Further, $DI_{j,k} = [t_{j,k}^{meet}, t_{j,k}^{dev}]$ represents the dwelling interval of v_k in the coverage of f_j, where $t_{j,k}^{meet}$ and $t_{j,k}^{dev}$ represent the time stamps when v_k and f_j meet and deviate, respectively.

We consider that tasks are offloaded in different periods, which are denoted by the set $T = \{t_1, t_2, \cdots, t_{|T|}\}$. Denote the computation rate for task b_i executing in fog node f_j as $r_{i,j}$. We also denote a binary variable $x_{i,j,t}$ to indicate the offloading strategy, where $x_{i,j,t} = 1$ represents task b_i is offloaded to fog node f_j at period t; otherwise, $x_{i,j,t} = 0$.

3.2 Multi-period Distributed Task Offloading Problem

In this section, we formulate the multi-period distributed task offloading (MPDTO) problem. Concretely, we introduce the constraints of resources and delay, respectively, and formulate the MPDTO into a integer linear problem (ILP).

During the process of task offloading, each $f_j \in F$ has two resource queues, such as the computation resource queue RQ_j, and the storage resource queue SQ_j. At period t, the resource queues may increase due to the completion of current tasks, where the increased computation resource is denoted by $r_{j,t}^+$ and the increased storage resource is denoted by $s_{j,t}^+$. Within duration $[1, t]$, the total required computation and storage resources of all tasks offloaded to f_j can not exceed the maximum available resource, which can be expressed as follows:

$$\sum_{t_0 \in [1,t]} \sum_{b_i \in B} r_{i,j} \cdot x_{i,j,t_0} \leq \sum_{t_0 \in [1,t]} r_{j,t_0}^+, \forall f_j \in F, \forall t \in T \tag{1}$$

$$\sum_{t_0 \in [1,t]} \sum_{b_i \in B} s_i \cdot x_{i,j,t_0} \leq \sum_{t_0 \in [1,t]} s_{j,t_0}^+, \forall f_j \in F, \forall t \in T \tag{2}$$

In addition to demand for resources, each $b_i \in B$ needs to be completed within the lifetime LP_i and the result needs to be returned to corresponding client vehicle before leaving the coverage of fog node. Based on this insight, we define two concepts of the maximum tolerated delay and the response delay.

Definition 1 Maximum tolerated delay. *It refers to the maximum time stamp for f_j to complete b_i, which is denoted by $t_{i,j}^{max}$. The maximum tolerated delay must be no more than the deadline of b_i and cannot exceed the time that client vehicle leaves the coverage of f_j (i.e., $t_{j,k}^{dev}$), computed as:*

$$t_{i,j}^{max} = min\{dt_i, t_{j,k}^{dev}\}, \forall b_i \in B, f_j \in F, \exists k, a_{ik} = 1 \tag{3}$$

Definition 2 Response Delay. *It refers to the time stamp of completing the b_i and returning the result back to corresponding client vehicle when b_i is offloaded to f_j at t (i.e., $x_{i,j,t} = 1$), which is denoted by $t_{i,j,t}^{res}$. The response delay include the waiting delay, transmission delay and computation delay, defined as follows:*

$$t_{i,j,t}^{res} = t_{i,j,t}^{wait} + t_{i,j,t}^{trans} + t_{i,j,t}^{proc} \tag{4}$$

We neglect the time overhead of the result returned back to the corresponding client vehicle, due to the size of the result in general is much smaller than the size of task data. The waiting delay can be computed by $t_{i,j,t}^{wait} = t - 1$, since b_i may not be offloaded instantly and needs to wait at the client vehicle before being transmitted to f_j. Specifically, the other two related concepts will be introduced as follows:

Definition 3 Transmission Delay. *It refers to the time duration for offloading b_i from client vehicle to f_j at period t, denoted by $t_{i,j,t}^{trans}$, which depends on the size of b_i and the transmission rate between b_i and f_j, computed as follows:*

$$t_{i,j,t}^{trans} = \frac{s_i}{\gamma_{i,j,t}}, \tag{5}$$

According to Shannon theory [18], the transmission rate $\gamma_{i,j,t}$ can be expressed as follows:

$$\gamma_{i,j,t} = W_j \cdot log_2(1 + \frac{p_{i,j} \cdot g_{i,j}}{\sigma^2}), \tag{6}$$

where W_j is the bandwidth of f_j, and $p_{i,j}$ is the transmission power for b_i offloading to f_j. Further, $g_{i,j}$ is the channel gain from client vehicle to f_j and σ^2 is the power of the receiver additive white Gaussian noise (AWGN), which is assumed to be identical between any client vehicle and fog node.

Definition 4 Processing Delay. *It refers to the time duration by executing b_i in f_j at period t, denoted by $t_{i,j,t}^{proc}$. We assume that the $t_{i,j,t}^{proc}$ is equal to the required number of CPU cycles (i.e., c_i) divided by the allocated computation rate of task from f_j (i.e., $r_{i,j}$), defined as follows:*

$$t_{i,j,t}^{proc} = \frac{c_i}{r_{i,j}}, \tag{7}$$

In order to quantitatively measure the system performance, we define a metric, namely, system service ratio.

Definition 5 System Service Ratio (SSR). *Given an offloading strategy x, SSR is defined as the ratio of summation of completed tasks to the total number of tasks generated during service interval $[1, |T|]$, which is computed as follows:*

$$SSR(\boldsymbol{x}) = \frac{\sum\limits_{t \in [1,T]} \sum\limits_{b_i \in B} \sum\limits_{f_j \in F} x_{i,j,t}}{|B|} \tag{8}$$

With the above definition, we can formulate the problem formally, as follows:

Given the service interval $[1, |T|]$, a set of tasks $B = \{b_1, b_2, \cdots, b_{|B|}\}$, and a set of fog nodes $F = \{f_1, f_2, \cdots, f_{|F|}\}$, the objective of Multi-period Distributed Task Offloading (MPDTO) problem is to search the optimal task offloading strategy $\mathbf{x} = \{x_{i,j,t} | \forall b_i \in B, \forall f_j \in F, \forall t \in T\}$ in a distributed way under the constraints of vehicular mobility, deadline of tasks and fog resources to maximize the SSR, which is expressed as follows:

Objective:

$$\max \frac{\sum\limits_{t \in [1,T]} \sum\limits_{b_i \in B} \sum\limits_{f_j \in F} x_{i,j,t}}{|B|} \tag{9}$$

Subject To:

$$\sum_{f_j \in F} \sum_{t \in LP_i \cap DI_{j,k}} x_{i,j,t} \leq 1, \ \forall b_i \in B, \exists k, a_{ik} = 1, \tag{10}$$

$$\sum_{f_j \in F} \sum_{t \in T - LP_i \cap DI_{j,k}} x_{i,j,t} = 0, \ \forall b_i \in B, \exists k, a_{ik} = 1, \tag{11}$$

$$\sum_{t_0 \in [1,t]} \sum_{b_i \in B} r_{i,j} \cdot x_{i,j,t_0} \leq \sum_{t_0 \in [1,t]} r_{j,t_0}^+ \leq R_j,$$

$$\forall f_j \in F, \forall t \in T, \tag{12}$$

$$\sum_{t_0 \in [1,t]} \sum_{b_i \in B} s_i \cdot x_{i,j,t_0} \leq \sum_{t_0 \in [1,t]} s_{j,t_0}^+ \leq S_j,$$

$$\forall f_j \in F, \forall t \in T, \tag{13}$$

$$(t - 1 + \frac{s_i}{\gamma_{i,j,t}} + \frac{c_i}{r_{i,j}}) \cdot x_{i,j,t} \leq t_{i,j}^{max},$$

$$\forall b_i \in B, \ \forall f_j \in F, \ \forall t \in T, \tag{14}$$

$$x_{i,j,t} \in \{0,1\}, \forall b_i \in B, \ \forall f_j \in F, \ \forall t \in T. \tag{15}$$

The conditions (10) (11) imply that each task can be offloaded to no more than one fog node at only one period t, and the offloading period t of each task must be in the lifetime of task and the dwelling interval of client vehicle in the coverage of fog node. The condition (12) indicates that the total required computation rates of all tasks offloaded to each fog node from period 1 to t cannot exceed the cumulative available computation resource. Similarly, the condition (13) represents that the total size of tasks offloaded to each fog node from period 1 to t should be within its cumulative available storage capacity. The condition (14) guarantees that each task offloaded to fog node should be completed before the maximum tolerated delay. The condition (15) is the binary variable constraint.

Theorem 1. *The MPDTO problem that find the scheduling strategy to maximum the system service ratio is NP-hard.*

Proof. We prove the NP-hardness of MPDTO by constructing a polynomial-time reduction from the maximum cardinality bin packing (MCBP) problem, which is a well-known NP-hard problem [15]. First, the general instance of the MCBP is described as: Given a set $N = \{1, 2, \cdots, n\}$ of items with sizes $\rho_k, 1 \leq k \leq n$ and set $M = \{1, 2, \cdots, m\}$ of bins with identical capacity C. The objective is to maximum the number of items assigned to bins without violating the capacity constraint. Mathematically, we can formulate the problem as followers:

Objective:

$$\max \sum_{k=1}^{n} \sum_{l=1}^{} \alpha_{k,l} \tag{16}$$

Subject To:

$$\sum_{k \in N} \rho_k \alpha_{k,l} \leq C, \quad \forall l \in M, \tag{17}$$

$$\sum_{l \in M} \alpha_{k,l} \leq 1, \quad \forall k \in N \tag{18}$$

$$\alpha_{k,l} \in 0, 1, \forall k \in N, \forall l \in M \tag{19}$$

For simplification, we only consider the special instance of MPDTO, that is, the service interval is set to a short period (i.e., $T = [1,1]$) so that the fog nodes and the client vehicles are relatively stationary. Consider the lifetime of all the tasks in this scene are large enough that we can ignore the constraints of response delay. It is obvious the special instance of MPDTO belongs to NP problem. Then ,we need to construct a polynomial transformation from MCBP to MPDTO to prove that our problem is NP-hard, which consists of two steps: (1) mapping any instance of MCBP to an instance of MPDTO, and (2) showing the results of MPDTO are the results of MCBP under the above mapping.

For the first step, restrict to MCBP by allowing only instances in which $|B| = |N|$, $r_{i,j} = s_i = \frac{1}{2}\rho_i, \forall b_i \in B, \forall f_j \in F$ and $R_j = S_j = \frac{1}{2}C, \forall f_j \in F$. Then, the set of tasks B and the set of fog nodes F in MPDTO corresponds to the set of items N and the set of bins M in MCBP, respectively. Both the size s_i and the computation rate $r_{i,j}$ for all $b_i \in B$ and $f_j \in F$ in MPDTO correspond to the size items $\rho_i, 1 \leq i \leq n$.

For the second step, we will show that the results of MPDTO are the results of MCBP under the above mapping. If there is a scheduling strategy $\mathbf{x} = \{0,1\}^{|B| \times |F| \times |T|}$ of MPDTO satisfies the constraints $\sum\limits_{t_0 \in [1,t]} \sum\limits_{b_i \in B} r_{i,j} \cdot x_{i,j,t_0} = \sum\limits_{b_i \in B} r_{i,j} \cdot x_{i,j,1} \leq R_j$ and $\sum\limits_{t_0 \in [1,t]} \sum\limits_{b_i \in B} s_i \cdot x_{i,j,t_0} = \sum\limits_{b_i \in B} s_i \cdot x_{i,j,1} \leq S_j$, then $\sum\limits_{b_i \in B} (r_{i,j} + s_i) \cdot x_{i,j,1} = \sum\limits_{i \in N} \rho_i \cdot \alpha_{i,j} \leq R_j + S_j = C$ when $\alpha_{i,j} = x_{i,j,1}, \forall i \in N, \forall j \in M$. By this, we can ensure that as long as a task b_i offloaded to its assigned fog node f_j achieves the constraints of computation and storage resources, for an item $i \in N$, the total sizes of the items on its assigned bin $j \in M$ will not violate the capacity constraint C. Therefore, if we have an algorithm that can find the maximum number of completed tasks, then we can also obtain the optimal solution to the MCBP problem.

The above proves that the MCBP problem can be reduced to a special instance of MPDTO. Therefore, MPDTO is an NP-hard problem.

4   A Heuristic Distributed Algorithm

In this section, we propose the Iterative Distributed Algorithm Based on Dynamic Programming (IDA_DP), which is decomposed into two subprocess in each period and schedules task offloading by coordinating fog nodes and clients in a distributed manner. The first subprocess executed at each fog node makes task offloading decisions based on Dynamic Programming. The second subprocess executed at each client vehicle selects the final fog node according to the exploitation method. Executing the above subprocesses iteratively, we can obtain an approximate task offloading strategy. Then, we will illustrate how the proposed algorithm works in detail.

4.1   Task Offloading Decision-Making Process

The basic idea of this subprocess is to offload tasks as much as possible to each fog node, which is executed at each fog node locally. At period $t^* \in T$, we use one fog node $f_{j^*} \in F$ as an example and the steps are illustrated below.

Step 1 Initialization: At period t^*, compute the available resources of f^*, i.e., $S_{j^*,t^*} = \sum_{t_0 \in [1,t^*]} s^+_{j^*,t_0}$ and $R_{j^*,t^*} = \sum_{t_0 \in [1,t^*]} r^+_{j^*,t_0}$. Within the communication range of f_{j^*}, there are some client vehicles, which generate a lot of tasks that need to be offloaded to f_{j^*} for executing, denoted by B_{j^*,t^*}. According to Eq. (10), Eq. (11) and Eq. (14), each task $b_i \in B_{j^*,t^*}$ can be offloaded to f_{j^*}, only if it satisfies the following four conditions: (1) t^* is in the lifetime of b_i (i.e., $t \in LP_i$), (2) the client vehicle of b_i is in the communication range of f_{j^*} in period t^*, (3) b_i is not offloaded at early periods (i.e., $t' < t^*$), and (4) the response delay for f_{j^*} to complete b_i is no more than the maximum tolerated delay of b_i (i.e., t^{max}_{i,j^*}). Initially, $x_{i,j^*,t^*} = 0$ for $\forall b_i \in B_{j^*,t^*}$, and

$$
B_{j^*,t^*} = \{b_i | t^* \in LP_i \bigcap DI_{j^*,k}, t^{res}_{i,j^*,t^*} \le t^{max}_{i,j^*},
$$
$$
\sum_{t' \in [1,t^*]} \sum_{f_j \in F} x_{i,j,t'} = 0, \forall b_i \in B, \exists k, a_{i,k} = 1\} \tag{20}
$$

Step 2 Result Table Construction: At period t^*, construct a three-dimensional result table $dp_{j^*,t^*}(k,r,s)$ to record the maximum number of tasks offloaded to f_{j^*}, where the first dimension k represents a task $b_k \in B_{j^*,t^*}$, the second dimension r represents the computation resource occupied by the previous k^{th} tasks, and the last dimension s represents the storage capacity occupied by the previous k^{th} tasks. Each entry of the table records the optimal solution of all possible offloading scheme meeting the following conditions: (1) All the tasks from b_1 to b_k in B_{j^*,t^*} have been considered whether be selected to f_{j^*}. (2) The total computation resource occupied by the selected tasks is no more than r. (3) s is large enough to accommodate all the selected tasks. (4) The response delay of each selected tasks from b_1 to b_k is no more than its maximum tolerated delay. If there are no solutions satisfying all the conditions, then $dp_{j^*,t^*}(k,r,s) = 0$. To trace back an optimal solution simplicity, we construct a table $P_{j^*,t^*}(k,r,s)$ to record lists of selected tasks after the table $dp_{j^*,t^*}(k,r,s)$ is updated.

Step 3 Result Table Update: During the update process, the first unarranged task in B_{j^*,t^*} will be considered whether be offloaded to f_{j^*}. It indicates that all the tasks before current considered task b_k have been optimally determined. Calculating each entry $dp_{j^*,t^*}(k,r,s)$ consists two cases:

Case 1: b_k cannot be offloaded to f_{j^*} due to any of the reasons as following: (1) the maximum tolerated response delay t_{k,j^*}^{max} is smaller than t_{k,j^*,t^*}^{res}, (2) r is smaller than the allocated computation rate r_{k,j^*} . (3) s is smaller than the size s_k. In this situation, $dp_{j^*,t^*}(k,r,s) = dp_{j^*,t^*}(k-1,r,s)$ and $P_{j^*,t^*}(k,r,s) = P_{j^*,t^*}(k-1,r,s)$.

Case 2: b_k can be offloaded to f_{j^*}, while the scheduling strategy satisfies the following conditions: (1) t_{k,j^*,t^*}^{res} is no more than t_{k,j^*}^{max}; (2) r is no less than r_{k,j^*}; and (3) s is no less than the size s_k. In this case, we need to further consider how to arrange b_k to get the maximum number of offloaded task. If b_k is selected by f_{j^*}, the number of offloaded tasks is denoted by $dp_{j^*,t^*}(k-1,r-r_{k,j^*},s-s_k)+1$; otherwise, it is denoted by $dp_{j^*,t^*}(k-1,r,s)$. Then, we will select the strategy with bigger value as the result of $dp_{j^*,t^*}(k,r,s)$ and the placement is stored in $P_{j^*,t^*}(k,r,s)$.

For each $f_j \in F$ and $t \in T$, the recursive function is formulated as Eq. (21).

$$dp_{j,t}(k,r,s)= \begin{cases} 0, & \text{if } k=0 \text{ or } r=0 \text{ or } s=0 \\ dp_{j,t}(k-1,r,s), & \text{if } k>0 \text{ and } (t_{k,j,t}^{res} > t_{k,j}^{max} \\ & \text{or } r < r_{k,j} \text{ or } s < s_k \\ max\{dp_{j,t}(k-1,r,s), & \\ dp_{j,t}(k-1,r-r_{k,j},s-s_k)+1\}, & \text{if } k>0 \text{ and } t_{k,j,t}^{res} \le t_{k,j}^{max} \\ & \text{and } r \ge r_{k,j} \text{ and } s \ge s_k \end{cases} \quad (21)$$

Step 4 Allocation Result: Iteratively execute step 3 until all the tasks in B_{j^*,t^*} have been determined whether be offloaded to f_{j^*} or not. At last, the maximum number of selected tasks for f_{j^*} at t^* is stored in $dp_{j^*,t^*}(|B_{j^*,t^*}|, R_{j^*,t^*}, S_{j^*,t^*})$. There exists an optimal assignment $\mathbf{x}_{j^*,t^*} = \{x_{k,j^*,t^*} | \forall b_k \in B_{j^*,t^*}\}$ found by tracing back $P_{j^*,t^*}(|B_{j^*,t^*}|, R_{j^*,t^*}, S_{j^*,t^*})$.

Similarly, each fog node $f_j \in F$ follows the same procedure as f_{j^*} to obtain an optimal task offloading scheme.

4.2 Final Node Selection Process

After the above process, each fog node return the offloading scheme back to corresponding client vehicle. However, a task may be selected by multiple fog nodes, which violates the rules defined in Eq. (10) and Eq. (11). Therefore, we need to design a method for each client to select the final fog node for each task. Based on $x_{i,j,t^*}, \forall f_j \in F$ at t^*, there are three cases for each $b_i \in B$:

Case 1: b_i was not selected by any fog node, i.e., $\sum_{f_j \in F} x_{i,j,t^*} = 0$. Then, b_i will be offloaded at the next periods (i.e., $t > t^*$).

Case 2: b_i was selected by only one fog node, i.e., $\sum_{f_j \in F} x_{i,j,t^*} = 1$. In this case, we need to offloaded b_i to f_{j^*} where $x_{i,j^*,t^*} = 1$.

Case 3: b_i was selected by multiple fog nodes, i.e., $\sum_{f_j \in F} x_{i,j,t^*} > 1$. Thus, the corresponding client vehicle needs to choose one fog node to offload b_i. The exploitation method is designed to select the one with the minimum response delay among all the possible fog nodes in $\{f_j | x_{i,j,t^*} = 1, \forall f_j \in F\}$, i.e., f_{j^*} where $t_{i,j^*,t^*}^{res} = min\{t_{i,j,t^*}^{res} | x_{i,j,t^*} = 1, \forall f_j \in F\}$.

We can obtain the scheduling strategy $\mathbf{X}_{t^*} = \{x_{i,j,t^*} | \forall b_i \in B, \forall f_j \in F\}$ at period t^* after the above subprocess. Further, an approximate task offloading scheme $\mathbf{X} = \{\mathbf{X}_t | \forall t \in T\}$ can be obtained by executing the above two subprocesses iteratively during the service time.

5   Performance Evaluation

5.1   Experimental Setup

In this section, we built the simulation model according to the system architecture described in Sect. 2 to evaluate the performance of the proposed algorithm. Specifically, the simulation model integrates the real-world map extracted from 1.5 km × 1.5 km areas of Chengdu, China (as shown in Fig. 2), the vehicular trajectory database on August 20, 2014 is adopted to simulate vehicular mobility, and the scheduling module as well as the proposed IDA_DP implemented by MATLAB.

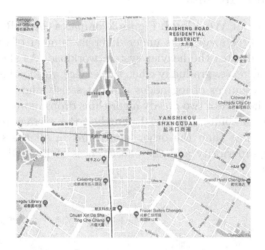

Fig. 2. 2.25 km^2 areas of Chengdu downtown

In the default setting, the service interval is set to [1, 300] and the number of stationary fog node is set to 4 with the fixed locations. The vehicles severed as mobile fog node is randomly selected. To differentiate the resources and capability, the attributes of each fog node are generated at random. For each stationary fog node, the storage capability is randomly generated from [250, 500], the computation rate is randomly generated from [10, 20] $\times 10^8$ CPU cycles/unit time, the bandwidth is randomly generated from [20, 40] and the communication range is randomly generated from [500, 1000] m. For each mobile fog node, the storage capability is randomly generated from [100, 250], the computation rate is randomly generated from [5, 10] $\times 10^8$ CPU cycles/unit time, the bandwidth is randomly generated from [10, 20] and the communication range is randomly generated from [300, 600] m. Further, the number of client vehicles is set to 90, which randomly selected from the database. For each client vehicle, the number of generated tasks is randomly generated in the range of [1, 6] and the attributes of each task are generated at random. The generate time is randomly generated from [1, 100] and the deadline is randomly generated from [100, 300]. The size of task is randomly generated from [10, 25], and the required computation resources of service requests is randomly generated in the range of [10, 25] $\times 10^9$ CPU cycles.

For performance comparison, we implement two algorithms: (1) "greedy_early" with first in first out (FIFO) scheduling policy, and (2) "gready_short" with earliest deadline first (EDF) scheduling policy. During simulation, we collect the following statistic: the number of completed tasks executed in f_j during simulation, denoted by B_j^{comp}. We have defined SSR (i.e., system service ratio) in Sect. 3, which is the primary metric for performance evaluation.

5.2   Experimental Results

(1) **Effect of System Workload:** Figure 3 compares the SSR and the average response delay of three algorithms under different number of tasks. The larger number of tasks indicates the heavier system workload. As shown, the SSR of three algorithms decreases gradually when the service workload is getting higher, since the limited resources of fog nodes can not satisfy the increasing requirements. However, the SSR of two greedy algorithms drops more drastically, which demonstrates the scalability of IDA_DP. Further, the average response delay of IDA_DP is always less than the other algorithms in all the scenarios, which demonstrates the superiority and the effectiveness of IDA_DP under different system workloads.

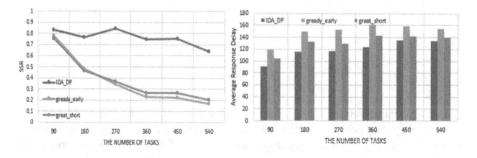

Fig. 3. Performance evaluation under different number of tasks

(2) **Effect of Network Scale:** Figure 4 compares the SSR and the average response delay of the three algorithms under different number of mobile fog nodes. Obviously, the greater number of fog nodes brings a lager network scale.

With the increasing network scale, the SSR of the IDA_DP maintains the best performance, which the result is close to 100% and better than the other greedy algorithms. The SSR decreases gradually when the number of mobile fog nodes is 20 and 30, since the increasing network scale makes the coordination of fog nodes more challenging. Further, the average response delay of the IDA_DP is less than the two greedy algorithms in all the scenarios. The above results show the scalability and adaptiveness of the IDA_DP against network scales.

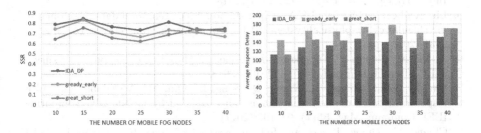

Fig. 4. Performance evaluation under number of mobile fog node

6   Conclusion

In this paper, we presented a distributed VFC architecture in vehicular network, where each fog node served as a distributed scheduler and made the offloading scheme for all the tasks submitted by the clients within its communication coverage. On this basis, we formulated the MPDTO problem aiming at maximizing the SSR, where comprehensively investigated the effects of multiple critical factors, such as different resource and delay requirements of tasks, heterogeneous resources of fog nodes and the dwelling time of client vehicles. Then, we proved that the MPDTO is NP-hard by constructing a polynomial-time reduction from the known NP-hard problem named MCBP, and we designed a heuristic algorithm based on dynamic programming called IDA_DP to adaptively achieve approximate optimal performance in a distributed way. Finally, we built the simulation model by integrating vehicular trajectory database, real-world map and a scheduling module. We also implemented the proposed algorithm IDA_DP as well as two competitive greedy algorithms for performance evaluation. The extensive simulation results demonstrated the effectiveness and scalability of the proposed IDA_DP under a wide range of scenarios.

Acknowledgement. This work was supported in part by the National Natural Science Foundation of China under Grant No. 61872049, and in part by the Fundamental Research Funds for the Central Universities under Project No. 2020CDCGJ004.

References

1. Nightingale, J., Salva-Garcia, P., Calero, J.M.A., Wang, Q.: 5G-QOE: QOE modelling for ultra-hd video streaming in 5G networks. IEEE Trans. Broadcast. **64**(2), 621–634 (2018)

2. Okuyama, T., Gonsalves, T., Upadhay, J.: Autonomous driving system based on deep Q learning. In: 2018 International Conference on Intelligent Autonomous Systems (ICoIAS), pp. 201–205. IEEE (2018)

3. Wang, X., Ning, Z., Wang, L.: Offloading in internet of vehicles: a fog-enabled real-time traffic management system. IEEE Trans. Industr. Inf. **14**(10), 4568–4578 (2018)

4. Liu, K., Feng, L., Dai, P., Lee, V.C., Son, S.H., Cao, J.: Coding-assisted broadcast scheduling via memetic computing in SDN-based vehicular networks. IEEE Trans. Intell. Transp. Syst. **19**(8), 2420–2431 (2018)

5. Liu, K., Lee, V.C.S., Ng, J.K.Y., Chen, J., Son, S.H.: Temporal data dissemination in vehicular cyber-physical systems. IEEE Trans. Intell. Transp. Syst. **15**(6), 2419–2431 (2014)

6. Abbas, N., Zhang, Y., Taherkordi, A., Skeie, T.: Mobile edge computing: a survey. IEEE Internet Things J. **5**(1), 450–465 (2018)

7. Liu, J., Wan, J., Zeng, B., Wang, Q., Song, H., Qiu, M.: A scalable and quick-response software defined vehicular network assisted by mobile edge computing. IEEE Commun. Mag. **55**(7), 94–100 (2017)

8. Chiang, M., Zhang, T.: Fog and IoT: an overview of research opportunities. IEEE Internet Things J. **3**(6), 854–864 (2016)

9. Xu, X., Liu, K., Xiao, K., Ren, H., Feng, L., Chen, C.: Design and implementation of a fog computing based collision warning system in VANETs. IEEE International Symposium on Product Compliance Engineering-Asia (2018)

10. Liu, K., Ng, J.K.Y., Wang, J., Lee, V.C., Wu, W., Son, S.H.: Network-coding-assisted data dissemination via cooperative vehicle-to-vehicle/-infrastructure communications. IEEE Trans. Intell. Transp. Syst. **17**(6), 1509–1520 (2016)

11. Qiao, G., Leng, S., Zhang, K., He, Y.: Collaborative task offloading in vehicular edge multi-access networks. IEEE Commun. Mag. **56**(8), 48–54 (2018)

12. Hou, X., Li, Y., Chen, M., Wu, D., Jin, D., Chen, S.: Vehicular fog computing: a viewpoint of vehicles as the infrastructures. IEEE Trans. Veh. Technol. **65**(6), 3860–3873 (2016)

13. Zhu, C., et al.: FOLO: latency and quality optimized task allocation in vehicular fog computing. IEEE Internet Things J. **6**(3), 4150–4161 (2018)

14. Zhang, Y., Wang, C.Y., Wei, H.Y.: Parking reservation auction for parked vehicle assistance in vehicular fog computing. IEEE Trans. Veh. Technol. **68**(4), 3126–3139 (2019)

15. Liu, C., Liu, K., Guo, S., Xie, R., Lee, V.C.S., Son, S.H.: Adaptive offloading for time-critical tasks in heterogeneous internet of vehicles. IEEE Internet Things J. 1 (2020). IEEE

16. Wu, Y., Zhu, Y., Li, B.: Trajectory improves data delivery in vehicular networks. In: 2011 Proceedings IEEE INFOCOM, pp. 2183–2191. IEEE (2011)

17. Pathirana, P.N., Savkin, A.V., Jha, S.: Location estimation and trajectory prediction for cellular networks with mobile base stations. IEEE Trans. Veh. Technol. **53**(6), 1903–1913 (2004)

18. Wyner, A.: Recent results in the shannon theory. IEEE Trans. Inf. Theory **20**(1), 2–10 (1974)

Template-Enhanced Aspect Term Extraction with Bi-Contextual Convolutional Neural Networks

Chenhui Liu[1], Zhiyue Liu[1], Jiahai Wang[1(✉)], and Yalan Zhou[2]

[1] Department of Computer Science, Sun Yat-sen University, Guangzhou, China
wangjiah@mail.sysu.edu.cn
[2] College of Information, Guangdong University of Finance and Economics,
Guangzhou, China

Abstract. Opinion always has a target that usually appears in the form of an aspect. Thus, aspect term extraction is a fundamental subtask in opinion mining. Although deep learning models have achieved competitive results, they have neglected the effectiveness of the local context feature in extracting aspect terms. This paper proposes a simple yet effective bi-contextual convolutional neural network model to capture the local context feature. Besides, the lack of annotated training data restricts the potential of deep model performance. Thus, this paper proposes a template-driven data augmentation strategy to alleviate the influence of data insufficiency. Experimental results demonstrate that the data augmentation strategy can effectively enhance the performance of the proposed model, which outperforms the existing state-of-the-art methods.

Keywords: Deep learning · Aspect term extraction · Local context · Data augmentation.

1 Introduction

Aspect term extraction (ATE) is a fundamental subtask in opinion mining research and has many applications [8]. The goal of ATE is to identify opinion targets (or aspects) from opinion texts. In product reviews, aspect refers to product features. For example, in laptop review *"Incredible graphics and brilliant colors."*, *"graphics"* and *"colors"* will be extracted as aspect terms.

The majority of researches formulate ATE task as a supervised sequence labeling problem. Traditionally, conditional random field [3,12] is widely used in sequence labeling problem. However, it fails to capture the deep semantic feature that leads to inferior performance. Recently, deep learning models have achieved competitive performance in ATE by taking the advantage of automatic feature representation.

Aspect terms should co-occur with opinion terms. To date, most progress made in aspect term extraction has focused on the co-extraction of aspect and

© Springer Nature Singapore Pte Ltd. 2020
H. Zhang et al. (Eds.): NCAA 2020, CCIS 1265, pp. 475–487, 2020.
https://doi.org/10.1007/978-981-15-7670-6_39

opinion terms [5,14,15] because opinion information is crucial to aspect term extraction. However, these methods require additional works on opinion term annotation. Opinion terms usually appear around aspect terms, so as the other important features. This paper introduces an effective feature that is neglected by previous researches, called the local context. The local context in this paper refers to the neighbors of the aspect terms (Fig. 1). The local context is the most straightforward feature to distinguish aspect terms from other words. Modeling on the local context will benefit ATE in two ways: (1) Opinion information can be capture effectively without additional annotation. (2) The token representation will be more accurate.

Opinion words generally appear in the local context. For example, in product review *"It's the best keyboard ever!"*, *"keyboard"* is the aspect term, and *(the best, ever)* is the local context, in which the word *"best"* is the opinion word targeting the aspect term *"keyboard"*. Besides, local context can be considered as a kind of opinion expression patterns. Such a pattern is shared among different product reviews. For example, in another product review, *"They offer the best service in this town."*, *"service"* is the aspect term, and its local context *(the best, in)* is similar to the previous example. If the model were aware of the local context pattern in the first example, the aspect term *"service"* in the second example would be easily recognized.

Token representation is the output of the neural network hidden layer. It contains word meanings and useful information for label inference. Specifically, word meanings vary with their local context. Taking the aspect term *"hard drive"* as an example, generally *"hard"* means difficult and *"drive"* is a verb. *"hard drive"* is considered as an aspect term in the laptop domain when the two words are coupled together. Obviously, *"hard"* and *"drive"* are local context to each other. Modeling on the local context helps the neural network model understand words in sentences and further improves the aspect term extraction performance.

Convolutional Neural Network (CNN) was first introduced in image processing. Recently, CNN has been used in ATE task [17] and achieved competitive performance. However, due to the difference in the form of text and image data, CNN is not able to capture local features from text data as effective as image data. To better capture the local context feature from product reviews, this paper further separates the local context into the left and right local context and proposes a bi-contextual convolutional neural network.

Moreover, deep learning models require larger training data than the traditional models for automatic feature learning. Deep learning models often overfit the training data. The size and quality of the training data determine the model generalization performance. Manual labels require an expensive data annotation process. [2] proposed a ruled-based model as a weak supervision strategy to label a huge amount of auxiliary data. However, the quality of the labeled data relies on the parsing result, which will not be precise on many user-generated texts. Data augmentation is a solution for data insufficiency problem in image processing [11]. However, it is not a common practice in natural language processing.

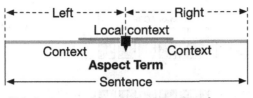

This laptop provides the best **trackpad** ever .

Fig. 1. Local context. Context, local context, and aspect terms are in grey, red, and boldface, respectively. (Color figure online)

The existing text data augmentation strategies [4, 16, 18] focus on text classification task. Such strategies can not work properly on ATE as a sequence labeling task where the labels are on token level. The random transformations and text generation techniques that are performed in the existing data augmentation strategies can not maintain the correctness of label sequences. To effectively alleviate the data insufficiency problem in ATE task, a template concept based on the local context idea is introduced in this paper. Sentences that contain similar local contexts can be abstracted as a template. For example, if replacing the aspect term *"keyboard"* with a placeholder *[TARGET]* in review *"It's the best keyboard ever !"*, the sentence becomes a template *"It's the best [TARGET] ever !"*. The template can transform into a new review sentence *"It's the best hard drive ever !"* by filling a different aspect term *"hard drive"* in the placeholder. Therefore, this paper proposes a template-driven data augmentation strategy to automatically expand the training data with reasonable texts.

In summary, our contributions are as follow:

- A bi-contextual convolutional neural network (BiCNN) model is proposed to extract features from local context and construct accurate token representation.
- A template-driven data augmentation (TDDA) strategy is designed to alleviate the data insufficiency problem by generating new training data.
- Extensive experiments are conducted on three benchmark datasets. Experimental results demonstrate the effectiveness of the proposed model and the augmentation strategy.

2 Related Work

The mainstream research formulates ATE as a sequence labeling problem. Traditionally, conditional random field [3, 12] is the most commonly used model. However, it fails to capture the deep semantic features and requires intensive labor for constructing rules and handcrafted features. Recently, deep learning models have drawn huge attention. [9] proposed a recurrent neural network-based model with word embeddings outperformed traditional conditional random field-based models. Recent progress of ATE researches is made by the co-extraction of aspect

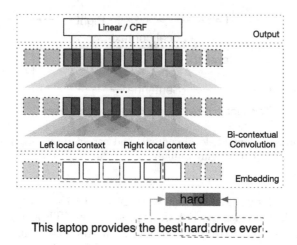

Fig. 2. BiCNN model, with left context convolution printed in red and right context convolution in blue (Color figure online)

and opinion terms. Dependencies between the aspect and opinion terms are used to enhance the performance of aspect term extraction. [14] proposed a recursive neural network based on a dependency tree to learn the representation for the aspect and opinion terms. However, the syntactic or dependency structures of many user-generated texts may not be precise. Therefore, more and more methods are proposed without parsers. [15] proposed a multi-layer attention model. [6] equipped a long short-term memory (LSTM) network with extended memories. [5] presented a new framework to exploit the opinion summary and aspect detection history. Although the co-extraction model achieved competitive performance, they require additional works on opinion term annotation. There are some other methods that do not explicitly model on the opinion terms. [17] coupled general-purpose embeddings and domain-specific embedding together to encode domain information. [10] exploited the overall meaning of the sentence with a sequence-to-sequence framework. Although they achieved competitive performance, they did not realize the fact that useful opinion expression patterns are contained in the local context. Different from the previous researches, this paper proposes a BiCNN model to capture the local context feature effectively.

Data augmentation is a common strategy in computer vision researches [11] and has become a standard technique to train robust deep networks for image processing. In natural language processing, text data augmentation strategies such as Synonym replacement, random insertion, random swap, random deletion [16], back translation [18] and text generation [4,19] are proposed for text classification task and has achieved considerable enhancement on certain tasks.

While writing this paper, we noticed that the local context has recently been applied to tackle unsupervised aspect extraction problem [7]. This paper is different from [7] in three aspects. Firstly, the problem definition is different.

This paper formulates ATE as a supervised sequence labeling problem while [7] formulated ATE as an unsupervised clustering problem. Secondly, the modeling strategy is different. This paper proposes a bi-contextual convolution layer while [7] proposed an LSTM layer to capture the local context feature. Experimental results in this paper show that the bi-contextual convolution layer is more effective in local context modeling. Lastly, this paper extends the local context idea to the concept of template and designs a data augmentation strategy.

3   Methodology

3.1   Problem Definition and Overview

Given a sequence $X = \{x_1, ..., x_N\}$ of N words, this paper formulates ATE as a sequence tagging problem. Model output the corresponding tag sequence $Y = \{y_1, ..., y_N\}$, where each $y_i \in \{B, I, O\}$. B, I, and O denote the **B**egining, **I**nside of an aspect, and **O**ther words, respectively.

The proposed method consists of two parts, BiCNN model and TDDA augmentation strategy. Suppose there is a human-annotated training data D_{train}. D_{train} contains a certain amount of product reviews, in which the aspect terms are annotated. BiCNN is trained on D_{train}, and then the trained model can be used on unseen data. TDDA is an augmentation strategy that aims at expanding the training data. TDDA constructs augmented data D_{aug} from D_{train} following certain rules. Then, D_{aug} is used to enhance the performance of BiCNN.

3.2   BiCNN Model

As is depicted in Fig. 2, the proposed BiCNN model has one embedding layer, multiple bi-contextual convolution layers, and one output layer. Assume the input sequence is $X = \{x_1, ..., x_N\}$.

First of all, the embedding layer maps each token x_i to a unique continuous representation e_i. Then, the bi-contextual convolution layer is implemented to capture the local context feature. Local context is composed of two parts, the left local context and the right local context (Fig. 1). The left and right local context contribute differently to the aspect term recognition. For example, in product review "*This laptop provides the best trackpad ever.*" and "*It's the best keyboard in 2015*", (*the best, ever*) and (*the best, in*) are local context to the aspect term "*trackpad*" and "*keyboard*", respectively. The two local context patterns share the same left but different right local context, and the left local context is more useful to the aspect term recognition. Therefore, the bi-contextual convolution layer implements two types of convolution kernels, namely the left local context convolution and the right local context convolution. They are designed to dynamically reshape word representations with consideration of the left and right local context. Either of the convolution has many 1D-convolution filters with fixed kernel size k, and performs the following convolution operation with

ReLU activation:

$$c_{i,r}^L = \max(0, (\sum_{j=-k}^{0} W_{j,r}^L e_{i+j}) + b_r^L), \tag{1}$$

$$c_{i,r}^R = \max(0, (\sum_{j=0}^{k} W_{j,r}^R e_{i+j}) + b_r^R), \tag{2}$$

where $c_{i,r}^L$, and $c_{i,r}^R$ indicates the output of the left and right local context convolution for each token x_i. W and b represent the convolution kernel weight and bias, respectively. r represents the r-th filter.

The bi-contextual layer applies each filter to all positions and concatenates the output of the left and right local context convolution together:

$$c_i = c_i^L \oplus c_i^R, \tag{3}$$

where $c_{i,r}$ is considered as a token representation of x_i that contains valuable semantic information extracted from the local context.

Lastly, this paper investigates two types of output layers, namely the linear layer and the conditional random field layer.

Linear Layer: The obtained token representation is directly fed to linear layer with softmax activation function:

$$P(y_i|x_i) = \text{softmax}(W_o c_i + b_o) \tag{4}$$

where W_o and b_o are learnable parameters of linear layer.

Conditional Random Fields: CRF is commonly used to solve the sequence labeling tasks with neural networks. A linear-chain CRF layer is adopted here on top of bi-contextual convolution layer. Different from the linear layer which aims at maximizing the token-level likelihood $p(y_i|x_i)$, linear-chain CRF aims at constructing the globally most possible label sequence. The sequence score $s(X, Y)$ and the sequence likelihood $p(Y|X)$ are calculated as follows:

$$s(X, Y) = \sum_{i=0}^{N} M_{y_i,y_{i+1}}^A + \sum_{i=1}^{N} M_{i,y_i}^P, \tag{5}$$

$$p(Y|X) = \text{softmax}(s(X, Y)) \tag{6}$$

where M^A is a randomly initialized transition matrix and M^P is the emission matrix linearly transformed from the token representation c_i. Given a text sequence X as input, the softmax function is conducted over all of the possible tag sequences. During inference, model output the best tag sequence with the viterbi search algorithm.

3.3 TDDA Augmentation Strategy

The lack of training data restricts deep learning models from achieving better results. TDDA is a data augmentation strategy to expand the training data,

Algorithm 1. TDDA Strategy

Input: D, ω

Output: D_{aug}.

1: D_a, D_o = Separate(D)

   /***Replacing***/

2: D_a^r = NewList()

3: collect aspect terms A from D_a

4: **for** each review r_i in D_a **do**

5:    **for** each aspect term a_{ij} in r_i **do**

6:       randomly pick new aspect term $\hat{a_{ij}}$ from A

7:       $\hat{r_i}$ = Replace(r_i, a_{ij}, $\hat{a_{ij}}$)

8:       append $\hat{r_i}$ to D_a^r

9:    **end for**

10: **end for**

   /***Trimming***/

11: D_a^{rt} = NewList()

12: **for** each review r_i in D_a^r **do**

13:    locate all aspect terms T_i in r_i

14:    **for** each aspect term span $(from, to)$ in T_i **do**

15:       r_i^c = Trim(r_i, from, to, ω)

16:       append r_i^c to D_a^{rt}

17:    **end for**

18: **end for**

19: **Return** $D_{aug} = D + D_a^{rt}$

---

thus alleviates the data insufficiency problem. The core idea of TDDA is that sentences contain similar local contexts can be abstracted as a template. Templates are general opinion expression patterns that can be applied to different aspect terms. Following the procedure in Algorithm 1, the input of TDDA is the training data D and the local context window size ω. The output is the augmented data D_{aug}.

D is separated into D_a and D_o. D_a denotes the texts that contain aspect terms while D_o denotes the rest that does not contain aspect terms. Then, the following process is divided into two steps, replacing and trimming.

Replacing: Product reviews can be abstracted as a template by replacing its aspect terms with placeholders. Templates are general opinion expression patterns that can be applied to different aspect terms. Thus, replacing aspect terms in one sentence with another will not affect the sentence structure and the expressed sentiment. Aspect terms are collected from D_a and stored in set A. Aspect terms should be categorized into two types, noun and verb. Aspect terms should be replaced by aspect terms with similar part-of-speech. For example, in sentence "*I can't believe how quick this machine boots up.*", "*boots up*" is a verb aspect term. If the "*boots up*" is replaced by a noun aspect term, the generated sentence will be strange. Besides, aspect terms are usually phrases in different sizes. The label sequences of aspect terms change according to their size, e.g., "*B I I*" is a label sequence of size 3. Thus, to improve the diversity of the label

Table 1. A sample of augmented text with $\omega = 3$. The aspect terms are in boldface and words in the local context window are in italic.

Text	*The* **service** *was superb* , they treat you like family .
Augmented Text	*The* **lunch buffet** *was superb* ,

Table 2. Datasets statistics. The second and third columns refer to the number of product review and the amount of aspect terms.

	Dataset		Sentence	Aspect
Data	SE14-LAPT	Train	3045	2358
		Test	800	654
	SE14-REST	Train	3041	3693
		Test	800	1134
	SE16-REST	Train	2000	1743
		Test	676	612
Augmented Data	SE14-LAPT Train		4982	4613
	SE14-REST Train		5992	7214
	SE16-REST Train		3462	3350

sequences, the new aspect term $\hat{a}_{ij}$ should be randomly picked from aspect term set A with different sizes. Then, the original aspect terms a_{ij} in review sentences will be replaced by the new aspect term $\hat{a}_{ij}$. $\hat{D}_a^r$ is composed of the processed review sentences.

Trimming: Opposite to the local context, remote context refers to context away from the aspect term. Remote context is usually unrelated to aspect term recognition. For example, in the product review "... *has excellent battery life* ...", a human can recognize "*battery life*" as an aspect term in the laptop domain simply by reading these 4 four words. A longer sentence contains more redundant information, which may affect the model judging. Therefore, the trimming step is to trim sentences $\hat{D}_a$ into chunks and remove redundant words from remote context according to the local context window size ω. For each review r_i in $\hat{D}_a$, r_i has already been tokenized, and the aspect term span *(from, to)* is a token level span. Then, r_i is trimmed into chunks according to the window size ω. For reviews that contain multiple aspects, if the distance between two adjacent aspect terms is smaller than ω, they will not be separated into different chunks.

Finally, the augmented data D_{aug} is the combination of the original training data D and the processed templates D_a^{rt}. An example of the augmented text is depicted in Table 1. The augmented text in the second row replaces the original aspect term "*service*" with "*lunch buffet*" and trims the redundant words "... *they treat you like family*".

Table 3. Experimental results. BiCNN and BiCNN-CRF are trained on the original SemEval datasets. BiCNN-TDDA and BiCNN-CRF-TDDA are trained on the augmented SemEval datasets. F1-score is reported.

Model	SE14-LAPT	SE14-REST	SE16-REST
IHS_RD	74.55	79.62	-
DLIREC	73.78	84.01	-
BiLSTM-CRF	77.50	84.57	69.57
RNCRF	78.42	84.93	-
CMLA	77.80	85.29	-
MIN	77.58	-	73.44
HAST	79.52	85.61	73.61
DECNN	81.59	-	74.37
Seq2SeqATE	80.31	-	75.14
RINANTE	81.37	86.76	-
BiCNN	82.56	86.11	75.17
BiCNN-CRF	82.99	86.49	74.12
BiCNN-TDDA	82.86	86.74	**75.45**
BiCNN-CRF-TDDA	**83.52**	**86.83**	74.88

4 Experiments

4.1 Datasets and Experimental Settings

Experiments are conducted on three benchmark datasets from SemEval2014[1] and SemEval 2016[2] challenges. These three datasets contain human-annotated aspect terms from the laptop and restaurant domain. Table 2 lists the statistics of these datasets, where SE14-LAPT, SE14-REST, and SE16-REST are used to represent SemEval14-Laptop, SemEval14-Restaurant, and SemEval16-Restaurant dataset, respectively. Dataset is split by randomly holding out 150 samples as validation data and the rest as training data.

For the TDDA process, the training data is input to TDDA as D and obtains the corresponding augmented data. If there is no additional statement, the parameter ω of TDDA is set to 3. Statistics of the augmented data is presented in the Table 2.

In preprocessing, sentences are tokenize with NLTK[3]. The embedding layer implements the double embedding process [17]. BiCNN contains 4 layers of bi-contextual convolution layers, and the kernel sizes of the 4 bi-contextual convolution layers are set to [3, 5, 5, 5], respectively. Each bi-contextual convolution layer has 256 filters. The dropout rate is 0.55 for each layer. During training, we

---

[1] http://alt.qcri.org/semeval2014/task4/.

[2] http://alt.qcri.org/semeval2016/task5/.

[3] http://www.nltk.org/.

form the batch size with 128 training samples, and the learning rate of Adam optimizer is set to 0.0001. We train our model 5 times with different random seeds and report the average F1-score.

4.2   Baseline Models

The proposed model is compared with the following baselines, which are categorized into three groups. The first group contains conditional random field-based models.

IHS_RD [1] and **DLIREC** [13] : IHS_RD and DLIREC are conditional random field-based models, and the best performing systems at SemEval 2014.

BiLSTM-CRF: It is a vanilla bi-directional LSTM with a conditional random field.

The second group contains the aspect and opinion term co-extraction models.

RNCRF [14]: It integrates recursive neural networks and conditional random fields into a unified framework for explicit aspect and opinion terms co-extraction.

CMLA [15]: CMLA implements multiple layers of attention networks to propagate information between aspect terms and opinion terms dually.

MIN [6]: MIN equips two LSTMs and neural memory operations for aspect extraction.

HAST [5]: A truncated history-attention and a selective transformation network are proposed to improved aspect extraction.

The third group contains the current state-of-the-art models.

DECNN [17]: DECNN couples two types of pre-trained embedding with multiple layers of CNN for aspect term extraction.

Seq2Seq4ATE [10]: It formulates ATE as a sequence-to-sequence task where the source sequences and target sequences are words and labels, respectively.

RINANTE [2]: RINANTE is a neural network model that adds mined rules as weak supervision. It is trained on a huge amount of auxiliary data and a small amount of human-annotated data.

4.3   Main Results

The main results are shown in Table 3. For the first group of baselines, IHS_RD, DLIREC, and BiLSTM-CRF are based on the conditional random field. IHS_RD and DLIREC use handcrafted features with shallow semantic, while BiLSTM-CRF extracts deep semantic automatically from the training data. BiLSTM-CRF outperforms IHS_RD and DLIREC, which shows the importance of deep semantic to ATE study.

For the second group of baselines, co-extraction models are taking advantage of the additional annotated opinion terms. However, BiCNN does not require annotation on the opinion terms. Comparing with the best performance in this group, BiCNN obtains 3.04%, 0.5%, and 1.56% absolute gains on SE14-LAPT, SE14-REST, and SE16-REST, respectively. It is because the opinion information is usually located in the local context, and is effectively modeled by BiCNN.

Besides, co-extraction models are mainly based on the recurrent neural network structure that aims at capturing the long-term dependencies. Thus they are not good at modeling the local context features.

For the third group of baselines, BiCNN outperforms DECNN on SE14-LAPT and SE16-REST with 0.97% and 0.8%, respectively. Although CNN is capable of extracting local features, it is not originally designed for text modeling. BiCNN separates local context into two parts and achieves a better performance than the CNN-based model. Seq2SeqATE is aware of the importance of adjacent words of aspect terms. However, Seq2SeqATE is based on a recurrent neural network which is not good at local feature extraction. Besides, in the sequence labeling problem, restrictions among the adjacent labels are also considered important. For example, label I will not follow right behind label O in label sequences. BiCNN-CRF implements CRF layer on the output layer to capture such label dependencies. The result from Table 3 shows that the performance of BiCNN-CRF improves 0.43% and 0.38% from BiCNN on SE14-LAPT and SE14-REST dataset, respectively. BiCNN-CRF decreases nearly 1% from BiCNN on SE16-REST dataset, probably because SE16-REST is just a half size of SE14-LAPT and SE14-REST, thus it is insufficient to train an additional CRF layer.

Comparing BiCNN and BiCNN-CRF with their data augmentation editions, the enhancement of TDDA strategy is obvious. BiCNN-CRF-TDDA achieves state-of-the-art performance on SE14-LAPT and SE14-REST, and BiCNN-TDDA achieves state-of-the-art performance on SE16-REST. It is worth noticing that BiCNN-CRF-TDDA obtains 1.93% improvement over DECNN on SE14-LAPT. RINANTE is trained with a huge amount of auxiliary data annotated by mined rules. Compared with RINANTE, BiCNN-TDDA and BiCNN-CRF-TDDA achieve better performance without additional auxiliary data. Experimental results demonstrate the effectiveness of the proposed BiCNN and BiCNN-CRF model and the promising enhancement of the TDDA augmentation strategy.

4.4   Effect of Window Size ω

Table 4. Performance of BiCNN-TDDA with different ω. "w/o trim" denotes TDDA strategy without the trimming process. F1-score is reported.

ω	SE14-LAPT	SE14-REST	SE16-REST
1	82.62	86.01	74.48
3	**82.86**	**86.74**	**75.45**
5	82.49	86.68	74.93
w/o trim	82.28	86.70	74.92

Parameter ω in TDDA strategy controls the window size of the local context in the trimming process. A smaller ω means a shorter local context on both sides.

This paper wonders how does the performance of BiCNN-TDDA change with different ω. The experiment in this section is conducted on BiCNN-TDDA model. As is shown in Table 4, ω is set to 1, 3, and 5. "w/o trim" refers to the ablation experiment of TDDA without the trimming process. The experimental result demonstrates that a proper ω leads to better performance. When $\omega = 3$, BiCNN-TDDA achieves the best performance on all three datasets. This experiment demonstrates that a smaller window size of local context may lose important information, while a bigger window size of local context may distract model training.

5   Conclusion

This paper introduces the local context as a critical feature for the supervised ATE task. To effectively capture the local context feature, BiCNN is proposed with the ability to consider the left and right local context separately. This paper further introduces a template concept, based on which the TDDA augmentation strategy is proposed to expand the training data automatically, thus alleviates the data insufficiency problem. Extensive experiments are conducted on three benchmark datasets. Experimental results demonstrate the effectiveness of BiCNN model and the promising enhancement of TDDA augmentation strategy for the ATE task.

Acknowledgement. This work is supported by the National Key R&D Program of China (2018AAA0101203), the National Natural Science Foundation of China (61673403, U1611262), and the Natural Science Foundation of Guangdong Province (2018A030313703).

References

1. Chernyshevich, M.: IHS RandD Belarus: cross-domain extraction of product features using CRF. Proc. SemEval **2014**, 309–313 (2015)
2. Dai, H., Song, Y.: Neural aspect and opinion term extraction with mined rules as weak supervision. Proc. ACL **2019**, 5268–5277 (2019)
3. Jakob, N., Gurevych, I.: Extracting opinion targets in a single and cross-domain setting with conditional random fields. Proc. EMNLP **2010**, 1035–1045 (2010)
4. Kusner, M.J., Hernández-Lobato, J.M.: GANS for sequences of discrete elements with the gumbel-softmax distribution. arXiv preprint arXiv:1611.04051 (2016)
5. Li, X., Bing, L., Li, P., Lam, W., Yang, Z.: Aspect term extraction with history attention and selective transformation. Proc. IJCAI **2018**, 4194–4200 (2018)
6. Li, X., Lam, W.: Deep multi-task learning for aspect term extraction with memory interaction. Proc. EMNLP **2017**, 2886–2892 (2017)
7. Liao, M., Li, J., Zhang, H., Wang, L., Wu, X., Wong, K.F.: Coupling global and local context for unsupervised aspect extraction. Proc. EMNLP-IJCNLP **2019**, 4571–4581 (2019)
8. Liu, B.: Sentiment analysis and opinion mining. Synthesis lectures on human language technologies pp. 1–184 (2012)

9. Liu, P., Joty, S., Meng, H.: Fine-grained opinion mining with recurrent neural networks and word embeddings. Proc. EMNLP **2015**, 1433–1443 (2015)
10. Ma, D., Li, S., Wu, F., Xie, X., Wang, H.: Exploring sequence-to-sequence learning in aspect term extraction. Proc. ACL **2019**, 3538–3547 (2019)
11. Shorten, C., Khoshgoftaar, T.M.: A survey on image data augmentation for deep learning. J. Big Data **1**, 1–48 (2019)
12. Shu, L., Xu, H., Liu, B.: Lifelong learning CRF for supervised aspect extraction. Proc. ACL **2017**, 148–154 (2017)
13. Toh, Z., Wang, W.: DLIREC: aspect term extraction and term polarity classification system. Proc. SemEval **2014**, 235–240 (2014)
14. Wang, W., Pan, S.J., Dahlmeier, D., Xiao, X.: Recursive neural conditional random fields for aspect-based sentiment analysis. Proc. EMNLP **2016**, 616–626 (2016)
15. Wang, W., Pan, S.J., Dahlmeier, D., Xiao, X.: Coupled multi-layer attentions for co-extraction of aspect and opinion terms. Proc. AAAI **2017**, 3316–3322 (2017)
16. Wei, J.W., Zou, K.: EDA: Easy data augmentation techniques for boosting performance on text classification tasks. Proc. IJCNLP **2019**, 6381–6387 (2019)
17. Xu, H., Liu, B., Shu, L., Yu, P.S.: Double embeddings and CNN-based sequence labeling for aspect extraction. Proc. ACL **2018**, 592–598 (2018)
18. Yu, A.W., et al.: QANET: combining local convolution with global self-attention for reading comprehension. arXiv preprint arXiv:1804.09541 (2018)
19. Yu, L., Zhang, W., Wang, J., Yu, Y.: SeqGAN: sequence generative adversarial nets with policy gradient. Proc. AAAI **2017**, 2852–2858 (2017)

Privacy Sensitive Large-Margin Model for Face De-Identification

Zhiqiang Guo[1], Huigui Liu[1], Zhenzhong Kuang[1(✉)], Yuta Nakashima[2],
and Noboru Babaguchi[2]

[1] Key Laboratory of Complex Systems Modeling and Simulation, School of Computer
Science and Technology, Hangzhou Dianzi University, Hangzhou 310018, China
zzkuang@hdu.edu.cn
[2] Osaka University, Osaka, Japan

Abstract. There is an increasing concern of face privacy protection
along with the wide application of big media data and social networks
due to free online data release. Although some pioneering works obtained
some achievements, they are not sufficient enough for sanitizing the sensitive identity information. In this paper, we propose a generative approach to de-identify face images yet preserving the non-sensitive information for data reusability. To ensure a high privacy level, we introduce
a large-margin model for the synthesized new identities by keeping a safe
distance with both the input identity and existing identities. Besides, we
show that our face de-identification operation follows the ϵ-differential
privacy rule which can provide a rigorous privacy notion in theory. We
evaluate the proposed approach using the vggface dataset and compare
with several state-of-the-art methods. The results show that our approach outperforms previous solutions for effective face privacy protection
while preserving the major utilities.

Keywords: Face image · Privacy protection · Generative model ·
Large margin.

1 Introduction

Due to the blooming popularity of surveillance cameras, camera-equipped personal devices, and social networks, the amount of online face images has witnessed significant increase [13,26]. In most cases, such images contain lots of sensitive personal information (*e.g.*, identity, location, health, and religion), which
may lead to the disclosure of privacy or social relations and may bring unde-
sirable troubles. Besides, these public information may also be used by some
malicious attackers to get some illegal profits [1,31]. Thus, it is a trend to sani-
tize the sensitive data for privacy protection before releasing it to public.

There are many different attempts to de-identify face images to protect identity information, such as pixelation, blurring, and k-same method [9,10,21]. However, standard de-identification methods are no longer effective due to the rapid

development of recognition techniques [17] because: (a) deep learning can easily identify the de-identified images [15]; (b) the naïve de-identification methods have poor smoothness in sanitized faces and may stimulate users' curiosity to find the real identity by background attacking.

Recently, generative adversarial network (GAN) has been adopted to generate new faces [6,15,18,30] which can be divided into two categories: partial tampering and global synthesis. The former focuses on changing the local facial parts (*e.g.*, AnonymousNet [15] and PFTD [16]), while the latter generates new faces by keeping visual utilities (*e.g.*, k-same-net [18] and PPGAN [30]). But, existing methods may have high identification for human and lack of rigorous theoretical guarantee on privacy.

In this paper, we present a novel privacy sensitive large-margin (PSLM) solution to de-identify the released face images by focusing on: (1) how to balance between data usability and privacy; (2) how to anonymize the face without being identified as the others; and (3) how to improve and analyze the privacy level. The main contributions of the paper are summarized as follows:

- We develop a novel large-margin algorithm for face privacy protection by keeping a safe distance with both the input face and the known identities.
- We use $N+k$ classifiers to explicitly distinguish the newly generated identities from the existing ones by following the k-same anonymity mechanism.
- We extend the concept of ϵ-differential privacy as one of the evaluation criteria to provide a rigorous guarantee for face de-identification. To the best of our knowledge, this is the first GAN-based approach to relate ϵ-differential privacy with face de-identification.
- The experimental results have verified the effectiveness of our approach on standard dataset.

2 Related Work

Face de-identification is one of the most important privacy protection strategies and, initially, it is achieved by replacing the face region using naive obscuring methods, such as pixelation, blurring, masking-out, adding-noise, and blending [2,4,26]. However, this kind of anonymization is visually not natural and cannot ensure the anonymity in theory [6,22]. Higher-level privacy protection can be achieved by replacing a face with a different one. Bitouk *et al.* [5] proposed to swap the original face with a real face from a database by adjusting its attributes. Lin *et al.* [32] proposed a privacy-preserving image sharing method and Pavel *et al.* [12] introduced a morphing method to protect visual privacy.

To guarantee a certain security level, many different models have been studied under different conditions to alleviate privacy disclosure [29,30]. k-anonymity is one of the most popular theoretical backbones in developing new privacy protection algorithms [2,29] and the k-same framework has received lots of attention to synthesize face images with a low probability of $1/k$ to recognize the original face [21,26,28]. But this family of algorithms would degrade the visual quality and may result in ghosting effects due to misalignment between multiple faces.

In the data security field, the ϵ-differential privacy plays an important role to provide rigorous guarantee in privacy that is provable in theory, but there are limited discussions on face de-identification because it is challenging to apply on non-aggregated data [10]. Although some studies tried to use ϵ-differential privacy to provide high-level privacy guarantee, they mainly focus on traditional privacy protection methods, such as pixelation [10], and they also suffer from the problem of low visual quality which may hurt the data usability.

The recent development of GANs has derived another inspiration for face de-identification by synthesizing realistic faces [3,19]. Ren *et al.* [25] introduced a video face anonymizer to realize a minimal effect on action detection, but the face quality is not guaranteed. Karla *et al.* [6] built a GAN-based model to generate a full body for de-identification. Wu *et al.* [30] designed a verifier and a regulator for de-identified face synthesis. Blaž *et al.* [18] developed a k-same-net approach and Li *et al.* [15] proposed AnonymousNet. Majumdar *et al.* [16] proposed a partially tamper method to spoil face recognition. Although recent works could preserve identity privacy in some sense, they may: (1) have high identification performance; (2) raise the possibility of identity attacks if the synthesized faces are similar to certain real people; and (3) not provide a rigorous criterion for evaluating the privacy level in theory.

This paper concentrates on developing a generative model to synthesize de-identified face, which is different from existing methods in: (1) it does not rely on group-wise anonymous strategy by assigning the same anonymous face to each group as with the k-same framework [18]; (2) it provides an apparent constraints for keeping data utility and alleviating identity switch; and (3) it incorporates the ϵ-differential privacy concept to evaluate the de-identification results instead of adding (Laplace) noise to the original data [10,29].

3   Large Margin Face De-Identification

In this section, we present a novel privacy sensitive large margin (PSLM) approach for face de-identification to fool face identifiers and provide a safer mechanism for image release and reuse. According to Fig. 1, our approach consists of four main components: (a) an encoder E to map the input face image x into a vector $z = E(x) \in \mathbb{R}^{K_{id}}$ in the identity feature space; (b) a generator G to synthesize a de-identified face $x' = G(z)$ for given z; (c) a discriminator D to classify whether a face image is real or fake for adversarial learning; and (d) the classifier C to classify an image into real and fake identities to guarantee the synthesized face is anonymous through training. The networks D and C are only used for training. We suppose that some effective face detection algorithm is known [3,7,15,16] so that we do not need to study how to obtain it.

3.1   Requirements

Let $\mathbb{I} = \{I_i | i = 1, 2, \ldots, N\}$ be the identity domain and $X = \{x | u_{id}(x) \in \mathbb{I}, x \in \Omega\}$ be the face image set, where Ω is the set of all possible images. We also

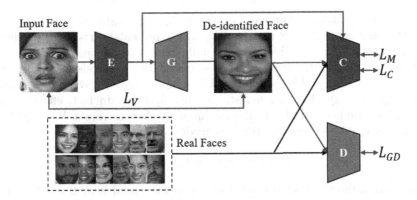

Fig. 1. The pipeline of our approach for face de-identification. Given an input face, we sample a conditioned vector by encoder E and then feed it to generator G to synthesize a de-identified face. We optimize our model by using large-margin losses associated with generator G, classifier C, and Discriminator D.

define attribute labels $u_c(x)$ associated with face image x, where c is in the set of attributes (*e.g.*, $c \in \{\mathrm{id, gender, age, ethnic}\}$) under consideration. Our goal is to learn a function $x' = \phi(x)$ to anonymize the sensitive identity of any given face x yet preserving its non-sensitive attributes. That is, x and x' should satisfy:

$$u_{\mathrm{id}}(x) \neq u_{\mathrm{id}}(x'),\ u_{\mathrm{id}}(x') \notin \mathbb{I},\ u_c(x') = u_c(x) \text{ for } c \neq \mathrm{id}. \tag{1}$$

3.2 Encoder

The recent improvements in deep neural nets have enabled a much higher performance in face recognition [3,7]. Following these works, we train the encoder network E with face-identity pairs $\{x, u_{\mathrm{id}}(x)\}$ and the softmax cross-entropy loss. Let f_{id} be a face identity classifier network and $f_{\mathrm{id}}^{u_{\mathrm{id}}(x)}(x)$ the f_{id}'s output corresponding to label $u_{\mathrm{id}}(x)$. To train f_{id} We use loss function

$$L_E = \mathbb{E}_{x \sim \mathbb{P}_r}[-\log f_{\mathrm{id}}^{u_{\mathrm{id}}(x)}(x)], \tag{2}$$

where $\mathbb{P}_r$ is the empirical distribution of x (*i.e.*, X). The output of the second last layer of the network is used as the identity feature vector $z = E(x)$ for x.

3.3 Objective Model

With trained encoder E, identity I_i can be described by the set of identity vectors of all face images of the identity and their average given by:

$$E_i = \frac{1}{|X_i|} \sum_{x'} E(x'), \tag{3}$$

where the summation is computed over $x' \in X_i = \{x|u_{id}(x) = I_i, x \in X\}$ and $|X_i|$ is the number of the face images in X_i. This representation is a natural choice because the face images of identity I_i usually exhibit intra-class variations.

It is hard for the traditional face de-identification methods to obtain realistic faces [9,26], e.g., like pixelation. Fortunately, GANs have shown significant progress in synthesizing realistic images [3,19,24]. In fact, effective face de-identification is a complex task involving multiple constraints. We consider a sum of multiple losses to optimize our encoder and generator:

$$L = L_{\text{GD}} + L_V + L_M + L_C, \tag{4}$$

where L_{GD} is the traditional GAN loss, L_V is the visual loss, L_M is the large margin de-identification loss, and L_C is the classification loss.

Adversarial Loss. L_{GD} is designed based on conditional GAN (cGAN) [20] by adversarially learning a generator G and a discriminator D. The output z of E serves as the input of G. The training strategy is to play a minimax game between G and D through solving the following loss function.

$$L_{\text{GD}} = \mathbb{E}_{x \sim \mathbb{P}_r} \left[\log D(x) \right] + \mathbb{E}_{x \sim \mathbb{P}_r} \left[\log(1 - D(G(E(x)))) \right]. \tag{5}$$

Visual Loss. L_V penalizes the inconsistent reconstruction by G. Ideally, synthesized face images should be compared with ground-truth anonymous face images, which are not available. Instead, we approximate it by using the ℓ_1 loss

$$L_V = \alpha \mathbb{E}_{x \sim \mathbb{P}_r} \left[\|G(E(x)) - x\|_1 \right] \tag{6}$$

to push the synthesized face $G(E(x))$ close to the input x, where α is the weight to control the contribution of the visual loss. With this loss function, the generator can learn to keep the utility in the visual space.

Large Margin De-identification Loss. L_M is a newly added loss defined as

$$L_M = \beta_1 L_{\text{M1}} + \beta_2 L_{\text{M2}}. \tag{7}$$

L_{M1} penalizes the synthesized face $x' = G(E(x))$ close to the input x in the identity feature space with a large margin m:

$$L_{\text{M1}} = \mathbb{E}_{x \sim \mathbb{P}_r}[\max(0, m - \|E(x) - E(x')\|_2)]. \tag{8}$$

With L_{M1}, a synthesized image x' is not likely to lie close to the input face image in the identity feature space, although it may still around the input due to L_V. At this point, no other constraints are imposed to x', which implies that $E(x')$ can be close to any identity in $\mathbb{I}$. If $E(x')$ is close to a certain identity I_i, x' may look similar to $I'i$, leading to *identity switch*. To alleviate this problem, we enforce $E(x')$ to keep a certain distance with all identities in $\mathbb{I}$ by using

$$L_{M2} = \mathbb{E}_{x \sim \mathbb{P}_r} \left[\sum_i \max(0, m - \|E(x') - E_i\|_2) \right], \tag{9}$$

where $x' = G(E(x))$ is a function of x. This formulation can improve the identity security and ensure that x' would not have conflicts with any other identities.

Pseudo Class Loss. L_C is another newly added term to distinguish synthesized identities from real ones in $\mathbb{I}$. We use another classifier C that classifies an real image x into one of identities in $\mathbb{I}$ and a synthesized image $x' = G(E(x))$ into one of k *pseudo classes*. The choice of the pseudo classes could be arbitrary. The loss is given as a standard softmax cross-entropy:

$$L_C = \mathbb{E}_{x \sim \mathbb{P}_r}[-\log C^{u_{\mathrm{id}}(x)}(x)] + \mathbb{E}_{x \sim \mathbb{P}_r}[-\log C^k(x')]. \tag{10}$$

This loss may reduce the chance of identity switch, described by Eq. (1).

3.4 Training

Encoder E and classifier C can share a same set of parameters except for their top layers since C has extra k outputs. Their lower layers' parameters are fixed once they are pretrained. We borrow G and D from DCGAN [24]. The top layer of classifier C can be trained together with G and D by optimizing $\min_{G,C} \max_D L$.

4 Evaluation Criteria

Evaluation of face de-identification (DeID) methods is challenging due to its subjectivity. We employ three objective criteria and user study.

4.1 Rank-1 Identification Rate

The rank-1 identification rate directly related to how well face DeID works. Specifically, this criterion shows whether or not a synthesized face image x' is recognized as its corresponding input face image x, which is related to our DeID requirement in Eq. (1). We depend on feature distance to find the rank-1 identity. If the rank-1 identity is different from the identity of the input face x, we say rank-1 identification fail (or DeID success).

4.2 Identity Switch Rate

By the IDS requirement in Eq. (1), the Identity switch (IDS) shows the false presence of existing identities. We follow [30] to evaluate the IDS rate to measure the possibility of a DeID algorithm attacking the other identities, where the feature distance between the synthesized face image x' and the original image x is used to determine if there happens IDS.

4.3   Differential Privacy

The reusability of data attributes is another important topic in face DeID. Being related to the requirement in Eq. (1), we formulate the face image reusability with differential privacy (DP) [8,10].

In the area of statistical databases, privacy protection is widely applied in query-based data releasing, which has been extended to to handle image-level privacy protection [10,11]. DP relies on the notion of adjacent databases that differ only in one record. Let B and B' denote two adjacent datasets, and a be any (numerical) property obtained from in B and B'. A privacy protection mechanism Ψ is ϵ-differentially private if the probability distributions $p(\Psi(B)(a) = \hat{v})$ and $p(\Psi(B')(a) = \hat{v})$ differ on a at most by e^{ϵ}, i.e.,

$$p(\Psi(B)(a) = \hat{v}) < e^{\epsilon} \times p(\Psi(B')(a) = \hat{v}), \tag{11}$$

where $\Psi(\cdot)(a)$ denotes the value of property a after privacy protection, and ϵ denotes the degree of privacy offered by Ψ.

In this work, we redefine DP to enable evaluation of the reusability of face images in semantic level. We first regard each face image as a database and then rewrite Eq. (11) with a non-sensitive attribute classifier $f_c(x)$

$$f_c^{u_c(x)}(x) < e^{\epsilon(x,x')} \times f_c^{u_c(x')}(x') \tag{12}$$

where $x' = G(E(x))$ is a synthesized face image that only hides face identity and $\epsilon(x, x')$ the distinguishability level between x and x'. From the viewpoint of DP, x and x' are an adjacent pair that only differs in a single property, i.e., identity ($u_{\text{id}}(x) \neq u_{\text{id}}(x')$) and has the same values for all other properties ($u_c(x) = u_c(x')$ for $c \neq$ id). That is, instead of applying DP for face DeID in the low-level domain [10], we synthesize an anonymous face using our encoder and generator and utilize $\epsilon(x, x')$ to show our method' performance on preserving utilities as well as on protecting privacy. Suppose $f_c(x)$ is the probability of face x having possible values of property c. Based on Eq. (12), we use

$$\epsilon(x, x') = \log f_c^{u_c(x)}(x) / f_c^{u_c(x')}(x') \tag{13}$$

as our criterion. For the DeID capability (i.e., $c = $ id), a larger value of $\epsilon(x, x')$ is the better [8,10]. For other properties, such as, ethnic, age, and gender, a smaller value of $\epsilon(x, x')$ implies better utility preservation.

5   Experiments

We evaluate our approach on the publicly VGGFace2 dataset [7], which contains 3.31 million images from 9,131 identities. The dataset suits to practical face-related research as it consists of various international faces. For systematical analysis, we adopt a pair of train and test set containing 1,000 identities, and we choose the commonly used facial attributes, i.e., ethnic, age, and gender, to show the performance of data reusability as with existing works did [6,9,15,26], where the identity discribution on face attributes is almost uniform. We group ethnic into 5 categories by *Black, Black2Yellow, Yellow, Yellow2White* and *White*, and gender into *Female* and *Male*. Since we have no ideas about the age of each face image, we roughly divide them into *Young* and *Old*.

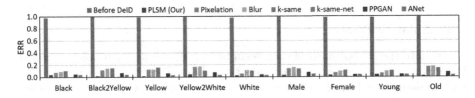

Fig. 2. Comparison of the rank-1 identification rate before and after de-identification.

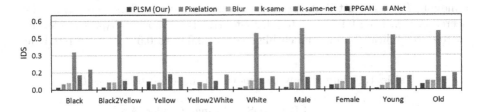

Fig. 3. Comparison of the IDS performance of after de-identification.

We use the VGG network [27] to implement E and C. To synthesize 128×128 face images, we employ G and D from DCGAN [24] because of the good stability for face generation [3]. Before using the standard mini-batch SGD and Adam solver for training our model, we set the learning rate $l = 0.0002$, $\beta_1 = 10$ and $\beta_2 = 5$. As with existing works [3,7], we also use the representative VGG classifier to identify the attribute information for verifying the data reusabiity.

5.1 De-Identification Performance

Fig. 2 shows the DeID performance with respect to different attribute-based image groups. Before DeID, the identity classifier f_{id} achieved very high recognition accuracy rate around 98.7% for all the utilities. After DeID, the recognition accuracy rate of our PLSM method dropped to very low around 0.03. This implies that our method can well de-identify faces. All other methods for face DeID worked well, although pixelation, blur, and the k-same method gave relatively high rank-1 identification rates. Interestingly, the rate for k-same-net [18] is very low, the reason of which lies in random face mapping using delegate faces.

5.2 Identity Switch Performance

According to Fig. 3, the IDS rates for different methods on each attribute are low except for the k-same method. The IDS rate of our method over all identities is 2.8%, which is low compared to other face DeID methods. This means that our method has less risk of false identification.

5.3 Ablation Study

To validate the effectiveness of each term in Eq. (4), we perform ablation study by removing one term at a time except for L_{GD}. Table 1 reports the results. By comparing our PSLM with others, we observe the following: (i) By removing L_V, the DeID performance became better, but the attribute preservation performance decrease a lot, e.g., from 0.882 to 0.491 for ethnic. (ii) By removing L_M (i.e., both L_{M1} and L_{M2}), although the DeID and IDS performances increase (i.e., close to 0), the attribute preservation performance has dropped. (iii) By removing L_{M1}, the attribute preservation performance suffers from significant decrease, e.g., from 0.933 to 0.655 for age. (iv) By removing L_{M2}, the identity switch possibilities become much higher from 0.034 to 0.091. (v) By removing

Table 1. Ablation study by removing a single loss term from L.

	ERR	IDS	Ethnic	Gender	Age
w/o L_V	**0.001**	0.037	0.491	0.697	0.731
w/o L_M	**0.001**	**0.002**	0.390	0.504	0.260
w/o L_C	0.044	0.048	0.851	0.932	0.931
w/o L_{M1}	0.012	0.014	0.666	0.815	0.655
w/o L_{M2}	0.020	0.091	**0.882**	**0.937**	**0.947**
PSLM (ours)	0.030	0.034	**0.882**	0.931	0.933

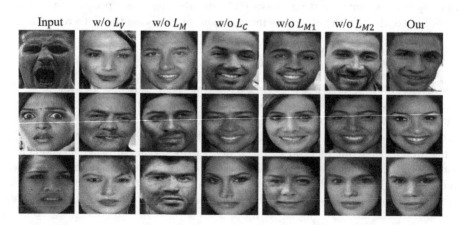

Fig. 4. Some results on ablation study by removing one loss term at a time.

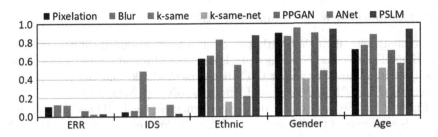

Fig. 5. Comparison of PLSM with competing face de-identification methods.

Table 2. De-identification evaluation with DP criteria, where higher ϵ indicates better for identity, while lower ϵ indicates better for ethnic, gender, and age.

	Identity	Ethnic	Gender	Age
Pixelation	4.46	0.48	0.11	0.33
Blur	3.88	0.42	0.15	0.27
k-same	0.50	0.19	0.05	0.13
k-same-net	**9.21**	1.83	0.92	0.67
PPGAN	2.75	0.59	0.11	0.35
ANet	6.62	0.87	0.86	0.71
PLSM	7.33	**0.14**	**0.06**	**0.07**

L_C, the DeID and IDS performances become worse, implying that classification into the identities in $\mathbb{I}$ works well but the pseudo class do not.

The above results show that the terms in both feature and visual space (L_M and L_V) could produce complementary effects to ensure high level security. Although PSLM is not the best on all criteria because the choice of loss terms exhibits trade-offs, it can offer a good balance among all criteria. We also found that L_{M1} and L_{M2} may spoil the DeID performance, and L_C may spoil both DeID and ethnic preserving performance, which is predictable from their inherent nature. Yet, compared to the performance gain on the other items, this performance deterioration should be acceptable. Figure 4 illustrates some visual results when a single term in L is removed. It is obvious that: (i) without L_V and L_M, the generated faces have much higher probability for changing utility classes (*e.g.*, gender); (ii) PSLM can be seen as an adjustment of the results of L_C for anonymizing the identity; (iii) without L_{M1}, DeID may not preserve the attributes; and (iv) although the synthesized faces without L_{M2} look as good as our final results, they may suffer from identity switch.

5.4   Comparison with Competing Methods

We compare our algorithm with the most related works, including both the traditional low-level methods (face pixelation, face blur and k-same based on face fusion) [9,10,21] and the recent GAN-based methods (k-same-net [18], privacy-protective GAN (PPGAN) [30] and AnonymousNet (ANet) [15]).

Figure 2 and Fig. 3 also contain the performances the DeID and IDS of these methods with respect to different items. Figure 5 plots the overall barchart results after de-identification on ERR rate, IDS rate and atrribute preservation accuracy on ethnic, gender and age. We observe that both PSLM and the GAN-based methods have significant superior de-identification performance than the traditional methods. They also has very low IDS performance, which indicates much better security of the GAN-based methods in terms of false identification. Although the k-same-net method works well on both DeID and IDS, it has relatively poor performance on attribute preservation without using additional annotations for network learning. Although PPGAN outperforms PSLM on IDS, PSLM has much lower DeID rate and better attribute preservation performance.

In Table 2, we present the performance in the DP criterion. It is clear that ours has achieved a high privacy level (*i.e.* 7.33) and it got the best performance on utility preservation, *i.e.*, the smallest $\epsilon(x, x')$ values for ethnic, gender, and age. Figure 6 visually compares anonymous faces by different methods and we observe that: (i) the naturalness of the traditional methods is poor; (ii) some example face images for PPGAN look similar to the corresponding original faces; and (iii) ours has comparable or better visual performance.

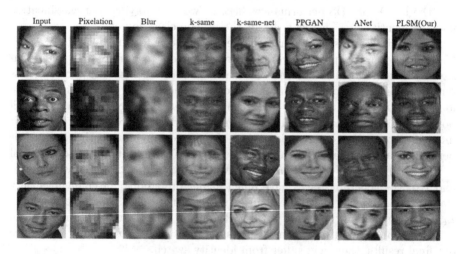

Fig. 6. Comparison of PSLM with competing face de-identification methods.

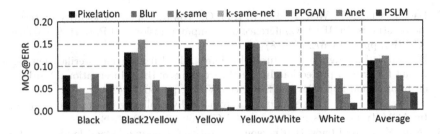

Fig. 7. User study results on ERR.

5.5 User Study

Given a set of de-identified faces, we perform user study to evaluate the subjective de-identification performance by asking 30 participants. We ask the participants if the synthesized face identity is the same with the original face identity and using the popular mean opinion score (MOS) as the quantitive measure [14,23]. Our comparison result is shown in Fig. 7 and one can see that our algorithm has achieved very competitive results compared with both deep generative methods and traditional anonymous methods.

6   Conclusion

This paper has introduced a novel face de-identification approach by synthesizing new faces for protecting identity information from presenting to the public. We model de-identification as a large margin generative problem and it has worked positively towards privacy sensitive face synthesis while preserving the major face attributes to keep data usability. Our experimental results have verified its effectiveness for privacy protection with very competitive results. Although we have achieved some good results, our approach is still restricted to the face quality without ground truth targets. Thus, in the future work, we would explore more effective ways for improving the image quality.

Acknowledgements. This work was supported by the National Natural Science Foundation of China under Grant No. 61806063, 61772161, 61622205. The authors would like to thank the reviewers who have provided insightful comments and valuable suggestions.

References

1. Agarwal, S., et al.: Protecting world leaders against deep fakes. In: IEEE Conference on Computer Vision and Pattern Recognition Workshops (CVPRW), pp. 38–45 (2019)
2. Ayala-Rivera, V., et al.: A systematic comparison and evaluation of k-anonymization algorithms for practitioners. Trans. Data Privacy **7**(3), 337–370 (2014)

3. Bao, J., Chen, D., Wen, F., Li, H., Hua, G.: Towards open-set identity preserving face synthesis. In: IEEE Conference on Computer Vision and Pattern Recognition (CVPR), pp. 6713–6722 (2018)

4. Bhattarai, B., Mignon, A., Jurie, F., Furon, T.: Puzzling face verification algorithms for privacy protection. In: The Proceedings of IEEE International Workshop on Information Forensics and Security (WIFS), pp. 66–71 (2014)

5. Bitouk, D., et al.: Face swapping: automatically replacing faces in photographs. In: ACM Transactions on Graphics (ToG). vol. 27, p. 39 (2008)

6. Brkic, K., et al.: I know that person: generative full body and face de-identification of people in images. In: IEEE Conference on Computer Vision and Pattern Recognition Workshops (CVPRW), pp. 1319–1328. IEEE (2017)

7. Cao, Q., et al.: Vggface2: a dataset for recognising faces across pose and age. In: IEEE International Conference on Automatic Face & Gesture Recognition (FG 2018), pp. 67–74 (2018)

8. Chatzikokolakis, K., et al.: Broadening the scope of differential privacy using metrics. Privacy Enhancing Technologies, pp. 82–102 (2013)

9. Du, L., Yi, M., Blasch, E., Ling, H.: Garp-face: balancing privacy protection and utility preservation in face de-identification. In: IEEE International Joint Conference on Biometrics, pp. 1–8 (2014)

10. Fan, L.: Image pixelization with differential privacy. In: IFIP Annual Conference on Data and Applications Security and Privacy (2018)

11. Fan, L.: image obfuscation with quantifiable privacy. In: CV-COPS (2019)

12. Korshunov, P., Ebrahimi, T.: Using face morphing to protect privacy. In: IEEE International Conference on Advanced Video and Signal Based Surveillance, pp. 208–213 (2013)

13. Kuang, Z., Li, Z., Lin, D., Fan, J.: Automatic privacy prediction to accelerate social image sharing. In: IEEE Third International Conference on Multimedia Big Data (BigMM), pp. 197–200. IEEE (2017)

14. Ledig, C., et al.: Photo-realistic single image super-resolution using a generative adversarial network. In: IEEE Conference on Computer Vision and Pattern Recognition (CVPR), pp. 4681–4690 (2017)

15. Li, T., Lin, L.: Anonymousnet: natural face de-identification with measurable privacy. In: IEEE Conference on Computer Vision and Pattern Recognition Workshops (CVPRW) (2019)

16. Majumdar, P., et al.: Evading face recognition via partial tampering of faces. In: IEEE Conference on Computer Vision and Pattern Recognition Workshops (CVPRW) (2019)

17. Mcpherson, R., et al.: Defeating image obfuscation with deep learning. In: arXiv:1609.00408v2 (2016)

18. Meden, B., et al.: k-same-net: k-anonymity with generative deep neural networks for face deidentification. Entropy 20(1), 60 (2018)

19. Mirza, M., Osindero, S.: Conditional generative adversarial nets (2014)

20. Mirza, M., Osindero, S.: Conditional generative adversarial nets. arXiv: Learning (2014)

21. Newton, E.M., Sweeney, L., Malin, B.: Preserving privacy by de-identifying face images. IEEE Trans. Knowl. Data Eng. (TKDE) 17(2), 232–243 (2005)

22. Oh, S.J., et al.: Faceless person recognition: privacy implications in social media. In: European Conference on Computer Vision (ECCV), pp. 19–35 (2016)

23. van den Oord, A., et al.: Parallel wavenet: fast high-fidelity speech synthesis. In: Proceedings of Machine Learning Research, pp. 3918–3926 (2018)

24. Radford, A., et al.: Unsupervised representation learning with deep convolutional generative adversarial networks. arXiv:1511.06434 (2015)
25. Ren, Z., Jae Lee, Y., Ryoo, M.S.: Learning to anonymize faces for privacy preserving action detection. In: European Conference on Computer Vision (ECCV), pp. 620–636 (2018)
26. Ribaric, S., Ariyaeeinia, A., Pavesic, N.: De-identification for privacy protection in multimedia content: a survey. Signal Process. Image Commun. **47**, 131–151 (2016)
27. Simonyan, K., Zisserman, A.: Very deep convolutional networks for large-scale image recognition. arXiv preprint arXiv:1409.1556 (2014)
28. Sun, Z., et al.: Distinguishable de-identified faces. In: IEEE International Conference and Workshops on Automatic Face and Gesture Recognition (FG), vol. 4, pp. 1–6 (2015)
29. Sweeney, L.: k-anonymity: a model for protecting privacy. Int. J. Uncertainty, Fuzziness Knowl.-Based Syst. **10**(05), 557–570 (2002)
30. Wu, Y., Yang, F., Ling, H.: Privacy-protective-gan for face de-identification. J. Comput. Sci. Technol. (JCST) **31**(1), 47–60 (2019)
31. Yang, X., Li, Y., Lyu, S.: Exposing deep fakes using inconsistent head poses. In: The Proceedings of IEEE International Conference on Acoustics, Speech and Signal Processing (ICASSP), pp. 8261–8265 (2019)
32. Yuan, L., et al.: Privacy-preserving photo sharing based on a secure JPEG. In: IEEE Conference on Computer Communications Workshops, pp. 185–190 (2015)

Balancing of Bike-Sharing System via Constrained Model Predictive Control

Yuguang Chen[1,2], Yan Zeng[1,2], Junjun Wu[1,2], Qian Li[1,2], and Zhou Wu[1,2(✉)]

[1] Key Laboratory of Dependable Service Computing in Cyber Physical Society of Ministry of Education, Chongqing University, Chongqing, China
wuzhsky@gmail.com
[2] College of Automation, Chongqing University, Chongqing 400044, China

Abstract. With the development of social urbanization, bike-sharing was born and developed rapidly. Many cities around the world believe that Shared bikes promote environmental protection and development towards sustainable society. In this paper, we consider two problems: bike-sharing system (BSS) redistribution efficiency maximization and system rebalance synchronize. Firstly, we describe BSS dynamic model with a linear form based on graph theory. Then, we propose a quantitative representation to measure the operational efficiency of BSS. We present a model predictive control (MPC) method to solve operational efficiency problem with system constraints. Both the dynamic state and the constraints in the redistribution are considered in MPC algorithm. Then, we verify the effectiveness of our proposed algorithm on different connection type BSS network. According to experimental results, the operational efficiency is maximized and BSS network can reach an equilibrium state during dynamic optimization. Compared to other methods, MPC approach is shown more effective.

Keywords: Bike-sharing system · System equalization · Model predictive control · Rebalancing

1 Introduction

With rapid economic development, the level of urbanization throughout the country has gradually increased. As a sustainable means of transportation, the use of bike sharing is often connected to subways and buses, solving the "the last mile" travel problem of people traveling. It provides benefits to people from economy, environmental protection, health and other aspects [1]. Globally, the use of BSS reduces the amount of pollutants emitted by 170,000 motor vehicles a year [2]. BSS has been widely used spread in more than 1,560 cities and are being planned or under construction in more than 400 cities [3]. Public transportation sharing programs stand out as one of the most promising initiatives to promote sustainable cycling in cities [4].

In the maintenance of bicycle sharing systems, the cost of rebalancing accounts for a big part [5]. In general, BSS has two equilibrium ideas: truck-based reallocation and price-based rebalancing [6]. They proposed a novel approach, pricing-based rather than

© Springer Nature Singapore Pte Ltd. 2020
H. Zhang et al. (Eds.): NCAA 2020, CCIS 1265, pp. 502–512, 2020.
https://doi.org/10.1007/978-981-15-7670-6_41

vehicle-based, to the BSS to reach equilibrium. In terms of static rebalancing problem, researchers study the rebalancing one-truck routing problem and propose algorithms for solving the shortest distance problem [7–9]. Raviv and Kolka studied the BSS rebalancing problem of small dispatch fleets, and selected the weight function of the total dispatch distance and the number of rejected users as the equilibrium objective function [10]. Based on the establishment of BSS dynamic scheduling model, Vélib redistributed the scheduling route, solved the problem of BSS scheduling path optimization [11]. Since some kind of manual balancing is always required, most of rebalancing literature is focused on optimal truck routing. The main purpose of achieving dynamic redistribution of bicycles is to obtain higher user satisfaction and reduce dispatch costs [12]. Dell et al. proposed a single-dispatch vehicle routing problem (1-PDVRPD) algorithm that is suitable for maximizing user usage time, which solves the problem of SBRP's path optimization problem while ensuring maximum user usage time [13]. On a summary of [14–17], they all concentrate on vehicle routing problem, and analyze the optimal dispatching routing and shortest time cost problems. "Travel entropy" models are often used to analyze such dispatching efficiency problems. On the other hand, users demand should also be considered in the model. The ideal state of BSS is that each station of the system is in an occupied state between empty and full to ensure that users can renting and returning [18].

The remainder structure of this paper is as follows. To depicting the dynamic bicycle flow, we combine graph network with BSS rebalancing problem and set up a dynamic BSS model in Sect. 2. In Sect. 3, we propose a description to measure the effective states of each station in BSS network, and present a metric to measure the system efficiency. We conclude that the system can get maximum operation efficiency when it reaches equilibrium state. In Sect. 4, we present a model predictive control method to control BSS network rebalance. To verify our MPC rebalance approach, different connection style network are experimented in Sect. 5. Finally, we give our concluding remarks in Sect. 6.

2 Problem Description and Mathematical Formulation

In this section, we present a system dynamic model that combines graph network theory analysis with BSS's historical phenomenon of usage behavior.

In a network with general circumstances, the network characteristics of each node are consistent. It is different from the diversified station of the analytical bike-sharing system. To deal with this difference, we give each node different characteristics to represent different sites in our BSS system. Each station has its own degree of necessity coefficient. Necessity coefficient illuminate historical station bike flow usage. To formulate BSS dynamic model, the information about the flow of bicycles in the BSS should be specifically described.

Notations	
n	Total number of stations
G	Bike-sharing network
$d(i,j)$	Distance between station i to j
x_i	The storage state of bicycle station i
C_i	Dock capacity of station i (i = 1, ..., n)
v_{sk}, v_{rk}	Sender and receiver nodes in BSS
B	BSS Network connection matrix
α_i	Weight of item i, which represents the degree of importance in system operation efficiency
u_{ij}	Scheduling variables which indicates the number of schedules from station i to station j, $i = 0,..., $ n; j = 1, ..., n and $i \neq j$
σ	Orientation assign that indicates the traffic flow
χ	Constraints of network states
υ	Constraints of network bike transport flow $u(t)$

2.1   System Dynamics Model

BSS network is defined as $G = (V, E)$. u represents global control information during redistribution operation. The distance $d(i,j)$ represents the connection distance between i and j. Then, we describe the system dynamics model and constraints. The discrete-time dynamic integral model of the ith station is established as follows:

$$C_i x_i(t+1) = C_i x_i(t) + \omega_i(t) \quad \text{for i} = 1...\text{n} \tag{1}$$

where C_i is the docks capacity of station i, $C_i x_i(t)$ is current number of exist bicycles at a bike-sharing station. $\omega_i(t)$ is the total bicycles rent or return from users. Equation (1) shows the dynamic equilibrium of the number of bicycles at the ith station.

To depicting renting and returning information, the dynamic model of the ith station becomes

$$C_i x_i(t+1) = C_i x_i(t) + \sum_{j \in V} u_{ji}(t) - \sum_{j \in V} u_{ij}(t) \tag{2}$$

where $V = \{j | (i,j) \in E\}$ is a collection of stations connected to station i. The last two terms represent station renting and returning detail.

We define that the direction of the bicycle flow of u_{ij} is opposite to the direction u_{ji}. For convenience, we define a traffic flow direction function $e(k)$ low direction of the bicycles flow is i to j or j to i when e is positive or negative, respectively. That is, in the system $G = (V, E)$, $e(k) = (i,j) \in E$ represents the direction from i to j, which is $+1$ or -1.

By defining the direction function e (k), we can get a clearer new dynamic model of the system. The dynamic equation of the i th station is

$$C_i x_i(t+1) = C_i x_i(t) + B_i u \tag{3}$$

where $B_i^T \in R^m$ is the ith column of the direction decision matrix B of the bicycle flow. B is incidence matrix in BSS network. In connected bike-sharing station network $G = (V, E)$ with entries

$$B_{ik} = \begin{cases} e_i(k) & \text{if } e(k) = (i,j) \text{ for some } j \in V \\ 0 & \text{otherwise} \end{cases} \tag{4}$$

The dynamics equation of BSS network is

$$x(t+1) = x(t) + C^{-1} Bu \tag{5}$$

where $x = [x_1 \cdots x_n]^T \in R^n$ is exist station state in BSS, and $C = diag([C_1 \cdots C_n]) \in R^{n \times n}$ is the diagonal matrix of station maximal capacities.

2.2 BSS Constraints

In the redistribution process $t \in N$, the BSS network states must satisfy the constraints $x(t) \in \chi \subset R_n$ where

$$\chi = \{x \in R_n | 0 \le x_i(t) \le 1 \text{ for } i \in 1 \ldots n\}. \tag{6}$$

The constraints on the network bike transport flow $u(t)$ is based on the capacity of dispatching truck. We have $u(t) \in \upsilon \subset R_n$ for all $t \in N$ where

$$\upsilon = \{u \in R_n | u_{\min} \le u_i(t) \le u_{\max} \text{ for } i \in 1 \ldots n\} \tag{7}$$

3 Rebalance and BSS Operation Efficiency

In this section, we propose a multi-objective metric for measuring the operation efficiency of a bike-sharing system.

3.1 System Operation Efficiency of Equalization

The station state x_i is defined as the bicycle capacity normalized by C_i. It is worth mentioning that in the operation of BSS, we assume that there is no bicycle damage in the entire system operation process, no new shared bicycle delivery, and the user can only use it within the system. The total number of bicycles in BSS remains unchanged. Station normalized state $x_i \in [0, 1]$ can be formulated as

$$x_i = \frac{1}{x_{up} - x_{low}} (\widehat{x}_i - x_{low}) \tag{8}$$

where $\widehat{x}_i$ is exist i-th station state. x_{low} and x_{up} are the minimum and maximum thresholds stored by the station.

In one hand, we wish to satisfy the users renting and returning demands as much as possible. The station with the least number of bicycles, most difficult station to meet the needs of the user, is

$$\min_{i \in v} C_i \alpha_i x_i \tag{9}$$

Where α_i is the importance parameter of each station. The station with the largest number of bicycles, make the most difficult transportation for suppliers, is

$$\min_{i \in v} C_i(1 - \alpha_i x_i) \tag{10}$$

We define the operation efficiency of each station in the BSS network as maximal difficulty station bicycle number to satisfy bicycles plus the largest demand to do redistribution.

$$f_i^{sta} = \min_{i \in v} C_i \alpha_i x_i + \min_{i \in v} C_i(1 - \alpha_i x_i) \tag{11}$$

The operation efficiency maximization problem is to maximize the satisfaction of user's needs (the first term in (11)) and minimize the cost for transportation bicycles (the second term).

We define the states that can maximize the redistribution operation efficiency objective (11) as

$$R_\infty^*(x_0) = \arg \max_{x \in R_\infty(x_0)} \sum_i f_i^{sta}(x). \tag{12}$$

Then, we define the problem of operation efficiency maximization as select a redistribution control policy $u(t)$ for the system such that the flow $u(t)$ is feasible $u(t) \in v$;

3.2   System Balancing and Efficiency Maximization

In this section, we prove that when BSS network reach the equalization states $\bar{R}_\infty(x_0)$, the operational efficiency can also get maximal states $R_\infty^*(x_0)$. The equilibrium set be stable means if the network state convergence to equilibrium $\bar{x}$, then it should remain in this equilibrium state.

The BSS network equalization states is defined as

$$\bar{R}_\infty(x_0) = \{x \in R_\infty(x_0) | x_i = x_j \forall i, j \in G\} \tag{13}$$

Theorem 1. If the states of system (5) are balanced, the operation efficiency in this station of the system can be maximized in the case.

Proof. Let $x = \bar{x} \in \bar{R}_\infty(x_0) \subset R_\infty(x_0)$. If $\bar{x}$ can maximizes $\bar{e}(x)$ in (11), then $\bar{x} \in R_\infty^*$ and theorem can be proved. The function $\bar{e}(x)$ at $\bar{x}$ is

$$
\begin{aligned}
\bar{e}(x) &= \min_{i \in V}\{C_i \alpha_i x_b\} + \min_{i \in V}\{C_i(1 - \alpha_i x_b)\} \\
&= \min_{i \in V}\{C_i\}
\end{aligned}
\tag{14}
$$

where x_b is the balanced state of each station in BSS. In Theorem 1, we find that when each station is in an equilibrium state, the state efficiency in (11) of the system can be maximized. Therefore, we transform the operation efficiency maximization problem into the equilibrium problem in the BSS network $= (V, E)$.

4 Equilibrium Control Algorithm

One-way rides always create station imbalances throughout the system, causing some users have no bicycles to ride. Then, some stations get bicycles redundancy. This is not only harmful to people's travel, but also can cause road congestion.

Here we propose a MPC control algorithm to solve the BSS network equilibrium and the operation efficiency maximization problem. In Theorem 1, it has been proved that the BSS is most efficient, when each station reaches its balanced state. The closed-loop MPC proposed in this chapter detects and corrects the interference of the shared bicycle system in real time to achieve the goal of optimizing dispatch efficiency, including: (1) maximizing the operating efficiency of the BSS; (2) satisfying all BSSs in each sampling range Operational constraints; (3) The closed-loop system is stable with respect to the dynamic bicycle flow; (4) The BSS can reach the maximum efficiency state and stabilize in this state within a specified time. The predicted equation is calculated by

$$
\begin{bmatrix}
x(t+1|t) \\
x(t+2|t) \\
\cdots \\
x(t+N|t)
\end{bmatrix}
= \bar{A}x(t) + \bar{D}
\begin{bmatrix}
u(t) \\
u(t+1) \\
\cdots \\
x(t+N-1)
\end{bmatrix}
\tag{15}
$$

Where

$$
\bar{A} =
\begin{bmatrix}
A \\
A^2 \\
\cdots \\
A^N
\end{bmatrix}
\tag{16}
$$

$$\bar{B} = \begin{bmatrix} D & 0 & \cdots & 0 \\ AD & D & \cdots & 0 \\ \vdots & \vdots & \ddots & \vdots \\ A^{N-1}D & A^{N-2}D & \vdots & D \end{bmatrix} \tag{17}$$

where $A = I_N$, $D = C^{-1}B$ in BSS system in Eq. (5).

Let $U(t) = [u(t|t), u(t+1|t), \ldots, u(k+N-1|t)]^T$ and $X(t+1) = [x(t+1|t), \ldots, x(t+N-1|t), x(t+N|t)]^T$. The objective function over the prediction horizon in the MPC can be formulated as

$$\text{maximize }_{u(t)} \sum_{i \in V} \sum_{j=0}^{N-1} f_i^{sta}(t+j+1|t) - \rho \sum_{j=0}^{N-1} \lfloor u(t+j|t) \rfloor \tag{18}$$

s.t.

$$x_i(t+j|t) \in \chi, j = 0 \ldots N-1 \tag{19}$$

$$u(t+j|t) \in \upsilon, j = 0 \ldots N-1 \tag{20}$$

$$Cx(t+1) = Cx(t) + Bu + w \tag{21}$$

where (19), (20) ensure that in each time period, the system state x and input u satisfy the constraints proposed in (6), (7). Constraint (21) ensure the update of network states The process of MPC algorithm is shown in Algorithm 1.

Algorithm 1 the proposal MPC algorithm flow

1:**Input**:*BSS redistribution control bicycle flow uij(t)(i, j ∈ 1...n, i ≠ j);*
2:**Output**:*BSS station state xi(t)(i ∈ 1...n);*
3:*Analyse BSS network characteristics, generate the connection matrix* **B** *and capacity matrix* **C**;
4:*Set t=0;*
5:**while** *t<to* **do**;
6: *Solve the maximization operation efficiency; problem (18) subject to Constraint (19-21);*
7: *Update system state variables based on function (5);*
8:*t=t+1;*
9:**end while**

---

5 Numerical Example

In this section analyzes BSS network instances of different scales and different connection methods in simulation in order to verify the proposed operational efficiency maximization MPC algorithm. Then, experiments on a series of 10, 20, 50 stations are conducted on the experimental prototype.

The network connections of 20-stations BSS is showed in Fig. 1. We build the data sets for a medium sized network with 20 stations in it. In Fig. 1, the blue circle represents the station, and the straight line represents the connection between the stations. We distribute corresponding degree of necessity coefficient α to each station. In this example, we define

$$\alpha_i = n_p/n * C_i \tag{22}$$

where n_p is the total quantity of bicycles that operator have putted in. n is the number of station in BSS. The characteristics of instance networks are reported in Table 1, where we depict the storage capacity of each station, the distance parameter and the number of connections edges.

We use tuning parameters $\rho = 10^{-3}$ since for this example. The period of network equilibrium rate within the specified time is set to 100. The simulations of system (5) in close loop with Algorithm 1 are shown in Fig. 2. The different method simulation result data comparison is shown in Table 2.

Table 1. Characteristics of 20 stations instance networks

Specifications of 20 stations bike sharing system	
Initial state $x_1(0) \sim x_{20}(0)$	0.1; 0.9; 0.2; 0.8; 0.4; 0.6; 0.5; 0.5; 0.75; 0.25; 0.45; 0.55; 0.675; 0.325; 0.452; 0.575; 0.35; 0.65; 0.33; 0.67
ρ	$2 * 10^{-3}$
$C_i\ i \in 1...20$	100
n_1	20 stations
m_1	36 edges

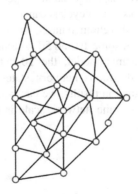

Fig. 1. The BSS network with 20 stations (Color figure online)

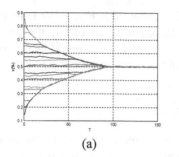 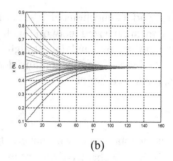

(a)   (b)

Fig. 2. Dynamic equalization for 20-vertices BSS network: (a) MPC approach; (b) feedback control

It can be seen from the comparison of two methods that MPC has an advantage over feedback intuitive method in terms of equilibrium time. However, Fig. 2 shows that the state of the BSS storage stations G in different distribution situation converged to an imbalanced state. During this BSS operation, Algorithm (18) can control the BSS network to an efficiency maximization state.

Table 2. Performance comparison of different method in 20 stations BSS network

Method	Convergence time (period)	Equilibrium rate (%)
MPC	94	98.87
Feedback	135	94.53

We still do experiences about two different actual BSS systems, including small scale (10-stations) and large scale (50-stations). Represented by historical data of people's bicycle using habit, customers always ride around nearby several stations around themselves. To cater to this usage habit of people, we present two sizes of simply connected BSS networks i.e. bicycles can only be transported to nearby 3 stations. Based on the Paris and Tehran dataset, we created a set of 2 benchmark problems using actual locations of some 10 and 50 stations respectively located in these cities. Based on the BSS dynamic model, the 10-stations network states profiles equilibrium process and the 50-stations network states profiles are also plotted as shown in Fig. 3. From the simulation results, we can conclude that when the network connection method is simpler and the network scale is larger, the convergence time required is longer.

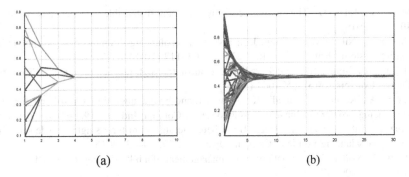

(a)   (b)

Fig. 3. MPC for dynamic equalization: (a) 10-vertices simply connected BSS network; (b) 50-vertices simply connected BSS network

6 Conclusion

In this study, a BSS network redistribution equalization method based on model predictive control is proposed. To analyzing the dynamic process within the BSS, a dynamic model of the bike sharing system is formulated based on the graph network theory. The importance of the location of stations were taken into account during the redistribution process. A description to measure the effective states of each station and a metric to measure the system efficiency are presented. We further prove that the system can get maximum operation efficiency, when it reaches balanced states. To cope with frequent change states and constraints of BSS, a closed-loop model predictive control approach is proposed. In experiment, we show our MPC approach can deal with different connection style BSS network. By comparing to other control method, the simulation results verify that our method have less balancing time and less redistribution cost.

Here are some areas to be perfected in the future. In the process of modeling a shared bicycle network, problems such as bicycle damage, re-delivery of bicycles, bicycle maintenance, and bicycle interworking between different systems should also be considered. When formulating the scheduling optimization objective function, more objectives should be taken into account, such as accurate user satisfaction. Finally, accurate station demand prediction is also a vital part during rebalancing process. Future research may be able to develop in these areas.

References

1. Si, H., Shi, J., Wu, G.: Mapping the bike sharing research published from 2010 to 2018: a scientometric review. J. Cleaner Prod. **213**, 415–427 (2019)
2. Ricci, M.: Bike sharing: a review of evidence on impacts and processes of implementation and operation. Res. Transp. Bus. Manag **15**, 28–38 (2015)
3. Chen, R.: "Bike litter" and obligations of the platform operators: lessons from China's dockless sharing bikes. Comput. Law Secur. Rev. **35**(5), 105317 (2019)

4. Mi, Z., Coffman, D.: The sharing economy promotes sustainable societies. Nat. Commun. **10**, 1–3 (2019)

5. Haider, Z., Nikolaev, A., Kang, J.E.: Inventory rebalancing through pricing in public bike sharing systems. Eur. J. Oper. Res. **270**(1), 103–117 (2018)

6. He, J.: Multi-objective model-predictive control for high-power converters. IEEE Trans. Energy Convers. **28**(3), 652–663 (2013)

7. Kadri, A., Kacem, I., Labadi, K.: A branch-and-bound algorithm for solving the static rebalancing problem in bicycle-sharing systems. Comput. Ind. Eng. **95**, 41–52 (2016)

8. Cruz, F., Subramanian, A., Bruck, B.: A heuristic algorithm for a single vehicle static bike sharing rebalancing problem. Comput. Oper. Res. **79**, 19–33 (2016)

9. Raviv, T., Kolka, O.: Optimal inventory management of a bike-sharing station. IIE Trans. **45** (10), 1077–1093 (2013)

10. Benchimol, M., Benchimol, P., Chappert, B.: Balancing the stations of a self-service bike hire system. RAIRO-Oper. Res. **45**(1), 37–61 (2011)

11. Caggiani, L., Camporeale, R., Ottomanelli, M.: A modeling framework for the dynamic management of free-floating bike-sharing systems. Transp. Res. Part C: Emerg. Technol. **87**, 159–182 (2018)

12. Dell'Amico, M., Iori, M., Novellani, S.: A destroy and repair algorithm for the bike sharing rebalancing problem. Comput. Oper. Res. **71**, 149–162 (2016)

13. Erdoan, G., Battarra, M., Wolfler, C.R.: An exact algorithm for the static rebalancing problem arising in bicycle sharing systems. Eur. J. Oper. Res. **245**(3), 667–679 (2015)

14. O'Mahony, E., Shmoys, D.B.: Data analysis and optimization for (citi)bike sharing. In: Twenty-Ninth AAAI Conference on Artificial Intelligence. AAAI Press (2015)

15. Benjamin, L.: Dynamic repositioning strategy in a bike-sharing system; how to prioritize and how to rebalance a bike station. Eur. J. Oper. Res. **272**(2), 740–753 (2019)

16. Aritra, P., Zhang, Y.: Free-floating bike sharing: Solving real-life large-scale static rebalancing problems. Transp. Res. Part C: Emerg. Technol. **80**, 92–116 (2017)

17. Repoux, M., Burak, B., Geroliminis, N.: Simulation and optimization of one-way car-sharing systems with variant relocation policies. In: 94th Annual Meeting of the Transportation Research Board (2015)

18. Wang, K., Gulsah, A.: Gender gap generators for bike share ridership: Evidence from Citi Bike system in New York City. J. Transp. Geograp. **76**, 1–9 (2019)

A Tree-Structure Convolutional Neural Network for Temporal Features Exaction on Sensor-Based Multi-resident Activity Recognition

Jingjing Cao[(✉)], Fukang Guo, Xin Lai, Qiang Zhou, and Jinshan Dai

School of Logistics Engineering, Wuhan University of Technology, Wuhan, China
bettycao@whut.edu.cn

Abstract. With the propagation of sensor devices applied in smart home, activity recognition has ignited huge interest and most existing works assume that there is only one habitant. While in reality, there are generally multiple residents at home, which brings greater challenge to recognize activities. In addition, many conventional approaches rely on manual time series data segmentation ignoring the inherent characteristics of events and their heuristic hand-crafted feature generation algorithms are difficult to exploit distinctive features to accurately classify different activities. To address these issues, we propose an end-to-end Tree-Structure Convolutional neural network based framework for Multi-Resident Activity Recognition (TSC-MRAR). First, we treat each sample as an event and obtain the current event embedding through the previous sensor readings in the sliding window without splitting the time series data. Then, in order to automatically generate the temporal features, a tree-structure network is designed to derive the temporal dependence of nearby readings. The extracted features are fed into the fully connected layer, which can jointly learn the resident labels and the activity labels simultaneously. Finally, experiments on CASAS datasets demonstrate the efficiency of our model in multi-resident activity recognition compared to state-of-the-arts techniques.

Keywords: Multi-resident activity recognition · CNN · Feature generation

1 Introduction

Human Activity recognition (HAR) at smart home has emerged as a popular topic to ubiquitous computing with the vast proliferation of sensor devices and Internet of Things [1]. Among all kinds of sensors for smart home information collection, ambient sensor can collect effective data without intruding the lives of residents and efficiently protecting privacy. In practical scenarios, there are

This work was supported by National Natural Science Foundation of China.

H. Zhang et al. (Eds.): NCAA 2020, CCIS 1265, pp. 513–525, 2020.
https://doi.org/10.1007/978-981-15-7670-6_43

more than one resident at home, which means that we need to capture the complexity nature of joint activities. However, ambient sensors can receive very limited information with binary information about on or off which poses a great challenge to recognize multi-resident activities.

In conventional approaches, there are three main steps in HAR: data pre-processing, human activity feature generation and activity classification. The time series data will be first segmented into slices and some statistical features will be extracted. They view each slice as an event ignoring the inherent continuity of events. Advanced heuristic hand-crafted features are generated based on processed features. Inputting those advanced features vectors, a classifier will be trained through machine learning. However, artificially cutting the data into sequences is easy to lead to the disappearance of advanced information. And methods such as Markov models [2] and support vector machines (SVMs) [3] which rely on activity recognition basic signal statistics and traits to manually generate features is limited by human domain knowledge [4]. There is currently no systematic feature extraction framework to effectively capture distinctive features of human activity.

Recent years have witnessed the rapid development of deep learning, achieving unparalleled performance in HAR for their favorable advantages of automatically feature learning [5]. Temporal convolutional neural networks (TCNNs) recently have been used for time series-processing in sensor-based HAR [6,7]. TCNNs first segment the time series data into preprocessing unit, and then treat each unit as event to yield its high-order representation via stacking there processing units, so that TCNNs can characterize the saliency in different scales. TCNNs share a convolution kernel and has little pressure on high-dimensional data processing. The input series information can be retained with the original position relationship after the convolution operation. However, sharing parameters across temporal series is incapable for exploiting all of the correlations between inputting events and local connectivity constrains the output to a small number of adjacent events. Another deep neural network recurrent neural networks (RNNs) learn the sequence transfer pattern of the sensor signal well, thus they are widely used to capture temporal dependency for HAR [8,9]. While RNNs only extracts information from the current input, and there is only one shared parameter matrix and difficult to extract deeper representation than TCNN. In addition, little has been learnt in multi-resident field and the existing work rarely make full use of temporal information. Hence, in this work, we proposed a Tree-Structure CNN framework for Multi-Resident Activity Recognition (TSC-MRAR). This approach treat each sample as an event to do predict task, and it can automatically captures neighboring temporal features through previous events and generates deep temporal representation in different dimensions with parameters adaptedly learning in each sliding window. The contributions of this work are as follows:

- A novel end-to-end framework TSC-MRAR for multi-resident activity recognition from binary ambient sensor is proposed, which can automatically generate temporal features leveraging previous events and enable us to classify both residents and activities' labels simultaneously.

- We put forward a tree-structure network for exploiting deep discriminative features with temporal dependence. Each two features is stacked together and then prorogated to the next layer, which can transmit temporal information to the top embedding, so that we can finally obtain an deep feature representation from neighboring events.

- Experimental results utilizing multi-resident dataset demonstrate that our approach outperforms other techniques in multi-resident activity recognition tasks, exspecially in resident recognition.

The rest of the paper is as follows: In Sect. 2, we review the related work of human activity recognition. In Sect. 3, we elaborate the details of our proposed approach, especially the tree-structure CNN model. We thoroughly evaluate the performance of the proposed framework in Sect. 4. Finally, we conclude the paper and the future directions in Sect. 5.

2 Related Work

2.1 Deep Learning for HAR

Deep learning has significant impact on HAR, which benefits for automatically feature extraction procedures for whose lack of the robust professional basis knowledge [10]. In most of existing work on CNNs for HAR, 1D convolution operation is utilized to exact local dependency along the temporal dimension of individual time series. In reference [11], they convolve the sequence information on different axes separately to obtain feature representations, and then concatenate them as the input of the top fully connected layer. However, this work only made use of a one-layered CNN architecture, which in turn did not exploit the latent hierarchical natures of activities. In [12], similarly, this model uses finite adjacent events as input units, and obtains the final representation through multiple convolutions. Yang et al. [6] redefined CNN operators and pooling filters. CNN operators with the same parameters are applied on local signals at different time segments for temporal information. Additionally, the huge number of feasible CNN operation designs leading to time wasting and resulting in suboptimal solutions, Baldominos et al. [7] proposed a method to automatically infer the topology of the CNN by an evolutionary algorithm to improve the efficiency. In [13], in order to obtain local dependency along both temporal and spatial domains, a framework with 2D convolution and pooling operations in multivariate time series data is proposed. However, as mentioned above, CNNs are restricted by the size of convolutional kernels for unadaptable to a wide range of activity recognition configurations.

RNNs perform well in capturing time dependence in lots of researches. In [14], the RNN model with the LSTM block is introduced to identify individual activities through automaticly learning of representative features and encoding of time information in the feature learning process. Pawlyta et al. [8] proposed models first cut time series into variable-length segments and treat each sequence as an event to classify activities by utilizing RNNs to read variable-length sequences of input samples. Francisco et al. [9] proposed a framework combined CNN and LSTM. The input series data with a slidding window is introduced to CNN layers and output representation is fed into LSTM model to model the temporal dependencies. To exploit multiple residents' activities, Tran et al. [15] employ multi-label RNN for modelling activities of multiple residents which can recognize activities for each resident respectively. However, RNN is limited in its structure that make it difficult to exploit deeper representation than CNNs.

2.2 Multi-resident Activity Recognition

Compare to individual activity recognition, multi-resident activity recognition is more realistic and complex. Solutions designed to deal wtih the issue of sensor-based multiple resident activity recognition can be categorized as knowledge-driven and data-driven models [16].

Knowledge-driven models utilize symbolic representation and ontology to analyze the semantic relations of activities [17]. Ye et al. [18] integrate a knowledge base and the statistical techniques to segment a sensor sequence into partitions. Based on the well-established ontology, this model can evaluate the semantic similarity between sensor events and connect the sensor with the activity ontology. Alam et al. [19] present a model for the recognition of complex daily activities. A loosely-coupled hierarchical dynamic Bayesian network is designed to recognize coarse-grained activities via fine-grained atomic actions. In [20], the ontological correlation between sequential behavior patterns is extracted which is well organized in a graphical knowledge base, without the intervention of domain experts.

Data-driven models rely on data and construct statistical and probabilistic models, such as hidden Markov models (HMMs) [2], conditional random fields (CRFs) [21]. There are two ways to deal with multi-resident issue based on the HMM models. One approach is to utilize single HMM for cooperative activities, ie. considering the activities of different residents as a random variable, named coupled HMM. Another approach is to model the activities of multiple residents, termed parallel HMM. Chiang et al. [22] propose a dynamic Bayesian network model. In order to improve the performance of the model, they classify the sensor observations based on data association and some domain knowledge to model the multi-resident activity model. Benmansour et al. [23] combine activity tags and observation tags of different residents to generate corresponding activity sequences and corresponding observation sequences based on conventional HMM.

However, little has been learnt in multi-resident field to the best of our knowledge in deep learning for HAR. Hence, in this work, we propose an end-to-end framework to automatically capture features for multi-resident activity classification.

3 TSC-MRAR Model

3.1 Problem Statement

HAR aims to recognize human activity which enable the computing systems to proactively assist users based on their requirement. Let S be the set of N sensors $S = \{S_1, S_2, \ldots, S_N\}$. There is a sequence of sensor reading that captures the activity information $x = (x_1, x_2, \ldots, x_t)$, where x_t denotes the sensor reading at time t, which is an event in this work and $x_t \in R^N$. $Y_t = \left(y_1^t, y_2^t, \ldots y_i^t, \ldots, y_m^t, y_{m+1}^t, \ldots y_j^t, \ldots, y_{m+n}^t\right)$ is the label referring to x_t, where m, n are the number of residents and activities, respectively. y_i^t is 1 or 0, 1 denotes the certain resident, otherwise, 0 means not. Similarly, y_j^t is 0 or 1, and 1 denoted the certain activity. Hence the HAR problem can be defined as: given a sequence of sensor reading x, we train a network to jointly predict resident and activity label Y_t for the event x_t.

3.2 Framework Overview

In this section, we elaborate the framework of the proposed approach. TSC-MRAR mainly contains three components: sampling, tree-structure convolutional layer and fully-connected layer for classification recognition. The overall framework is shown in Fig. 1. The input time series data is sampled in a fix sliding window. Those events are embedded into initial vectors, and then are fed into a tree-structure convolutional network, which will output high-order information representation. Each convolutional layer is constructed into a residual block. Finally, top fully connected layer finally combine there embedding to classify of the labels of residents and activities in the meanwhile.

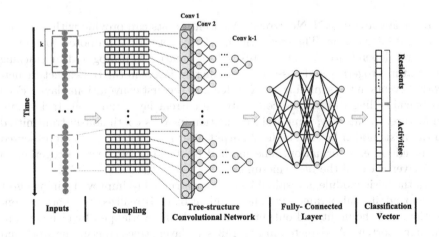

Fig. 1. Algorithm flow chart

3.3 Tree-Structure Convolutional Network

Sampling. Traditional deep learning approaches normally segment series data into several slice, and view each slice as an event, while in this work, we treat each ON as an event to do predict task. For each task, we conduct a sample window with fixed size to get previous informations of event in time series so that we can extract the sample feature of this task. Sampling set is denoted as $X_t = \left(x_{t-k}^0, \ldots, x_{t-1}^0, x_t^0\right)$, specifically, k is fixed sliding window size. To further describe the proposed model, we define the original input vector as x_t^0 and output for each layer of this network denote as x_t^i. X_t is fed to our network, and output an feature embedding considering to event x_t.

Convolutional Operators. In this section, we elaborate convolutional operators in each layer. There are $k - 1$ layers of convolution operations in total. In ith layer, the convolution operator is denoted as Conv i. Set input channel of the first layer convolutional operator Conv 1 to be 1, output channel to be 16, which represents 32 feature maps and convolution kernel size to be 3. Input channel of Conv 2 is 16 and output channel is set to be 32. The third-layer convolution operator Conv 3 leverages the output feature vectors of Conv 2 with output channel to be 64. The latter $k - 4$ convolution operators are the same as Conv 3. A feature map j in i th convolutional layer, denoted by $z_{i,j}$ is computed as:

$$z_{i,j} = \sigma\left(\sum_{m=1}^{M} w_m z_{i-1,j+m} + b_{i-1,j}\right) \tag{1}$$

where $\sigma(\cdot)$ is the rectified linear unit (ReLU), M is the size of convolution kernel, w_m is the weight matrix and $b_{i-1,j}$ is the bias for this feature map.

Tree-Structure CNN Network. Adjacent events can provide with sequence temporal information. The tree-structure CNN network is proposed to capture temporal dependence. Events contextual information is prorogated by stacking two features together and pass between different layers in the tree-structure network. t events in the sampling set X_t is fed into the first convolutional layer. Each event embedding will first be convoluted in current layer and then stacked into an feature. In this way, temporal information in X_t is continuously transmitted to the next hidden layer. The final output after $k - 1$ layer x_t^{k-1} is represented with high-order information. The method of stacking two feature together in each layer is called the basic module.

In the basic module, a residual block is constructed to improve training effect without additional parameters. The Residual block is realized by shortcut connection, and the input and output of this block are overlapped by element-wise through shortcut. As depicted in Fig. 2, it's a 5 layers tree-structure network and symbol $\oplus$ in the figure represents element-level concatenation. The two inputting events embedding x_{t-1}^0 and x_t^0 in the sampling set X_t is fed into the basic module. x_{t-1}^0 and x_t^0 is convoluted with Conv 1 and mapped out to be $F\left(x_{t-1}^0\right)$

and $F\left(x_t^0\right)$, respectively. These two feature map are concatenated into a feature vector $h_1(x)$ in an element-wise level, and then fed into the residual block. In the residual block, $h_1(x)$ will be convoluted for three time with Conv 1 and go through function ReLu to map out as $F(h_3(x))$. $h_1(x)$ is shortcut connected with $F(h_3(x))$ as the final feature vector x_t^1 in this basic module.

Every two features in the tree-structure network are concatenated in the way as depicted in the basic module. Hence, temporal information can be transmitted from bottom to top.

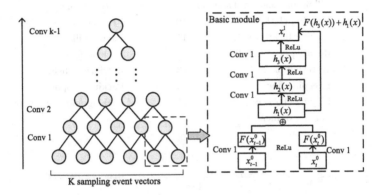

Fig. 2. Tree-structure CNN network

3.4   Fully-Connected Layer

In our end-to-end network, features in the previous convolutional layer will be directly fed into a fully-connected layer. t events in the sampling set X_t stack into an feature embedding with high information via tree-structure network. After convolutional layers have extracted relevant features from inputting sampling set, these features can be introduced to a classifier. Taking the extracted feature vectors as the input of CNN feature extracted layer, the fully connected layer trains the data, and classifies different activity types through the learning model. The final fully connected layer classifies an input batch into one predefined actions and residents classification vector Y_t. Here we utilize softmax function as classifier. Once one iteration of forward propagation is done, we will have the error value, with the cross entropy loss function. We are able to use optimizer Adam to update each edge w for the fully connected layer.

4   Experiments

4.1   Experimental Setup

Dataset. The dataset utilized in this experiment are collected by CASAS project in a smart workplace [24]. There are 15 activities in this activity recognition task

with 2 residents, and activities are as shown in Table 1. The test bed is equipped
with 37 different sensors set in the ceiling as depicted in Fig. 3. Each record is
regarded as an event including the following information(Date, Time, SensorID,
Value, ResidentID, ActivityID). The notation of collected sensor events are manually labeled.

Table 1. The list of 15 activities.

Activity ID	Description	Activity ID	Description
1	Filling medication dispenser	9	Setting table
2	Hanging clothes	10	Reading books (R1)
3	Moving furniture	11	Paying bills
4	Reading books (R2)	12	Packing food
5	Watering plants	13	Retrieving dishes
6	Sweeping floor	14	Packing supplies
7	Playing checkers	15	Packing and bring supplies
8	Making dinner		

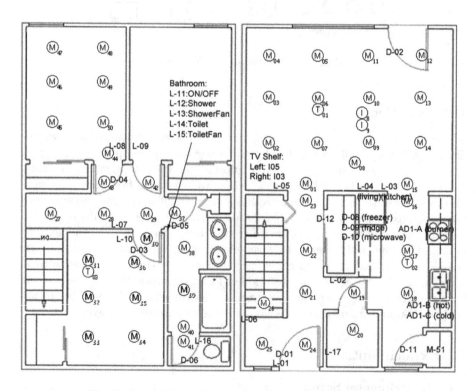

Fig. 3. Sensor layout displayed in the CASAS smart home

Baselines. We compare our proposed model with the following methods: k-Nearest Neighbor (KNN) and Decision Tree (DT) are traditional classification technology for benchmarking the performance; CNN [25] is deep learning methods which can only predict activity; RNN is classical deep learning model which is able to predict resident and activty at the same time.

For machine methods, we use the default hyper parameters provided in scikit learn package, then we train them for ten times and keep the best result. For deep learning method, we utilize the same hyper parameters as proposed model and feed them with data preprocessed in the same way.

Parameter Settings. Specifically, there are 26 files in the dataset that we utilize. The annotation experiment was conducted by WSU team for 26 times and each experiment was recorded in a single file. We divide 70% of them into training set and the remaining into testset, i.e, 18 files are randomly selected as the training set with 6197 events, and the remaining 8 are test sets with 2743 events.

We only utilize the events with sensor values ON, since there is an time interval in turning on and off the sensor, and these intervals will result in noisy for analysis. Sensors are represented by numbers 0–37 as shown in Table 2 and one hot technology is applied to get initial embedding of each event. Here we set the parameter k in our approach to be 8.

Table 2. Tags of sensors.

Sensor ID	Tags	Sensor ID	Tags	Sensor ID	Tags	Sensor ID	Tags
1	M01	11	M11	21	M21	31	D09
2	M02	12	M12	22	M22	32	D10
3	M03	13	M13	23	M23	33	D11
4	M04	14	M14	24	M24	34	D12
5	M05	15	M15	25	M25	35	D13
6	M06	16	M16	26	M26	36	D14
7	M07	17	M17	27	M51	37	D15
8	M08	18	M18	28	I04		
9	M09	19	M19	29	I06		
10	M10	20	M20	30	D07		

In detail, we utilize Pytorch 1.2.1 running on python 3.7.4 to construct our networks. We trained our model on server with Ryzen 7 3700x processer and Nvida GTX 1080ti GPU, and the operating system is Ubuntu 18.04 LTS with CUDA 10.0 and cuDNN 7.6.4.

Initially, we adopt 15 epochs for hyper parameters tuning to find appropriate hyper parameters combination of networks. Then we exploit 25 epochs and best hyper parameters combination to train our network, in each epoch, we feed all files in train set into our network and find the average loss to be this epoch's loss output.

4.2   Experimental Results

We first tune hyperparameters to rearch the best hyperparameters of our model. Ten-fold cross-validation for evaluation is adopted in our experiment, and we compare the performance of our proposed model with baseline and other competitive methods mentioned above. To evaluate the performance of aforementioned models, we leverage binary classification metrics for resident classification, including accuracy, F1-score and precision, and multi-classification metric accuracy for activity classification.

Sensitivity of Hyperparameters. We characterise the effects of the key hyperparameters of this model, including the batch size α, L2 weight β and learning rate γ. The approach of hyperparameters tuning is to set a few fixed values for the parameters, and then train model to find the optimal parameter combination. Parameter α is preset several discrete values, 64, 128 and 256. β and γ are both set to range from 0.0001 to 0.001 with step 0.0002. Metrics of max loss, average loss and min loss is utilized to measure batch size performance. From Fig. 4(a) we can see that $\alpha = 128$ performs best on its average loss and max loss, while this point is a little bit worse than 64 on min loss. In summary, we pick $\alpha = 128$. Then we fix α to tune β and γ. Figure 4(b) shows that β performance in each learning rate setting. When γ is set to be 0.0002, β can achieve best performance in most case except $\beta = 0.0008$ on the overall trend. β with value 0.0004 outperform than other values when $\gamma = 0.0002$. In conclusion, we select $\alpha = 128$, $\beta = 0.0004$ and $\gamma = 0.0002$ as our final hyperparameters.

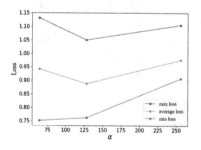   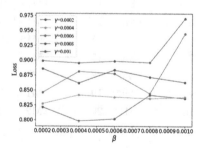

(a) Effect of batch size on loss   (b) Effect of L2 and learning rate on loss

Fig. 4. Effects of hyper-parameters on loss

Results Comparison. We further compare the best convolutional framework with aforementioned methods as shown in Table 3. For resident classification, among all the models, TSC-MRAR performs the best for all three criterias and DT performs second best. In this case, we can draw a conclusion that deep learning methods are not necessarily better than machine learning methods. Noting that when under F1-score and Precision, there is a very large improvement of the proposed model compared with other methods. Considering the complexity of the data with multiple residents recognition, all of the algorithms need huge improvement. CNN get the highest accuracy but cannot recognize resident and activity at the same time, which is extremely important in smart home environment.

Table 3. Performance results of TSC-MRAR model and RNN.

Method	Resident			Activity
	Accuracy	Precision	F1-score	Accuracy
KNN	0.658	0.6517	0.6463	0.4502
DT	0.7073	0.6993	0.7	0.4765
CNN	-	-	-	0.6593
RNN	0.7044	0.6974	0.6931	0.3802
TSC-MRAR	0.7922	0.9259	0.8541	0.5848

In conclusion, under multiple residents' activities scenario, these results corroborate that our approach outperform than other machine learning methods and deep learning approaches.

5 Conclusion and Future Works

In this paper, we consider the scenerio of multi-resident activity recognition. An end-to-end framework TSC-MRAR to automatically capture temporal features is proposed. Each sample is treated as an event without series data segmentation. This model can automatically extract temporal features in hidden layers and temporal information is propagated though the tree-structure network. Furthermore, features extraction and classification are unified in one model to classify multi-resident activities. Experimental results demonstrate that the proposed CNN method outperforms other state-of-the-art approaches, and we therefore believe that the proposed approach can serve as a competitive tool of multi-resident activity classification.

In the future, we plan to broaden and deepen this work in several directions. First we will try to utilize other multi-resident data to confirm the wide applicability. More deep learning framework to explore multi-resident features will be concerned to improve the accuracy of multi-resident activity recognition.

References

1. Alam, M.R., Reaz, M.B.I., Ali, M.A.M.: A review of smart homes–past, present, and future. IEEE Trans. Syst. Man Cybern. Part C **42**(6), 1190–1203 (2012)
2. Crandall, A.S., Cook, D.J.: Using a hidden Markov model for resident identification. In: 2010 Sixth International Conference on Intelligent Environments, pp. 74–79, IEEE (2010)
3. Lu, H., Yang, J., Liu, Z., Lane, N.D., Choudhury, T., Campbell, A.T.: The jigsaw continuous sensing engine for mobile phone applications. In: Proceedings of the 8th ACM Conference on Embedded Networked Sensor Systems, pp. 71–84 (2010)
4. Plötz, T., Hammerla, N.Y., Olivier, P.L.: Feature learning for activity recognition in ubiquitous computing. In: Twenty-Second International Joint Conference on Artificial Intelligence (2011)
5. Wang, J., Chen, Y., Hao, S., Peng, X., Hu, L.: Deep learning for sensor-based activity recognition: a survey. Pattern Recogn. Lett. **119**, 3–11 (2019)
6. Yang, J., Nguyen, M.N., San, P.P., Li, X.L., Krishnaswamy, S.: Deep convolutional neural networks on multichannel time series for human activity recognition. In: Twenty-Fourth International Joint Conference on Artificial Intelligence (2015)
7. Baldominos, A., Saez, Y., Isasi, P.: Evolutionary design of convolutional neural networks for human activity recognition in sensor-rich environments. Sensors **18**(4), 1288 (2018)
8. Pawlyta, M., Hermansa, M., Szczęsna, A., Janiak, M., Wojciechowski, K.: Deep recurrent neural networks for human activity recognition during skiing. In: Gruca, A., Czachórski, T., Deorowicz, S., Hareżlak, K., Piotrowska, A. (eds.) ICMMI 2019. AISC, vol. 1061, pp. 136–145. Springer, Cham (2020). https://doi.org/10.1007/978-3-030-31964-9_13
9. Ordóñez, F.J., Roggen, D.: Deep convolutional and lstm recurrent neural networks for multimodal wearable activity recognition. Sensors **16**(1), 115 (2016)
10. Chen, K., Zhang, D., Yao, L., Guo, B., Yu, Z., Liu, Y.: Deep learning for sensor-based human activity recognition: Overview, challenges and opportunities. arXiv preprint arXiv:2001.07416 (2020)
11. Zeng, M., et al.: Convolutional neural networks for human activity recognition using mobile sensors. In: 6th International Conference on Mobile Computing, Applications and Services, pp. 197–205. IEEE (2014)
12. Ronao, C.A., Cho, S.-B.: Deep convolutional neural networks for human activity recognition with smartphone sensors. In: Arik, S., Huang, T., Lai, W.K., Liu, Q. (eds.) ICONIP 2015. LNCS, vol. 9492, pp. 46–53. Springer, Cham (2015). https://doi.org/10.1007/978-3-319-26561-2_6
13. Ha, S., Choi, S.: Convolutional neural networks for human activity recognition using multiple accelerometer and gyroscope sensors. In: 2016 International Joint Conference on Neural Networks (IJCNN), pp. 381–388. IEEE (2016)
14. Ding, J., Wang, Y.: Wifi CSI-based human activity recognition using deep recurrent neural network. IEEE Access **7**, 174257–174269 (2019)
15. Tran, S.N., Zhang, Q., Smallbon, V., Karunanithi, M.: Multi-resident activity monitoring in smart homes: a case study. In: 2018 IEEE International Conference on Pervasive Computing and Communications Workshops (PerCom Workshops), pp. 698–703, IEEE (2018)
16. Benmansour, A., Bouchachia, A., Feham, M.: Multioccupant activity recognition in pervasive smart home environments. ACM Comput. Surv. (CSUR) **48**(3), 1–36 (2015)

17. Riboni, D., Pareschi, L., Radaelli, L., Bettini, C.: Is ontology-based activity recognition really effective? In: 2011 IEEE International Conference on Pervasive Computing and Communications Workshops (PERCOM Workshops), pp. 427–431. IEEE (2011)
18. Ye, J., Stevenson, G., Dobson, S.: KCAR: a knowledge-driven approach for concurrent activity recognition. Pervasive Mob. Comput. **19**, 47–70 (2015)
19. Alam, M.A.U., Roy, N., Misra, A., Taylor, J.: CACE: exploiting behavioral interactions for improved activity recognition in multi-inhabitant smart homes. In: 2016 IEEE 36th International Conference on Distributed Computing Systems (ICDCS), pp. 539–548. IEEE (2016)
20. Hao, J., Bouzouane, A., Gaboury, S.: Recognizing multi-resident activities in nonintrusive sensor-based smart homes by formal concept analysis. Neurocomputing **318**, 75–89 (2018)
21. Hsu, K.-C., Chiang, Y.-T., Lin, G.-Y., Lu, C.-H., Hsu, J.Y.-J., Fu, L.-C.: Strategies for inference mechanism of conditional random fields for multiple-resident activity Recognition in a Smart Home. In: García-Pedrajas, N., Herrera, F., Fyfe, C., Benítez, J.M., Ali, M. (eds.) IEA/AIE 2010. LNCS (LNAI), vol. 6096, pp. 417–426. Springer, Heidelberg (2010). https://doi.org/10.1007/978-3-642-13022-9_42
22. Chiang, Y.T., Hsu, K.C., Lu, C.H., Fu, L.C., Hsu, J.Y.J.: Interaction models for multiple-resident activity recognition in a smart home. In: 2010 IEEE/RSJ International Conference on Intelligent Robots and Systems, pp. 3753–3758. IEEE (2010)
23. Benmansour, A., Bouchachia, A., Feham, M.: Modeling interaction in multi-resident activities. Neurocomputing **230**, 133–142 (2017)
24. Singla, G., Cook, D.J., Schmitter-Edgecombe, M.: Recognizing independent and joint activities among multiple residents in smart environments. J. Ambient Intell. Humaniz. Comput. **1**(1), 57–63 (2010)
25. Singh, D., Merdivan, E., Hanke, S., Kropf, J., Geist, M., Holzinger, A.: Convolutional and recurrent neural networks for activity recognition in smart environment. In: Holzinger, A., Goebel, R., Ferri, M., Palade, V. (eds.) Towards Integrative Machine Learning and Knowledge Extraction. LNCS (LNAI), vol. 10344, pp. 194–205. Springer, Cham (2017). https://doi.org/10.1007/978-3-319-69775-8_12

Author Index

Printed in the United States
By Bookmasters